Stochastic Control in Discrete and Continuous Time

Atle Seierstad

Stochastic Control in Discrete and Continuous Time

 Springer

Atle Seierstad

Department of Economics
University of Oslo
1095 Blindern
0317 Oslo
Norway
Atle.Seierstad@econ.uio.no

ISBN 978-1-4419-4569-3   e-ISBN 978-0-387-76617-1
DOI: 10.1007/978-0-387-76617-1

Mathematics Subject Classification (2000): 90C40, 93E20

Printed on acid-free paper

springer.com

This book is dedicated to a lynx I saw in Maridalen in the vicinity of Oslo in April 2006.

Without having read this book (so far), yet maybe it hunts according to an optimized piecewise deterministic process.

Preface

This book contains an introduction to three topics in stochastic control: discrete time stochastic control, i.e., stochastic dynamic programming (Chapter 1), piecewise deterministic control problems (Chapter 3), and control of Ito diffusions (Chapter 4). The chapters include treatments of optimal stopping problems. An Appendix recalls material from elementary probability theory and gives heuristic explanations of certain more advanced tools in probability theory.

The book will hopefully be of interest to students in several fields: economics, engineering, operations research, finance, business, mathematics. In economics and business administration, graduate students should readily be able to read it, and the mathematical level can be suitable for advanced undergraduates in mathematics and science. The prerequisites for reading the book are only a calculus course and a course in elementary probability. (Certain technical comments may demand a slightly better background.)

As this book perhaps (and hopefully) will be read by readers with widely differing backgrounds, some general advice may be useful: Don't be put off if paragraphs, comments, or remarks contain material of a seemingly more technical nature that you don't understand. Just skip such material and continue reading, it will surely not be needed in order to understand the main ideas and results.

The presentation avoids the use of measure theory. At certain points, mainly in Chapter 4, certain measure theoretical concepts are used, they are then explained in a heuristic manner, with more detailed but still heuristic explanations appearing in Appendix. The chosen manner of exposition is quite standard, except in Chapter 3, where a slightly unusual treatment is given that can be useful for the elementary types of problems studied there. In all chapters, problems with terminal restrictions are included.

One might doubt if Ito-diffusions can at all be presented in a useful way without using measure theory. One then has to strike a balance between being completely intuitive and at least giving some ideas about where problems lie hidden, how proofs can be constructed, and directions in which a more advanced treatment must move. I hope that my choices in this respect are not too bad.

A small chapter (Chapter 2) treats deterministic control problems and has been included because it makes possible a very simple exposition of ideas that later on reappear in Chapters 3 and 4. In addition, and more formally, certain proofs in Chapter 3 make use of proven results in Chapter 2.

The level of rigor varies greatly. Formal results should preferably be stated will full rigor, but certain compromises have been unavoidable, especially in Chapter 4, but in fact in all chapters. The degree of rigor in the proofs varies even more. Some proofs are completely heuristic (or even omitted), other ones are nearly, or essentially, rigorous. Quite frequently, first nonrigorous proofs are presented, and then, perhaps annoyingly often, some comment on what is lacking in the proofs, or how they might be improved upon, are added.

Hopefully, what might be called introductory proofs of the most central results are easy to read. Other proofs of more technical material may be quite compact, and then more difficult to read. So the reader may feel that readability varies a lot. In a book of this type and length, it was difficult to avoid this variability in the manner of exposition.

Altogether, there are a great number of remarks in the text giving refinements or extension of results. On (very) first reading, it is advisable to skip most of the remarks and concentrate on the main theory and the examples. Asterisks, usually one (*), are used to indicate material that can be jumped over at first reading; when two asterisks (**) are used, it indicates in addition that somewhat more advanced mathematical tools are used.

Solved examples, examples with analytical (or closed form) solutions, play a big role in the text. The aim is to give the reader a firmer understanding of the theoretical results. It will also equip him or her with a better knowledge of how to solve similar, simple problems, and an idea of how solutions may look like in slightly more complicated problems where analytical solutions cannot be found. The reader should get to know, however, that most problems cannot be solved analytically, they need numerical methods, not treated in the current book.

On the whole, sufficient conditions for optimality are proved with greater rigor (sometimes even with full rigor) compared with proofs of necessary conditions. Even if the latter proofs are heuristic, the necessary conditions that they seem to provide can be compared with the sufficient conditions established, and the former (slightly imprecise) conditions can tell how useful the latter conditions are, in other words how frequently we can hope that the sufficient conditions can help us solve the optimization problems we consider.

A number of exercises, with answers provided (except for a few theoretical problems), have been included in the book.

For the reader wanting to continue studying some of the themes of this book, or who wants to consult alternative expositions of the theory, a small selection of books and articles are provided in the References. These works are also referred to at the ends of the chapters. The few titles provided can only give some hints as to where one can seek more information; for more extensive lists of references, one should look into more specialized works.

Sections 1.1 and 1.2 have, with some changes, appeared in the book by K. Sydsæter et al. (2005), *Further Mathematics for Economic Analysis* by Prentice Hall.

Early versions of the chapters have been tried out in courses for PhD students in economics, from whom useful feedback is gratefully acknowledged.

Peter Hammond has read early versions of the chapters in the book and given a tremendous amount of useful advice, much more than he surely remembers.

Knut Sydsæter and Arne Strøm have read and given comments on parts of the book and also helped in technical matters related to presentation and layout. Also, Tore Schweder has helped me at certain points. For all this help, I am extremely grateful.

All errors and other shortcomings are entirely my own.

The Department of Economics at University of Oslo has over the years made available an excellent work environment and Knut Sydsæter and Arne Strøm, my co-mathematicians at this department, have carried out more than their share of work related to teaching and advising students of all types, which has given me more time to work on this book, among other things. Thanks again.

Finally, I am very grateful for the excellent technical support provided by Springer concerning the preparation of the final version of the manuscript.

Oslo, Norway   *Atle Seierstad*
October 2007

Contents

Chapter 1
Stochastic Control over Discrete Time

This chapter describes the optimal governing of certain discrete time stochastic processes over time. First, solution tools for finite horizon problems are presented, the most important being the dynamic programming equation, but also a stochastic maximum principle is rendered. In three sections, infinite horizon problems are treated. Optimal stopping problems are discussed, where when to stop is *a* — or *the* — central question, both for a finite and infinite horizon. Problems of incomplete observations, where we learn more the longer the process runs, are also discussed, and we end this part by presenting stochastic control with Kalman filtering. Finally, some approximation methods are briefly discussed, and an extension to stochastic time periods is presented.

1.1 Stochastic Dynamic Programming

What is the best way of controlling a system governed by a difference equation that is subject to random disturbances? Stochastic dynamic programming is a central tool for tackling this question.

In deterministic dynamic programming, the state develops according to a difference equation $x_{t+1} = f(t_1, x_t, u_t)$, controlled by appropriate choices of the control variables u_t. In the current chapter, the function f is also influenced by random disturbances, so that x_{t+1} is a stochastic quantity. Following common practice, we often (but not always) use capital letters instead of lower-case letters for stochastic quantities, e.g., X_t instead of x_t.

Suppose then that the *state equation* is of the form

$$X_{t+1} = f(t, X_t, u_t, V_{t+1}), \quad X_0 = x_0, V_0 = v_0, x_0, v_0 \text{ given}, u_t \in U, \qquad (1.1)$$

where, for each t, V_{t+1} is a random variable that takes values in a finite set $\mathcal{V}$. The probability that $V_{t+1} = v \in \mathcal{V}$ is written $P_t(v|v_t)$; it is assumed that it may depend on the outcome v_t at time t, as well as explicitly on time t. We may allow V_{t+1} to

A. Seierstad, *Stochastic Control in Discrete and Continuous Time*,
DOI: 10.1007/978-0-387-76617-1_1,
© Springer Science+Business Media, LLC 2009

be a continuous variable that takes values anywhere in $\mathbb{R}$. Then the distributions of V_{t+1} are often given by densities $p_t(v|v_t)$, separately piecewise continuous in v and v_t. Mostly, we speak as if the V_t's are discrete random variables. However, the solution tools presented can also be used for continuous stochastic variables. We assume that $t = 0, 1, \ldots, T$, T a given positive integer, that x_t belongs to $\mathbb{R}^n$, and that u_t is required to belong to a given subset U of $\mathbb{R}^r$. The vectors u_t are subject to choice, and these choices, as well as the stochastic disturbances V_{t+1} determine the development of the state X_t.

Example 1.1. Suppose that $Z_1, Z_2, \ldots$ are independently distributed stochastic variables that take a finite number of positive values (or a continuum of positive values) with specified probabilities independent of both the state and the control. The state X_t develops according to:

$$X_{t+1} = Z_{t+1}(X_t - u_t), \qquad u_t \in [0, \infty). \tag{i}$$

Here u_t is consumption, $X_t - u_t$ is investment, and Z_{t+1} is the return per invested dollar. Moreover, the utility of the terminal state x_T is $\beta^T B x_T^{1-\gamma}$ and the utility of the current consumption is $\beta^t u_t^{1-\gamma}$ for $t < T$, where β is a discount factor, and $0 < \gamma < 1$. The development of the state x_t is now uncertain (stochastic). The objective function to be maximized is the sum of expected discounted utility, given by

$$\sum_{t=0}^{T-1} \beta^t E u_t^{1-\gamma} + \beta^T B E X_T^{1-\gamma}. \tag{ii}$$

$\square$

Let us, for a moment, consider a two-stage decision problem. Assume that one wants to maximize the criterion:

$$E\{f_0(0, X_0, u_0) + f_0(1, X_1, u_1)\} = f_0(0, X_0, u_0) + E f_0(1, X_1, u_1),$$

where E denotes expectation and f_0 is some given function. Here the initial state $X_0 = x_0$ and an initial outcome v_0 are given and X_1 is determined by the difference equation (1.1), i.e., $X_1 = f(0, x_0, u_0, V_1)$. To find the maximum, the following method works: We can first maximize with respect to u_1, and then with respect to u_0. When choosing u_1, we simply maximize $f_0(1, X_1, u_1)$, assuming that X_1 is known before the maximization is carried out. The maximum point u_1^* becomes a function $u_1^*(X_1)$ of X_1. Imagine that this function is inserted for u_1 in the criterion, and that the two occurrences of X_1 are replaced by $f(0, x_0, u_0, V_1)$. Then the criterion becomes equal to

$$f_0(0, X_0, u_0) + E\{f_0(1, f(0, x_0, u_0, V_1), u_1^*(f(0, x_0, u_0, V_1)))\},$$

i.e., u_0 occurs in both terms in the criterion. A maximizing value of u_0 is then chosen, taking both these occurrences into account.

When there are more than two stages, this process is continued backwards, as we shall see.

To see why it matters that we can observe X_1 before choosing u_1, consider the following problem: Let $T = 1$, $f_0(0,x_1,u_1) = 0$, $f_0(1,x_1,u_1) = X_1u_1$, $X_1 = V_1$, where V_1 takes the values 1 and -1 with probabilities $1/2$, and where u can take the values 1 and -1. Then, $EX_1u_1 = 0$ if we have to choose u_1 before observing X_1 (hence a constant u_1), but if we can first observe X_1, then we can let u_1 depend on X_1. If we choose $u_1 = u_1(X_1) = X_1$, then $EX_1u_1 = 1$, which yields a better value of the objective. In Sections 1.1–1.6, we shall assume that X_t, in fact both X_t and V_t, can be observed before choosing u_t.

Let us turn to the general problem. The process X_t, determined by (1.1) and the random variables V_t, is to be controlled in the best possible manner by appropriate choices of the variables u_t. The *objective function* is now the expectation

$$\sum_{t=0}^{T} E[f_0(t,X_t,u_t(X_t,V_t))]. \tag{1.2}$$

Here several things have to be explained. Each control u_t, $t = 0,1,\ldots,T$ is a function, $u_t(x_t,v_t)$, of the current state x_t and the outcome v_t. Such a function is called a *policy* (or more specifically a *Markov policy* or a *Markov control*). For a large class of stochastic optimization problems, including the one we are now studying, this is the natural class of controls to consider in order to achieve an optimum. Both V_t and X_t are random variables, the X_t's arising from the state equation when the functions $u_s(X_s,V_s)$ are inserted in the equation. The letter E, as before, denotes expectation. For completeness, a detailed description of its calculation follows in the next paragraph, but because it will not be much used later on, readers may want to skip reading it.

To compute the expectation requires specifying the probabilities that are needed in the calculation of the expectation. Given v_0, recall that the probability for the events $V_1 = v_1$ and $V_2 = v_2$ jointly to occur equals the conditional probability for $V_2 = v_2$ to occur, given $V_1 = v_1$, times the probability for $V_1 = v_1$ to occur, given $V_0 = v_0$. Hence it equals $P_1(v_2|v_1)$ times $P_0(v_1|v_0)$. Similarly, given v_0, the probability of the joint event $V_1 = v_1$, $V_2 = v_2, \ldots, V_t = v_t$, is given by

$$p^*(v_1,\ldots,v_t) := P_0(v_1|v_0) \cdot P_1(v_2|v_1) \cdot \ldots \cdot P_{t-1}(v_t|v_{t-1}). \tag{1.3}$$

(This is actually a conditional probability, v_0 given.) Now, given the policies $u_t(x_t,v_t)$, the sequence X_t, $t = 1,\ldots,T$, in (1.2) is the solution of (1.1), found by calculating, successively, $X_1,X_2,\ldots$, when, successively, $V_1,V_2,\ldots$. and $u_1 = u_1(X_1,V_1)$, $u_2 = u_2(X_2,V_2),\ldots$ are inserted. Hence X_t depends on $V_1, \ldots, V_t$ and, for each t, the expectation $Ef_0(t,X_t,u_t(X_t,V_t))$ is calculated by means of the probabilities specified in (1.3).

Though not always necessary, we shall assume that f_0 and f are continuous in (x,u), (in (x,u,v) if V_t takes values in a nondiscrete set).

The optimization problem is to find a sequence of policies $u_0^*(x_0,v_0),\ldots$, $u_T^*(x_T,v_T)$ that gives the expression in (1.2) the largest possible value. Such a policy sequence is called an *optimal policy sequence*.

We now define the *optimal value function*

$$J(t,x_t,v_t) = \max E\left[\sum_{s=t}^{T} f_0(s,X_s,u_s(X_s,V_s)) \mid x_t,v_t\right], \qquad (1.4)$$

where the maximum is taken over all policy sequences $u_s = u_s(x_s,v_s), s = t,\ldots,T$, given v_t and given that we "start the equation" (1.1) at the state x_t at time t, as indicated by "$\mid x_t,v_t$" in (1.4) and apply the controls u_s from the sequence when using (1.1) to calculate all the X_s's. The computation of the expectation in (1.4), (or expectations, when E is taken inside the sum), is now based on conditional probabilities of the form $p^*(v_{t+1},\ldots,v_s|v_t) = P_t(v_{t+1}|v_t)\cdot\ldots\cdot P_{s-1}(v_s|v_{s-1})$.

The central tool in solving optimization problems of the type (1.1), (1.2) is the following *optimality* (or *dynamic programming*) equation:

$$J(t-1,x_{t-1},v_{t-1}) = \max_{u_{t-1}}\{f_0(t-1,x_{t-1},u_{t-1}) + E[J(t,X_t,V_t) \mid x_{t-1},v_{t-1}]\} \quad (1.5)$$

where $X_t = f(t-1,x_{t-1},u_{t-1},V_t)$ is to be inserted. The "x_{t-1}" in the symbol "$|x_{t-1},v_{t-1}|$" is just a reminder that x_{t-1} occurs in the expression to be inserted. After the insertion, the equation becomes $J(t-1,x_{t-1},v_{t-1}) =$

$$\max_{u_{t-1}}\{f_0(t-1,x_{t-1},u_{t-1}) + E[J(t,f(t-1,x_{t-1},u_{t-1},V_t),V_t) \mid v_{t-1}]\},$$

$t = 1,\ldots,T$. Moreover, at time T, we have

$$J(T,x_T,v_T) = J(T,x_T) = \max_{u_T} f_0(T,x_T,u_T). \qquad (1.6)$$

The equations (1.5), (1.6) are, essentially, both necessary and sufficient. They are sufficient in the sense that *if* $u_{t-1}^*(x_{t-1},v_{t-1})$ maximizes the right-hand side of (1.5) for $t = 1,\ldots,T$ and the right-hand side of (1.6) for $t = T+1$, then $u_{t-1}^*(x_{t-1},v_{t-1})$, $t = 1,\ldots,T+1$, are optimal policies. On the other hand, they are necessary in the sense that, for every x_{t-1},v_{t-1}, an optimal control $u_{t-1}^*(x_{t-1},v_{t-1}),t = 1,\ldots,T$, yields a maximum on the right-hand side of (1.5), and, for $t = T+1$, on the right-hand side of (1.6). To be a little more precise, it is necessary that the optimal control $u_{t-1}^*(x_{t-1},v_{t-1})$ yields a maximum on the right-hand side of (1.5), (1.6) for all values of x_{t-1},v_{t-1} that can occur with positive probability, given $\{u_s^*\}_s$.

The solution method is thus as follows: The relation (1.6) is used to find the functions $u_T^*(x_T,v_T)$ and $J(T,x_T,v_T)$, and then (1.5) is used to find first $u_{T-1}^*(x_{T-1},v_{T-1})$ and $J(T-1,x_{T-1},v_{T-1})$ (then $J(T,x_T,v_T)$ is needed), and then $u_{T-2}^*(x_{T-2},v_{T-2})$ and $J(T-2,x_{T-2},v_{T-2})$ (then $J(T-1,x_{T-1},v_{T-1})$ is needed), and so on, going backwards in time until $u_0(x_0,v_0)$ and $J(0,x_0,v_0)$ have been constructed. At any time t, the optimal control to use, given that (x_t,v_t) has been observed, is then $u_t(x_t,v_t)$.

The intuitive argument for (1.5) is as follows: Suppose the system is in a given state x_{t-1}, and v_{t-1} is given. For a given u_{t-1}, the "instantaneous" reward is $f_0(t,x_{t-1},u_{t-1})$. In addition, the maximal expected sum of rewards at all later times is $E[J(t,X_t,V_t)|x_{t-1},v_{t-1}]$ when $X_t = f(t-1,x_{t-1},u_{t-1},V_t)$. When using u_{t-1}, the total expected maximum value gained over all future time points (now including even $t-1$), is the sum in (1.5). The largest expected gain comes from choosing u_{t-1} to maximize this sum.

Note that when $P_t(v|v_t)$ does not depend on v_t, then v_t can be dropped in the functions $J_t(x_t,v_t), u_t(x_t,v_t)$, and in (1.5),(1.6). Then in (1.5) the conditioning on v_{t-1} drops out, and $J(t-1,x_{t-1},v_{t-1})$, and the maximizing vector $u_{t-1} = u_{t-1}(x_{t-1},v_{t-1})$ will not depend on v_{t-1}. (In some later sections, $f_0(t,.,.)$ will depend also on v_t and then this simplification does not hold.) In examples below, this simplification is employed.

Example 1.2. Consider the following example

$$\max E\left[\sum_{t=0}^{T-1}(1/2)^t((1-u_t)x_t)^{1/2} + (1/2)^T 2^{1/2}(X_T)^{1/2}\right],$$

subject to

$$X_{t+1} = u_t X_t V_{t+1}, \; X_0 = 1, V_t \in \{0,8\}, \; \Pr[V_t = 8] = 1/2, \; u_t \in [0,1].$$

This problem is closely related to Example 1.1.

Solution. Evidently, $J(T,x_T) = (1/2)^T 2^{1/2}(x_T)^{1/2}$.

Next, let us find the optimal $u = u_{T-1}$ and $J(T-1,x_{T-1})$, where $J(T-1,x_{T-1}) =$

$$\max_u\{(1/2)^{T-1}((1-u)x_{T-1})^{1/2} + E[(1/2)^T 2^{1/2}(ux_{T-1}V_T)^{1/2}]\} =$$

$$\max_u\{(1/2)^{T-1}((1-u)x_{T-1})^{1/2} + (1/2)^T 2^{1/2}(1/2)8^{1/2}(ux_{T-1})^{1/2}\} =$$

$$\max_u\{(1/2)^{T-1}(x_{T-1})^{1/2}[(1-u)^{1/2} + u^{1/2}]\}.$$

When differentiating to obtain the maximum point (we have a concave function in u), we get

$$(1/2)^{T-1}(x_{T-1})^{1/2}[-(1/2)(1-u)^{-1/2} + (1/2)u^{-1/2}] = 0,$$

which gives $(1-u)^{-1/2} = u^{-1/2}$, or $1-u = u$, i.e., $u = u_{T-1} = 1/2$. Inserting in the maximand, we get

$$J(T-1,x_{T-1}) = (1/2)^{T-1}(x_{T-1})^{1/2}2(1/2)^{1/2} = (1/2)^{T-1}2^{1/2}(x_{T-1})^{1/2}.$$

We now guess that, generally, $J(t,x_t) = (1/2)^t 2^{1/2}(x_t)^{1/2}$. Let us try this guess, hence let us find $J(t-1,x_{t-1})$ and the optimal $u = u_{t-1}$ from the optimality equation (we now see that we can repeat the above calculations for T replaced by t):

$$J(t-1,x_{t-1}) = \max_u\{(1/2)^{t-1}((1-u)x_{t-1})^{1/2} + E[(1/2)^t 2^{1/2}(ux_{t-1}V_t)^{1/2}]\}$$

$$= \max_u\{(1/2)^{t-1}(x_{t-1})^{1/2}[(1-u)^{1/2} + u^{1/2}]\}$$

$$= (1/2)^{t-1}2^{1/2}(x_{t-1})^{1/2},$$

the last equality because when differentiating to obtain the maximum point, we get $(1/2)^{t-1}(x_{t-1})^{1/2}[(-1/2)(1-u)^{-1/2} + (1/2)u^{-1/2}] = 0$, which gives $(1-u)^{-1/2} = u^{-1/2}$, i.e., $u = u_{t-1} = 1/2$ again.

In this example, incidentally, u_{t-1} came out as independent of x_{t-1}.   □

In the next example, the outcome of the stochastic variable depends on its value one period earlier.

Example 1.3. We want to solve the problem

$$\max E[X_T + V_T], \ X_{t+1} = u_t X_t V_{t+1} + (1-u_t)X_t(1-V_{t+1}),$$

$$X_0 = 1, u_t \in [0,1],$$

$$V_t \in \{0,1\}, \Pr[V_{t+1} = 1|V_t = 1] = 3/4, \Pr[V_{t+1} = 1|V_t = 0] = 1/4.$$

Solution. Formally, we need to work with a second state variable, say Y_t governed by $Y_{t+1} = V_{t+1}$. Then, $f_0(T,x_T,y_T) = x_T + y_T$, while $f_0(t,.,.)$ vanishes for $t < T$. However, below we write $x_T + v_T$ and $J(t,x_t,v_t)$ instead of $x_T + y_T$ and $J(t,x_t,y_t,v_t)$. Note that, by necessity, $X_t \geq 0$ for all t.

Now, $J(T,x_T,v_T) = x_T + v_T$. Let us next find $J(T-1,x_{T-1},v_{T-1})$. For $v_{T-1} = 1, J(T-1,x_{T-1},v_{T-1}) =$

$$\max_u E\{ux_{T-1}V_T + (1-u)x_{T-1}(1-V_T) + V_T|v_{T-1} = 1\}$$

$$= \max_u\{(3/4)ux_{T-1} + 3/4 + (1/4)(1-u)x_{T-1}\} = (3/4)x_{T-1} + 3/4.$$

Here, $u = u_{T-1} = 1$ is optimal.

For $v_{T-1} = 0, J(T-1,x_{T-1},v_{T-1}) =$

$$\max_u E\{ux_{T-1}V_T + (1-u)x_{T-1}(1-V_T) + V_T|v_{T-1} = 0\}$$

$$= \max_u\{(1/4)ux_{T-1} + 1/4 + (3/4)(1-u)x_{T-1}\} = (3/4)x_{T-1} + 1/4.$$

Here, $u = u_{T-1} = 0$ is optimal.

Let us now find $J(T-2,x_{T-2},v_{T-2})$. We can write $J(T-1,x_{T-1},v_{T-1}) = (3/4)x_{T-1} + 3v_{T-1}/4 + (1-v_{T-1})/4$.

For $v_{T-2} = 1, J(T-2,x_{T-2},v_{T-2}) =$

$$\max_u\{E[(3/4)(ux_{T-2}V_{T-1} + (1-u)x_{T-2}(1-V_{T-1})) + 3V_{T-1}/4$$

$$+ (1-V_{T-1})/4|v_{T-2}]\}$$

$$= \max_u\{(3/4)[(3/4)ux_{T-2} + 3/4] + (1/4)[(3/4)(1-u)x_{T-2} + 1/4]\}$$

$$= (3/4)^2 x_{T-2} + (3/4)^2 + (1/4)^2$$
$$= (3/4)^2 x_{T-2} + 10/16.$$

Here, $u = u_{T-2} = 1$ is optimal.

For $v_{T-2} = 0$, $J(T-2, x_{T-2}, v_{T-2}) =$

$$\max_u \{ E[(3/4)(ux_{T-2}V_{T-1} + (1-u)x_{T-2}(1-V_{T-1})) + 3V_{T-1}/4$$
$$+ (1-V_{T-1})/4 | v_{T-2}] \}$$
$$= \max_u \{ (1/4)[(3/4)ux_{T-2} + 3/4] + (3/4)[(3/4)(1-u)x_{T-2} + 1/4] \}$$
$$= (3/4)^2 x_{T-2} + 6/16.$$

Here, $u = u_{T-2} = 0$ is optimal.

We now guess that $J(t, x_t, v_t)$ is of the form $J(t, x_t, v_t) = (3/4)^{T-t} x_t + a_t$ when $v_t = 1$, $J(t, x_t, v_t) = (3/4)^{T-t} x_t + b_t$ when $v_t = 0$. We can write $J(t, x_t, v_t) = (3/4)^{T-t} x_t + a_t v_t + b_t (1 - v_t)$. Then, $J(t-1, x_{t-1}, v_{t-1}) =$

$$\max_u E\{ (3/4)^{T-t}(ux_{t-1}V_t + (1-u)x_{t-1}(1-V_t)) + a_t V_t + b_t(1-V_t) | v_{t-1} \}.$$

For $v_{t-1} = 1$ this expression equals

$$\max_u \{ (3/4)[(3/4)^{T-t} ux_{t-1} + a_t] + (1/4)[3/4]^{T-t}(1-u)x_{t-1} + b_t] \}$$
$$= (3/4)^{T-(t-1)} x_{t-1} + (3/4)a_t + (1/4)b_t,$$

with $u = u_{t-1} = 1$ optimal, and for $v_{t-1} = 0$, we get $J(t-1, x_{t-1}, v_{t-1}) =$

$$\max_u \{ (1/4)[(3/4)^{T-t} ux_{t-1} + a_t] + (3/4)[3/4]^{T-t}(1-u)x_{t-1} + b_t] \}$$
$$= (3/4)^{T-(t-1)} x_{t-1} + (1/4)a_t + (3/4)b_t,$$

with $u = u_{t-1} = 0$ optimal.

Note that for all t, the optimal u_t equals v_t.

The entities a_t and b_t are governed by the backwards difference equations $a_{t-1} = (3/4)a_t + (1/4)b_t$, $b_{t-1} = (1/4)a_t + (3/4)b_t$, $a_T = 1$, $b_T = 0$, and so are known. In fact, it is easy to find a formula for them. Adding the right-hand side of the equations, we see that $a_{t-1} + b_{t-1} = a_t + b_t$, so using $a_T + b_T = 1$ yields $a_t + b_t \equiv 1$. So $a_{t-1} = (1/2)a_t + 1/4$, which has the solution $a_t = (1/2)^{T-(t-1)} + 1/2$, while $b_t = 1 - a_t = 1/2 - (1/2)^{T-(t-1)}$. □

Remark 1.4 (State- and time-dependent control region*). The theory above holds also if the control region depends on t, x in the manner that $U = U(t,x) = \{ u : h_i(t,x,u) \geq 0, i = 1, \ldots, i^* \}$, for some given functions h_i's that are continuous in (x,u). If $U(t,x)$ is empty, then, by convention, the maximum over $U(t,x)$ is set equal to $-\infty$. Hence, now $u_t(x_t, v_t)$ has to take values in $U(t, x_t)$, and the maximization in (1.5), respectively (1.6), is carried out over $U(t-1, x_{t-1})$, respectively, $U(T, x_T)$ □

An additional comment is perhaps needed to make quite clear what the problem now is: A maximum of the criterion is sought in the set of all pairs of sequences $\{X_s\}_s$, $\{u_s(x, v)\}_s$ that satisfy the state equation and the condition $u_s(X_s, V_s) \in U(s, X_s)$ a.s. for $s = 0, \dots, T$. If the set of such pairs is empty, the problem has no solution.

Example 1.5. Let us solve the problem in Example 1.1:

$$\max E\left[\sum_{t=0}^{T-1} \beta^t u_t^{1-\gamma} + \beta^T B X_T^{1-\gamma}\right], \tag{i}$$

$$X_{t+1} = Z_{t+1}(X_t - u_t), \qquad u_t \in (0, x_t), \tag{ii}$$

where $0 < \gamma < 1, 0 < \beta < 1, B > 0$, and $Z_t, t = 0, 1, \dots$ are independently distributed non-negative random variables, $EZ_t^{1-\gamma} < \infty$.

Solution. Here $J(T, x_T) = \beta^T B x_T^{1-\gamma}$. To find $J(T-1, x_{T-1})$, we use the optimality equation

$$J(T-1, x_{T-1}) = \max_u \left(\beta^{T-1} u^{1-\gamma} + E\left[\beta^T B(Z_T(x_{T-1} - u))^{1-\gamma}\right]\right). \tag{iii}$$

The expectation must be calculated by using the probability distribution for Z_T. Now the expectation in (iii) is equal to

$$\beta^T B D_T (x_{T-1} - u)^{1-\gamma}, \qquad D_t = E\left[Z_t^{1-\gamma}\right]. \tag{iv}$$

Hence, the expression to be maximized in (iii) is $\beta^{T-1} u^{1-\gamma} + \beta^T B D_T(x_{T-1} - u)^{1-\gamma}$. If we put $u = wx$, $w \in (0, 1)$, and let $\varphi(w) := w^{1-\gamma} + h(1-w)^{1-\gamma}$, where $h = \beta B D_T$, then $J(T-1, x_{T-1}) = \beta^{T-1} x^{1-\gamma} \max_w \varphi(w)$, and we see that we need to solve the maximization problem

$$\max_{w \in (0,1)} \varphi(w) = \max_{w \in (0,1)} \left[w^{1-\gamma} + h(1-w)^{1-\gamma}\right]. \tag{v}$$

We find the maximum of the concave function φ, by solving

$$\varphi'(w) = (1-\gamma)w^{-\gamma} - (1-\gamma)h(1-w)^{-\gamma} = 0,$$

which yields $w^{-\gamma} = h(1-w)^{-\gamma}$. Solving for w yields

$$w = \frac{1}{1 + h^{1/\gamma}}. \tag{vi}$$

Inserting this in φ gives its maximal value

$$\max_w \varphi(w) = 1/(1 + h^{1/\gamma})^{1-\gamma} + h[h^{1/\gamma}/(1 + h^{1/\gamma})]^{1-\gamma} = (1 + h^{1/\gamma})^\gamma. \tag{vii}$$

Define $C_T := B, C_{T-1}^{1/\gamma} := 1 + (\beta B D_T)^{1/\gamma} = 1 + h^{1/\gamma}$, and generally,

$$C_t^{1/\gamma} := 1 + (\beta C_{t+1} D_{t+1})^{1/\gamma}. \tag{viii}$$

Then, the optimal $u_{T-1} = w x_{T-1} = x_{T-1}/C_{T-1}^{1/\gamma}$ and $J(T-1,x_{T-1}) = \beta^{T-1} x_{T-1}^{1-\gamma} \max \varphi(w) = \beta^{T-1} C_{T-1} x_{T-1}^{1-\gamma}$. As $J(T-1,x_{T-1})$ has the same form as $J(T,x_T)$, then, to find the optimal u_{T-2} and $J(T-2,x_{T-2})$, (vi) and (vii) are used for $h = \beta C_{T-1} D_{T-1}$. This yields $u_{T-2} = x_{T-2}/C_{T-2}^{1/\gamma}$ and $J(T-2,x_{T-2}) = \beta^{T-2} C_{T-2} x_{T-2}^{1-\gamma}$. This continues backwards, so evidently we obtain generally $u_t = x_t/C_t^{1/\gamma} \in (0,x_t)$, $(C_t^{1/\gamma} > 1)$, and $J(t,x_t) = \beta^t C_t x_t^{1-\gamma}$.

Note that C_t is a known sequence; it is determined by $C_T = B$ and backwards recursion, using (viii). $\square$

1.2 Infinite Horizon

Suppose that $P_t(v_{t+1}|v_t)$ and f are independent of t, and that f_0 can be written $f_0(t,x,u) = g(x,u)\alpha^t$, $\alpha \in (0,1]$ (f and g continuous). The problem is often called stationary, or autonomous, if these properties hold. Put $\pi = ((u_0(x_0,v_0), u_1(x_1,v_1)\ldots)$. The problem is now

$$\max_{\pi} E\left[\sum_{t=0}^{\infty} \alpha^t g(X_t, u_t(X_t,V_t))\right], \ u_t(X_t,V_t) \in U, \quad \Pr[V_{t+1} = v|v_t] = P(v|v_t), \tag{1.7}$$

where X_t is governed by the stochastic difference equation

$$X_{t+1} = f(X_t, u_t(X_t,V_t), V_{t+1}), \tag{1.8}$$

with X_0, V_0 given, $(g, f, P(v|v_t)$ given entities, the control u_t subject to choice in U, U given). Again, V_{t+1} can also be allowed to be a continuous random variable, governed by a density $p(v|v_t)$, separately piecewise continuous in v and v_t, but independent of t. The maximum in (1.7) is sought when considering all sequences $\pi := (u_0(x_0,v_0), u_1(x_1,v_1), \ldots)$ and selecting the best. We base our discussion upon condition (1.11) below, or P., or N. in Remark 1.6 below, implying that the infinite sum in (1.7) always exists in $[-\infty,\infty]$, for condition (1.11), the sum belongs to $(-\infty,\infty)$. For a given sequence $\pi := (u_0(x_0,v_0), u_1(x_1,v_1), \ldots)$, let us write

$$J_\pi(s,x_s,v_s) = E\left[\sum_{t=s}^{\infty} \alpha^t g(X_t, u_t(X_t,V_t)) \mid x_s, v_s\right], \tag{1.9}$$

where we now start the difference (state) equation at $X_s = x_s$. Let $J(s,x_s,v_s) = \sup_\pi J_\pi(s,x_s,v_s)$. We now prove that $J(1,x_0,v_0) = \alpha J(0,x_0,v_0)$. The intuitive argument is as follows. Let $J_\pi^k(x,v) = \sum_{t=k}^{\infty} E[\alpha^{t-k} g(X_t, u_t(X_t,V_t))|x,v]$ and let $J^k(x,v) =$

$\sup_\pi J_\pi^k(x,v)$. Then $J^k(x,v)$ is the maximal expected present value of future rewards discounted back to $t = k$, given that the process starts at (x,v) at time $t = k$. When starting at (x,v) at time $t = 0$, and discounting back to $t = 0$, the corresponding maximal expected value is $J^0(x,v) = J(0,x,v)$. Because time does not enter explicitly in $P(v|v_t), g$ and f, the future looks exactly the same at times $t = 0$, and $t = k$, hence $J^k(x,v) = J(0,x,v)$. As $J_\pi(k,x_0,v_0) = \alpha^k J_\pi^k(x_0,v_0)$ (in the definition of $J_\pi(k,x_0,v_0)$ we discount back to $t = 0$) and hence $J(k,x_0,v_0) = \alpha^k J^k(x_0,v_0) = \alpha^k J(0,x_0,v_0)$ and in particular $J(1,x_0,v_0) = \alpha J(0,x_0,v_0)$.

The heuristic argument for the optimality equation can be repeated in the infinite horizon case. So (1.5) still holds. Using (1.5) for $t = 1$, and then inserting $\alpha J(0,x,v)) = J(1,x,v)$ and writing $J(x,v) = J(0,x,v), x = x_0, v = v_0$, gives the following *optimality equation*, or *equilibrium optimality equation* or *Bellman equation*

$$J(x,v) = \max_u \{g(x,u) + \alpha E[J(X_1,V_1) \mid x,v]\}, \tag{1.10}$$

where $X_1 = f(x,u,V_1)$.

Observe that (1.10) is a "functional equation," an equation (hopefully) determining the unknown function J (J occurs on both sides of the equality sign). Once J is known, the optimal Markov control is obtained from the maximization in the optimality equation. Evidently, the maximization yields a control function $u(x,v)$, not dependent on t, and this is what we should expect of an optimal control function: If we have observed x,v at time 0 and time t, the optimal choice of control should be the same in the two situations, because then the future looks exactly the same at these two points in time.

It can be shown that the optimal value function $J(x,v) = \sup_\pi J_\pi(0,x,v)$ is defined and satisfies the equilibrium optimality equation in three cases to be discussed below, (1.11), as well as P. and N. in Remark 1.6. (At least this is so when "max" is replaced by "sup" in the equation.) Let us first consider the following case.

$$M_1 \leq g(x,u) \leq M_2 \text{ for all } (x,u) \in \mathbb{R}^n \times U, \tag{1.11}$$

where M_1 and M_2 are given numbers. In case of the boundedness condition (1.11), it is known that the equilibrium optimality equation has a unique bounded solution $J(x,v)$ (when "max" is replaced by "sup," if necessary). Furthermore, $J(x,v)$ is automatically the optimal value function in the problem, and a control $u(x,v)$ giving maximum in the optimality equation, given $J(x,v)$, is the optimal control.

Remark 1.6 (Alternative boundedness conditions*). Complications arise when the boundedness condition (1.11) fails to hold. Then we cannot know for sure that the optimal value function is bounded, so we may have to look for unbounded solutions of the Bellman equation. But then false solutions can occur (bounded or not bounded), not equal to the optimal value function. Even in the case where (1.11) holds, allowing unbounded solutions may lead to nonunique solutions, even a plethora of solutions (see Exercise 1.48 and the following even simpler problem, where $g = 0$, $f = x/\alpha$, and where $J(x) = ax$ satisfies the Bellman equation for all a).

We shall consider two cases, P. and N., where some results can be obtained. In both cases, we must allow for infinite values for the optimal value function $J(x,v)$, $+\infty$ in case P, and $-\infty$ in case N.

P. *Either $g(x,u) \geq 0$ for all $(x,u) \in \mathbb{R}^n \times U$, and $\alpha = 1$, or for some negative number γ, $g(x,u) \geq \gamma$ for all $(x,u) \in \mathbb{R}^n \times U$ and $\alpha \in (0,1)$.*

Let $J^u(x,v)$ be the value function arising from using $u(.,.)$ all the time. In the current case, if $u(x,v)$ yields the maximum in the Bellman equation when $J^u(x,v)$ is inserted, then $u(.,.)$ is optimal. (In other words, if we have been able to find a control $u(x,v)$ such that the pair $(u(x,v), J^u(x,v))$ behaves in this way, then $u(x,.,.)$ is optimal.)

Most often, it can be imagined that first the Bellman equation were solved and a pair $(u(x,v), \hat{J}(x,v))$ satisfying it were found (in particular, then, $u(x,v)$ yields the maximum in the equation). Next, if we are lucky enough to be able to prove that $J^u(x,v) = \hat{J}(x,v)$, then all is well.

Sometimes it is useful to know the fact that if $J^u(s,x,v,T)$ is the value function arising from using $u = u(x,v)$ all the time from s until $t = T$ when starting at (s,x,v), then $J^u(0,x,v,T) \to J^u(x,v)$ as $T \to \infty$. Also $J(0,x,v,T) \to J(x,v)$ as $T \to \infty$, where $J(s,x,v,T)$ is the optimal value function in the problem with finite horizon T, and where we start at (s,x,v).

Note that, in the current case, the optimal value function J is $\leq \hat{J}$ for any other solution $\hat{J}$ of the Bellman equation for which $\hat{J} \geq \hat{M}(1-\alpha)$ for some $\hat{M} \leq 0$.

N. *Either $g(x,u) \leq 0$ for all $(x,u) \in \mathbb{R}^n \times U$, and $\alpha = 1$, or for some positive number γ, $g(x,u) \leq \gamma$ for all $(x,u) \in \mathbb{R}^n \times U$ and $\alpha \in (0,1)$.*

In this case, it is known that if $u(.,.)$ satisfies the Bellman equation with the optimal value function inserted, then $u(.,.)$ is optimal. It is also known that if U is compact, then $J(0,x,v,T) \to J(x,v)$ as $T \to \infty$. (In fact, for this result, we now do need that f and g are continuous and $\mathcal{V}$ is finite. For other assumptions on $\mathcal{V}$, see Section 1.6 below.)

How can this information be used? Assume that we have found a function $\hat{J}(x,v)$ satisfying $\hat{J} \leq \hat{M}(1-\alpha)$ for some $\hat{M} \geq 0$, together with a function $u(x,v)$ satisfying the Bellman equation. If we are able to prove that the Bellman equation has only one such solution $\hat{J}(x,v)$, then this is the optimal value function $J(x,v)$ (because $J(x,v)$ is known to satisfy the Bellman equation, both in case P. and N.), and then $u(x,v)$ is optimal. Another possibility is the following: Suppose that U is compact and that we can apply the limit result $\lim_{T\to\infty} J(0,x,v,T) = J(x,v)$ mentioned above. If we then find that $\lim_{T\to\infty} J(0,x,v,T) = \hat{J}(x,v)$, then $\hat{J}(x,v)$ is the optimal value function and $u(x)$ is optimal.

Note that in case N., the optimal value function J is $\geq \hat{J}$ for any other solution $\hat{J}$ of the Bellman equation for which $\hat{J} \leq \hat{M}(1-\alpha)$ for some $\hat{M} \geq 0$. (This fact lies behind the uniqueness argument in the last paragraph.) $\qquad\square$

Remark 1.7 (Modified boundedness conditions*). The boundedness condition (1.11) and the conditions in P. and N. need only hold for x in $\mathcal{X}(x_0) := \cup_s \mathcal{X}_s(x_0)$, where

$\mathcal{X}_s(x_0)$ is the set of states that can be reached at time s, when starting at x_0 at time 0, considering all outcomes and all controls.

The conclusions drawn in the case where (1.11) is satisfied also hold if the following alternative condition holds: There exist positive constants M, M^*, β, and δ such that for all $x \in \mathcal{X}(x_0)$, all $u \in U$ and all V, $|f(x,u,V)| \leq M + \delta|x|$ and $|g(x,u)| \leq M^*(1+|x|^\beta)$, with $\alpha\delta^\beta < 1$, and $\alpha \in (0,1)$. Moreover, the conclusions in case P. (respectively, case N.) hold if the next to last inequality is replaced by $g(x,u) \geq -M^*(1+|x|^\beta)$ (respectively, $g(x,u) \leq M^*(1+|x|^\beta)$). Note that J needs to be defined only for x in $\mathcal{X}(x_0)$, this set having the property that if x belongs to the set, also $f(x,u,v)$ belongs to it.   □

Example 1.8. Consider the problem

$$\max_{u_t \in (0,1)} E\left[\sum_{t=0}^{\infty} \beta^t x_t^{1-\gamma} u_t^{1-\gamma}\right] \tag{i}$$

$$x_{t+1} = V_{t+1}(1-u_t)x_t, \quad x_0 \text{ is a positive constant.} \tag{ii}$$

Here, $V_1, V_2, \ldots$ are identically and independently distributed non-negative stochastic variables, with $D = EV^{1-\gamma} < \infty$, where V is any of the V_t's. We may think of x_t as the assets of, say, some timeless institution. At each point in time an amount $u_t x_t$ is spent on some useful purpose, and the total effect is measured by the expectation in (i). (For a comment on (ii), see Example 1.1.) It is assumed that

$$\rho = (\beta D)^{1/\gamma} < 1, \quad \beta \in (0,1), \quad \gamma \in (0,1) \tag{iii}$$

Solution. In the notation of problem (1.7), (1.8), $g(x,u) = x^{1-\gamma}u^{1-\gamma}$ and $f(x,u,V) = V(1-u)x$. The equilibrium optimality equation (1.10) yields

$$J(x) = \max_{u \in (0,1)} \left[x^{1-\gamma}u^{1-\gamma} + \beta EJ(V(1-u)x)\right] \tag{iv}$$

We guess that $J(x)$ had the form $J(x) = kx^{1-\gamma}$ for some constant k (the optimal value function had a similar form in the finite horizon version of this problem discussed in the previous section). Then, canceling the factor $x^{1-\gamma}$, (iv) reduces to

$$k = \max_{u \in (0,1)} \left[u^{1-\gamma} + \beta kD(1-u)^{1-\gamma}\right], \tag{v}$$

where $D = EV^{1-\gamma}$. Using the result from Example 1.5 (the maximization of φ) gives that the maximum in (v) is obtained for $u =$

$$u_* = \frac{1}{1+\rho k^{1/\gamma}}, \quad \rho = (\beta D)^{1/\gamma} \tag{vi}$$

and the maximum value in (v) equals $(1+\rho k^{1/\gamma})^\gamma$, so k is determined by the equation

$$k = (1+\rho k^{1/\gamma})^\gamma.$$

Raise each side to the power $1/\gamma$, and solve for $k^{1/\gamma}$ to obtain $k^{1/\gamma} = 1/(1-\rho)$, or $k = (1-\rho)^{-\gamma}$. Hence, the solution is $J(x) = (1-\rho)^{-\gamma} x^{1-\gamma}$, with $u = 1 - \rho$.

In this example, the boundedness condition (1.11) is not satisfied for $x \in \mathcal{X}(x_0)$. One method out is to use the transformation $y_t = x_t/z_t$, $z_{t+1} = V_{t+1} z_t$, $z_0 = 1$, which gives that $y_{t+1} = (1-u)y_t$, $y_0 = x_0$. Replacing x_t by $y_t z_t$, as $Z_t = V_1 \cdot \ldots \cdot V_t$, taking the expectation inside the sum in the criterion (using actually what is called the monotone convergence theorem), the problem can be transformed into a deterministic one. The deterministic difference equation $y_{t+1} = (1-u)y_t$, $y_0 = x_0$ is the state equation, we have a new discount factor $\hat{\beta} = \beta E V^{1-\gamma}$ and a new g-function equal to $y^{1-\gamma} u^{1-\gamma} \in [0, x_0^{1-\gamma}]$ for all $y \in \mathcal{X}(y_0) \subset [0, x_0]$. In this problem, the modified boundedness condition in Remark 1.7 is satisfied. Another way out is the following: Let us use P. in Remark 1.6: Then we need to know that $J^{u_*}(x) = J(x)$. It is fairly easy to carry out the explicit calculation of $J^{u_*}(x)$, by taking the expectation inside the sum and summing the arising geometric series. But we don't need to do that. Noting that $x_t = x_0 \rho^t V_1 \cdot \ldots \cdot V_t$, evidently, we must have that $J^{u_*}(x_0) = k x_0^{1-\gamma}$, for some k. We must also have that $J^{u_*}(x_0)$ satisfies the equilibrium optimality equation with $u = u_*$ and the maximization deleted, (in the problem where $U = u_*$, u_* is optimal!). But the only value of k for which this equation is satisfied we found above. Thus the test in P. works and u_* as specified in (vi) is optimal.   □

1.3 State and Control-Dependent Probabilities

Suppose that the state equation is still of the form

$$X_{t+1} = f(t, X_t, u_t, V_{t+1}), \quad x_0, v_0 \text{ are given} \tag{1.12}$$

where V_{t+1} takes values in a finite set $\mathcal{V} = \{\bar{v}_0, \ldots, \bar{v}_m\}$, whose elements have probabilities $Pr[V_{t+1} = \bar{v}_0] = P^{(0)}(t, x_t, u_t, v_t), \ldots, Pr[V_{t+1} = \bar{v}_m] = P^{(m)}(t, x_t, u_t, v_t)$, respectively, hence these probabilities are conditional ones, also written $P_t(v|x_t, u_t, v_t)$, $v \in \mathcal{V}$. Thus, the probability $Pr[V_{t+1} = v] = P_t(v|x_t, u_t, v_t)$ of the event $V_{t+1} = v$, is supposed to depend on the time t, the outcome v_t, the state x_t, and the control u_t we select at time t. We may allow $\mathcal{V}$ instead to be all $\mathbb{R}^{\hat{n}}$, for some $\hat{n}$, thus allowing the V_t's to be continuous stochastic variables. Then the distribution of V_{t+1} is often given by a density $p_t(v|x_t, u_t, v_t)$, separately piecewise continuous in each component of v, x_t, u_t and v_t. In the main theoretical discussions, we mostly stick to discrete random variables. However, the solution tools presented can also be used for continuous stochastic variables. Again it is assumed that x_t belongs to $\mathbb{R}^n$, that u_t belongs to a given subset U of $\mathbb{R}^r$, and that $t = 0, \ldots, T$.

Example 1.9. A machine is supposed to be in one of three states. Either "as good as new," denoted by (2), or "functioning" (1), or "broken" (0). After having been used all day, the machine is checked in the evening and its state is determined. The following table describes the "transition probabilities" of the state from one evening to the next one (it is an example of a so-called Markov process).

The state next evening

(1)		0	1	2
The state when the	0	1	0	0
machine is checked	1	0.4	0.6	0
	2	0.2	0.4	0.4

The table should be understood as follows. The first column lists the three possible states of the machine, when it is checked in the evening. The uppermost row shows the possible states of the machine after it has run for one day. If the machine is, say, "as good as new," i.e., in state 2, then the last row says that upon checking the machine the next evening there is a probability of 0.2 of finding that it is "broken" (state 0), a probability of 0.4 of finding that it is functioning (state 1), and a probability of 0.4 of finding that it is as good as new (state 2). The other two rows below the bar are read similarly. If we use the symbols above, we let $\bar{v}_0 = 0$, $\bar{v}_1 = 1$, $\bar{v}_2 = 2$, $X_{t+1} = f(t, x_t, u_t, V_{t+1}) = V_{t+1}$, where $x_{t+1} \in \{0, 1, 2\}$ and $v_t \in \{0, 1, 2\}$. Moreover, $u_t \in \{0, 1\}$, $u_t = 0$ means that we do not repair the machine after the evening check, whereas $u_t = 1$ means that we repair it. The above table describes the situation the next evening if we do not repair the machine. The elements in the matrix in (1) are hence the probabilities $P^{(i)}(t, x_t, 0) = P_t^{(i)}(t, x_t, u_t = 0)$, $i = 0, 1, 2$, $x_t = 0, 1, 2$, where i gives the column number and x_t the row number.

If the machine is repaired one evening, then it is simply assumed that it is as good as new (in state 2) the next evening. Thus, $P_t^{(2)}(t, x_t, 1) = 1$ and $P_t^{(i)}(t, x_t, 1) = 0$ for $i = 0, 1$, regardless of x_t.   □

Let us return to the general problem. The process determined by (1.12) and the random events V_1, V_2, ..., is to be controlled in the best possible manner by appropriate choices of the variables u_t. The criterion to be maximized is the expectation

$$E\left[\sum_{t=0}^{T} f_0(t, X_t, u_t(X_t, V_t), V_t)\right]. \tag{1.13}$$

Again, each control u_t, $t = 0, 1, 2, \ldots, T$ should be a function, $u_t(x_t, v_t)$, $t = 0, \ldots, T$, of the current state x_t and the current outcome v_t. To compute the expectation in (1.13), i.e., to calculate $E[f_0(t, X_t, u_t(X_t, V_t), V_t)]$ for any given t, requires specifying the probabilities that lie behind the calculation of this expectation. Let us consider the case where V is discrete. Given that the policies $u_0(x_0, x_0)$, ..., $u_T(x_T, v_T)$ are used, note first that $X_s = X_s(V_1, \ldots, V_s)$ in other words, X_s depends on the outcomes of $V_1, \ldots, V_s$. The probability of the joint event $V_1 = v_1$, $V_2 = v_2$, ..., $V_t = v_t$, is given by $p^*(v_1, \ldots, v_t) =$

$$P_0(v_1|x_0, u_0, v_0) \cdot P_1(v_2|x_1, u_1, v_1) \cdot \ldots \cdot P_{t-1}(v_t|x_{t-1}, u_{t-1}, v_{t-1}) \tag{1.14}$$

where $u_0 = u_0(x_0, v_0)$, $u_1 = u_1(x_1, v_1), \ldots, u_{T-1} = u_{T-1}(x_{T-1}, v_{T-1})$ and where each $x_s = x_s(v_1, \ldots, v_s)$ (the x_s's forming a solution sequence of the state equation for the specified control sequence), so the expression in (1.14) is a function (only) of

$(v_1, \ldots, v_t)$. Similarly, when inserting $X_t = X_t(V_1, \ldots, V_t)$ in $f_0(t, X_t, u_t(X_t, V_t), V_t)$, this function becomes a function only of $(V_1, \ldots, V_t)$ and the probabilities for the various outcomes $(v_1, \ldots, v_t)$ we have already specified, so $E[f_0(t, X_t, u_t(X_t, V_t), V_t)]$ can be calculated. Thus, the expression in (1.13) is equal to

$$\left(\sum_{t=0}^{T} \sum_{v_1, \ldots, v_t} f_0(t, x_t, u_t(x_t, v_t), v_t) \right) p^*(v_1, \ldots, v_t), \qquad (1.15)$$

where the inner sum is taken over all combinations of values $(v_1, \ldots, v_t)$. The probabilities $p^*(v_1, \ldots, v_t)$, and hence the expected value, depend on the policies chosen, so sometimes we write $E_{u_0, \ldots, u_T}$ instead of E in (1.13).

Though not always necessary, we shall assume that f_0 and f are continuous in (x, u), even in (x, u, v) if $\mathcal{V}$ is nondiscrete.

The optimization problem is to find a sequence of policies $u_0^*(x_0, v_0), \ldots, u_T^*(x_T, v_T)$, which gives the expression in (1.13) the largest possible value, subject to the difference equation (1.12).

We now define

$$J(t, x_t, v_t) = \sup E_{u_t, \ldots, u_T} \left[\sum_{s=t}^{T} f_0(s, X_s, u_s(X_s, V_s), V_s) \; \middle| \; x_t, v_t \right], \qquad (1.16)$$

where the supremum is taken over all policy sequences $u_s = u_s(x_s, v_s), s = t, \ldots, T$, given v_t and given that we start at the state x_t at time t, as indicated by "$| x_t, v_t$." The computation of the expectation is now based on conditional probabilities of the form

$$P_t(v_{t+1} | x_t, u_t(x_t, v_t), v_t) \cdot \ldots \cdot P_{T-1}(v_T | x_{T-1}, u_{T-1}(x_{T-1}, v_{T-1}), v_{T-1}).$$

In (1.16), and in these probabilities, given $u_t(\cdot, \cdot), \ldots, u_{T-1}(\cdot, \cdot)$, for $s = t + 1, \ldots, T$, x_s is a function of $(v_{t+1}, \ldots, v_s)$ (and the given v_t), as well as of the given start value x_t, again determined by the difference equation (1.12).

(We seek a maximum in (1.16), and in a similar definition in Section 1.1, we wrote max and not sup. When we write max we indirectly say that a maximum exists, and being a little more formal in this section, we don't want to include such an assumption in the definition. A similar remark pertains to the optimality equation (1.17), (1.18) below.)

Again, the central tool in solving optimization problems of the type (1.12)–(1.13) is the following *optimality* equation (we write also here sup instead of max). For $t < T$

$$J(t-1, x_{t-1}, v_{t-1}) = \sup_{u_{t-1}} \Big\{ f_0(t-1, x_{t-1}, u_{t-1}, v_{t-1})$$

$$+ \sum_{v_t \in \mathcal{V}} P_{t-1}(v | x_{t-1}, u_{t-1}, v_{t-1}) J(t, f(t-1, x_{t-1}, u_{t-1}, v_t), v_t) \Big\}. \quad (1.17)$$

Of course also here, if possible, we want to maximize, and in the maximization, the vector u_{t-1} is constrained to lie in U. The equation can be written more concisely as

$$J(t-1,x_{t-1},v_{t-1}) = \sup_{u_{t-1}}\{f_0(t-1,x_{t-1},u_{t-1},v_{t-1})$$

$$+ E_{u_{t-1}}[J(t,X_t,V_t) \mid x_{t-1},v_{t-1}]\}. \qquad (1.18)$$

Of course, this version is also valid for continuous stochastic variables. Moreover, when $t = T$, we must have

$$J(T,x_T,v_T) = \sup_{u_T} f_0(T,x_T,u_T,v_T). \qquad (1.19)$$

The intuitive argument for (1.17) is exactly as before: Suppose the system is in state x_{t-1}. For a given u_{t-1}, the "instantaneous" reward is equal to $f_0(t,x_{t-1},u_{t-1},v_{t-1})$. In addition, the sum of rewards at all later times is at most $J(t,x_t)$ if $x_t = f(t-1,x_{t-1},u_{t-1},v)$, and the probability of this event is $P_{t-1}(v|x_{t-1},u_{t-1},v_{t-1})$. When using u_{t-1}, the total expected maximum value gained over all future time points (now including even $t-1$) is the sum in (1.17). The largest expected gain comes from choosing u_{t-1} to maximize this sum.

A formal proof is presented later on. In connection with the proof, certain theoretical questions are discussed. In particular, it can be shown that the maximal value of the criterion cannot be increased by allowing policies that depend on past states as well as on the present state.

Remark 1.10 (Criterion to be minimized). Suppose that we want to minimize the value of the criterion. Then, to obtain the optimal value functions $J(t,x_t,v_t)$ a minimization is carried out instead of a maximization. In the optimality equation, "max" (or "sup") must then be replaced by "min" ("inf").

To see that this is correct, recall that to minimize a criterion is the same as maximizing (-1) times the criterion. Thus, we can apply the above "maximization theory" to a problem where f_0 is replaced by $-f_0$. From this it is easy to see that the "min"-version of the optimality equation follows.   □

Example 1.11. Consider Example 1.9 again. In this example, the values of f_0 will be costs, rather than rewards. Let the values of the function f_0 for all t be given by the table

		u	
		0	1
(2)	0	2	5
x	1	0	1
	2	0	1/2

From the table, we see for instance that $f_0(t,x_t,u_t) = f_0(x_t,u_t) = 5$ when $x_t = 0$, $u_t = 1$. The costs in the table may be interpreted as follows: A broken machine leads to lost sales. But, if it is repaired, then that will add to the costs (see the numbers

2 and 5 in the table). Repair carried out on a machine in better shape costs less, as indicated by the last column. We are going to use the machine in a production run over a period of three days. Before we start (i.e., at time $t = 0$), the machine is in state 1. Because we are going to minimize costs, we replace sup with inf, (or min) in (1.18) and (1.19), see Remark 1.10. For $J(3, X_3)$ we get:

(3)
$$J(3,0) = 2 \qquad J(3,1) = 0 \qquad J(3,2) = 0$$
$$u_3^* = 0 \qquad u_3^* = 0 \qquad u_3^* = 0$$

We naturally choose $u_3^* = 0$ because we shall not produce anything the next day. Let us compute $J(2,x_2)$ for $x_2 = 0, 1, 2$.

First let $x_2 = 0$. If $u = 0$ is chosen, then the expected cost is $f_0(0,0) + 1 \cdot J(3,0) + 0 \cdot J(3,1) + 0 \cdot J(3,2) = 2 + 1 \cdot 2 + 0 \cdot 0 + 0 \cdot 0 = 4$, where the factors 1, 0, 0 make up the first row in the matrix (1) in Example 1.9. If $u = 1$ is chosen, the expected cost is $f_0(0,1) + 1 \cdot J(3,2) = 5 + 1 \cdot 0 = 5$. (Recall that a newly repaired machine is still as good as new $(x = 2)$ after one day's use.) The minimum of the numbers 4 and 5 is 4, attained by $u = 0$, so $J(2,0) = 4$.

Next, let $x_2 = 1$. If $u = 0$ is chosen, the expected cost is $f_0(1,0) + 0.4 \cdot J(3,0) + 0.6 \cdot J(3,1) + 0 \cdot J(3,2) = 0 + 0.4 \cdot 2 + 0.6 \cdot 0 + 0 \cdot 0 = 0.8$, where the factors 0.4, 0.6, and 0 make up the second row in table (1) in Example 1.9. If $u = 1$ is chosen, the expected cost is $f_0(1,1) + 1 \cdot J(3,2) = 1$. The minimum of the numbers 0.8 and 1 is 0.8, attained for $u = 0$.

Finally, put $x_2 = 2$. If $u = 0$ is chosen, we get $f_0(2,0) + 0.2 \cdot J(3,0) + 0.4 \cdot J(3,1) + 0.4 \cdot J(3,2) = 0 + 0.2 \cdot 2 + 0.4 \cdot 0 + 0.4 \cdot 0 = 0,4$. If $u = 1$ is chosen, we get $f_0(2,1) + 1 \cdot J(3,2) = 0,5 + 1 \cdot 0 = 0.5$. The minimum of the numbers 0.4 and 0.5 is 0.4, attained for $u = 0$. We summarize our calculations thus:

(4)
$$J(2,0) = 4 \qquad J(2,1) = 0.8 \qquad J(2,2) = 0.4$$
$$u_2^* = 0 \qquad u_2^* = 0 \qquad u_2^* = 0$$

Let us compute $J(1,x_1)$, $x_1 = 0, 1, 2$, in the same way.

Let $x_1 = 0$. If $u = 0$ is chosen, we get $f_0(0,0) + 1 \cdot J(2,0) + 0 \cdot J(2,1) + 0 \cdot J(2,2) = 2 + 1 \cdot 4 + 0 \cdot 0.8 + 0 \cdot 0.4 = 6$. If $u = 1$ is chosen, we get $f_0(0,1) + 1 \cdot J(2,2) = 5 + 1 \cdot 0.4 = 5.4$. The minimum, 5.4, is attained for $u = 1$.

Next, let $x_1 = 1$. If $u = 0$ is chosen, the expected cost is $f_0(1,0) + 0.4 \cdot J(2,0) + 0.6 \cdot J(2,1) + 0 \cdot J(2,2) = 0 + 0.4 \cdot 4 + 0.6 \cdot 0.8 + 0 \cdot 0.4 = 2.08$. If $u = 1$ is chosen, the expected cost is $f_0(1,1) + 1 \cdot J(2,2) = 1 + 0.4 = 1.4$. The minimum, 1.4, is attained for $u = 1$.

Finally, let $x_1 = 2$. If $u = 0$ is chosen, we obtain $f_0(2,0) + 0.2 \cdot J(2,0) + 0.4 \cdot J(2,1) + 0.4 \cdot J(2,2) = 0 + 0.2 \cdot 4 + 0.4 \cdot 0.8 + 0.4 \cdot 0.4 = 1.28$. If $u = 1$ is chosen, we get $f_0(2,1) + 1 \cdot J(2,2) = 0.5 + 1 \cdot 0.4 = 0.9$. The minimum, 0.9, is attained for $u = 1$.

This gives the following table:

(5)
$$J(1,0) = 5.4 \qquad J(1,1) = 1.4 \qquad J(1,2) = 0.9$$
$$u_1^* = 1 \qquad u_1^* = 1 \qquad u_1^* = 1$$

From (5), we now conclude that if two production days remain, we always repair the machine, whatever its state. If only one production day is left, then it is too expensive to repair the machine for such a short spell of time.   □

In the next example, we go back to a very simple probability structure. (Recall that any minimization problem can be rewritten as a maximization problem by changing the sign of the criterion function, and in case of minimization, we get minimization also in the optimality equation.)

Example 1.12 (Linear quadratic multidimensional problem). Let H' be the transpose of the matrix H, and call a symmetric $n \times n$ matrix positive definite if, for all $x \in \mathbb{R}^n$, $x \neq 0$, $x'Hx > 0$, and positive semidefinite if $x'Hx \geq 0$. Consider the following problem with n state variables and r control variables:

$$\min_{u_0,\ldots,u_T} E \left[\sum_{0 \leq t \leq T} x_t'R_t x_t + u_t'Q_t u_t \right], \qquad (1.20)$$

where R_t and Q_t are given symmetric positive definite square matrices. The minimization is subject to the condition (equation)

$$x_{t+1} = A_t x_t + B_t u_t + \varepsilon_t, \quad u_t \in \mathbb{R}^r, \quad x_0 \text{ given in } \mathbb{R}^n, \qquad (1.21)$$

where A_t and B_t are given $n \times n$ and $n \times r$ matrices, respectively, and where the random variables ε_t are independently distributed with mean zero and finite covariance matrices, their distributions being independent of history.

Solution. We will need the following result: Let Q be a symmetric and positive definite $r \times r$–matrix, let C be a symmetric and positive semidefinite $n \times n$–matrix, let A be a $n \times n$–matrix, and let B be an $n \times r$–matrix. The following equality is obtained by a completing-the-square argument presented below:

$$h(u) := u'Qu + (Ax + Bu)'C(Ax + Bu) = (u' + x'H')K(u + Hx) + x'Jx, \quad (*)$$

where $K = Q + B'CB$, $H = K^{-1}B'CA$, $J = A'CA - H'KH = A'CA - A'CB(Q + B'CB)^{-1}B'CA$ (K is symmetric and positive definite).
The equality (*) follows from $h(u) =$

$$u'Qu + u'B'CBu + x'A'CAx + x'A'CBu + u'B'CAx$$
$$= u'Ku + x'A'CAx + x'A'CBK'^{-1}K'u + u'KK^{-1}B'CAx$$
$$= u'Ku + u'KHx + x'H'Ku + x'H'KHx + x'Jx$$
$$= (u' + x'H')K(u + Hx) + x'Jx.$$

The minimum point and minimal value of $h(u)$ are evidently given by

$$u = -Hx, \quad \min_u h(u) = x'Jx. \qquad (**)$$

Define the symmetric, positive definite matrix C_t by the (backwards) Riccati equation

$$C_t = R_t + A_t'C_{t+1}A_t - (A_t'C_{t+1}B_t)(Q_t + B_t'C_{t+1}B_t)^{-1}(B_t'C_{t+1}A_t), \qquad (1.22)$$

$C_{T+1} = 0$. As a backwards induction hypothesis, assume that $J(t,x)$ is of the form $x'C_tx + d_t$ for t replaced by $t+1$ and let us prove that then the formula is also correct for t (it *is* correct for $t = T$, for $C_T = R_T, d_T = 0$). Using the induction hypothesis, the optimality equation is:

$$J(t,x) = \min_u\{x'R_tx + u'Q_tu + E(A_tx + B_tu + \varepsilon_t)'C_{t+1}(A_tx + B_tu + \varepsilon_t)\} + d_{t+1}$$

Now,

$$E[(A_tx + B_tu + \varepsilon_t)'C_{t+1}(A_tx + B_tu + \varepsilon_t)] = (A_tx + B_tu)'C_{t+1}(A_tx + B_tu)$$
$$+ E[(A_tx + B_tu)'C_{t+1}\varepsilon_t + \varepsilon_t'C_{t+1}(A_tx + B_tu)] + E(\varepsilon_t'C_{t+1}\varepsilon_t), \qquad (1.23)$$

where the second term on the right-hand side vanishes. Only the first of the three terms is relevant to the minimization, because the third one is independent of u, so

$$J(t,x) = \min_u\{x'R_tx + u'Q_tu + (A_tx + B_tu)'C_{t+1}(A_tx + B_tu)\} + d_t,$$

where $d_t = E(\varepsilon_t'C_{t+1}\varepsilon_t) + d_{t+1}$. Using $(**)$, we have that the optimal control $u = u_t$ satisfies $u_t = -D_tx$, where $D_t = (Q_t + B_t'C_{t+1}B_t)^{-1}B_t'C_{t+1}A_t$. Moreover, using $(**)$ and (1.22), we get $J(t,x) = x'C_tx + d_t$, where d_t satisfies the backwards recursion

$$d_t = d_{t+1} + \Sigma_{i,j}N_t^{ij}C_{t+1}^{ij}, N_t = Cov(\varepsilon_t), d_T = 0,$$

(the top indices ij indicating elements in the matrices). We have obtained results in conformity with the so-called "certainty equivalence principle," namely that the control is the same as that obtained by taking expectation on the right-hand side of the state equation, i.e., by putting $\varepsilon_t = 0$, as if there were no uncertainty. This is a rather exceptional result, completely dependent on the particular structure of the problem.   □

Proof of the optimality equation (1.18), (1.19)

The proof is provided for the specially interested reader and we assume that $\mathcal{V}$ is finite. Write $z_t = (x_t, v_t)$. For simplicity, $f_0(s,.,.,.)$ is assumed to be independent of v_s. We define, as before,

$$J(t,z_t) = \sup_{u_t,\dots,u_T} E_{u_t,\dots,u_T}\left[\sum_{s=t}^T f_0(s,X_s,u_s(Z_s)) \,\Big|\, z_t\right], \qquad (1.24)$$

for $t < T$, with

$$J(T,z_T) = \sup_{u_T} f_0(T,x_T,u_T). \qquad (1.25)$$

The optimality equation to be proved is

$$J(t-1,z_{t-1}) = \sup_{u_{t-1}}\{f_0(t-1,x_{t-1},u_{t-1}) + E_{u_{t-1}}[J(t,Z_t) \mid z_{t-1}]\}. \qquad (1.26)$$

In the proof, we shall consider a larger class of control policies, namely the general history-dependent controls $u_t(z_1,\ldots,z_t)$. Thus the controls are allowed to depend on all previous events v and states x. The proof to be presented makes it possible to answer the following question: Is it possible to achieve even better results if we are allowed to select policies from this larger collection of policies?

The argument below uses the following iterated expectation rule that can be found in standard texts on probability theory (see also the Appendix):

$$E[Y \mid X_1,\ldots,X_m] = E[E[Y \mid X_1,\ldots,X_n] \mid X_1,\ldots,X_m], m < n.$$

Let us write $J(t,z_{\to t})$ for the value that results when the policies in (1.24) are chosen from the class of policies $u_s(z_{\to s}) := u_s(z_1,\ldots,z_s)$, where the symbol $z_{\to s}$ means the sequence $(z_1,\ldots,z_s)$, and where we condition on $z_{\to t}$ rather than on just z_t.

Write $E^{s-1} := E_{u_{s-1},\ldots,u_T}$ (if the probabilities P_t do not depend on the controls, drop the superscripts $s-1$ and s on E below, the reader may want to concentrate on this slightly simpler case). The following sequence of equalities will be explained shortly:

$$
\begin{aligned}
J(s-1,z_{\to s-1}) &= \sup_{t \geq s-1} E^{s-1}\left[\sum_{\tau=s-1}^{T} f_0(\tau,X_\tau,u_\tau(Z_{\to \tau})) \mid z_{\to s-1}\right] \\
&= \sup_{u_{s-1}(\cdot)}\left[f_0(s-1,x_{s-1},u_{s-1}(z_{\to s-1}))\right. \\
&\qquad \left. + \sup_{t \geq s} E^{s-1}\left\{\sum_{\tau=s}^{T} f_0(\tau,X_\tau,u_\tau(Z_{\to \tau})) \mid z_{\to s-1}\right\}\right] \\
&= \sup_{u_{s-1}(\cdot)}\left[f_0(s-1,x_{s-1},u_{s-1}(z_{\to s-1}))\right. \\
&\qquad \left. + \sup_{t \geq s} E^{s-1}\left[E^s\left\{\sum_{\tau=s}^{T} f_0(\tau,X_\tau,u_\tau(Z_{\to \tau})) \mid Z_{\to s}\right\} \mid z_{\to s-1}\right]\right] \\
&= \sup_{u_{s-1}(\cdot)}\left[f_0(s-1,x_{s-1},u_{s-1}(z_{\to s-1}))\right. \\
&\qquad \left. + E^{s-1}\left\{\sup_{t \geq s} E^s\left[\sum_{\tau=s}^{T} f_0(\tau,X_\tau,u_\tau(Z_{\to \tau})) \mid Z_{\to s}\right] \mid z_{\to s-1}\right\}\right] \\
&= \sup_{u_{s-1}}\left[f_0(s-1,x_{s-1},u_{s-1}) + E^{s-1}[J(s,Z_{\to s}) \mid z_{\to s-1}]\right]. \qquad (1.27)
\end{aligned}
$$

Here $\sup_{t \geq s-1}$ means the supremum over all policy sequences $u_{s-1}(\cdot), \ldots, u_T(\cdot)$, and $\sup_{t \geq s}$ has a corresponding interpretation. In the last line, $u_{s-1}(.)$ and $u_{s-1}(z_{\to s-1})$ are abbreviated to u_{s-1}. The first equality repeats the definition. The second equality follows from a general rule stating that joint suprema equal iterated suprema (so we can first take supremum over $u_s(\cdot), \ldots, u_T(\cdot)$, and afterwards over u_{s-1}). The third equality is the iterated expectation rule. The fifth (last) equality follows from the definition of $J(s, z_{\to s})$. The fourth equality requires a few more words: Consider the problem of finding the supremum of a sum $\sum a_i(w_i)q_i, q_i \geq 0$, where the variable w_i, for every i, can be chosen from some set W_i. Evidently, $\sup \sum a_i(w_i)q_i = \sum \sup a_i(w_i)q_i$, to "maximize" the sum, make each term as large as possible, a rule that works when there is no interdependence between the choices of the different w_i's. The supremum in the fifth line is of this type: Given $z_{\to s-1}$, the inner expectation depends on X_s, V_s, and $u_s(.), \ldots, u_T(.)$, the dot indicating here (as elsewhere) functions (i.e., the functional forms of these controls), when we calculate this expectation, we then imagine that the X_τ's occurring both in $f_0, u_\tau, \tau > s$, and in the probabilities appearing as factors in the products are expressed by means of X_s, V_s and the controls, using the difference equation. Write $\bar{z}_i = (f(s-1, x_{s-1}, u_{s-1}(z_{\to s-1}), \bar{v}_i), \bar{v}_i)$, When i runs through $\{1, \ldots, m\}$, these are the possible values (X_s, V_s) takes. The outer expectation is a sum of the type $\sum a_i(w_i)q_i$ where q_i is the probability $P_{s-1}(\bar{v}_i | x_{s-1}, u_{s-1}, v_{s-1})$ and where $w_i = (u_s(z_{\to s-1}, \bar{z}_i), u_{s+1}(z_{\to s-1}, \bar{z}_i, z_{s+1}), \ldots, u_T(z_{\to s-1}, \bar{z}_i, z_{s+1}, \ldots, z_T))$. Here $z_{\to s-1}$ is fixed and $w_i = w_i(z_{s+1}, \ldots, z_T)$, $i = 1, \ldots, m$, can be chosen independently of each other. (This point in the proof makes essential use of the history-dependence of the controls.) But then, as was just seen, the "sup" can be taken inside the sum, in other words, the fourth equality is valid. $\qquad\square$

Note that the supremum in the last line in (1.27) is evidently the same whether we consider u_{t-1} to be a vector ranging through U or a function ranging through the set of history-dependent controls; it is the latter version of (1.27) that we obtained from the arguments above, see the last line in (1.27), where, by the way, E^{s-1} evidently can be replaced by $E_{u_{s-1}(z_{\to s-1})}$.

Having justified (1.27), we shall prove that for all s, $J(s, z_{\to s}) = J(s, x_s, v_s)$. The proof goes by backwards induction. Consider first the case where $s = T$. By definition, $J(T, z_{\to T}) = \sup_{u_T} f_0(T, x_T, u_T)$. The supremum and hence $J(T, \ldots)$ depend only on x_T. By induction, assume that we can write $J(s, z_{\to s}) = J(s, x_s, v_s)$. Take a look at the last line in (1.27). By the induction hypothesis, we can replace $J(s, Z_{\to s})$ by $J(s, X_s, V_s)$. Disregarding u_{s-1}, the expression $E_{u_{s-1}}[J(s, X_s, V_s) \mid z_{\to s-1}] =$

$$E_{u_{s-1}}[J(s, f(s-1, x_{s-1}, u_{s-1}, V_s), V_s) \mid z_{\to s-1}]$$

depends on $z_{\to s-1}$ only through (x_{s-1}, v_{s-1}) (remember that the probabilities needed to calculate this conditional expectation depend only on x_{s-1}, v_{s-1}). Hence, the supremum in the last line is a supremum of a function only depending on x_{s-1}, v_{s-1} (and s) in addition to the "maximization" variable u_{s-1}. Thus, this supremum depends only on x_{s-1}, v_{s-1} (and s), so we can write $J(s-1, z_{\to s-1}) = J(s-1, x_{s-1}, v_{s-1})$ and the induction argument is finished. From this it follows that (1.26) is satisfied.

Let us next consider the necessary and sufficient conditions connected with the optimality equation for a sequence of controls $u_t^*(x_t, v_t)$ to be optimal. First, write

$$J(t, x_t, v_t; u_t, \ldots, u_T) := E_{u_t, \ldots, u_T}\left[\sum_{s=t}^{T} f_0(s, X_s, u_s(Z_s)) \mid x_t, v_t\right],$$

and note that $J(t, x_t, v_t; u_t, \ldots, u_T)$ satisfies

$$J(t-1, x_{t-1}, v_{t-1}; u_{t-1}, \ldots, u_T) = f_0(t-1, x_{t-1}, u_{t-1}(z_{t-1}), v_{t-1})$$
$$+ E_{u_{t-1}(z_{t-1})}[J(t, X_t, V_t; u_t, \ldots, u_T) \mid x_{t-1}, v_{t-1}]. \qquad (1.28)$$

(This is a special case of the optimality equation, in which there is a single choice available for u_{t-1}, of course the equality follows by the double expectation rule again.)

First a proof of sufficiency: Suppose that the suprema in (1.26) are attained (i.e., sup can be replaced by max), that maximum is obtained by the vectors $u_{t-1}^*(x_{t-1}, v_{t-1})$ for all t, and likewise that max in (1.25) is obtained by $u_T^*(x_T) = u_T^*(x_T, v_T)$. Then, by definition, $u_T^*(x_T)$ gives us the optimal optimal value $J(T, x_T) = J(T, x_T, v_T)$. Now, letting $t = T$ in (1.26) and using $J(T, X_T, V_T) = J(T, X_T, V_T, u_T^*(x_T))$, it follows from (1.26) and (1.28) that $J(T-1, x_{T-1}, v_{T-1}) = J(T-1, x_{T-1}, v_{T-1}; u_{T-1}^*, u_T^*)$. (Remember that u_{T-1}^* gives maximum in (1.26) in this case.) As a backward induction hypothesis, assume

$$J(t, x_t, v_t) = J(t, x_t, v_t; u_t^*, \ldots, u_T^*). \qquad (1.29)$$

Substituting the latter expression for the former one in (1.26), and using the fact that u_{t-1}^* gives the maximum, by (1.28) we obtain (1.29) even for $t-1$. Thus the equality (1.29) holds for all t. This means that the policies u_t^* are optimal: The optimality equation is a sufficient condition for optimality.

Next, let us prove necessity. So let $u_0^*(., .), \ldots, u_T^*(., .)$ maximize the criterion (1.13). If, by contradiction, $u_s^*(x_s, v_s)$ for some x_s, v_s does not satisfy the optimality equation with $t-1$ replaced with s, then we know that $J(s, x_s, v_s)$ is $> J(s, x_s, v_s; u_s^*, u_{s+1}^*, \ldots, u_T^*)$. If we know that we can attain (x_s, v_s) with positive conditional probability, given x_{s-1}, v_{s-1} and using u_{s-1}^*, then, the optimality equation gives that also $J(s-1, x_{s-1}, v_{s-1}) > J(s-1, x_{s-1}, v_{s-1}; u_{s-1}^*, u_s^*, u_{s+1}^* \ldots, u_T^*)$. If we also can obtain a pair x_{s-1}, v_{s-1} with positive conditional probability, given x_{s-2}, v_{s-2} and using u_{s-2}^*, such that (x_s, v_s) can be attained with positive probability, given this x_{s-1}, v_{s-1}, then $J(s-2, x_{s-2}, v_{s-2}) > J(s-2, x_{s-2}, v_{s-2}; u_{s-2}^*, u_{s-1}^*, u_s^*, u_{s+1}^* \ldots, u_T^*)$. In this manner, we can continue backwards and obtain that $J(0, x_0, v_0) > J(0, x_0, v_0; u_0^*, \ldots, u_{s-1}^*, u_s^*, u_{s+1}^* \ldots, u_T^*)$ if there is a solution sequence $x_1^*, \ldots, x_s^*$ of (1.12), given $u_t^*, t < s$, such that (x_s^*, v_s) can be attained with a positive probability. A contradiction has arisen. This argument entails that the optimality equation is also necessary, at least for points (x_{t-1}, v_{t-1}), the occurrence of which has a positive probability, given $u_0^*, \ldots, u_T^*$. $\square$

In practice, we often search for control functions that satisfy the optimality equation for every x_{t-1}, v_{t-1}.

Let us write down in a more precise form what we have now proved (see Theorem 1.21 for a slightly more precise version in case $\mathcal{V}$ is nondiscrete).

Theorem 1.13. *If a sequence of pairs* $(u_t^*(x_t, v_t), J(t, x_t, v_t))$, $t = 0, \ldots, T$, *has been found such that the optimality equations (1.19) and (1.18) are satisfied for all* (x_T, v_T), *respectively,* (x_{t-1}, v_{t-1}), $u_t^*(x_t, v_t)$ *yielding maximum in the equations, then* $u_t^* = u_t^*(x_t, v_t)$, $t = 0, \ldots, T$, *are optimal. On the other hand, if* $u_t^* = u_t^*(x_t, v_t)$, $t = 0, \ldots, T$, *are optimal, then in the optimality equations (1.19), respectively (1.18), where now the optimal value functions appear, these controls must yield the suprema at any* (x_T, v_T), *respectively* (x_{t-1}, v_{t-1}), *that can be reached with positive probability from the given start point* x_0, *given* $u_t^* = u_t^*(x_t, v_t), t = 0, \ldots, T$.   □

Let us now return to history-dependent controls. We showed above that $J(t, x_t, v_t) = J(t, z_{\to t})$. Now, considering only the case where suprema are attained (the general case is only slightly more complicated), we now know from the sufficiency results above that the $u_t^*(x_t, v_t)$'s coming out of the optimality equations are optimal, hence, $J(t, x_t, v_t) = J(t, x_t, v_t; u_t^*, \ldots u_T^*) = J(t, z_{\to t})$. Thus, the more general history-dependent controls do not "achieve more" as measured by the criterion than Markov controls.

Remark 1.14 (Complete observability). Important questions in stochastic optimization over time are what it is possible to observe and what it is necessary to remember. We need to be able to observe the x_t, v_t's, but we do not need to store past values, in order to behave optimally. If the probabilities P_t and $f_0(t, ., ., .)$ do not depend on v_t, we have noted that it does not do any harm, if the v_t's are unobservable. If, however, say, the x_t's are only "partially observable," then the controls depending on these "partial observations" (even on their entire history) will frequently be inferior to the optimal ones in the inaccessible "completely observable" case. A particular type of such problems is discussed in Sections 1.7, 1.8 below.   □

Remark 1.15 (Control sets $U(t, x)$ and $f_0 = f_0(t, x_t, u_t, V_{t+1})$).* Let us note that even in the current setting, the control set U can be allowed to depend on t, x, we can allow sets of the form $U(t, x_t) = \{u \in U^* : h_i(t, x_t, u) \geq 0, i = 1, \ldots, i^*\}$, U^* some given subset of $\mathbb{R}^r$. Compare Remark 1.4. We can even allow h_i to be independent of u. (Then $U(t, x_t) = U^*$ if $h_i(t, x_t) \geq 0$ for all i, and $U(t, x_t) = \emptyset$ if not all $h_i(t, x_t) \geq 0$.)

Note that for $t < T$, the maximization in the optimality equation must actually be restricted to $U^*(t - 1, x_{t-1}) := \{u \in U(t - 1, x_{t-1}) : (f(t - 1, x_{t-1}, u, V_t), V_t) \in Q(t)$ a.s.$\}$, where $Q(t) = \{(x, v) : J(t, x, v) > -\infty\}$. This follows from the fact that if the property $(f(t - 1, x_{t-1}, u, V_t), V_t) \in Q(t)$ a.s. fails to hold for any given u, then for this u, the expectation term in the optimality equation equals $-\infty$, see, e.g., (1.17). If $U^*(t - 1, x_{t-1})$ is empty, $J(t - 1, x_{t-1}, v_{t-1})$ equals $-\infty$. (We assume $J(t, x, v) < \infty$ for all t, x, v.)

Sometimes, situations are encountered where the instantaneous reward equals $f_0(t, x_t, u_t, V_{t+1})$, i.e., the reward depends on a future random shock V_{t+1}. In equation (1.18) then $f_0(t - 1, x_{t-1}, u_{t-1}, v_{t-1})$ must be replaced by $E_{u_{t-1}}[f_0(t - 1, x_{t-1}, u_{t-1}, V_t) \mid x_{t-1}, v_{t-1}]$. Furthermore, $J(T, x_T, v_T) = \sup_{u_T} E_{u_T}[f_0(T, x_T, u_T, v_{T+1}) \mid x_T, v_T]$. (To see this, note that it is possible to rewrite such a problem on the form considered earlier, by putting $Z_{t+1} = f_0(t, X_t, u_t, V_{t+1})$, $Z_0 = 0$, with Z as a new state variable, and then maximizing $\sum_{t=0}^{T} Z_{t+1}$ instead of the original criterion.) $\square$

Remark 1.16 (Stochastic Euler equation). Sometimes, a problem of the following form is encountered:

$$\max E \left[\sum_{t=0}^{T} F(t+1, X_t, X_{t+1}, V_{t+1}) \right], \quad x_0 \text{ given}, \ x_1, \ldots, x_{T+1} \text{ free} \qquad (1.30)$$

(i.e., subject to choice in $\mathbb{R}$), V_{t+1} governed by a conditional probability distribution $P_t(v|x_t, v_t)$ or a conditional density $p(v|x_t, v_t)$, F a C^1-function.

Such problems can be solved by using the so–called Euler equations, which take two forms (F_2 and F_3 are derivatives): For $t = 0, \ldots, T-1$,

$$E[F_2(t+2, x_{t+1}, x_{t+2}(x_{t+1}, V_{t+2}), V_{t+2})|x_{t+1}, v_{t+1}]$$
$$+ F_3(t+1, x_t, x_{t+1}, v_{t+1}) = 0 \qquad (1.31)$$

(expectation with respect to V_{t+2}), and for $t = T$, the first term drops out, so:

$$F_3(T+1, x_T, x_{T+1}, v_{T+1}) = 0. \qquad (1.32)$$

The solution method is as follows: First, solve (1.32) for x_{T+1}, to obtain a function $x_{T+1} = x_{T+1}(x_T, v_{T+1})$, next, insert this function and $t = T-1$ in (1.31), solve this equation for x_T to obtain a function $x_T(x_{T-1}, V_T)$, insert this function and $t = T-2$ in (1.31), solve this equation for x_{T-1} to obtain a function $x_{T-1}(x_{T-2}, V_{T-1})$, and so on backwards. (Generally, the optimal choices of x_t depend on x_{t-1} and v_t, but on nothing more.)

To see that (1.31) is correct, note that if we replace X_{t+1} by u_{t+1} and X_t by u_t, and next put $Y_{t+1} = u_t$, then the sum in (1.30) equals $\sum_{t=0}^{T} F(t+1, Y_{t+1}, u_{t+1}, V_t) = \sum_{t=1}^{T+1} F(t, Y_t, u_t, V_t)$. Now, given a sequence of optimal policies $\{u_t(y_t, v_t)\}_t$, in the current case the term $E[J(t, Y_t, V_t)|y_{t-1}, v_{t-1}]$ in the optimality equation equals $E[\sum_{s=t}^{T+1} F(s, Y_s, u_s(Y_s, V_s), V_s))|y_{t-1}, v_{t-1}]$. Because only the first term in the sum depends on $u_{t-1}(= Y_t)$, then when differentiating in the optimality equation with respect to u_{t-1}, where we have inserted $y_t = u_{t-1}$, and using the envelope theorem (i.e., disregarding the term $u_t(y_t, v_t) = u_t(u_{t-1}, v_t)$), we get that

$$0 = F_3(t-1, y_{t-1}, u_{t-1}, v_{t-1}) + E[F_2(t, u_{t-1}, u_t(y_t, V_t), V_t)|y_{t-1}, v_{t-1}].$$

Replacing $(t-1, y_{t-1}, u_{t-1}, u_t(y_t, V_t))$ by $(t+1, x_t, x_{t+1}, x_{t+2}(x_{t+1}, V_{t+2}))$, we get (1.31). $\square$

Remark 1.17 (History dependence versus Markov dependence*). Why don't we need history-dependent controls? Some new arguments will now be presented. First, take a look at a much simplified situation: Suppose $T = 2$, $f \equiv 0$, $f_0(0, ., ., .) = 0$, $f_0(1, ., ., .) = 0$, x_t is a constant, which we then ignore. Insert a history-dependent control $u_2^*(V_1, V_2)$ in the criterion, which now is $E[f_0(2, u_2^*(V_1, V_2), V_2)]$. Can we find a function $u(V_2)$ giving as good a criterion value? Intuitively speaking yes: We might for example get a proposal for such a control by considering $\max_{V_1} f_0(2, u_2^*(V_1, V_2),$ $V_2)$. Supposing that a maximum point exists for all V_2, it would then be one of the function values of $u_2^*(.,.)$, and the value would be a function only of V_2. Hence a function $u_*(V_2)$ would exist such that $f_0(2, u_*(V_2), V_2) \geq f_0(2, u_2^*(V_1, V_2), V_2)$. The criterion value for $u_*(.)$ would be $\geq$ the one for $u^*(.,.)$.

A (much) refined argument of this type is used by Arkin and Evstigneev (1987) to show that Markov controls suffices.

Another argument, that we want to be "forward–looking," is the following. Define as before

$$J(t, z_{\to t}) = \sup_{u_t, \ldots, u_T} E_{u_t, \ldots, u_T} \left[\sum_{s=t}^{T} f_0(s, X_s, u_s(Z_{\to s})) \mid z_{\to t} \right], \qquad (1.33)$$

where the supremum is taken over all history-dependent controls. Let us show the equality $J(t, z_{\to t}) = J(t, z_t)$, i.e., let us show that when z_t is given, the supremum in (1.33) is independent of $z_{t'}, t' < t$. Let $\{u_s\}$ be any given sequence $u_t, \ldots, u_T$ of history-dependent controls, and let us denote by $J_{\{u_s\}}(t, z_{\to t})$ the value of the criterion in (1.33) obtained by deleting "sup." Let $z_{t'}^1$ and $z_{t'}^2$, $t' < t$ be two sequences of values of $z_{t'}, t' < t$. For $\hat{u}_s(z_t, Z_{t+1}, \ldots, Z_s) := u_s(z_0^1, \ldots, z_{t-1}^1, z_t, Z_{t+1}, \ldots, Z_s)$, we have that $J_{\{\hat{u}_s\}}(t, z_0^2, \ldots, z_{t-1}^2, z_t) = J_{\{u_s\}}(t, z_0^1, \ldots, z_{t-1}^1, z_t)$, because the controls $\hat{u}_s(.)$ and u_s give the same X_s's, $s > t$, given z_t and the $z_{t'}^1$'s. Then, surely, we have $J(t, z_0^2, \ldots, z_{t-1}^2, z_t) \geq J_{\{u_s\}}(t, z_0^1, \ldots, z_{t-1}^1, z_t)$, hence even $J(t, z_0^2, \ldots, z_{t-1}^2, z_t) \geq J(t, z_0^1, \ldots, z_{t-1}^1, z_t)$. A symmetric argument gives the opposite inequality, hence we have equality and so independence of $z_0, \ldots, z_{t-1}$. Then, moreover, for any given t, the supremum in (1.33) can be found by restricting the controls to be of the type $u_s^{(t)}(z_t, \ldots, z_s)$, in particular, $u_t^{(t)}$ is of the form $u_t^{(t)}(z_t)$. Assume that supremum in (1.33) is attained for a sequence of controls $u_s^{(t)}(z_t, \ldots, z_s), s \geq t$.

For any sequence of controls $u_s(z_{\to s}), s = 0, 1, 2, \ldots$, by optimality of $u_s^{(t)}$,

$$J_{\{u_s\}}(0, z_0) = E\left[\sum_{s=0}^{t-1} f_0(s, X_s, u_s(Z_{\to s})) | z_0 \right] + E[J_{\{u_s\}}(t, Z_{\to t}) | z_0]$$

$$\leq E\left[\sum_{s=0}^{t-1} f_0(s, X_s, u_s(Z_{\to s})) | z_0 \right] + E[J_{\{u_s^{(t)}\}}(t, Z_t) | z_0]. \qquad (1.34)$$

Define

$$u_t^+ = (u_0^{(0)}(z_0), u_1^{(1)}(z_1), \ldots, u_t^{(t)}(z_t), u_{t+1}^{(t)}(z_t, z_{t+1}), u_{t+2}^{(t)}(z_t, z_{t+1}, z_{t+2}), \ldots).$$

Let $\{u_s^{**}(z_{\to s})\}$ be any given sequence of history-dependent controls. Moreover, let "is better than" mean "has a criterion value no less than the criterion value of." Evidently, using (1.34) for $t = 0$ and $\{u_s\} = \{u_s^{**}\}$, we get that u_0^+ is better than u_s^{**}. Next, using (1.34) for $t = 1$ and $\{u_s\} = u_0^+$, we get that u_1^+ is better that u_0^+. Then, using (1.34) for $t = 2$ and $\{u_s\} = u_1^+$, we get that u_2^+ is better than u_1^+, and so on.

Hence, for any t, u_t^+ is better that $\{u_s^{**}\}$. This holds in particular for $t = T$, and u_T^+ consists only of Markov controls.

We should rather not assume that the maximum in (1.33) is attained. It is possible to work with a modification of the proof where $u_s^{(t)}(z_t, \ldots, z_s)$ only brings us within a distance of $\varepsilon/2^t$ from the supremum. Then u_0^+ has a criterion value $\geq -\varepsilon/2^0+$ the criterion value of u_s^{**}, u_1^+ has a criterion value $\geq -\varepsilon/2^1+$ the criterion value of u_0^+ and so on. Then, u_t^+ has a criterion value $\geq -\varepsilon/2^0 - \ldots - \varepsilon/2^t+$ the criterion value of u_s^{**}. Letting $\varepsilon \downarrow 0$ yields again the conclusion that the supremum of the criterion is not influenced by restricting the controls to be Markov ones, rather than history-dependent. (When $\mathcal{V}$ is nondiscrete, an argument is needed to show that $u_s^{(t)}(z_t, \ldots, z_s)$ can be chosen to be at least "minimally well-behaved," it is dropped here.)   □

Remark 1.18 (Retarded systems can be rewritten as nonretarded ones*). We now generalize the standard problem (1.12), (1.13). In a *retarded* system that we shall now consider, $P_t(v_{t+1}|.,.,.)$, f_0, and f depend also on past values of x and v. Thus, in conformity with earlier notation, write $x_{\to t} := (x_0, \ldots, x_t)$, $v_{\to t} := (v_0, \ldots, v_t)$, $P_t(v_{t+1}|x_{\to t}, u_t, v_{\to t})$, $f_0(t, x_{\to t}, u_t, v_{\to t})$ and $f(t, x_{\to t}, u_t, v_{\to t+1})$. Now, the controls must depend on the history, the controls have to be of the form $u_t(x_{\to t}, v_{\to t})$. Then the optimality equation reads:

$$J(t-1, x_{\to t-1}, v_{\to t-1}) = \max_{u_{t-1}}\{f_0(t-1, x_{\to t-1}, u_{t-1}, v_{\to t-1}))$$

$$+ E_{u_{t-1}}[J(t, x_{\to t-1}, X_t, v_{\to t-1}, V_t)|x_{\to t-1}, v_{\to t-1}]\}. \tag{1.35}$$

The sequence of equalities in (1.27) actually establishes the optimality equation also in the current problem. In such problems, controls $u_t(z_0, \ldots, z_t)$ can give better results than Markov controls.

This retarded problem can be rewritten as a nonretarded one, by changing the definition of the state. Define the new state vector to be $\hat{x}_t = (x_0, \ldots, x_t, 0, \ldots, 0) \in \mathbb{R}^{n(T+1)}$, $\hat{v}_t = (v_0, \ldots, v_t, 0, \ldots, 0) \in \mathbb{R}^{q(T+1)}$ (assuming that $\mathcal{V}$ is a subset of $\mathbb{R}^q$). Write $\hat{f}_0(t, \hat{x}_t, u_t, \hat{v}_t) = f_0(t, x_{\to t}, u_t, v_{\to t})$. Then the original problem can be expressed as:

$$\max E \sum_t \hat{f}_0(t, \hat{x}_t, u_t, \hat{v}_t) \tag{1.36}$$

subject to

$$\hat{x}_{t+1} = \hat{f}(t, \hat{x}_t, u_t, \hat{V}_{t+1}) := (x_0, \ldots, x_t, f(t, x_{\to t}, u_t, v_{\to t+1}), 0, \ldots, 0) \tag{1.37}$$

(the last vector having $n(T+1)$ components), where the component V_{t+1} in $\hat{V}_{t+1}$ is distributed according to $\hat{P}_t(v|\hat{x}_t, u_t, \hat{v}_t) := P_t(v|t, x_{\to t}, u_t, v_{\to t})$, (of course, the other components are fixed because $\hat{v}_t$ is given). Now, the Markov controls (policies) $u_t(.)$ depend on the state $(\hat{x}_t, \hat{v}_t)$, and one can apply the optimality equation to the $(\hat{x}_t, \hat{v}_t)$-system. The problem has become nonretarded.

Assume that $P_t(v|t, \ldots)$, f_0 and f also depend on past values of u_t, for example let $P_t(v|\ldots) = P_t(v|x_{\to t}, u_{\to t}, v_{\to t})$. Now, introduce a new state variable w_t, governed by $w_{t+1} = u_t$, (where w_0 is an arbitrary vector). Then we can write $P_t(v|\ldots) = P_t(v|r_{\to t}, u_t, v_{\to t})$, where $r_t = (x_t, w_t) = (x_t, u_{t-1})$, and similarly for f_0 and f. Hence, replacing x by r brings us back to a problem of the previous retarded type. □

Example 1.19 (Problem with a "hard" end constraint). Consider the problem

$$\max E\left[\sum_{0 \leq t \leq 2} -u_t^2/2\right],$$

subject to

$$x_{t+1} = x_t + u_t + V_{t+1}, \; x_2 \geq 2 \text{ a.s.}$$

(i.e., with probability 1), $u_t \in (-\infty, 2], x_0 = 0.5$. Here, V_t takes values in $\{-1, 0\}$, and $\Pr[V_t = 0] = 3/4$, all V_t being i.i.d.

Solution. Using Remarks 1.4, 1.14, as always, let us solve the problem backwards: $J(2, x_2) = 0$ if $x_2 \geq 2, J(2, x_2) = -\infty$ if $x_2 < 2, u_T = 0$. Next, consider the equation

$$J(1, x_1) = \max\{-u_1^2/2 + EJ(2, x_1 + u_1 + V_2)\}. \tag{a}$$

If possible, a u_1 must be chosen such that $J(2, x_1 + u_1 + V_2) > -\infty$ with probability 1, i.e.,

$$x_1 + u_1 - 1 \geq 2 \Leftrightarrow u_1 \geq 3 - x_1, \tag{b}$$

otherwise there is a positive probability (in fact at least $1/4$), that x_2 becomes <2 (which is forbidden), and which is taken care of in the current set-up by the fact that the expectation in (a) then has value $-\infty$. Thus u_1 must, if possible, be chosen such that (b) holds. The best u_1 that can be chosen lies at the left boundary of the set $U^*(1, x_1) := \{u : u \geq 3 - x_1\} \cap (-\infty, 2] = [3 - x_1, 2]$, if $3 - x_1$ is positive, otherwise $u_1 = 0$ is chosen, i.e., $u_1 = \max\{0, 3 - x_1\}$. If $3 - x_1 > 2$, i.e., $1 > x_1$, then $U^*(1, x_1) = \emptyset$ (here meaning that the end condition cannot be met). Hence, $J(1, x_1) = -(\max\{0, 3 - x_1\})^2/2$ if $x_1 \geq 1$ and $J(1, x_1) = -\infty$ if $x_1 < 1$. Finally, consider the equation

$$J(0, x_0) = \max\{-u^2/2 + EJ(1, x_0 + u_0 + V_1)\}. \tag{c}$$

At the first step, we don't need to calculate $J(0, x_0)$ for all x_0 values, only for the given start point $x_0 = 0.5$. If u_0 is chosen such that $x_1 = 0.5 + u_0 - 1 < 1, (u_0 < 1.5)$, then $J(1, 0.5 + u_0 + V_1)$ has value $-\infty$ with positive probability, so if possible, we avoid such a value. We restrict the value of u_0 to $U^*(0, x_0) := (-\infty, 2] \cap [1.5, \infty) = [1.5, 2]$. The right-hand side of (c) becomes $\{-u_0^2/2 - (3/4)(\max\{0, 2.5 - u_0\})^2/2 - (1/4)(\max\{0, 3.5 - u_0\})^2/2\} =: \phi(u)$ when $u_0 \in U^*(0, x_0)$. Now, ϕ is concave and at $u_0 = 1.5$ it has the derivative $\phi'(1.5) = -1.5 + (3/4)(2.5 - 1.5) + (1/4)(3.5 - 1.5) = -0.25$, which shows that $u_0 = 1.5$ yields maximum. Thus $u_0 = 1.5, u_1 = \max\{0, 3 - x_1\}, u_2 = 0$ are optimal. In fact, $u_1 = 3 - x_1$ in all actual runs of the process, as $3 - x_1$ never becomes negative.

(If $\Pr[V_t = 0] = 1/4$, then $\phi'(u_0) = 0$ for $u_0 = 14/8 \in (1.5, 2)$, so this is the optimal u_0 in this case.) □

In Example 1.28, the condition $x_2 \geq 2$ a.s. is replaced by $Ex_T \geq y$, and the problem is solved by means of the stochastic maximum principle.

Remark 1.20 (Technical note*). Certain problematic questions have been glossed over in the presentation so far. When $\mathcal{V}$ is finite, there are few problems, in particular the criterion (1.13) is always defined, but its supremum may be infinite (so $J(t,x_t,v_t)$ may be infinite). If $\mathcal{V}$ is discrete but nonfinite, for any given policy sequence the criterion value may be infinite, or the expectation may not exist. Now, the main use of (1.18), (1.19) are as parts of sufficient conditions, see the first assertion in Theorem 1.13, in which we implicitly *assume* finiteness of the suprema in (1.18), (1.19). When $\mathcal{V}$ is nonfinite, it follows from the assumption of finiteness of suprema that $E_u[J(t,f(t-1,x_{t-1},u,V_t),V_t)|x_{t-1},v_{t-1}] < \infty$ for all t, all u,x_{t-1},v_{t-1}, a strengthened version of this inequality is explicitly included as an assumption in the theorem to follow.

When $\mathcal{V}$ is nondiscrete, the same problems arise, and in addition some "behavioral conditions" on $J(t,x_t,v_t)$ and the control functions are needed. The sufficient condition in Theorem 1.13 is stated in more precise form in Theorem 1.21 below in this case.

Define a function of several real variables to be separably piecewise continuous (for short called sp-continuous) if it is piecewise continuous in each real variable, i.e., as a function of any of these variables it has a finite set of discontinuity points, it has one-sided limits everywhere, and it is right-continuous or left-continuous.

Theorem 1.21. *If we have found sp-continuous functions $J(t,x_t,v_t)$ satisfying $E_u[|J(t,f(t,x_{t-1},u,V_t),V_t)||x_{t-1},v_{t-1}] < \infty$ for $(x_{t-1},u,v_{t-1}) \in \mathcal{X}_{t-1}(x_0) \times U \times \mathcal{V}$, as well as (1.18) and (1.19), with sp-continuous functions $u^*_{t-1}(x_{t-1},v_{t-1})$ yielding maximum in (1.18), (1.19), then the controls $\{u^*_{t-1}(x_{t-1},v_{t-1})\}_{t-1}$ are optimal in the set of all sequences of sp-continuous controls $\{u_{t-1}(x_{t-1},v_{t-1})\}_{t-1}$.* $\qquad\square$

(From the assumptions in Theorem 1.21, certain properties not explicitly stated follow. For example, the sp-continuity of $J(t-1,x_{t-1},v_{t-1})$ puts certain restrictions on the dependence of the expectation in (1.18) on $(x_{t-1},u_{t-1}, v_{t-1})$, compare the stronger continuity conditions explicitly stated in case (c) and (d) below.)

Let us next discuss another question. Can conditions be stated that ensure at the outset that a finite maximum of the objective in (1.13) can be obtained, and that the above dynamic programming method does yield optimal controls? Below some examples of such conditions are given.

Assume that U is compact. Then a maximum of (1.13) is attained, the maxima in (1.18), (1.19) are attained for $x_{t-1} \in \mathcal{X}_{t-1}(x_0)$, and controls $u^*_{t-1}(x_{t-1},v_{t-1}),x_{t-1} \in \mathcal{X}_{t-1}(x_0)$ yielding maximum in (1.18), (1.19) exist in four cases now to be considered.

(a) $\mathcal{V}$ is a finite set.

(b) $\mathcal{V}$ is nonfinite but discrete, and $f_0(t,x,u,v)$ is bounded, (i.e., for some $K \geq 0, K \geq |f_0(t,x,u,v)|$ holds for all $u \in U, x \in \mathcal{X}_t(x_0), v \in \mathcal{V}$).

(c) $\mathcal{V}$ is nondiscrete, f_0 is bounded, and, for all bounded continuous $\phi(x,u,v)$, the function $\psi(x_t,u_t,v_t) := E_{u_t}[\phi(x_t,u_t,V_{t+1})|x_t,v_t)]$ is continuous in $w = (x_t,v_t,u_t) \in \mathbb{R}^n \times \mathcal{V} \times U$.

(d) V is nondiscrete, $(x_t, u_t, v_t) \rightarrow p_t(v|x_t, v_t, u_t)$ is continuous, $\int |v| \psi_t(v) dv < \infty$, where $\psi_t(.)$ is given in (1.38) below, and, for some k, $|f_0(t, x, u, v)| + |f(t, x, u, v)| \leq k(1 + |x| + |v|)$ for all t, all $(x, u, v) \in \mathbb{R}^n \times U \times V$. (Then boundedness of f_0 is not necessary.)

In the case (c), we can even let V_t have a probability distribution that is a mixture of a discrete one and one given by a density.

For the above conditions, the functions $J(t, x, v)$ become continuous in x, respectively in (x, v) in case V is nondiscrete. If a unique maximum point always exists in (1.18) and (1.19), even u_t is continuous. (For these continuity results, see the Maximum Theorem in Sydsaeter et al. (2005).) Now, if unique maximum points do not necessarily exist, then the u_t's usually become at least sp-continuous (in exceptional circumstances they exhibits only weaker well-behavior properties, namely so-called Borel measurability, see the Appendix).

When U is noncompact, results are not so neat. (Let us confine our attention to the nondiscrete case.) Perhaps $J(t, x_t, v_t)$ is not continuous. (For a simple calculus example, note that $\max_{u \in \mathbb{R}} \{-u^2 (\ln x)^2 - 2u \ln x\}$ equals 1 for $x \neq 1$, and equals 0 for $x = 1$.) Usually, however, $J(t, x_t, v_t)$ becomes what is called lower semicontinuous (lsc) in $\mathcal{X}_t(x_0) \times V$ (a function $h(y)$ is lsc if, for any $\bar{y}$ in its set of definition, $\liminf_{y_n \to \bar{y}} h(y_n) \geq h(\bar{y})$). Generally, the supremum of an infinite number of lsc-functions (hence in particular of continuous functions) is again lsc, and usually, taking the conditional expectation in the optimality equation also preserves the lsc-property, i.e., if $J(t, x_t, v_t)$ is lsc, then the conditional expectation of $J(t, f(t - 1, x_{t-1}, u_{t-1}, V_t), V_t)$ is lsc in $(x_{t-1}, u_{t-1}, v_{t-1})$. (Note that if $J(t, x_t, v_t)$ is lsc, then $(x_{t-1}, u_{t-1}, v_t) \rightarrow J(t, f(t, x_{t-1}, u_{t-1}, v_t), v_t)$ is lsc.) This surely holds if $f_0(t, x, u, v)$ is bounded for $(x, u, v) \in \mathcal{X}_t(x_0) \times U \times V$, $(x_t, u_t, v_t) \rightarrow p_t(v|x_t, v_t, u_t)$ is continuous and, for all $(x_t, v_t, u_t) \in \mathbb{R}^n \times V \times U$,

$$p_t(v|x_t, v_t, u_t) \leq \psi_t(v), \quad \psi_t(v) \quad \text{some integrable function.} \tag{1.38}$$

(Then surely the criterion (expectation) in (1.13) is defined, and the supremum of it is finite.)

Now, let us imagine that the only functions that we are able to integrate are sp-continuous functions, but not, say, lsc-functions. (Often, taking conditional expectation means finding a multiple integral, and we are able to integrate each iterative integral with respect to the one-variable sp-continuous function in question.) Most often, $J(t, x_t, v_t)$ and $J(t, f(t - 1, x_{t-1}, u_{t-1}, v_t), v_t))$ will not only be lsc but also be sp-functions, so it is not problematic to work with such an assumptions in a sufficient condition. Very often, the maximizing controls will also come out as such functions. So it does not subtract much from the usefulness of Theorem 1.21 that we work with such assumptions in that theorem.

As the reader can see now, when V is nonfinite, we have been slightly imprecise, or sloppy, at certain points above, and we will continue this praxis.

At one point above, we mentioned a measurability concept. It should now perhaps come as no surprise that more advanced books on stochastic dynamic programming use measure theory with a vengeance. □

Remark 1.22 (Reformulation of the problem*). When the probability distributions are allowed to depend on (x_t, v_t, u_t), for $z_t = (x_t, v_t)$, we write $f_0(t, z_t, u_t) = f_0(t, x_t, u_t, v_t)$ (in a sense we consider v_t also to be a state). There is a probability distribution of Z_{t+1}, denoted $Q_t(z_t, u_t)$, which is determined by the probability distribution of V_{t+1} and the state (difference) equation. Hence, the problem (1.13) can be rephrased as follows: We seek $\max E[\sum f_0(t, Z_t, u_t(Z_t))]$ subject to Z_{t+1} governed by the distribution Q_t. This is a frequently occurring formulation of dynamic programming problems.

□

1.4 Stochastic Maximum Principle in Discrete Time

Occasionally, it happens that one can find the solution of a problem more quickly by using the stochastic maximum principle set out below instead of the optimality equation. Essentially, the maximum principle can be *derived* from the optimality equation. Hence it contains "no new information," but it *can* be useful to start directly with this principle.

We shall only consider the problem (1.1)–(1.2) in Section 1.1, and again we write

$$\Pr[V_{s+1} = v | v_s] = P_s(v | v_s), \tag{1.39}$$

thus this probability does not depend on x_s, u_s, as it did in Section 1.3. We also assume that U is convex.

Assume that the components f_i, $i = 1, \ldots, n$ of f, f_0, and the partial derivatives f_{ix}, f_{iu}, $i = 0, \ldots, n$, (exist and) are continuous. Let $u_0^*(x_0, v_0), \ldots, u_T^*(x_T, v_T)$ be an optimal sequence of policies in problem (1.2), with corresponding solution sequence $X_0, X_1^*(V_1), \ldots, X_T^*(V_1, \ldots, V_T)$ of (1.1). In case $\mathcal{V}$ is nondiscrete, we assume the controls to be piecewise continuous in each component of their arguments x_t, v_t. This was called sp-continuity above. (Drop the sp-continuity if $\mathcal{V}$ is discrete, also in the theorem to follow.) The following necessary condition is satisfied:

Theorem 1.23 (Necessary condition for optimality). *There exists a sequence of sp-continuous functions $\psi(s, x_s, v_s)$ such that the following conditions hold. For all pairs (x_T, v_T) such that $\Pr[X_T^* = x_T, V_T = v_T] > 0$, (or in the nondiscrete case, that the corresponding density is positive at (x_T, v_T)), and for all $u \in U$,*

$$f_{0u}(T, x_T, u_T^*(x_T, v_T))(u - u_T^*(x_T, v_T)) \le 0. \tag{1.40}$$

For $s = 1, \ldots, T$, for all pairs (x_{s-1}, v_{s-1}) such that $\Pr[X_{s-1}^* = x_{s-1}, V_{s-1} = v_{s-1}] > 0$, (or in the nondiscrete case, that the corresponding density is positive at (x_{s-1}, v_{s-1})), and for all $u \in U$,

$$0 \ge \Big[f_{0u}(s-1, x_{s-1}, u_{s-1}^*(x_{s-1}, v_{s-1}))$$
$$+ E[\psi(s, f(s-1, x_{s-1}, u_{s-1}^*(x_{s-1}, v_{s-1}), V_s), V_s) \cdot$$
$$f_u(s-1, x_{s-1}, u_{s-1}^*(x_{s-1}, v_{s-1}), V_s) | v_{s-1}] \Big] (u - u_{s-1}^*(x_{s-1}, v_{s-1})), \tag{1.41}$$

where f_u is the Jacobian matrix $\{f_{iu_j}\}$ of partial derivatives with respect to u_j, $j = 1, \ldots, r$. The functions $\psi(s, x_s, v_s)$ are governed by (1.43) and (1.44) below. □

In the case where $\mathcal{V}$ is discrete, (1.41) can be written

$$
\begin{aligned}
\Big[f_{0u}(s - 1, x_{s-1}, u_{s-1}^*(x_{s-1}, v_{s-1})) \\
+ \sum_{v_s \in \mathcal{V}} \{P_{s-1}(v_s|v_{s-1})\psi(s, f(s-1, x_{s-1}, u_{s-1}^*(x_{s-1}, v_{s-1}), v_s), v_s) \\
\times f_u(s-1, x_{s-1}, u_{s-1}^*(x_{s-1}, v_{s-1}), v_s)\} \Big] (u - u_{s-1}^*(x_{s-1}, v_{s-1})) \le 0, \quad (1.42)
\end{aligned}
$$

where $u \in U$. The equations determining $\psi(s, x_s, v_s)$ recursively backwards are as follows:

$$
\begin{aligned}
\psi(s, x_s, v_s) = f_{0x}(s, x_s, u_s^*(x_s, v_s)) \\
+ E[\psi(s+1, f(s, x_s, u_s^*(x_s, v_s), V_{s+1}), V_{s+1}) f_x(s, x_s, u_s^*(x_s, v_s), V_{s+1})|v_s].
\end{aligned}
$$
$$(1.43)$$

Here f_x is the Jacobian matrix $\{f_{ix_j}\}$ of partial derivatives with respect to x_j, $j = 1, \ldots, n$. Moreover, the following terminal condition holds:

$$
\psi(T, x_T, v_T) = f_{0x}(T, x_T, u_T^*(x_T, v_T)). \qquad (1.44)
$$

In the case where $\mathcal{V}$ is discrete, (1.43) can be written

$$
\begin{aligned}
\psi(s, x_s, v_s) = f_{0x}(s, x_s, u_s^*(x_s, v_s)) \\
+ \sum_{v_{s+1} \in \mathcal{V}} \{P_s(v_{s+1}|v_s)\psi(s+1, f(s, x_s, u_s^*(x_s, v_s), v_{s+1}), v_{s+1}) \\
\times f_x(s, x_s, u_s^*(x_s, v_s), v_{s+1})\}. \qquad (1.45)
\end{aligned}
$$

The conditions (1.40)–(1.44) are called the stochastic maximum principle for discrete time. Note that these conditions make it possible to compute $\psi(t, x_t, v_t)$ and u_t^* recursively backwards. First, use (1.40) to find u_T^* as function of (x_T, v_T). Next, calculate $\psi(T, x_T, v_T)$ by means of (1.44). Then use (1.41) to find u_{T-1}^* as a function of (x_{T-1}, v_{T-1}) and then use (1.43) to find $\psi(T-1, x_{T-1}, v_{T-1})$. Then calculate u_{T-2}^* using (1.41) and then find $\psi(T-2, x_{T-2}, v_{T-2})$, using (1.43). And so on backwards. We often construct a solution of (1.40)–(1.43) without taking into consideration the conditions $\Pr[X_{s-1} = x_{s-1}, V_{s-1} = v_{s-1}] > 0$, $\Pr[X_T = x_T, V_T = v_T] > 0$. Hence, we disregard the fact that (1.40)–(1.44) need only be satisfied when strict inequalities for the probabilities specified prevail (or corresponding inequalities for densities in case of continuous variables V_s).

Note than when $u_{s-1}^*(x_{s-1}, v_{s-1})$ is an interior point in U, then (1.41) holds with equality when $(u - u_{s-1}^*(x_{s-1}, v_{s-1}))$ is deleted (this follows from the original version of (1.41)).

Using the envelop theorem, the formulas (1.40)–(1.43) can be deduced from the optimality equation, if we assume that J is C^1 in x. (Differentiating the equation with respect to x, using the envelop theorem, and letting $\psi(s, x_s, v_s) = J_x(s, x_s, v_s)$

gives (1.43), and using the last equality and first order conditions related to the fact that $u^*(t-1, x_{t-1}, v_{t-1})$ maximizes the right-hand side of the optimality equation gives (1.41).) Let us sketch another derivation of the stochastic maximum principle that does not make use of the C^1-assumption on J. (In what follows, for the sake of simplicity the reader may very well assume that $x \in \mathbb{R}$, in which case f_x is a number.)

Proof of Theorem 1.23 (sketch)

Write $z_{s-1} = (x_{s-1}, v_{s-1})$ and $Z_t^* = (X_t^*, V_t)$. Fix an integer s and choose a small control $w_{s-1}(z_{s-1})$ with the additional property that the control $u_{s-1}(z_{s-1}) :=$ $w_{s-1}(z_{s-1}) + u_{s-1}^*(z_{s-1})$ belongs to U. Let $u_t(z_t) = u_t^*(z_t)$ for $t < s-1$, and let $u_t := u_t^*(Z_t^*)$ for $t > s-1$. For $t > s-1$, u_t is not a Markov control, but a history-dependent control, dependent on $(x_0$ and$)$ $v_0, \ldots, v_{t-1}$ (x_t^* is a function of these variables, given $u_0^*, u_1^*, \ldots$). We saw above that a Markov control that is optimal among all Markov controls is also optimal among all history-dependent controls. Let us compute the effect on the criterion by changing from the sequence $\{u_s^* : s = 0, 1, \ldots, T\}$ to the sequence $\{u_s : s = 0, 1, \ldots, T\}$. Let $w_t(z) = 0, t \neq s-1$ and let $X_t(V_{\rightarrow t})$ be the path associated with u_t when the controls u_t are inserted in the state difference equation (1.1). This will be the same as the path obtained when $w_t(Z_t^*) + u_t^*(Z_t^*)$ is inserted, where $Z_t^* = (X_t^*(V_{\rightarrow t}), V_t)$. We then obtain that

$$\Delta := E \sum_{t=0}^{T} f_0(t, X_t, u_t^*(Z_t^*) + w_t(Z_t^*)) - E \sum_{t=0}^{T} f_0(t, X_t^*, u_t^*(Z_t^*))$$

$$= E\left[\sum_{t=s-1}^{T} f_0(t, X_t, u_t^*(Z_t^*) + w_t(Z_t^*)) - f_0(t, X_t^*, u_t^*(Z_t^*)) \right]. \quad (1.46)$$

From the state difference equation (1.1), we have as a first-order approximation that $x_s - x_s^* = f_u^*(s-1) \cdot w_{s-1}(z_{s-1}^*)$ where $f_u^*(s-1) := f_u(s-1, x_{s-1}^*, u_{s-1}^*(z_{s-1}^*), v_s)$.

Moreover, in a compact notation,

$$x_{s+1} - x_{s+1}^* \approx f_x(s, x_s^*, u_s^*(z_s^*), v_{s+1})(x_s - x_s^*) = f_x^*(s)(x_s - x_s^*)$$
$$x_{s+2} - x_{s+2}^* \approx f_x^*(s+1)(x_{s+1} - x_{s+1}^*) = f_x^*(s+1)f_x^*(s)(x_s - x_s^*)$$

and in general

$$x_{s+k+1} - x_{s+k+1}^* \approx f_x^*(s+k) \cdots f_x^*(s)(x_s - x_s^*),$$

where $x_s - x_s^* \approx f_u^*(s-1)w_{s-1}(z_{s-1}^*)$. The terms in the latter sum in (1.46) for $t \geq s$ approximately equal $f_{0x}^*(t)(x_t - x_t^*)$, $f_{0x}^*(t) := f_{0x}(t, X_t^*, u_t^*(Z_t^*))$. Using the approximations for the differences $(x_t - x_t^*)$ just calculated, in the first order we therefore get that Δ equals

$$E\Big[f^*_{0u}(s-1)w_{s-1}(Z^*_{s-1}) + f^*_{0x}(s)f^*_u(s-1)w_{s-1}(Z^*_{s-1})$$

$$+f^*_{0x}(s+1)f^*_x(s)f^*_u(s-1)w_{s-1}(Z^*_{s-1})$$

$$+\cdots+f^*_{0x}(T)f^*_x(T-1)\cdots f^*_x(s)f^*_u(s-1)w_{s-1}(Z^*_{s-1})\Big].$$

If we let $\bar{\psi}(s,x^*_s,v_{s\rightarrow}) :=$

$$f^*_{0x}(s) + f^*_{0x}(s+1)f^*_x(s) + \cdots + f^*_{0x}(T)f^*_x(T-1)\cdots f^*_x(s),$$

we can write this compactly as

$$\Delta \approx E\left[f^*_{0u}(s-1)w_{s-1}(Z^*_{s-1}) + \bar{\psi}(s,X^*_s,V_{s\rightarrow})f^*_u(s-1)w_{s-1}(Z^*_{s-1})\right]. \quad (1.47)$$

The symbol $\bar{\psi}(s,x^*_s,v_{s\rightarrow})$ means that $\bar{\psi}$ at time s depends on x^*_s, v_s, ..., v_T (given $u^*_s,\ldots,u^*_T$).

Fix a state $\bar{z}$, and choose $w_{s-1}(\bar{z}) = \bar{w}$, where $\bar{w}$ is small, $u^*_{s-1}(\bar{z}) + \bar{w} \in U$, and $w_{s-1}(z) = 0$, $z \neq \bar{z}$. (If $\mathcal{V}$ is nondiscrete, a similar perturbation is needed in a small interval around $\bar{z}$, the details here are omitted.) If we read $[\,]$ as the bracket in (1.47) we have $\Delta \approx E[\,] = P\cdot E\left\{[\,]\mid Z^*_{s-1} = \bar{z}\right\}$, where $P = \Pr\left[Z^*_{s-1} = \bar{z}\right]$. If $\Delta > 0$, then it would improve the expected value of the criterion to make the change from u^*_t to u_t for this choice of w_{s-1}. Therefore, $\Delta \leq 0$, because $\{u^*_t(X^*_t) : t = 0,1,\ldots\}$ is optimal. When $\Pr(Z^*_{s-1} = \bar{z}) \neq 0$, we can now show that for $\bar{z} = (x_{s-1},v_{s-1})$, $x^*_{s-1} = x_{s-1}$,

$$f^*_{0u}(s-1)w + E[\bar{\psi}(s,X^*_s,V_{s\rightarrow})f^*_u(s-1)\mid x_{s-1},v_{s-1}]w \leq 0 \quad (1.48)$$

for all $w \in U - u^*(\bar{z}) = \{u - u^*(\bar{z}) : u \in U\}$. The inequality (1.48) holds because we cannot have $PE\{[\,]\mid Z^*_t = \bar{z}\} > 0$ for $\bar{w} = \lambda w$, λ small and > 0, because in that case $PE\{[\,]\mid Z^*_t = \bar{z}\} \approx \Delta > 0$.

Note that

$$\bar{\psi}(s,x^*_s,v_{s\rightarrow}) = f^*_{0x}(s) + \{f^*_{0x}(s+1) + f^*_{0x}(s+2)f^*_x(s+1)$$

$$+\ldots+f^*_{0x}(T)f^*_x(T-1)\cdots f^*_x(s+1)\}f^*_x(s)$$

$$= f^*_{0x}(s) + \bar{\psi}(s+1,x^*_{s+1},v_{s+1\rightarrow})f^*_x(s). \quad (1.49)$$

Thus we have the following recursive formula (difference equation) for $\bar{\psi}$:

$$\bar{\psi}(s,x^*_s,v_{s\rightarrow}) = f^*_{0x}(s) + \bar{\psi}(s+1,x^*_{s+1},v_{s+1\rightarrow})f^*_x(s) \quad (1.50)$$

with end condition

$$\bar{\psi}(T,x^*_T,v_T) = f^*_{0x}(T). \quad (1.51)$$

Now define $\psi(s,x^*_s,v_s) = f^*_{0x}(s) + E_s\left[\bar{\psi}(s+1,X^*_{s+1},V_{s+1\rightarrow})f^*_x(s)\right]$, where E_s means conditional expectation given x^*_s,v_s. By (1.50), we obtain that

$$\psi(s,x_s^*,v_s) = f_{0x}^*(s) + E_s\left[f_{0x}^*(s+1) + \bar{\psi}(s+2,X_{s+2}^*,V_{s+2\rightarrow})f_x^*(s+1)f_x^*(s)\right]$$
$$= f_{0x}^*(s) + E_s\{f_{0x}^*(s+1)$$
$$+ E\left[\bar{\psi}(s+2,X_{s+2}^*,V_{s+2\rightarrow})f_x^*(s+1)|X_{s+1}^*,V_{s+1}\right]\cdot f_x^*(s)\}$$
$$= f_{0x}^*(s) + E_s\left[\psi(s+1,X_{s+1}^*,V_{s+1})f_x^*(s)\right]. \tag{1.52}$$

Now, for $x_s = x_s^*$, (1.43) is the same as (1.52). Next (in a compact notation), by (1.52),

$$E_{s-1}\{\bar{\psi}(s,\ldots)f_u^*(s-1)\} = E_{s-1}\{\left[f_{0x}^*(s) + \bar{\psi}(s+1,\ldots)f_x^*(s)\right]f_u^*(s-1)\}$$
$$= E_{s-1}\{f_{0x}^*(s) + \left(E_s\left[\bar{\psi}(s+1,\ldots)f_x^*(s)\right]\right)f_u^*(s-1)\}$$
$$= E_{s-1}\{\psi(s,\ldots)\}f_u^*(s-1).$$

Hence, by (1.48), (1.41) holds.   □

In the theorem to follow, a sufficient condition is presented.

Theorem 1.24 (Sufficient condition). *Let $u_0^*(x_0,v_0),\ldots,u_T^*(x_T,v_T)$ be a sequence of policies that, together with a sequence of adjoint functions $\psi(0,x_0,v_0),\ldots,$ $\psi(T,x_T,v_T)$ satisfy conditions (1.40)–(1.43), ((1.40) for all pairs (x_T,v_T) and (1.41) for all pairs (x_{s-1},v_{s-1})). Then $u_0^*(x_0,v_0),\ldots,u_T^*(x_T,v_T)$ are optimal, provided the function $(x,u) \rightarrow f_0(t,x,u) + \psi(t+1,x_{t+1}^*,v_{t+1})f(t,x,u,v_{t+1})$ is concave for all possible values of x_{t+1}^*,v_{t+1} and for all $t=0,\ldots,T$, with $\psi(T+1,.,.) = 0$.* □

(In case $\mathcal{V}$ is nondiscrete, add the assumptions that $u_t(.,.)$ and $\psi(t,.,.)$ are sp-continuous.) If X_t is required to belong to given convex sets A_t, $t=1,\ldots,T$, then concavity need only hold for $x \in X_t$. For a proof, see Exercise 1.23 below.

Example 1.25. Let us consider Example 1.5 in Section 1.1 again, and let us solve the problem by means of the maximum principle. Now, by (1.44), $\psi(T,x_T,z_T) = (1-\gamma)\beta^T Bx_T^{-\gamma}$ and, by (1.43)

$$\psi(s,x_s,z_s) = E\left[\psi(s+1,Z_{s+1}\{x_s - u_s(x_s)\},Z_{s+1})\cdot Z_{s+1}\right]. \tag{ix}$$

This equation has number (ix) and (i)–(viii) can be found in Example 1.5. Now, $u_s(x_s) := u_s^*(x_s)$ is determined by the equation

$$(1-\gamma)\beta^s u_s^{-\gamma} + E\left[\psi(s+1,Z_{s+1}(x_s - u_s),Z_{s+1})(-Z_{s+1})\right] = 0. \tag{x}$$

We obtain (x) from (1.41) by replacing $s-1$ with s and assuming an interior maximum point $u_s := u_s^*$.

Let us solve the problem backwards. Because we know $\psi(T,x_T,Z_T)$, then by (x), we can find u_{T-1}: We have

$$(1-\gamma)\beta^{T-1}u_{T-1}^{-\gamma} + E\left[(1-\gamma)\beta^T B\{Z_T(x_{T-1} - u_{T-1})\}^{-\gamma}(-Z_T)\right] = 0, \tag{xi}$$

which yields

$$(1-\gamma)\beta^{T-1}u_{T-1}^{-\gamma} - (1-\gamma)\beta^T \cdot BD_T(x_{T-1} - u_{T-1})^{-\gamma} = 0, \tag{xii}$$

where $D_T = E\left[Z_T^{1-\gamma}\right]$. This is a first-order condition for maximum of the expression $\beta^{T-1}u^{1-\gamma} + \beta^T BD_T(x_{T-1} - u)^{1-\gamma}$ that we maximized in Example 1.5, and the solution u_{T-1} is therefore given by (vi) in Example 1.5, i.e., $u_{T-1} = x_{T-1}/C_{T-1}^{1/\gamma}$, where $C_{T-1}^{1/\gamma} = 1 + (\beta BD_T)^{1/\gamma}$. Because u_{T-1} is known, $\psi(T-1, x_{T-1}, z_{T-1})$ can be found from (ix). Using $(C_{T-1}^{1/\gamma} - 1)^\gamma = \beta BD_T$, we obtain

$$\psi(T-1, x_{T-1}, z_{T-1}) = E\left[(1-\gamma)\beta^T B(Z_T(x_{T-1} - u_{T-1}))^{-\gamma} \cdot Z_T\right]$$

$$= (1-\gamma)\beta^{T-1} \cdot \beta BD_T\left(x_{T-1} - \frac{x_{T-1}}{C_{T-1}^{1/\gamma}}\right)^{-\gamma}$$

$$= (1-\gamma)\beta^{T-1}(C_{T-1}^{1/\gamma} - 1)^\gamma(C_{T-1}^{1/\gamma} - 1)^{-\gamma}C_{T-1}^{1/\gamma})^\gamma x_{T-1}^{-\gamma}$$

$$= (1-\gamma)\beta^{T-1}C_{T-1}x_{T-1}^{-\gamma},$$

where C_{T-1} is given in (viii) as before, with $C_T = B$. This means that $\psi(T-1, x_{T-1})$ is of the same form as $\psi(T, x_T)$. Hence, u_{T-2} is given by $u_{T-2} = x_{T-2}/C_{T-2}^{1/\gamma}$ where $C_{T-2}^{1/\gamma} = 1 + (\beta C_{T-1}D_{T-1})^{1/\gamma}$ and $\psi(T-2, x_{T-2}, z_{T-2})$ is given by $\psi(T-2, x_{T-2}, z_{T-2}) = (1-\gamma)\beta^{T-2}C_{T-2}x_{T-2}^{*}{}^{-\gamma}$. By continuing backwards, we again obtain that u_t is given by $u_t = x_t/C_t^{1/\gamma}$, where $C_t^{1/\gamma} = 1 + (\beta C_{t+1}D_{t+1})^{1/\gamma}$ and $\psi(t, x_t, z_t) = (1-\gamma)\beta^t C_t x_t^{-\gamma}$. In the above presentation, we have disregarded that we actually need to observe the inequality $X_t \geq u_t$, however the control u_t came out as satisfying this inequality (strictly), moreover sufficient conditions apply, see Theorem 1.31. (We could have used the control w of Example 1.5, in which case no state-dependent inequality appears; this, however, leads to slightly more complicated calculations.) $\qquad\square$

Remark 1.26 (Growth conditions*). For the above theory to hold in the case where $\mathcal{V}$ is nonfinite, we actually need that the system satisfies certain "growth conditions." These can for example be: For some functions $\phi(t, V_j)$, for all $i = 0, \ldots, n$, $|f_i(t, 0, u, V_j)| \leq \phi(t, V_j)$, $|f_{ix_j}(t, x, u, V_j)| \leq \phi(t, V_j)$, $|f_{iu_j}(t, x, u, V_j)| \leq \phi(t, V_j)$ for all $x, u \in U, V_j \in \mathcal{V}$, where

$$\sup_{V_{j-1} \in \mathcal{V}} E[\phi(t, V_j)|V_{j-1}] < \infty.$$

If $\mathcal{V}$ is nondiscrete, sp-continuity of $(v, v_t) \to p_t(v|v_t)$ is assumed. $\qquad\square$

Remark 1.27 (Terminal conditions). Let us make the following change in the problem: There is a mixture of soft and hard end constraints

(a) $x_T^i = \bar{x}^i$ for $i = 1, \ldots, l$ with probability 1
(b) $x_T^i \geq \bar{x}^i$ for $i = l+1, \ldots, m$ with probability 1
(c) $Ex_T^i = \bar{x}^i$ for $i = m+1, \ldots, m'$
(d) $Ex_T^i \geq \bar{x}^i$ for $i = m'+1, \ldots, m''$.

(The constraints (a), (b) are called hard, the constraints (c), (d) are called soft.) We call a control sequence admissible if it takes values in U, and if its corresponding solution sequence satisfies all the end constraints. (In addition, if $\mathcal{V}$ is nondiscrete, the controls are required to be sp-continuous.) Now, we may need to work with general history-dependent controls $u_t(v_{\rightarrow t})$, they may lead to strictly better criterion values, at least when soft constraints of the form (c) or (d) appear in the problem, see Remark 1.29 below.

Let $u_t^*(v_{\rightarrow t})$, $t = 0, \ldots, T$, be optimal, and write the corresponding solution as $x_t^*(v_{\rightarrow t})$. Now define the Hamiltonian $H(s,y,u,\psi,v_{s+1}) := \psi_0 f_0(s,y,u) + \psi f(s,y,u,v_{s+1})$, ψ_0 some number in $\{0,1\}$. The following maximum condition holds: There exists a sequence of adjoint functions $\psi(s,v_{\rightarrow s})$ and a number ψ_0 in $\{0,1\}$, such that, for all $u \in U$, all $v_{\rightarrow s-1}$ that occur with positive probability (positive density in the continuous case), $0 \geq$

$$E[H_u(s-1,x_{s-1}^*(v_{\rightarrow s-1}),u_{s-1}^*(v_{\rightarrow s-1}),\psi(s,V_{\rightarrow s}),V_s)|v_{\rightarrow s-1}](u - u_{s-1}^*(v_{\rightarrow s-1})) \tag{1.53}$$

The adjoint equation is now

$$\psi(s,v_{\rightarrow s}) = E[H_x(s,x_s^*(v_{\rightarrow s}),u_s^*(v_{\rightarrow s}),\psi(s+1,v_{\rightarrow s+1}),V_{s+1})|v_{\rightarrow s}]. \tag{1.54}$$

Finally, the following transversality condition on $\psi(T,v_{\rightarrow T})$ holds:

$$\psi(T,v_{\rightarrow T}) = \psi_0 f_{0x}(T,x^*(v_{\rightarrow T}),u^*(v_{\rightarrow T})) + p_T(v_{\rightarrow T}). \tag{1.55}$$

Here $p_T(v_{\rightarrow T})$ belongs to $\mathbb{R}^n$, the components $p_T(v_{\rightarrow T})^i$ for $i > m$ are constants, $p_T(v_{\rightarrow T})^i = 0$, $i > m''$, and, a.s.,

$$p_T(v_{\rightarrow T})^i \geq 0 (= 0 \text{ if } x_T^{*i}(V_{\rightarrow T}) > \bar{x}^i \text{ a.s.}) \text{ for } i = l+1,\ldots,m, \text{ and}$$
$$p_T(v_{\rightarrow T})^i \geq 0 (= 0 \text{ if } Ex_T^{*i}(V_{\rightarrow T}) > \bar{x}) \text{ for } i = m'+1,\ldots,m''.$$

Finally, $(\psi_0, p_T(V_{\rightarrow T}))$ is different from zero with probability 1.

In case $\mathcal{V}$ is nondiscrete, add the assumption that $u_t^*(.,.)$ are sp-continuous, and the property that (normally) $\psi(t,.,.)$ and $p_T(.)$ are sp-continuous. Conditions of the type of Remark 1.26 are again needed for the above result to hold if $\mathcal{V}$ is nonfinite.

If the variables V_t take values in a finite set $\mathcal{V}$, then the above maximum problem can be reduced to a deterministic nonlinear programming problem in Euclidean space, with a finite number of constraints, and from which the above necessary conditions can be deduced. If $\mathcal{V}$ is nonfinite, e.g., if the variables V_t are continuously distributed, there are an infinite number of constraints, and one has to work in an infinite dimensional space, at least if $m \geq 1$ ($m = 0$ means no hard constraints). And, in this case ($m \geq 1$), certain "constraints qualifications" may be needed, see Arkin and Evstigneev (1987) in a similar situation. (Thus, in the case where $\mathcal{V}$ is nondiscrete, completely precise necessary conditions are not stated.)

For $\psi_0 = 1$, the conditions are sufficient if $(y,u) \rightarrow H(s,y,u,\psi(s+1,v_{\rightarrow s+1}),v_{s+1})$ is concave for all $v_{\rightarrow s+1}$.

Arkin and Evstigneev (1987) have shown that if only hard end restrictions appear in the problem, then as mentioned before, it suffices to consider Markov controls.

Imagine that a given problem is to be solved. One can start by trying to solve the problem by assuming that $u^*(.,.)$ and $\psi(.,.,.)$ are as in Theorem 1.23 (i.e., are Markov functions), except that $\psi(T, x_T, v_t) = \psi(T, x_T^*, v_T)$ has to satisfy (1.54), for a $p_T(v \to T) = p_T(x_T^*, v_T)$ satisfying the inequality and complementary slackness conditions above. Occasionally, such a try may fail, but if it succeeds, then of course one candidate for optimality that satisfies the necessary conditions (1.53)–(1.55) has been found. To prove uniqueness of the candidate, (hopefully this will be the case), then one may have to use (1.53)–(1.55) in full generality. This is not needed when sufficient conditions applies. The current comment is used in the example to follow. □

Example 1.28 (Problem with a soft end constraint). Consider the problem

$$\max E\left[\sum_{0 \le t \le T} -u_t^2/2\right],$$

subject to

(i) $x_{t+1} = x_t + u_t + V_{t+1}$, x_0 given,
(ii) $Ex_T \ge y$, $u_t \in (-\infty, 2]$,

V_{t+1} takes values in $\{-1, 0\}$, $\Pr[V_{t+1} = 0] = p \in (0, 1)$, y a fixed number. We assume $2T > y - x_0 + T(1-p)$.

Solution. Evidently, $u_T = 0$. Let $\psi_0 = 1$. Now, $\psi(T)$ equals a non-negative constant a that is 0 if $Ex_T > y$. The first-order necessary condition (1.41) yields

(iii) $-u_{t-1} + E[\psi(t, x_{t-1} + u_{t-1} + V_t, V_t) \cdot 1 | v_{t-1}] = 0$ if $u_{t-1} \in (-\infty, 2)$, replace $=$ by $\ge$ if $u_{t-1} = 2$.

The equation for $\psi(t, x_t, v_t)$ is

(iv) $\psi(t-1, x_{t-1}, v_{t-1}) = E[\psi(t, x_{t-1} + u_{t-1} + V_t, V_t) \cdot 1 | v_{t-1}]$.

From the last equation, used backwards from time T, we see that ψ will only depend on time, so we write it ψ^t. If we try the value $a = 0$, we get $\psi^{T-1} = \psi^{T-2} = \dots = \psi^0 = 0$. Then (iii) yields $u_t = 0$ for all $t < T$. Then (i) gives $x_T(V_1, \dots, V_T) = x_0 + \sum_{1 \le t \le T} V_t$, which has expected value $x_0 - T(1-p)$. If $x_0 - T(1-p) \ge y$, then $u_0 = u_1 = \dots = u_T = 0$ is a candidate for optimality. If this inequality fails to hold, then we have to look for a suitable positive value of a: Then (iv) gives $\psi^{T-1} = \psi^{T-2} = \dots = \psi^0 = a$, and (iii) gives $u_t = a, t < T$, if interior solutions. Then (i) yields $x_T = x_T(V_1, \dots, V_T) = x_0 + Ta + \sum_{1 \le t \le T} V_t$. A positive a requires $Ex_T = y$, which yield $x_0 + Ta - T(1-p) = y$, i.e., $a = (y - x_0 + T(1-p))/T > 0$. Fortunately, we have assumed $2T > y - x_0 + T(1-p)$, so $u_t = a < 2$. Sufficient conditions give optimality. (As far as the necessary conditions go, trying the alternative $\psi_0 = 0$ gives $a \ne 0$, i.e., $a > 0$ and $EX_T = y$. But, at the same time, the maximum condition implies $u_t = 2$ and hence $EX_T = x_0 + 2T - T(1-p) > y$, a contradiction. Formally, at this point, the general versions of (1.53)–(1.55) should be used.) □

Remark 1.29 (Nonexistence of an admissible Markov control*). In softly end constrained problems, in certain cases history-dependent admissible controls may exist,
but not Markov controls. An example is given in Exercise 1.24. It was said in Sec-
tion 1.1 that when $P_t(v|v_t)$ does not depend on v_t, then it suffices to consider Markov
controls $u_t(x_t)$, (more general controls do not give a higher value of the criterion).
In a trivial sense, this property fails in this example. Moreover, in Section 1.1 it was
also said that when $P_t(v|v_t)$ does depend on v_t, then it suffices to work with Markov
controls $u_t(x_t, v_t)$. This is no longer necessarily the case in softly end constrained
problems, as the above-mentioned exercise shows.   □

Remark 1.30 (Soft terminal constraints, another method*). Problems with soft terminal constraints (c) and (d) ($m = 0$ above) can sometimes also be solved by means
of the optimality equation. Then the reward function $f_0(T, ., .)$ at time T has to be
modified by adding to it a term $\sum_{1 \leq i \leq m''} \lambda^i(x_T^i - \bar{x}^i)$, and using this in a free end
problem, denoted P^λ, $\lambda = (\lambda^1, \ldots, \lambda^{m''})$. The solution obtained from the optimality equation will then depend on the numbers λ_i, and by varying these parameters
a solution is sought such that (c) and (d) are satisfied. If the λ^i's also satisfy the
transversality conditions connected with (d), for $p_T(v_{\rightarrow T})^i$ replaced by λ^i, then an
optimal solution to the soft end constrained problem has been found. This method
represents a sufficient condition for finding an optimal solution in such problems,
it is not in general possible to show that the method is also necessary (i.e., that
there must exist λ^i's yielding such a solution). Without being very precise, it frequently fails in nonconcave problems. Compare the similar situation in nonlinear
programming, where it is not always the case that the Lagrangian is maximized,
though the first-order conditions related to such a maximization are satisfied by an
optimal point. In the current setting, these first-order conditions take the form of
the stochastic maximum principle. See also Exercise 1.22. Recall that the stochastic maximum principle is a necessary condition. (When it works, the above use of
the optimality equation to find an optimal Markov control yields optimality among
all history-dependent controls in P^λ as seen before, and then also in the original
problem.)   □

Infinite Horizon

Consider the problem

$$\max \lim_{T \to \infty} E\left[\sum_{t=0}^{T} f_0(t, X_t, u_t(X_t, V_t))\right], \quad x_t \in \mathbb{R}^n, \tag{1.56}$$

subject to

$$X_{t+1} = f(t, X_t, u_t(X_t, V_t), V_{t+1}), \quad x_0, v_0 \text{ given}, \quad u_t \in U \subset \mathbb{R}^r, U \text{ convex}. \tag{1.57}$$

Now, f_0, f and $P_t(v|v_{t+1})$ can depend explicitly on t, as written. Assume that the
conditions in Remark 1.26 are satisfied in case $\mathcal{V}$ is nonfinite, and that the limit in

the criterion (1.56) exists for any given sequence $u_0(X_0, V_0), u_1(X_1, V_1), \ldots$ We state a sufficient condition for such problems.

Theorem 1.31 (Sufficient condition). *Suppose that the pair of sequences* $(X_1^*(V_1),$ $X_2^*(V_1, V_2), X_3^*(V_1, V_2, V_3), \ldots, u_0^*, u_1^*(X_1, V_1), u_2^*(X_2, V_2), \ldots,$ *satisfies (1.57) and (1.41) for some function* $\psi(s, x_s, v_s)$ *satisfying (1.43). Assume that* $(x, u) \rightarrow f_0(t, x, u) +$ $\psi(t + 1, x_{t+1}^*, v_{t+1}) f(t, x, u, v_{t+1})$ *is concave for all possible values of* $x_{t+1}^*, v_{t+1},$ *and for all t. Then the pair of sequences is optimal, provided the following condition is satisfied:*

$$\lim_{t \to \infty} \inf E[\psi(t, X_t^*(V_1, \ldots, V_t), V_t)(X_t(V_1, \ldots, V_t) - X_t^*(V_1, \ldots, V_t))] \geq 0 \quad (1.58)$$

for all admissible pairs of sequences $(X_1(V_1), X_2(V_1, V_2), X_3(V_1, V_2, V_3), \ldots, u_0,$ $u_1(X_1, V_1), u_2(X_2, V_2), \ldots,$ *i.e., all pairs satisfying (1.57).* □

Here (as well as in the sufficient finite horizon condition of Theorem 1.24), it is permitted to have restrictions of the form $h_i(t, X_t, u_t) \geq 0$ a.s., for all $t, i = 1, \ldots, i'$ (h_i continuous and quasiconcave in (x, u)) on the sequence pair $(X_1(V_1), X_2(V_1, V_2),$ $X_3(V_1, V_2, V_3), \ldots, u_0, u_1(X_1, V_1), u_2(X_2, V_2), \ldots$ in order for the pair to be called admissible. The concavity is only needed in $\{(x_t, u_t) : h_i(t, x_t, u_t) \geq 0 \text{ for all } i\}$. (Again, assumptions of sp-continuity of $u_s^*(x, v)$ and $\psi(s, x_s, v_s)$ should be added in case $\mathcal{V}$ is nondiscrete.)

1.5 Optimal Stopping

Consider the problem

$$\max E_{u_0, \ldots, u_\tau} \left[\sum_{s=0}^{\tau} f_0(s, X_s, u_s(X_s, V_s), V_s) \right] \quad (1.59)$$

when $X_0 = x_0$ is given, and

$$X_{t+1} = f(t, X_t, u_t(X_t, V_t), V_{t+1}), \quad \Pr[V_{t+1} = v | x_t, u_t, v_t] = P_t(v | x_t, u_t, v_t) \quad (1.60)$$

where $u_t, t = 0, \ldots, \tau$ are policies, $u_t = u_t(X_t, V_t)$. Compared with the problem in Section 1.3, the only difference is the τ on top of the summation sign instead of T.

Let us now assume that we can vary the length of the period during which we let the process run. Hence in the maximization problem (1.59), τ is also subject to choice, in addition to $u_0, \ldots, u_\tau$.

We first assume that τ is restricted to the range $0 \leq \tau \leq T - 1$, with T given. By a couple of redefinitions, this new problem can be changed to a problem where we have a fixed horizon T. Let us add a new control value u^s ("stop") to those in U. When we use u^s we end up in a special state x^s outside of $\mathbb{R}^n$, which is sometimes called the coffin state. When x^s is reached, the process continuously stays in x^s

and no instantaneous rewards are earned. Thus, we extend the definitions of f_0 and f such that $x^s = x_{t+1} = f(t, x_t, u^s, v_t)$ for all t, x_t, v_t, while $x^s = f(t, x^s, u, v_t)$, and $f_0(t, x^s, u, v_t) = 0$ for all t, v_t, $u \in U \cup \{u^s\}$. (If the reader feels uneasy about what kind of objects u^s and x^s are, he can enlarge the state space and control space by adding one dimension to each, letting x^s and u^s be the vector $(0, \ldots, 0, 1)$ in $\mathbb{R}^{n+1}$ and $\mathbb{R}^{r+1}$, respectively. By redefinition, U then consists of vectors with last components equal to zero and the values of f are $(n+1)$-vectors, with last component equal to zero for all u different from u^s, $x \neq x^s$.)

We have now a fixed horizon T, and an extra control u^s, which, when used, takes us to a state x^s that we never can leave again, and where we do not get any rewards. It is required to use u^s at time $t = T$. Thus, u is chosen from $U \cup \{u^s\}$ for $t < T$ and from $\{u^s\}$ (no other choice!) for $t = T$. This "stopping problem" now incorporates the problem (1.59)–(1.60), provided $f_0(t, x_t, u^s, v_t) = 0$.

So far, we have assumed that no reward is obtained by using u^s, and particularly this is so at time T, at which we have to use u^s. However, it is now possible to change that feature of the problem: At any time t, there may be a reward connected with the use of u^s; that reward is $f_0(t, x_t, u^s, v_t)$ (still, $f_0(t, x^s, u^s, v_t) = 0$).

Because we are now dealing with a problem with a fixed horizon, the optimality equation and the methods associated with it are still valid. (The stochastic maximum principle must be modified because we no longer have differentiability of f_0 and f with respect to x.)

Let us now consider "pure" stopping problems where the only options available are either to stop (use $u = u^s$) or to continue (use $u = u^c$). In our original formulation, we can think of U as containing only the point u^c. Then we can write the optimality equation thus: $J(t-1, x_{t-1}, v_{t-1})$ is equal to the maximum of the two numbers in curly brackets below:

$$J(t-1, x_{t-1}, v_{t-1}) = \max\{f_0(t-1, x_{t-1}, u^s, v_{t-1}), f_0(t-1, x_{t-1}, u^c, v_{t-1})$$
$$+ E[J(t, f(t-1, x_{t-1}, u^c, V_t), V_t)|x_{t-1}, v_{t-1}]\}, \quad (1.61)$$

where $E = E_{u^c}$. (When using u^s, we do not get any expectation term in addition to $f_0(t-1, x_{t-1}, u^s, v_{t-1})$ because we end up in the coffin state x^s where no rewards are earned.) Evidently,

$$J(T, x_t, v_t) = f_0(T, x_T, u^s, v_T). \quad (1.62)$$

Let us look at a standard example in labor economics.

Example 1.32 (Job search). A student has a period of T days in which to find a one-month summer job. She receives one job offer with wage v_t each day. She cannot turn back to past offers. The decision whether to say yes ($u = u^s$) or no ($u = u^c$) is made on the basis of the wage offered. The wages $v = v_t$ offered are sampled from a single known probability distribution $\varphi(v)$ over an interval $[0, R]$, the v_t being i.i.d. (independently and identically distributed). If she turns down an offer one day, it costs her an amount c to obtain a job offer the next day. While still searching for a job, the student is in state x^c, but once she has accepted a job, she is in state x^s. On the last day she takes whatever is offered her, so $J(T, x^c, v) = v$

and $f_0(t,x^c,u^s,v) = v$, $f_0(t,x^c,u^c,v) = -c$, $f_0(t,x^s,u^s,v) = f_0(t,x^s,u^c,v) = 0$, and $f(t,x^c,u^c,v) = x^c$, $f(t,x,u^s,v) = x^s$.

When writing $J(t,v) = J(t,x^c,v)$, the optimality equation is

$$J(t-1,v_{t-1}) = \max\{v_{t-1}, -c + EJ(t,V_t)\} = \max\{v_{t-1}, v^t\}, \qquad (i)$$

where $v^t := -c + EJ(t,V_t)$. In particular, $v^T := -c + EV_T$. We shall assume $v^T \geq 0$.

Let us write $\Phi(v) = \int_0^v \varphi(v)\,dv$. Consider for a moment the task of computing $E\max[V,v^t]$. In fact, $E\max[V,v^t] = \int_0^R \max[v,v^t]\varphi(v)\,dv = v^t\int_0^{v^t}\varphi(v)\,dv + \int_{v^t}^R v\varphi(v)\,dv = v^t\Phi(v^t) + |_{v^t}^R v(\Phi(v)-1) - \int_{v^t}^R(\Phi(v)-1)\,dv = v^t + \int_{v^t}^R(1-\Phi(v))\,dv$. Here we have used that $\Phi(R) = 1$ and integration by parts, $((\Phi(v)-1)' = \varphi(v))$.

Now, $J(t-1,V_{t-1}) = \max\{V_{t-1},v^t\}$, so, by definition, $v^{t-1} = -c + E\max\{V_{t-1},v^t\}$. Thus, we get

$$v^{t-1} = -c + v^t + \int_{v^t}^R(1-\Phi(v))\,dv. \qquad (ii)$$

Starting from $v^T = -c + EV_T$, each value v_t is therefore determined by (ii) using backwards recursion. The stopping rule is: Stop if $v_{t-1} \geq v^t$, else keep searching. (We choose the convention that if $v_{t-1} = v^t$ we stop, though we could equally well have continued.) We shall now show that the sequence v^t is nondecreasing. One way of showing this is to use the definition of v^t and the easily proven fact that $J(t-1,v) \geq J(t,v)$: The latter property is a simple consequence of the definition of $J(t,v)$, because the set of options at time $t-1$ includes all options available at time t moved back one period in time. (The reader might want to formalize this argument. Try it!)

Let us instead infer the asserted monotonicity from (ii). Observe that the right-hand side of (ii), denoted for the moment by $g(v^t)$, is nondecreasing in v^t (calculate the derivative). Because $v^T = g(0) \geq 0$, then $v^{T-1} = g(v^T) \geq g(0) = v^T$, and then $v^{T-2} = g(v^{T-1}) \geq g(v^T) = v^{T-1}$, and so it continues backward: $v^T \leq v^{T-1} \leq v^{T-2} \leq v^{T-3} \ldots$.

If $R = 4$, $\varphi \equiv 1/4$ and $c = 1$, then $\Phi(v) = v/4$ and $v^T = 1$, and we find that v^t is recursively given by $v^{t-1} = 1 + \frac{1}{8}(v^t)^2$.   □

Some pure stopping problems have a structure that makes it possible to reach the solution quickly. Define the sets B_t by

$$B_t := \{(x_t,v_t) : f_0(t,x_t,u^s,v_t) \geq f_0(t,x_t,u^c,v_t)$$
$$+E[f_0(t+1,f(t,x_t,u^c,V_{t+1}),u^s,V_{t+1}) \mid x_t,v_t]\} \qquad (1.63)$$

for $t < T$, and for $t = T$, let $B_T = \mathbb{R}^n \times V$. The sets B_t consists of those point (x_t,v_t) for which it is better to stop at once than to continue one stage more and then stop. ("Better" here equals "at least as good.") It is easy to describe the optimal policies if it so happens that the sets B_t satisfy the property that

$$(x_t,v_t) \in B_t \Rightarrow (f(t,x_t,u^c,V_{t+1}),V_{t+1}) \in B_{t+1} \text{ with probability 1}, \qquad (1.64)$$

more precisely, with *conditional* probability equal to 1, given (x_t,v_t). When (1.64) is satisfied, the sets B_t are said to be *absorbing*, and in this case we can show that the optimal $u_t^*(x_t, v_t)$ is defined by: $u_t^*(x_t, v_t) = u^s \Leftrightarrow (x_t, v_t) \in B_t$. It is trivial that u_T^* is optimal. By backwards induction from T, we shall now show that if the above policy is the optimal one at time $t+1$, then this policy is also optimal at time t: First, assume that $(x_t, v_t) \in B_t$. Then consider the optimality equation: $J(t, x_t, v_t) =$

$$\max\{f_0(t, x_t, u^s, v_t), f_0(t, x_t, u^c, v_t) + E[J(t+1, f(t, x_t, u^c, V_{t+1}), V_{t+1}) \mid x_t, v_t]\}$$

Because $(f(t, x_t, u^c, V_{t+1}), V_{t+1}) \in B_{t+1}$ with conditional probability 1, then with conditional probability 1:

$$J(t+1, f(t, x_t, u^c, V_{t+1}), V_{t+1}) = f_0(t+1, f(t, x_t, u^c, V_{t+1}), u^s, V_{t+1})$$

(by the induction hypothesis that the proposed policy is the optimal one at time $t+1$, we do stop at $t+1$ in this case). Hence, $J(t, x_t, v_t) =$

$$\max\{f_0(t, x_t, u^s, v_t), \; f_0(t, x_t, u^c, v_t)$$
$$+ E[f_0(t+1, f(t, x_t, u^c, V_{t+1}), u^s, V_{t+1}) \mid x_t, v_t]\}$$

$= f_0(t, x_t, u^s, v_t)$ because $(x_t, v_t) \in B_t$. Thus, by the optimality equation, u^s is an optimal choice of control at time t when $(x_t, v_t) \in B_t$.

If, on the other hand, $(x_t, v_t) \notin B_t$, then

$$f_0(t, x_t, u^s, v_t) < f_0(t, x_t, u^c, v_t) + E[f_0(t+1, f(t, x_t, u^c, V_{t+1}), u^s, V_{t+1}) \mid x_t, v_t].$$

But the expression on the right-hand side is

$$\leq f_0(t, x_t, u^c, v_t) + E[J(t+1, f(t, x_t, u^c, V_{t+1}), V_{t+1}) \mid x_t, v_t],$$

(as we always have $f_0(t+1, x_{t+1}, u^s, v_{t+1}) \leq J(t+1, x_{t+1}, v_{t+1})$). Hence, by the optimality equation $u^*(x_t, v_t) = u^c$ when $(x_t, v_t) \notin B_t$. This completes the proof of the induction step.

Recall that checking whether $(x_t, v_t) \in B_t$ is equivalent to comparing the gain from stopping at once with the gain obtained from continuing one period more before stopping. Therefore $u^*(x_t, v_t)$ is called a one-stage look-ahead policy. Let us consider two problems with this structure:

Example 1.33 (The thief's problem). A thief avoids being caught with probability $p \in (0, 1)$ when out trying to steal something. The stolen objects have values $A_t \geq 0$, $A_1, A_2, \ldots$ being i.i.d. with mean $EA_t = a \in (0, \infty)$. The wealth of the thief is x_t, which is the sum of the values of all objects he has successfully stolen so far. If he is caught he loses all his wealth. Using $w = 0$ as representing being caught, whereas $w = 1$ represents not being caught, we can write $x_{t+1} = W_{t+1}(x_t + A_{t+1})$, where $\Pr[W_{t+1} = 1 \mid w_t = 1] = p$, $\Pr[W_{t+1} = 0 \mid w_t = 1] = 1 - p$, and $\Pr[W_{t+1} = 0 \mid w_t = 0] = 1$ (if he is caught, he never returns to his thievery). The variables A_t and W_t are stochastically independent. Moreover, the values of $f_0(t, x, u, w)$ are as follows: $f_0(t, x_t, u^s, 1) = x_t$, $f_0(t, x_t, u^c, 0) = 0$, $f_0(t, x_t, u^s, 0) = 0$, and $f_0(t, x_t, u^c, 1) = 0$ (his fortune is not guaranteed until he stops). Here the optimality equation becomes, for $w_t = 1$,

$$J(t, x_t, 1) = \max\{x_t, pEJ(t+1, x_t + A_{t+1}, 1)\}, \qquad (i)$$

(the variable A can be dropped in J). Instead of using (1), we shall take a short-cut and see if a one-stage look-ahead policy is optimal. This means that we check whether the sets B_t given by (1.63) are absorbing, $(B_T = \mathbb{R} \times \{0, 1\})$. The inequality in (1.63) in this example reduces to $x_t \geq p(x_t + a)$ if $w_t = 1$, and to $0 \geq 0$ if $w_t = 0$. Hence $(x_t, 0) \in B_t$ for all x_t, and $(x_t, 1) \in B_t \Leftrightarrow x_t \geq p(x_t + a) \Leftrightarrow x_t \geq pa/(1 - p)$. (As $V = (w, A)$ here, formally, B_t should have consisted of triples (x, w, A), not pairs (x, w), but the inequalities defining the sets B_t put no restrictions on $A \in [0, \infty)$, so it is dropped.) After noting this we are able to check whether the B_t's are absorbing, in other words, we prove that if $(x_t, w_t) \in B_t$, then $(X_{t+1}, W_{t+1}) \in B_t$ with probability 1: Consider first the case $w_t = 0$. Then $W_{t+1} = 0$ with probability 1 and $(X_{t+1}, W_{t+1}) \in B_{t+1}$. Next assume $w_t = 1$. If $W_{t+1} = 0$, then $(X_{t+1}, W_{t+1}) \in B_{t+1}$ automatically. Next, if $W_{t+1} = 1$, then $x_{t+1} = x_t + A_{t+1} \geq x_t \geq pa/(1 - p)$, which implies $(X_{t+1}, W_{t+1}) \in B_{t+1}$.

The optimal policy is therefore: Stop at the first time when $(x_t, w_t) \in B_t$. If $w_t = 0$ at this point, this condition is automatically satisfied (after all he has then been caught). If $w_t = 1$ he is not caught at the t-th theft. Then he stops if and only if $x_t \geq pa/(1 - p)$.

If it so happens that for all $t < T$, no pair (x_t, w_t) with $w_t = 1$ satisfies the last inequality, then the thief stops at the horizon T if he has not already been caught. $\square$

Example 1.34 (The house sale). Ole is going to sell his house, and obtains an offer every day. The offer on the t-th day is denoted by v_t. He has the option of recalling earlier offers to consider them anew. It costs him C every day his house remains unsold. The entity that Ole considers at day t in order to decide if to continue collecting offers or not is the maximum, $\max\{v_1, \ldots, v_t\} = \tilde{v}_t$, of all the offers received so far. This is his state variable. (It is possible to describe the situation in such a way that it exactly corresponds with the format in the theory above: Note that $\tilde{v}_t$ is governed by the difference equation $\tilde{v}_{t+1} = \max\{\tilde{v}_t, V_{t+1}\}$.) The offers V_t are i.i.d. and follow a known probability distribution given by $\Pr[v_t = \tilde{v}_i] = p_i$, $i = 0, \ldots, i^*$. Ole's objective is to maximize

$$E[\tilde{v}_\tau - \tau C],$$

where $\tau \geq 1$ is subject to choice. Let us see if a one-stage look-ahead policy is optimal here.

Here, $B_t = \{\tilde{v}_t : \tilde{v}_t \geq E[\tilde{v}_{t+1}] - C\}, t < T, B_T = \{\tilde{v}_i\}_i$. Now, $\tilde{v}_{t+1} = \max\{v_{t+1}, \tilde{v}_t\}$ and

$$E[\tilde{v}_{t+1}] = \sum_i \max\{\tilde{v}_i, \tilde{v}_t\} p_i = \sum_{\tilde{v}_i \leq \tilde{v}_t} \tilde{v}_t p_i + \sum_{\tilde{v}_i > \tilde{v}_t} \tilde{v}_i p_i.$$

The inequality defining B_t can be written

$$\sum_i \tilde{v}_t p_i = \tilde{v}_t \geq \sum_{\tilde{v}_i \leq \tilde{v}_t} \tilde{v}_t p_i + \sum_{\tilde{v}_i > \tilde{v}_t} \tilde{v}_i p_i - C,$$

i.e.,

$$C \geq \sum_{\tilde{v}_i \leq \tilde{v}_t} \tilde{v}_t p_i - \sum_i \tilde{v}_t p_i + \sum_{\tilde{v}_i > \tilde{v}_t} \tilde{v}_i p_i = -\sum_{\tilde{v}_i > \tilde{v}_t} \tilde{v}_t p_i + \sum_{\tilde{v}_i > \tilde{v}_t} \tilde{v}_i p_i$$

or

$$C \geq \sum_{\tilde{v}_i > \tilde{v}_t} (\tilde{v}_i - \tilde{v}_t) p_i.$$

We know that $\tilde{v}_{t+1} \geq \tilde{v}_t$. But the right-hand side of the inequality decreases when $\tilde{v}_t$ is replaced by $\tilde{v}_{t+1}$, hence if $\tilde{v}_t$ satisfies the inequality, then so does $\tilde{v}_{t+1}$. Thus, $\tilde{v}_t \in B_t \Rightarrow \tilde{v}_{t+1} \in B_{t+1}$.

The sets B_t therefore satisfy condition (1.64) (are absorbing), and the optimal policy is: Stop at the first time when

$$\tilde{v}_t \in \{\tilde{v}_t : \tilde{v}_t \geq E\tilde{v}_{t+1} - C\} = \{\tilde{v}_t : C \geq \sum_{\tilde{v}_i > \tilde{v}_t} (\tilde{v}_i - \tilde{v}_t) p_i\}.$$

If no $\tilde{v}_t$ is found to lie in B_t before T, Ole stops at $t = T$. The above policy is the optimal one when recall is allowed. A glance at the policy reveals that Ole never recalls an offer, except at $t = T$, where he may make use of all the offers received. □

Remark 1.35 (Markov and non-Markov stopping times*). In the stopping problems considered above, so-called Markov stopping times are sufficient for optimality (i.e., there is no need to consider more general ones). Any Markov stopping time τ can described as follows. There are given a sequence of sets $C_t \subset \mathbb{R}^n \times \mathcal{V}$ such that τ equals the first time (X_t, V_t) belongs to C_t. (Such controls come out of the optimality equation (1.61).) In more general problems, it may be that more general (history-dependent) stopping times τ are needed, where τ equals the first time $Z_{\to t}$ belongs to given sets C_t consisting of entities $z_{\to t}$ (recall $z_t = (x_t, v_t)$). □

1.6 Infinite Horizon

Let us again turn to infinite horizon problems, but let us now suppose that we have state- and control-dependent probabilities. Again, assume that P_t, and f are independent of t, and that f_0 can be written $f_0(t, x, u, v) = g(x, u, v)\alpha^t$, $\alpha \in (0, 1]$, (f and g continuous). Put $\pi = (u_0(x_0, v_0), u_1(x_1, v_1), \ldots)$. The problem is

$$\max_{\pi} E_{\pi} \left[\sum_{t=0}^{\infty} \alpha^t g(X_t, u_t(X_t, V_t), V_t) \right], \tag{1.65}$$

where X_t is governed by the stochastic difference equation,

$$X_{t+1} = f(X_t, u(X_t, V_t), V_{t+1}), \ \Pr[V_{t+1} = v_{t+1} | x_t, u_t, v_t] = P(v_{t+1} | x_t, u_t, v_t) \tag{1.66}$$

with X_0 given. The maximum in (1.65) is sought by considering all sequences $\pi = (u_0(x_0, v_0), u_1(x_1, v_1), \ldots)$ and selecting the best. We base our discussion upon conditions (see (1.70) below) implying that the infinite sum in (1.65), as well as its expectation, always exists in $[-\infty, \infty]$. For a given sequence π, let us write

$$J_\pi(s, x_s, v_s, T) = E_\pi \left[\sum_{t=s}^{T} \alpha^t g(X_t, u_t(X_t, V_t), V_t) \mid x_s, v_s \right], \tag{1.67}$$

where $\{X_t\}_t$ is the solution of the state difference equation in (1.66) corresponding to π, starting at (s, x_s, v_s). Define the function $J(s, x_s, v_s, T) = \sup_\pi J_\pi(s, x_s, v_s, T)$. Let $J_\pi(s, x_s, v_s, \infty)$ be the limit of $J_\pi(s, x_s, v_s, T)$ when $T \to \infty$, (again (1.70) secures existence), and write $J(s, x_s, v_s)$ for $\sup_\pi J_\pi(s, x_s, v_s, \infty)$. The equality $J(1, x_0, v_0) = \alpha J(0, x_0, v_0)$ holds, and it follows from the equality $J^1(x, v) = J^0(x, v)$ to be proved in a moment, these entities being defined as follows. Let

$$J_\pi^k(x, v) = \sum_{t=k}^{\infty} E_\pi [\alpha^{t-k} g(X_t, u_t(X_t, V_t), V_t) | x, v]$$

and let $J^k(x, v) = \sup_\pi J_\pi^k(x, v)$. Then $J^1(x, v)$ is the maximal expected present value of future rewards discounted back to $t = 1$, given that the process starts at (x, v) at time $t = 1$. When starting at (x, v) at time $t = 0$, and discounting back to $t = 0$, the corresponding maximal expected value is $J^0(x, v)$. It was said in Section 1.2 that $J^1(x, v) = J^0(x, v)$, "because the future looks exactly the same at times $t = 0$, and $t = 1$." The meaning of the last expression can be made more precise by the following consideration.

Given any sequence $\pi = (u_1(x_1, v_1), u_2(x_2, v_2), \dots)$, let us translate this sequence one step to the left, i.e., define a new sequence $\pi' = (u_0'(x_0, v_0), u_1'(x_1, v_1), \dots)$, where $u_0' = u_1$, $u_1' = u_2$, etc. Symmetrically, if π' is given, a translation to the right gives a policy π corresponding to π'. Then, due to the autonomy in the problem, $J_{\pi'}^0(x_0, v_0) = J_\pi^1(x_1, v_1)$ when $x_1 = x_0, v_1 = v_0$. Therefore, $J(0, x_0, v_0) := J^0(x_0, v_0) = \sup_{\pi'} J_{\pi'}^0(x_0, v_0) = \sup_\pi J_\pi^1(x_0, v_0) = J^1(x_0, v_0)$. Because $\alpha J_\pi^1(x_0, v_0) = J_\pi(1, x_0, v_0, \infty)$, $\alpha J(0, x_0, v_0) = J(1, x_1, v_1)$, when $x_1 = x_0, v_1 = v_0$.

Now, consider for a moment the optimality equation for a finite horizon T, (1.18), for $t = 1$. When we let $T \to \infty$, we get $J(0, x_0, v_0)$ on the left-hand side and, on the right-hand side, we get $J(1, X_1, V_1) = \alpha J(0, X_1, V_1)$, so most likely, the equality still holds. Thus, by this heuristic argument, we obtain the following Bellman equation or optimality equation (or "equilibrium optimality equation") for $J(x_0, v_0) := J(0, x_0, v_0)$:

$$J(x_0, v_0) = \sup_{u_0} \{ g(x_0, u_0, v_0) + \alpha E_{u_0}[J(X_1, V_1) \mid x_0, v_0] \}, \tag{1.68}$$

where $X_1 = f(x_0, u_0, V_1)$. For discrete $\mathcal{V}$, it can also be written

$$J(x_0, v_0) = \sup_{u_0} \left\{ g(x_0, u_0, v_0) + \alpha \sum_{v \in \mathcal{V}} P(v|x_0, u_0, v_0) J(f(x_0, u_0, v), v) \right\}. \tag{1.69}$$

We explain a little better below that the condition (1.68) *is* a valid necessary condition on the optimal value function if one or more of the following conditions are satisfied for all x, u and v.

$$(B) \quad M_2 \leq g(x,u,v) \leq M_1, \quad 0 < \alpha < 1 \quad \text{(Boundedness)}$$
$$(P) \quad 0 \leq g(x,u,v), \quad\quad\quad 0 < \alpha \leq 1 \quad \text{(Positivity)} \quad\quad (1.70)$$
$$(N) \quad g(x,u,v) \leq 0, \quad\quad\quad 0 < \alpha \leq 1 \quad \text{(Negativity)}$$

In case (P), the lower bound 0 on g can be replaced by a negative number M if $\alpha < 1$, and in case (N), the upper bound 0 on g can be replaced by a positive number M if $\alpha < 1$. We continue to call the cases (P) and (N), respectively, even if they have been modified in this way. Moreover, these conditions need only hold for $x \in \mathcal{X}(x_0)$. All the conclusions stated in the three cases (1.11), P. and N. in Section 1.2 still hold for the current setup. So recall the following results: Let $u^*(x,v)$ be a given control function, and let $J^{u^*}(x_0,v_0)$ be the value of the criterion for u^* (i.e., $J^{u^*}(x_0,v_0) = J_\pi(0,x_0,v_0,\infty)$, with $\pi = (u^*(x,v), u^*(x,v),\dots)$. With assumption (B) or (P) satisfied, it is sufficient for $u^*(x,v)$ to be optimal that $u^*(x,v)$ yields the supremum in (1.68) for J replaced by J^{u^*}. If (B) or (N) is satisfied, is suffices for $u^*(x,v)$ to be optimal that $u^*(x_0,v_0)$ gives the supremum of the right-hand side of (1.68) as written. Moreover, (1.68) has a unique bounded solution $\hat{J}(x_0,v_0)$ when (B) is satisfied, which equals the optimal value function $J(x_0,v_0)$. For further properties in case of (1.70), see Exercise 1.49.

An observation that is in particular useful in case (P) is that sometimes (B) holds for u^*, even if (P) fails when u is arbitrary. If (B) holds in this manner in case (P), and if the optimality equation (1.68) is satisfied by some function $\hat{J}(x,v)$ together with some control function u^* (giving maximum in (1.68) for $\hat{J}$ inserted) and such that (B) holds for u^*, then automatically $J = J^{u^*} = \hat{J}(x,v)$ and u^* is optimal. (Note that J^{u^*} has to satisfy (1.68) when the maximization is omitted, as J^{u^*} is the optimal value in a problem where the only admissible choice of u is $u = u^*$. Evidently, $\hat{J}(x,v)$ also satisfies this version of (1.68), hence by uniqueness $J^{u^*} = \hat{J}$.)

Finally, note that in all these cases, $J(0,x_0,v_0,T) \to J(x_0,v_0)$ when $T \to \infty$, in fact, in case (N), we need the assumption that U is compact, and some additional boundedness and continuity assumptions of the type of (C) or (D) mentioned in connection with (1.91) and (1.92) in Section 1.9. When these assumptions are added, we denote the case (N+). (In the same cases, for $\pi = (u,u,\dots)$, of course also $J_\pi(0,x_0,v_0,T) \to J_\pi^0(x_0,v_0)$, one way of seeing this is to imagine that we have a single choice of control, namely u, in the control problem.)

Note that it can be shown that history-dependent controls are "no better" than Markov controls (the supremum of the criterion is the same). Actually, the argument in Remark 1.17 works also for an infinite horizon, it still tells us that history-dependent controls are "no better" than Markov controls. Or at least, it comes close to to telling us that. (Using the notation of Remark 1.17, define $u_\infty^+ = (u_0^{(0)}, u_1^{(1)}, \dots)$. As u_{t+1}^+ was better than u_t^+ for all t, intuitively u_∞^+ must be the very best. But some limit theorems must be invoked to obtain this conclusion, see Exercise 1.37.)

Next, consider the sequence of equalities/inequalities in (1.27) in the proof of the optimality equation, for $s = 1$. Now, (1.27) also holds for $T = \infty$. And in the last line, (with $s = 1$), $J(1,Z_{\to s}) = J(1,Z_s) = \alpha J(0,Z_s)$, hence the Bellman equation (1.68) follows.

A few words about stopping problems are in order: As we saw above, these problems are only a special type of the general problems we have discussed, so the theory above is applicable to them. In particular, in cases (B), (N), and (P), as applied to the stopping problem (where the control region is augmented by u^s), the following optimality equation holds

$$J(x_0, v_0) = \sup_{u_0 \in U \cup \{u^s\}} \{g(x_0, u_0, v_0) + \alpha E_{u_0}[J(X_1, V_1) \mid x_0, v_0]\}, \tag{1.71}$$

which in a pure stopping problem reduces to

$$J(x_0, v_0) = \max\{g(x_0, u^s, v_0), g(x_0, u^c, v_0) + \alpha E_{u^c}[J(X_1, V_1) \mid x_0, v_0]\}. \tag{1.72}$$

Now there is no point in time at which we are forced to use u^s.

Of course, the comments connected to the cases (B), (N), and (P) also hold in the current particular case where all the time the control is chosen from $U = \{u^c, u^s\}$. In fact, when in the definition of the finite horizon optimal value function $J(0, x_0, v_0, T)$, u is allowed to be chosen from $\{u^c, u^s\}$ even at $t = T$, then we have that $J(0, x_0, v_0, T) \to J(x_0, v_0)$ in the three cases, in particular now only (N) (no further conditions) is needed in order to obtain this conclusion.

Define $J_T(0, x_0, v_0)$ as the supremum of the expected gain over all policies which surely stop at time $t = T$ (use u^s at T).

In a pure stopping problem, the equilibrium optimality equation is satisfied by the optimal value function $J(x, v)$ not only in cases (B), (P), and (N), but also if the following condition is satisfied: For some non-negative a, M', and M, for all x and v

$$g(x, u^c, v) < -a \leq 0, \quad -M' \leq g(x, u^s, v) \leq M, \quad \min\{-a, \alpha - 1\} < 0 \tag{1.73}$$

where we allow $M' = \infty$ iff $\alpha = 1$, and where $\min\{-a, \alpha - 1\} < 0$ means that the first inequality in (1.73) holds for $-a \leq 0$ if $\alpha < 1$, and for $-a < 0$ if $\alpha = 1$. (Of course, if $\alpha < 1$, (1.73) implies that (N) is satisfied.) Under condition (1.73), $J(x_0, v_0) = \sup_{\pi, \pi \text{ stops sometime}} J(0, x_0, v_0, \pi)$, where the supremum is taken only over stopping times π that stop sometimes with probability 1. Moreover, if the optimal value function is inserted in the equilibrium optimality equation, then maximizing the right-hand side of this equation gives us the optimal control (whether to stop or continue). Furthermore, at least when $\alpha = 1$, with probability 1, an optimal policy in the infinite horizon problem stops at some time point. Finally, when (1.73) holds and $M' < \infty$, then $J_T(0, x_0, v_0) \uparrow J(x_0, v_0)$ when $T \to \infty$.

In pure stopping problems, of course $J(0, x_0, v_0, T) = J_T(0, x_0, v_0)$ if

$$g(x, v, u^c) \leq g(x, v, u^s), \quad \text{for all} \quad x, v \tag{1.74}$$

so $J_T(0, x_0, v_0) \to J(x_0, v_0)$ if this inequality holds in case (P) and (N). Let us also mention that the sets B_t defined in connection with (1.64) now become independent of t. Using that $J_T(0, x_0, v_0) \to J(x_0, v_0)$, we can easily generalize the arguments associated with (1.64) to cover the situation with an infinite horizon. Thus, we have the following result. In the case (1.73) with $M' < \infty$, (or (N) or (P) with (1.74) added), assume that the set

$$B := \left\{ (x,v) : g(x,u^s,v) \geq g(x,u^c,v) + \alpha E_{u^c} \left[g(f(x,u^c,V_1),u^s,V_1) \mid x,v \right] \right\} \quad (1.75)$$

is absorbing in the sense that $(x,v) \in B \Rightarrow (f(x,u,V_1),V_1) \in B$ with conditional probability 1, given (x,v). Then the optimal policy is: stop the first time $(x_t,v_t) \in B$. (See Exercises 1.42, 1.43, and 1.44 below.)

Remark 1.36 (Weaker boundedness conditions). Remark 1.7 also holds in the current setting. (The bounds on g in that note must now hold also for all v.) For a proof, see Exercise 1.47.   □

Example 1.37. In the examples of the thief's problem (Example 1.33) and the house sale (Example 1.34) in the preceding section, we actually have examples of infinite horizon stopping problems if we simply drop the condition that the "play" stops at T (and the separate definition of B_t for $t = T$). In both cases the sets B_t are independent of t. Concerning the optimal policies, the solutions are the same, one can simply drop the remarks about what is done at the horizon T. In the house sale problem, the seller Ole never recalls earlier offers. Evidently, then, the policy described is also optimal in the problem where recall is not allowed. (To prove this intuitively evident fact in a formal manner, the reader may leave the "pure stopping" formulation we have used so far and introduce controls combining when to stop with which offer to pick if one stops. In this setting, the optimal policy found above is obviously optimal in the no recall case, because the set of policies in this latter problem is a subset of the set of policies in the recall case.)   □

Example 1.38. Consider Example 1.32, the job search, and assume in this example that the horizon is infinite. The equilibrium optimality equation becomes $J(v) = \max\{v, -c + E[J(V)]\}$. Denote $-c + E[J(V)]$ by v'. Then, taking expectations on both sides of the last equality, we get $E[J(V)] = E[\max\{V,v'\}]$. Calculating this expectation in the same way as $E[\max\{V,v'\}]$ in Example 1.32, we get $E[J(V)] = v' + \int_{v'}^R (1 - \Phi(v))\,dv$. Inserting this in the equality defining v', we get $c = \int_{v'}^R (1 - \Phi(v))\,dv$, a relationship that determines v'. We have already calculated the expected reward $E[\max\{V,v'\}]$ connected with the policy u^*: stop if and only if $v \geq v'$. By definition of v' we get that $\max\{v,v'\} = \max\{v, -c + E[\max\{V,v'\}]\} = \max\{v, -c + v' + \int_{v'}^R (1 - \Phi(v))\,dv\}$, hence the equilibrium optimality equation is satisfied for $J = \max\{v,v'\}$.

Observe that (1.73) is satisfied. Evidently, any upper bounded function J satisfying the equation must have the form $\max\{v,v'\}$, v' determined as above. Of course the optimal value function is bounded above by R. Thus, $\max\{v,v'\}$ must be the optimal value function. Hence, the policy described above is optimal.

Another argument telling us that $J = \max\{v,v'\}$ is the optimal value function is the following one. Let us prove that this function is the optimal value function by using the convergence property described subsequent to (1.72). Remember that v^t in Example 1.32 was nonincreasing. From (ii) in Example 1.32, we see that the sequence is bounded above by R (in fact, $v^{t-1} \leq g(R) \leq -c + R$ for all t, with g as defined subsequent to (ii) in that example). Hence, when t decreases indefinitely, v^t tends to a limit v', which coincides with the v' determined above, as we get the same relationship as that defining v' by taking limits in (ii). From the formula $J(t-1, v_{t-1}) = \max\{v_{t-1}, v^t\}$ we see that when $t \to -\infty$, $J(t-1, v) \to \max\{v, v'\}$, so the last expression *is* the optimal value function in the current problem. (To let t tend to $-\infty$ in a fixed horizon stationary problem is "the same" as letting the horizon tend to ∞.) $\qquad\square$

1.7 Incomplete Observations

So far, it has been assumed that x_t, and when needed, v_t, are observable. In such cases we speak of complete (or perfect) observability. Now that assumption is relaxed.

Consider the retarded system where P_t, f_0, and f also depend on past values of x, u, and v. Using an earlier notation, we write $P_t(v_{t+1}|x_{\to t}, u_{\to t}, v_{\to t})$, $f_0(t, x_{\to t}, u_{\to t}, v_{\to t})$ and $f(t, x_{\to t}, u_{\to t}, v_{\to t+1})$. Assume that x_t and v_t are not observable, but that there is another observable state z_t governed by $z_t = h(t, x_{\to t}, u_{\to t-1}, v_{\to t})$, $t > 0$. For simplicity, it is assumed that v_0 and x_0 are given, known entities. (The control u_t, whose values we choose, is naturally assumed to be observable.) Note that controls can only depend on what is observed. Hence, u_t can only be a function of $s_{\to t}$, where $s_t = (u_{t-1}, z_t)$, $(s_0 = z_0)$, so $u_t = u_t(s_{\to t})$, which means that u_t depends on the *vector* $(u_0, \ldots, u_{t-1}, z_0, \ldots, z_t)$. These are the types of controls considered in the following. The discussion below will be confined to the case where V is discrete.

There are now two methods for rewriting such a system as a perfectly observable one. Once that has been done, the problem can be solved by the familiar tools available in this case.

Method I

Noting that x_0 is fixed, a recursive use of the difference equation for x_t gives x_t as a function of $u_{\to t-1}, v_{\to t}$. Hence, one can write $z_t = \bar{h}(t, u_{\to t-1}, v_{\to t})$, and similarly, one can write $f_0(t, x_{\to t}, u_t, v_{\to t})$ as $\bar{f}_0(t, u_{\to t}, v_{\to t})$ and $f(t, x_{\to t}, u_t, v_{\to t+1})$ as $\bar{f}(t, u_{\to t}, v_{\to t+1})$, and, finally, $P_t(v|x_{\to t}, u_t, v_{\to t})$ as $\bar{P}(v; t, u_{\to t}, v_{\to t})$. Note that $Pr[V_{\to t+1} = v_{\to t+1}|u_{\to t}] = \bar{P}(v_1; 0, u_0, v_0) \cdot \ldots \cdot \bar{P}(v_{t+1}; t, u_{\to t}, v_{\to t}) =: P^*(v_{\to t+1}|t, u_{\to t})$.

The criterion to be maximized is now

$$\sum_{t=0}^{T} E \bar{f}_0(t, u_{\to t}(S_{\to t}), V_{\to t}),$$

where $u_{\to t}(S_{\to t}) = (u_0(S_0), u_1(S_{\to 1}), \ldots, u_t(S_{\to t}))$. (When the functions $u_0(.), \ldots, u_T(.)$ have been specified, then by induction, one may see that, for any t, $u_t(s_{\to t})$

simply depends on $v_0, \dots, v_t$, through the h-function, so $\bar{P}(v; t, u_{\to t}, v_{\to t}) = \check{P}_t(v|v_{\to t})$ depends also merely on $v_0, \dots, v_t$, the latter probabilities entering the calculation of the expectation in the criterion.) Consider the calculation of any of the terms $E\bar{f}_0(t, u_{\to t}(S_{\to t}), V_{\to t})$ in the criterion. By means of the rule of double expectation, the term can be written $EE[\bar{f}_0(t, u_{\to t}(S_{\to t}), V_{\to t})|S_{\to t}]$. Here $\hat{f}_0(t, s_{\to t}, u_t) := E[\bar{f}_0(t, u_{\to t}(s_{\to t}), V_{\to t})|s_{\to t}]$ is calculated by means of known conditional probabilities that are presented shortly. But note first that, because $s_{\to t}$ is fixed, when calculating this conditional expectation it does not matter whether u_t is considered to be a fixed vector or a function of $s_{\to t}$. The former assumption is made here, and this makes $\hat{f}$ dependent on the *vector* u_t. (When calculating the outer expectation, the dependence of u_t on $s_{\to t}$ is taken into account.) Let $1_A(v)$ be the indicator function of the set A (i.e., $1_A(v) = 1$ if $v \in A$, $= 0$ if $v \notin A$). The conditional probabilities mentioned are:

$$
\begin{aligned}
P^v(v_{\to t+1}|s_{\to t+1}) &:= Pr[V_{\to t+1} = v_{\to t+1}|s_{\to t+1}] \\
&= Pr[V_{\to t+1} = v_{\to t+1}|u_{\to t}, z_{\to t+1}] \\
&= Pr[V_{\to t+1} = v_{\to t+1} \& Z_{\to t+1} \\
&\quad = z_{\to t+1}|u_{\to t}]/Pr[Z_{\to t+1} = z_{\to t+1}|u_{\to t}] \\
&= Pr[V_{\to t+1} = v_{\to t+1}|u_{\to t}] \\
&\quad \times 1_A(v_{\to t+1})/Pr[V_{\to t+1} \in A|u_{\to t}], \quad (1.76)
\end{aligned}
$$

where

$$
A = \{v_{\to t+1} : \bar{h}(t'+1, u_{\to t'}, v_{\to t'+1}) = z_{t'+1}, 0 \le t' \le t\}.
$$

Here both probabilities occurring in the last fraction can be calculated by means of the P^*'s. When the probabilities P^v have been calculated, then also the expression $P^z(z_{t+1}|t, s_{\to t}, u_t)$ defined as follows is known: Write $B = \{v_{\to t+1} : \bar{h}(t+1, u_{\to t}, v_{\to t+1}) = z_{t+1}\}$ and define

$$
\begin{aligned}
P^z(z_{t+1}|t, s_{\to t}, u_t) &:= Pr[Z_{t+1} = z_{t+1}|u_{\to t}, z_{\to t}] \\
&= Pr[V_{\to t+1} \in B|u_{\to t}, z_{\to t}] \\
&= \sum_{v_{\to t}} Pr[V_{\to t+1} \in B|u_{\to t}, v_{\to t}] \\
&\quad \times Pr[V_{\to t} = v_{\to t}|u_{\to t}, z_{\to t}], \quad (1.77)
\end{aligned}
$$

where the sum extends over all $v_{\to t}$ satisfying $\bar{h}(t', u_{\to t'-1}, v_{\to t'}) = z_{t'} t' \le t$. The factors in the terms in the sum are calculated by means of $\bar{P}$ and P^v, respectively, ($Pr[V_{\to t} = v_{\to t}|u_{\to t}, z_{\to t}] = Pr[V_{\to t} = v_{\to t}|u_{\to t-1}, z_{\to t}]$ equals $P^v(v_{\to t}|s_{\to t})$).

The criterion to be maximized is now

$$
E\left[\sum_t \hat{f}_0(t, S_{\to t}, u_t(S_{\to t}))\right]. \quad (1.78)
$$

The maximization takes place subject to

$$S_{t+1} := (u_t, Z_{t+1}), \quad Pr[Z_{t+1} = z|s_{\to t}, u_t] = P^z(z|t, s_{\to t}, u_t). \tag{1.79}$$

Thus, as might be expected, Z_t plays the role of V_t in the completely observable case. The optimality equation is now:

$$J(t, s_{\to t}) = \max_u \{\hat{f}_0(t, s_{\to t}, u) + E_u[J(t+1, S_{\to t+1})|s_{\to t}]\}, \tag{1.80}$$

where E_u is calculated by means of $P^z(z|t, s_{\to t}, u)$.

The retarded system in Method I can of course be rewritten as a nonretarded one by means of the methods above, see Remark 1.18.

In the above discussion, no regard has been paid to the dimensionality (or complexity) of the resulting system. High dimensionality is a well-known obstacle to computing solutions numerically, a topic not touched in this chapter. Sometimes the next method is more conservative in this respect, especially if there are easy formulas for updating the parameters of the distributions.

Method II

This method is presented only for the case where $P_t(v|\ldots)$, f_0, f, and h are all "nonretarded," i.e., $f_0(t, ., ., .)$, $f(t, ., ., .)$, and $P_t(v|\ldots)$ depend only on the current values x_t, v_t, u_t of x, v, and u (v_{t+1} for f), whereas $h(t, ., .)$ depends only on x_t, v_t, and u_{t-1}. Thus we now consider the case where the observable variable satisfies $z_t = h(t, x_t, u_{t-1}, v_t)$, $t > 0$, where the state develops according to $x_{t+1} = f(t, x_t, u_t, V_{t+1})$, where the function f_0 in the criterion is $f_0 = f_0(t, x_t, u_t, V_t)$, and where the conditional probabilities for $V_{t+1} = v$ are of the form $Pr[V_{t+1} = v|x_t, u_t, v_t] = P_t(v|x_t, u_t, v_t)$ (so also explicitly dependent on t). Still x_0, v_0 are given and known. Let $y_t = (x_t, v_t)$. In Method II, the conditional probability Q_{t+1} defined by

$$y \to Q(y; \to t+1) := Pr[Y_{t+1} = y|u_{\to t}, z_{\to t+1}] \tag{1.81}$$

is taken as the state variable. Intuitively, this seems plausible: If we cannot observe y_t, it is reasonable to let u_t depend on the (updated) probability distribution of y_t. Using rules for calculating conditional expectations, the knowledge of the $P_t(v|\ldots)$'s, and the equations for x_t and z_t, it should be intuitively evident that once Q_t is known, u_t is chosen and z_{t+1} is observed, then it is possible to calculate Q_{t+1} from Q_t (and u_t and z_{t+1}), i.e., $Q_{t+1} = F(t, Q_t, u_t, Z_{t+1})$ for some known function F. Anyway, this is confirmed below. Furthermore, by the rule of double expectation, any term $f_0(t, X_t, u_t, V_t)$ in the sum in the expression $E[\sum f_0(t, X_t, u_t(Q_t), V_t)]$ can be replaced by $E[f_0(t, X_t, u_t, V_t)|Q_t]$, the conditional expectation calculated by means of $Q_t := Q(y; \to t)$. Because Q_t is fixed, when calculating the conditional expectation, it does not matter whether u_t is assumed to be a vector or a function of Q_t. For convenience, the former assumption is made. (When calculating the outer expectation in the expression $E \sum E[f_0|Q_t]$, the dependence of u_t on Q_t is taken into account.) Write $E[f_0(t, X_t, u_t, V_t)|Q_t] =: g(t, Q_t, u_t)$. Furthermore,

$$\Pr[Z_{t+1} = z | x_t, u_t, v_t]$$
$$= \Pr[V_{t+1} \in \{v : h(t+1, f(t, x_t, u_t, v), u_t, v) = z\} | x_t, u_t, v_t].$$

The latter probability can be calculated by means of $P_t(v | x_t, u_t, v_t)$. Hence, once the distribution $Q_t : y \to Q(y; \to t)$ is known, we can also calculate

$$S(z; t, Q_t, u_t) := \Pr[Z_{t+1} = z | Q_t, u_t] = \sum_{(x_t, v_t)} \Pr[Z_{t+1} = z | x_t, u_t, v_t] Q_t(x_t, v_t). \quad (1.82)$$

Then the problem has been reduced to maximizing

$$E \left[\sum_{0 \le t \le T} g(t, Q_t, u_t(Q_t)) \right] \quad (1.83)$$

subject to

$$Q_{t+1} = F(t, Q_t, u_t, Z_{t+1}) \quad \text{and} \quad \Pr[Z_{t+1} = z | Q_t, u_t] := S(z; t, Q_t, u_t), \quad (1.84)$$

where now Z_t plays the role of the random disturbance (instead of V_t).

In practical problems, it is often evident how to find the function F, so for solving such problems it is not necessary to read what follows, namely how we construct the function F in the general (but discrete probability) case. The function F is constructed as follows: Evidently,

$$R(v; t-1, y_{t-1}, u_{t-1}, z_t) := \Pr[V_t = v | y_{t-1}, z_t, u_{t-1}]$$
$$= \Pr[V_t = v \& Z_t = z_t | y_{t-1}, u_{t-1}] / \Pr[Z_t = z_t | y_{t-1}, u_{t-1}]$$
$$= \Pr[V_t \in \{v\} \cap A | y_{t-1}, u_{t-1}] / \Pr[V_t \in A | y_{t-1}, u_{t-1}], \quad (1.85)$$

where $A := \{v' : h(t, x_{t-1}, u_{t-1}, v') = z_t\}$. (Note that $R(v; t-1, ., ., .)$ has value zero, when $z_t \ne h(t, x_{t-1}, u_{t-1}, v)$.) Evidently, $R(v; t-1, ., ., .)$ can be calculated by means of the probabilities $P_{t-1}(v | x_{t-1}, u_{t-1}, v_{t-1})$.

Let (x, v) be arbitrarily given and define

$$R^*((x, v); t-1, y_{t-1}, u_{t-1}, z_t) := R(v; t-1, y_{t-1}, u_{t-1}, z_t) \cdot 1_{\{x : x = f(t-1, x_{t-1}, u_{t-1}, v)\}}(x)$$

(i.e., $R^*((x, v); y_{t-1}, u_{t-1}, z_t) = R(v; y_{t-1}, u_{t-1}, z_t)$ if $x = f(t-1, x_{t-1}, u_{t-1}, v)$ and $R^* = 0$, otherwise). Then, for $y = (x, v)$,

$$Q(y; \to t) := \Pr[Y_t = y | u_{\to t-1}, z_{\to t}]$$
$$= \sum_{y_{t-1}} \Pr[Y_t = y | y_{t-1}, u_{\to t-1}, z_{\to t}] \Pr[Y_{t-1} = y_{t-1} | u_{\to t-2}, z_{\to t-1}]$$
$$= \sum_{y_{t-1}} R^*(y_t; t-1, y_{t-1}, u_{t-1}, z_t) Q(y_{t-1}; \to t-1) \quad (1.86)$$

(by our current probability assumptions, $\Pr[Y_t = y | y_{t-1}, u_{\to t-1}, z_{\to t}] = \Pr[Y_t = y | y_{t-1}, u_{t-1}, z_t]$). Evidently, a recursive relation for $Q_t := y \to Q(y; \to t)$ has been obtained:

$$Q_t = F(t-1, Q_{t-1}, u_{t-1}, z_t) := \sum_{y_{t-1}} R^*(y_t; t-1, y_{t-1}, u_{t-1}, z_t) Q_{t-1}(y_{t-1}). \quad (1.87)$$

In this case, the optimality equations is:

$$J(t, Q_t) = \max_u \{ g(t, Q_t, u) + E_u[J(t+1, Q_{t+1})|Q_t] \}, \quad (1.88)$$

where $Q_{t+1} = F(t, Q_t, u_t, Z_{t+1})$ and the expectation with respect to Z_{t+1} is calculated by using the probability distribution $S(z; t, Q_t, u_t)$. Recall that $S(z; t, Q_t, u_t) := \Pr[Z_{t+1} = z | Q_t, u_t]$, where Q_t is the probability distribution of (X_t, V_t) obtained when z_t has been observed.

Example 1.39. We have just bought a machine that can be used to produce plastic cups. The state X of the machine can be "good" (state I) or "bad" (state II). The state does not change over time ($X_t = X$ for all t). The probability of a saleable product is a in state I and $b < a$ in state II. The revenue from a saleable cup is 1, whereas that from a defective cup is 0. The unit cost of production is $c > 0$, regardless of quality. At time t, the probability of state I is believed to be q_t. In other words, the probability P that the machine produces a saleable product takes one of two values a or b, and at time t we believe $\Pr[P = a] = q_t$. (P is a stochastic variable belonging to the two point set $\{a, b\}$.) Before $t = 0$, a random draw was made that determined P, with known probability λ of drawing a and $(1 - \lambda)$ of drawing b. (More prosaically, long experience in a variety of ways has told us that the frequency of getting a good machine when buying one is λ.) We assume that $b < c$, and that $\lambda a - c > 0$, which means that we produce at least one cup. We use the experience from producing cups by the machine bought to make inferences about the probability for the machine to be good. So there is "learning" in the sense of Bayesian updating, to which we return in a moment. At each time $t = 0, 1, 2, \ldots$, a cup is produced and its quality z_t is inspected to see whether it is saleable. If yes, $z_t = 1$, otherwise $z_t = 0$. Total profit is $\sum_{0 \leq t \leq \tau}(z_t - c)$, where the stopping time $\tau \in \{0, \ldots, T\}$ is subject to choice and T is fixed. This pure stopping problem will be solved by using Method II. Initially, a belief $q_{-1} = \lambda$ is given, before z_0 is observed, from which an updated value q_0 is determined. Similarly, q_t is our belief after z_t is observed. At the start of the time period of our problem, z_0 is known and so also q_0, and we wish to find $\max_{\tau(.)} E[\sum_{0 \leq t \leq \tau}(Z_t - c)]$, where the symbol $\tau(.)$ is a reminder that the stopping rules we consider depend on the states z_t, q_t. (Note that z_t is both an "observable" and a "state" determining the payoff. So, formally, in the framework of Method II, the updated distribution Q_t is actually the joint distribution of z_t and X_t, but as z_t is observed, the value of z_t is known exactly, so we can confine the interest to the distribution of $X_t = X$.) After observing z_{t-1}, and then inferring the updated q_{t-1}, we decide whether to continue or not. If we stop at time $t - 1$, we can pocket $z_{t-1} - c$. If we continue, we also get $z_{t-1} - c$ immediately and in addition the future expected profit $E[J_t(Z_t, q_t)|z_{t-1}, q_{t-1})]$. The optimality equation compares these two alternatives (we don't write Q_t but stick to a small letter, i.e., q_t, even when it is considered as stochastic):

$$J_{t-1}(z_{t-1},q_{t-1}) = \max\{z_{t-1}-c, z_{t-1}-c+E[J_t(Z_t,q_t)|z_{t-1},q_{t-1}]\}. \qquad (i)$$

First, let us find the difference equation for the state q_t (the updating formulas): Given q_{t-1}, let $\phi(q_{t-1}) := \Pr[P=a|Z_t=1]$ and let $\psi(q_{t-1}) := \Pr[P=a|Z_t=0]$. By standard rules for conditional probabilities, (dropping writing conditioning on q_{t-1}, e.g., writing $\Pr[Z_t=1]$ instead of $\Pr[Z_t=1|q_{t-1}]$),

$$\phi(q_{t-1}) = \Pr[P=a\,\&\,Z_t=1]/\Pr[Z_t=1]$$
$$= \frac{\Pr[Z_t=1|P=a]\Pr[P=a]}{\Pr[Z_t=1|P=a]\Pr[P=a]+\Pr[Z_t=1|P=b]\Pr[P=b]}$$

and

$$\psi(q_{t-1}) = \Pr[P=a\,\&\,Z_t=0]/\Pr[Z_t=0]$$
$$= \frac{\Pr[Z_t=0|P=a]\Pr[P=a]}{\Pr[Z_t=0|P=a]\Pr[P=a]+\Pr[Z_t=0|P=b]\Pr[P=b]}.$$

Hence,

$$q_t = aq_{t-1}/\{q_{t-1}a+(1-q_{t-1})b\} = \phi(q_{t-1}) \quad \text{when } z_t = 1 \text{ and}$$

$$q_t = (1-a)q_{t-1}/\{(1-a)q_{t-1}+(1-b)(1-q_{t-1})\} = \psi(q_{t-1}) \quad \text{when } z_t = 0. \quad (ii)$$

Thus, $q_t = F(z_t,q_{t-1})$, where the function F has the two values specified above for $z_t = 1$ and $z_t = 0$, respectively. Note also that, by (ii), $\phi(q_{t-1}) \geq aq_{t-1}/\{aq_{t-1}+a(1-q_{t-1})\} = q_{t-1} = (1-a)q_{t-1}/\{(1-a)q_{t-1}+(1-a)(1-q_{t-1})\} \geq \psi(q_{t-1})$, so q_{t-1} is revised upwards if a good cup is produced and downwards if a bad cup is observed. Define $\beta_{t-1}(q_{t-1}) = E[J_t(Z_t,q_t)|z_{t-1},q_{t-1}]$. This expectation depends only on q_{t-1}, as indicated by the notation $\beta_{t-1}(q_{t-1})$. Evidently, when calculating this expectation, the expression $q_t = F(z_t,q_{t-1})$ is inserted for q_t: The distribution of z_t depends only on q_{t-1} (e.g., see the expression for $\Pr[Z_t=1] = \Pr[Z_t=1|q_{t-1}] = q_{t-1}a+(1-q_{t-1})b$ used above).

Now, given q_{t-2}, q_{t-1} is a stochastic variable taking the value $\phi(q_{t-2})$ with probability $q_{t-2}a+(1-q_{t-2})b$ ($=\Pr[Z_{t-1}=1]$) and the value $\psi(q_{t-2})$ with the probability $(1-a)q_{t-2}+(1-b)(1-q_{t-2})$. Hence, taking the conditional expectation, given q_{t-2}, on both sides of the optimality equation, and using $\beta_{t-2}(q_{t-2}) = E[J_{t-1}(Z_{t-1},q_{t-1})|q_{t-2}]$ yields

$$\beta_{t-2}(q_{t-2}) = [\max\{1-c, 1-c+\beta_{t-1}(\phi(q_{t-2}))\}][aq_{t-2}+(1-q_{t-2})b]$$

$$+[\max\{-c, -c+\beta_{t-1}(\psi(q_{t-2}))\}][(1-a)q_{t-2}+(1-b)(1-q_{t-2})] \quad (iii)$$

This gives a recursive formula for $\beta_t(.)$: Note that $\beta_{T-1}(q_{T-1}) =$

$$E[J_T(Z_T,q_T)|z_{T-1},q_{T-1}] = aq_{T-1}+b(1-q_{T-1})-c(=1\Pr[z_T=1|q_{T-1}]-c).$$

Because the function $\beta_{T-1}(.)$ is known, one can calculate the function $\beta_{T-2}(.)$ by means of (iii) (with $t - 1 = T - 1$), then one can calculate the function $\beta_{T-3}(.)$ (for $t - 1 = T - 2$ in (iii)), and so on backwards. At each step, the complete β-function is needed. Using (iii), by backwards induction, it is easily seen that the functions β_t are increasing (see Exercise 1.38 below), with $\beta_t(0) = b - c, \beta_t(1) = [T - t](a - c)$. In fact, the equalities also follow immediately from the fact that if $q_t = 0$ we stop immediately, and when $q_t = 1$ we surely continue to $t = T$, and the expected earning at each stage from $t + 1$ on is $a - c$. Moreover, we must have $J_{t-1}(z_{t-1}, q_{t-1}) \geq J_t(z_t, q_t)$ when $(z_{t-1}, q_{t-1}) = (z_t, q_t)$, as there are more options available at time $t - 1$ than at time t. Using $\beta_t(0) < 0$ and $\beta_t(1) > 0$, it follows that points q^t exist, at which $\beta_t(q^t) = 0$. (From the formula for $\beta_{T-1}(q_{T-1})$, in particular, we get $q^{T-1} = (c - b)/(a - b)$.) Using $J_{t-1} \geq J_t$, $\beta_{t-1}(q_{t-1}) \geq \beta_t(q_t)$ when $q_{t-1} = q_t$. This entails that $q^{t-1} \leq q^t$ for all t. The optimal behavior is then as follows: As earlier stated, there is an original belief q_{-1}, and we observe z_0 (the saleability of the first cup produced). Using (ii), we can calculate q_0 for the appropriate case, depending on whether z_0 was 1 or 0. Production continues if and only if $q_0 > q^0$. If another cup is produced, inspecting it determines the value of z_1. Using this and q_0, we calculate q_1 from (ii), and production continues if and only if $q_1 > q^1$. And so on until T, if production has not already been stopped before.

Consider the situation where T is large. In the case where the machine is bad (the producer does not know this), then most often the updates q_t will be lower than the preceding one, and these estimates will eventually become less than q^t, so we will stop. Consider next the situation where the machine is good. Most often the updates q_t will be higher than the preceding one, so often (perhaps in the majority of possible production runs), we will not stop and the estimates q_t will increase toward 1. However, not infrequently, production runs may appear in which, to begin with, enough bad cups are produced to push the estimates q_t below q^t, causing the producer to stop production even in this case. Had he known that the machine was good, he would not have stopped: Then, at each point in time, the expected net income is $a - c > 0$. (Symmetrically, not infrequently, we may continue "too long" in cases where the machine is bad, but we never have such information for sure.) □

1.8 Control with Kalman Filter

Consider the linear-quadratic problem of Example 1.12 in Section 1.3. It will now be assumed that it is not possible to measure x exactly. The state is still governed by (1.21) and the criterion appears in (1.20). What we can observe is $z_t = H_t x_t + v_t$, where H_t is a known matrix, v_t is a random variable with all v_t, ε_t independent for all t. It is assumed that v_t has Gaussian density proportional to $\exp[-\frac{1}{2}\{v'G_t v\}]$ and that ε_t has Gaussian density proportional to $\exp[-\frac{1}{2}\{w'N_t w\}]$, G_t and N_t known matrices, $G_t^{-1} = \text{Covar}(v_t)$, $N_t^{-1} = \text{Covar}(\varepsilon_t)$. The current problem will be solved using the ideas that lie behind Method I. Let $s_t = (u_{t-1}, z_t)$.

The following results will be needed: Consider for the moment the situation where $z = Hx + v$, let ϕ be the normal density of x with $\text{Covar}(x) = M^{-1}$, $Ex = x^*$, and let ψ be the normal density of v (v and x independent), with $Ev = 0$, $\text{Covar}(v) = R^{-1}$. We will need expressions for $\hat{x} = E[x|z]$ and $\text{Covar}(x|z)$. The conditional density of z given x is $\psi(z - Hx)$. Then the conditional density of x given z is $\phi(x)\psi(z - Hx)/\psi^*(z)$, where $\psi^*(z) = \int_{-\infty}^{\infty} \phi(x)\psi(z - Hx)dx$. Let $z'' = z - Hx^*$. Using $\sim$ to mean "proportional," note that

$$
\begin{aligned}
&\phi(x)\psi(z - Hx) \\
&\sim \exp(-\tfrac{1}{2}\{(x - x^*)'M(x - x^*) + (z - Hx)'R(z - Hx)\}) \\
&\sim \exp(-\tfrac{1}{2}\{(x - x^*)'M(x - x^*) + (z'' - H(x - x^*))'R(z'' - H(x - x^*))\}) \\
&\sim \exp(-\tfrac{1}{2}\{(x - x^*)'(M + H'RH)(x - x^*) \\
&\quad - (z'')'RH(x - x^*) - (x - x^*)'H'Rz'' + (z'')'Rz''\}).
\end{aligned}
$$

The last expression is the same as $\exp(-\tfrac{1}{2}\{(x - x^* - (M + H'RH)^{-1}H'Rz'')'(M + H'RH)(x - x^* - (M + H'RH)^{-1}H'Rz'') + \alpha\})$, α a term in z (or in z'', i.e., independent of x). This can be checked by multiplying out in the last expression. The conditional density of x given z has the same form (the term in z is different). Hence, the conditional expectation of x given z is

$$\hat{x} := x^* + (M + H'RH)^{-1}H'Rz'' = x^* + (M + H'RH)^{-1}H'R(z - Hx^*) \qquad (a)$$

and the conditional covariance $\text{Covar}(x|z)$ is

$$\text{Covar}(x|z) = (M + H'RH)^{-1} = M^{-1} - M^{-1}H'(HM^{-1}H' + R^{-1})^{-1}HM^{-1}. \qquad (b)$$

To prove the last equality, first multiply from the left by H. Then the equality becomes

$$H(M + H'RH)^{-1} = HM^{-1} - (I + R^{-1}(HM^{-1}H')^{-1})^{-1}HM^{-1}.$$

Next, multiplying by $(M + H'RH)$ from the right, we get

$$H = H + HM^{-1}H'RH - (I + R^{-1}(HM^{-1}H')^{-1})^{-1}HM^{-1}(M + H'RH).$$

Canceling H and multiplying by $I + R^{-1}(HM^{-1}H')^{-1}$ from the left yields $0 = 0$, i.e., the equality in (b) is verified.

(Given the Gaussian assumptions, statistics texts tell us that the estimator $\hat{x}$ is a "maximum likelihood" and "minimum variance" estimator.)

We want to prove the formula

$$\text{Covar}(x|z)H'R = M^{-1}H'(HM^{-1}H' + R^{-1})^{-1}. \qquad (b')$$

Let $B = HM^{-1}H'$, $D = (B + R^{-1})^{-1}BR$. Note that, by (b),

$$\text{Covar}(x|z)H'R = M^{-1}H'R - M^{-1}H'D = (M^{-1}H'RD^{-1} - M^{-1}H')D$$
$$= (M^{-1}H'RR^{-1}B^{-1}(B+R^{-1}) - M^{-1}H')D$$
$$= (M^{-1}H' + M^{-1}H'B^{-1}R^{-1} - M^{-1}H')D$$
$$= M^{-1}H'B^{-1}R^{-1}(B+R^{-1})^{-1}BR$$
$$= M^{-1}H'[(B+R^{-1})RB]^{-1}BR = M^{-1}H'[BRB+B]^{-1}BR$$
$$= M^{-1}H'[BR(B+R^{-1})]^{-1}BR = M^{-1}H'(B+R^{-1})^{-1},$$

and (b') follows. Combining (a) and (b'), we get $\hat{x} =$

$$x^* + \text{Covar}(x|z)H'R(z - Hx^*) = x^* + M^{-1}H'(HM^{-1}H' + R^{-1})^{-1}(z - Hx^*). \quad (b'')$$

Let us now turn to the solution of the linear-quadratic problem. As before, let C_t be given by the following Riccati equation:

$$C_t = R_t + A_t'C_{t+1}A_t - (A_t'C_{t+1}B_t)(Q_t + B_t'C_{t+1}B_t)^{-1}(B_t'C_{t+1}A_t), \quad (1.89)$$

$C_{T+1} = 0$. Using the induction hypothesis that $J(t+1, s_{\to t+1})$ is equal to $E[x_{t+1}'C_{t+1}x_{t+1}|s_{\to t+1}] + d_{t+1}$, d_{t+1} some constant (it *is* correct for $t+1 = T$, with $C_T = R_T$, $d_T = 0$), we are going to prove that such a formula also holds for $t+1$ replaced by t. Using the induction hypothesis, the optimality equation reads (compare Method I):

$$J(t, s_{\to t}) = \min_{u_t} E\{x_t'R_t x_t + u_t'Q_t u_t + E[(A_t x_t + B_t u_t + \varepsilon_t)'C_{t+1}$$
$$(A_t x_t + B_t u_t + \varepsilon_t)|s_{\to t+1}] + d_{t+1}|s_{\to t}\}$$
$$= \min E\{x_t'R_t x_t + u_t'Q_t u_t + (A_t x_t + B_t u_t + \varepsilon_t)'C_{t+1}$$
$$(A_t x_t + B_t u_t + \varepsilon_t)|s_{\to t}\} + E\{d_{t+1}|s_{\to t}\}. \quad (1.90)$$

(Here the double expectation rule has been used.) Define $\hat{x}_t = E[x_t|s_{\to t}]$. The calculations leading to (b'') above tell us how to obtain this estimate of x_t, but let us postpone making this connection.

Now, $E[x_t'R_t x_t|s_{\to t}] = \hat{x}_t'R_t\hat{x}_t + E[(x_t - \hat{x}_t)'R_t(x_t - \hat{x}_t)|s_{\to t}]$ and

$$E[(A_t x_t + B_t u_t + \varepsilon_t)'C_{t+1}(A_t x_t + B_t u_t + \varepsilon_t)|s_{\to t}]$$
$$= E[(A_t\hat{x}_t + A_t(x_t - \hat{x}_t) + B_t u_t + \varepsilon_t)'C_{t+1}$$
$$\times (A_t\hat{x}_t + A_t(x_t - \hat{x}_t) + B_t u_t + \varepsilon_t)|s_{\to t}]$$
$$= (A_t\hat{x}_t + B_t u_t)'C_{t+1}(A_t\hat{x}_t + B_t u_t) + E[(A_t(x_t - \hat{x}_t) + \varepsilon_t)'C_{t+1}$$
$$\times (A_t(x_t - \hat{x}_t) + \varepsilon_t)|s_{\to t}],$$

see a similar calculation, namely (1.23) and the subsequent comment in Example 1.12. Using these expressions yields $J(t, s_{\to t}) =$

$$\min\{\hat{x}_t'R_t\hat{x}_t + u_t'Q_t u_t + (A_t\hat{x}_t + B_t u_t)'C_{t+1}(A_t\hat{x}_t + B_t u_t)\}$$
$$+ E[(A_t(x_t - \hat{x}_t) + \varepsilon_t)'C_{t+1}(A_t(x_t - \hat{x}_t) + \varepsilon_t)|s_{\to t}]$$
$$+ E[(x_t - \hat{x}_t)'R_t(x_t - \hat{x}_t)|s_{\to t}].$$

Using results from Example 1.12, the control $u_t = -D_t\hat{x}_t$, $D_t = (Q_t + B'_tC_{t+1}B_t)^{-1}$ $B'_tC_{t+1}A_t$, gives minimum in the optimality equation, see $(**)$ in that example. Moreover, for such a u_t,

$$\hat{x}'_tR_t\hat{x}_t + u'_tQ_tu_t + (A_t\hat{x}_t + B_tu_t)'C_{t+1}(A_t\hat{x}_t + B_tu_t) = \hat{x}'_tC_t\hat{x}_t, \qquad (c')$$

see Example 1.12 again.

Then, using (c') yields

$$J(t, s_{\rightarrow t}) = \hat{x}'_tC_t\hat{x}_t + E[(x_t - \hat{x}_t)'R_t(x_t - \hat{x}_t)|s_{\rightarrow t}]$$
$$+ E[(A_t(x_t - \hat{x}_t) + \varepsilon_t)'C_{t+1}(A_t(x_t - \hat{x}_t) + \varepsilon_t)|s_{\rightarrow t}] + d_{t+1}.$$

Similar to earlier calculations, inserting $\hat{x}_t = x_t + (\hat{x}_t - x_t)$ in $\hat{x}'_tC_t\hat{x}_t$ gives that $\hat{x}'_tC_t\hat{x}_t = E[\hat{x}'_tC_t\hat{x}_t|s_{\rightarrow t}]$ equals

$$E[x'_tC_tx_t|s_{\rightarrow t}] + E[(\hat{x}_t - x_t)'C_t(\hat{x}_t - x_t)|s_{\rightarrow t}].$$

Hence, $J(t, s_{\rightarrow t}) = E[x'_tC_tx_t|s_{\rightarrow t}] + d_t$, where

$$d_t = E[(\hat{x}_t - x_t)'C_t(\hat{x}_t - x_t)|s_{\rightarrow t}] + E[(x_t - \hat{x}_t)'R_t(x_t - \hat{x}_t)|s_{\rightarrow t}]$$
$$+ E[(A_t(x_t - \hat{x}_t) + \varepsilon_t)'C_{t+1}(A_t(x_t - \hat{x}_t) + \varepsilon_t)|s_{\rightarrow t}] + d_{t+1}.$$

The next to last addend equals $E[(A_t(x_t - \hat{x}_t))'C_{t+1}A_t(x_t - \hat{x}_t)|s_{\rightarrow t}] + E[(\varepsilon_t)'C_{t+1}\varepsilon_t)]$. We omit giving the explicit expression for d_t but we note that it can be expressed by means of d_{t+1} and the given matrices in the problem, (in particular each term of the form $E[(x^i_t - \hat{x}^i_t)a(x^j_t - \hat{x}^j_t)|s_{\rightarrow t}]$ is determined by the conditional covariance matrix $\text{Covar}(x_t|s_{\rightarrow t}) = P_t$ defined below).

It remains to describe the estimate $\hat{x}_t$, which determines the control u_t. Let $\hat{x}_t, \bar{x}_t$, M_t, H_t, G_t, and P_t play the roles of $\hat{x}, x^*, M^{-1}, H, R^{-1}$ and $\text{Covar}(x|z)$, respectively, in (b'') above. Note that $x^* = \bar{x}_t$ is the estimate at time t of the current state x_t, before z_t is observed. Evidently, $\bar{x}_t$ equals $A\hat{x}_{t-1} + B_{t-1}u_{t-1} = A\hat{x}_{t-1} - B_{t-1}D_{t-1}\hat{x}_{t-1}$. Then $\hat{x} = \hat{x}_t$ is the estimate of the current state x_t after observing z_t. As can be seen by induction, the current estimate $\hat{x}_t$, as well as the current estimate of $\text{Covar}(x_t)$, given $s_{\rightarrow t}$, in fact depend only on the entities in $s_{\rightarrow t}$, via $\bar{x}_t$ and z_t (and therefore on $\hat{x}_{t-1}$, z_t), as follows from (b) and (b''), see also $(d), (e), (f)$ below.

Altogether, for C_t as given in (1.89), we get the following formulas:

(c) $J(t, s_{\rightarrow t}) = E[x'_tC_tx_t|s_{\rightarrow t}] + d_t$

(d) $u_t = -D_t\hat{x}_t$, $D_t = (Q_t + B'_tC_{t+1}B_t)^{-1}B'_tC_{t+1}A_t$

(e) $\hat{x}_t = \bar{x}_t + M_tH'_t(H_tM_tH'_t + G_t)^{-1}(z_t - H_t\bar{x}_t)$, $\hat{x}_0 = E(x_0)$

(f) $\bar{x}_{t+1} = A\hat{x}_t + B_tu_t$

(g) $M_{t+1} = A_tP_tA'_t + N_t$, $M_0 = \text{Covar}(x_0)$

(h) $P_t = M_t - M_tH'_t(H_tM_tH'_t + G_t)^{-1}H_tM_t$

(i) $d_t = h(d_{t+1})$, $d_T = 0$,   the formula for $h(.)$ omitted.

The explanation of formula (g) is as follows: P_t is $\text{Covar}(x_t|s_{\rightarrow t})$, so M_{t+1} is evidently $\text{Covar}(x_{t+1}|s_{\rightarrow t})$. (So far, we have always assumed x_0 fixed and known, but the specification of $\hat{x}_0$ and M_0 indicates a generalization to the case where only these entities are known at the outset.) Note that (g) is solved forwards, using that P_t depends on M_t via (h). All values of M_t and all values of C_t (and then also of D_t) can be calculated before the initial time. Then at all times t, from (d) and (e), we see that the optimal control u_t can be calculated, given that z_t is observed.

The so-called Kalman filter consists of the algorithm that recursively generates the estimates $\hat{x}_t$, (i.e., $(d),(e),(f)$), as stated above, the update $\hat{x}_t$ depends only on z_t and the previous update $\hat{x}_{t-1}$.

1.9 Approximate Solution Methods in the Infinite Horizon Case*

Successive approximation. In this section, we shall give proofs of some results yielding approximate solutions to infinite horizon dynamic programming problems, as well as proofs of some of the results presented in the sections on an infinite horizon. The problem is as in Section 1.6, see (1.65), (1.66); to simplify the notation, we shall assume no dependence on v_t in $P_t(V_{t+1}|x_t, u_t)$ and in $g(t, ., .)$ in this section. Recall the definitions of $\mathcal{X}(x_0)$ (Remark 1.7 in Section 1.2) and $J(0, x_0, v_0, K)$ (now independent of v_0), see a definition subsequent to (1.67), where $K = T$. The optimal value function J depends now only on x.

One approximation result has already been mentioned in Section 1.6: Under certain conditions, $J(0, x_0, K) \to J(x_0)$ when $K \to \infty$. When K is large, then using the finite horizon optimal controls $u_0, \ldots, u_K$ at the stages $0, 1, \ldots, K$ and arbitrary controls afterwards yields approximately the infinite horizon optimal value (at least in the case (B) of (1.70)). Most often for t not too large, there will be no appreciable difference between the u_t's, so frequently, $u_0(x_0)$ is approximately optimal in the infinite horizon problem when K is large. Another method yielding approximately optimal solutions is the so-called successive approximation method, next to be described, with a little more precision.

Below, functions $h(x), h^*(.), \hat{h}(x), \tilde{J}(x)$ appear, assume them to be sp-continuous in case $\mathcal{V}$ is nondiscrete (no such assumption needed when $\mathcal{V}$ is discrete). For any given function $h(.)$, define $T(h)$ by

$$(Th)(x) = \sup_{u \in U}\{g(x, u) + \alpha E_u[h(f(x, u, V))|x]\},$$

and let $T^2(h) = T(Th), T^3(h) = T(T^2(h))$ and so on. Evidently, T is nondecreasing: $h \leq h^* \Rightarrow T(h) \leq T(h^*)$. Under certain conditions, we shall see later on that $\lim_{k \to \infty} T^k(0) = J$, where $J(x)$, as before is the optimal value function in the infinite horizon problem.

Let the control $u_k(x)$ be the one yielding the supremum (=maximum) when calculating $T^k(0)$, (recall: when $T^{k-1}(0)$ is known, a maximization is carried out to find $T^k(0)$). The control u_k is then usually approximately optimal for k large. In certain circumstances, it is easy to give bounds, telling clearly how close u_k is to being optimal, For example in case (B) of (1.70), $|J(x) - J_{u_k(.)}(x)| \leq 2\alpha^{k-1}M/(1-\alpha)^2$, where $M = \max\{|M_1|, |M_2|\}$ and $J_{u(.)}$ is the criterion value resulting from using the control $u(x)$ at all times. (See Exercise 1.45 below.)

Note that, formally, in the definition of T, h can be any extended real-valued function, i.e., taking values in $[-\infty, \infty]$, provided h is either bounded below or above (and is minimally well-behaved so the expectation can be calculated, with values in $[-\infty, \infty]$).

In this section, the cases (B), (P), and (N) denote the cases of (1.70), including the modifications described subsequent to (1.70) if $\alpha < 1$. Denote by (N+) the case where in addition to (N), the following conditions hold: U is compact and either (C) V is finite and $(x, u) \to P(v|x, u)$ is continuous, or (D) V is nondiscrete and the probabilities are given by piecewise continuous densities $v \to p(v|x, u)$ being continuous in (x, u) for each v, such that $p(v|x, u) \le \psi(v)$ for some integrable ψ, with $\int |v| \psi(v) dv < \infty$. In case (D), it is further assumed that, for some constants k_g and k_f, for all $x \in \mathcal{X}(x_0)$, all $v, u \in U$,

$$|g(x, u)| \le k_g(1 + |x|), |f(x, u, v)| \le k_f(1 + |x| + |v|), \qquad (1.91)$$

The probability assumptions in case (D) can be replaced by, for some k_v,

$$E_u[|V||x] \le k_v(1 + |x|) \text{ and } E_u[h(x, u, V)|x] \text{ is continuous in } (x, u)$$
$$\text{for any continuous function } h(x, u, v) \text{ satisfying } |h(x, u, v)|$$
$$\le k_h(1 + |x| + |v|) \text{ for some } k_h, \text{ for all } x \in \mathcal{X}(x_0), v, u \in U. \qquad (1.92)$$

(In case of (1.92), V can even be discrete but nonfinite.)

The following theorem will be proved.

Theorem 1.40. *In the cases (B), (P), (N+), $\lim_{K \to \infty} J(0, x, K) = J(x)$ and $\lim_{k \to \infty} T^k(0)(x) = J(x)$ for any $x \in \mathcal{X}(x_0)$.* $\qquad \square$

Proof. It turns out that the cases $M_2 \ge g, M_2 > 0, \alpha < 1$ and $M_1 \le g, M_1 < 0, \alpha < 1$, can be reduced to the cases $0 \ge g$, and $0 \le g$, respectively: In the second case, adding the positive number $-M_1$ to g changes the value of the criterion by $\sum \alpha^s(-M_1)$, in case $\alpha \in (0, 1)$. A similar remark pertains to the first case. Thus, it suffices to consider the cases in (1.70). Two lemmas are needed.

Lemma 1.41. *Assume $g \ge 0$, i.e., (P) in (1.70). Then for any non-negative $\tilde{J}$, if $\tilde{J} \ge T(\tilde{J})$, then $\tilde{J} \ge T^\infty(0)$. In particular, $J \ge T^\infty(0)$. Symmetrically, if $g \le 0$ (i.e., (N) in (1.70)) $\tilde{J} \le 0$ and $\tilde{J} \le T(\tilde{J})$, then $\tilde{J} \le T^\infty(0)$, in particular, $g \le 0 \Rightarrow J \le T^\infty(0)$.* $\quad \square$

Proof. Assume first $g \ge 0$. The inequality $\tilde{J} \ge 0$ implies the second of the inequalities $\tilde{J} \ge T(\tilde{J}) \ge T(0)$. Then, $\tilde{J} \ge T(\tilde{J}) \ge T^2(\tilde{J}) \ge T^2(0), \tilde{J} \ge T^3(\tilde{J}) \ge T^3(0)$, and so on, so $\tilde{J} \ge T^k(0)$ for any k. Hence $\tilde{J} \ge T^\infty(0)$. Because $J = T(J), J \ge T^\infty(0)$.

A symmetric proof works for the case $g \le 0$. $\qquad \square$

For any given control function $u(x)$, and any bounded function $h(x)$, define

$$T_{u(.)}(h)(x) = g(x, u(x)) + \alpha E_{u(x)}[h(f(x, u(x), V))|x].$$

Moreover, define $T_{u(.)}^k(h)$ recursively by $T_{u(.)}^k(h) = T_{u(.)}(T_{u(.)}^{k-1}(h))$.

Lemma 1.42. *Let $\hat{h}(x)$ be a bounded function. In the cases (B), (P), (N), for any given sequence of sp-continuous controls $\{u_s(.)\}_s$, for all k, for all $\hat{x}_0 \in \mathcal{X}(x_0)$,*

$$E\left[\sum_{0 \le s \le k} \alpha^s g(X_s, u_s(X_s)) + \alpha^{k+1}\hat{h}(X_{k+1})|\hat{x}_0\right] \le T^{k+1}(\hat{h})(\hat{x}_0), \qquad (1.93)$$

X_s the solution of the difference equation given $\{u_s(x)\}_s$ and $\hat{x}_0$ (starting at $t = 0$ at $\hat{x}_0$). Equality holds in (1.93) if $u_s(.) = u(.)$ for all s, $u(.)$ a given sp-continuous control, and $T^{k+1}(\hat{h})$ is replaced by $T_{u(.)}^{k+1}(\hat{h})$. $\qquad\square$

(Here, and below, we have omitted to indicate that the expectation depends on the control sequence $\{u_s\}_s$.)

Proof. Evidently, (1.93) holds if $k = 0$ (equality holds in case of $u_s(.) = u(.)$ and $T^1(0)$ replaced by $T_{u(.)}^1(0)$). Assume by induction that the inequality holds for a given k (equality in case $u_s(.) = u(.)$ and $T^k(0)$ replaced by $T_{u(.)}^k(0)$). Define $h(X_1) := E[\sum_{1 \le s \le k+1} \alpha^{s-1} g(X_s, u_s(x_s)) + \alpha^{k+1}\hat{h}(X_{k+2})|X_1]$. Then, using (1.93) for $\sum_{0 \le s \le k}$ replaced by $\sum_{1 \le s \le k+1}$ and $\hat{x}_0$ replaced by X_1,

$$E\left[\sum_{0 \le s \le k+1} \alpha^s g(X_s, u_s(X_s)) + \alpha^{k+2}\hat{h}(X_{k+2})|\hat{x}_0\right]$$

$$= E[g(\hat{x}_0, u_0(\hat{x}_0))|\hat{x}_0]$$

$$+ E\left[\alpha\left\{E\left[\sum_{1 \le s \le k+1} \alpha^{s-1} g(X_s, u_s(x_s))|X_1\right] + \alpha^{k+1}\hat{h}(X_{k+2})\right\}|\hat{x}_0\right]$$

$$= E[g(\hat{x}_0, u_0(\hat{x}_0)) + \alpha h(f(\hat{x}_0, u_0(\hat{x}_0), V))|\hat{x}_0]$$

$$\le T(h)(\hat{x}_0) \le T(T^{k+1}(\hat{h}))(\hat{x}_0) = T^{k+2}(\hat{h})(\hat{x}_0).$$

(Equality holds in case of $T^{k+2} = T_{u(.)}^{k+2}$, $u_s(.) = u(.)$.) Thus, the asserted inequality (equality) (1.93) holds for all k, all $\hat{x}_0 \in \mathcal{X}(x_0)$. $\qquad\square$

Lemma 1.42 has the evident implication that for $\hat{h} \equiv 0$

$$J(\hat{x}_0) \le T^\infty(0)(\hat{x}_0) \quad \text{for all } \hat{x}_0 \in \mathcal{X}(x_0), \qquad (1.94)$$

(with equality if $T^\infty(0)$ and $J(\hat{x}_0)$ are replaced by $T_{u(.)}^\infty(0)$ and $J_{u(.)}(\hat{x}_0)$).

To see this, by a monotone convergence theorem, see the Appendix, (1.93) holds also for $k = \infty$, next note that the supremum over all policy sequences of the left-hand side equals $J(\hat{x}_0)$.

Proof of $T^k(0)(x) \to J(x)$, $x \in \mathcal{X}(x_0)$, in cases (B) and (P)

Combine (1.94) with $J \ge T^\infty(0)$ (Lemma 1.42). $\qquad\square$

Proof of $T^k(0)(x) \to J(x)$, $x \in \mathcal{X}(x_0)$, in case (N+)

Note that $T^k(0) \downarrow T^\infty(0)$, so $T^{k+1}(0) = T(T^k(0)) \geq T(T^\infty(0))$. Fix an x in $\mathcal{X}(x_0)$, and (using compactness of U and continuity of g and f in u) choose u_k such that

$$T^{k+1}(0)(x) = g(x, u_k) + \alpha E_{u_k}[T^k(0)(f(x, u_k, V))|x]$$
$$= \max\{g(x, u) + \alpha E_u[T^k(0)(f(x, u, V))|x]\} \geq T^\infty(0)(x),$$

(it may be seen by induction and a Maximum Theorem, see Sydsaeter et al. (2005), that (1.91) and (1.92) imply continuity of the maximand). Because T^k is nonincreasing, then for $i \geq k$,

$$g(x, u_i) + \alpha E_{u_i}[T^k(0)(f(x, u_i, V))|x]$$
$$\geq g(x, u_i) + \alpha E_{u_i}[T^i(0)(f(x, u_i, V))|x] \geq T^\infty(0)(x).$$

Then also $g(x, u) + \alpha E_u[T^k(0)(f(x, u, V))|x] \geq T^\infty(0)(x)$, where u is a cluster point of u_i. Letting $k \to \infty$ (and using a monotone convergence theorem), gives $g(x, u) + \alpha E_u[T^\infty(0)(f(x, u, V))|x] \geq T^\infty(0)(x)$.

Define $U^k(x) = \{u \in U : g(x, u) + \alpha E_u[T^k(0)(f(x, u, V))|x] \geq T^\infty(0)(x)\}$, and note that we have show that $U^*(x) = \cap_k U^k(x) \neq \emptyset$. There exists a (minimally well-behaved) function $u_*(x) \in U^*(x)$ for all $x \in \mathcal{X}(x_0)$, such that

$$g(x, u_*(x)) + \alpha E_{u_*(x)}[T^\infty(0)(f(x, u_*(x), V))|x]$$
$$= T_{u_*(.)}(T^\infty(0))(x) \geq T^\infty(0)(x).$$

(On the existence of $u_*(x)$, see Remark 1.44 below.) It then follows that $T^2_{u_*(.)}(T^\infty(0))(x) \geq T^\infty(0)(x), T^3_{u_*(.)}(T^\infty(0))(x) \geq T^\infty(0)(x), \ldots, x \in \mathcal{X}(x_0)$.

Because $0 \geq T^\infty(0)$, then, using the "equality case" of (1.94), we get $J(x) \geq J_{u_*(.)} = \lim_k T^k_{u_*(.)}(0)(x) \geq \lim_k T^k_{u_*(.)}(T^\infty(0))(x) \geq T^\infty(0)(x)$. By (1.94), we have $J(x) \leq T^\infty(0)(x)$, so $J(x) = T^\infty(0)(x), x \in \mathcal{X}(x_0)$. $\qquad\square$

Proof of $\lim_{K \to \infty} J(0, x, K) = J(x)$, $x \in \mathcal{X}(x_0)$

Consider first the case (P), with $g \geq 0$. Note first, by non-negativity of g, for an arbitrary $\{u_s(.)\}_s$, that

$$E\left[\sum_{0 \leq s \leq K} \alpha^s g(X_s, u_s(X_s))|x\right] \leq E\left[\sum_{0 \leq s < \infty} \alpha^s g(X_s, u_s(X_s))|x\right] \leq J,$$

so $J(0, x, K) \leq J$ and $J(0, x, \infty) = \sup_K J(0, x, K) \leq J$. On the other hand,

$$E\left[\sum_{0 \leq s \leq K} \alpha^s g(X_s, u_s(X_s))|x\right] \leq J(0, x, K),$$

so

$$E\left[\sum_{0 \leq s < \infty} \alpha^s g(X_s, u_s(X_s))|x\right] \leq \sup_K J(0, x, K) := J(x, \infty)$$

and hence $J(x) \leq J(x, \infty)$.

In case (N+), with $g \leq 0$, evidently $J(0, x, K) \geq J(x)$ (J "contains more negative terms"), so $J(0, x, \infty) := \lim_K J(0, x, K) \geq J(x)$. On the other hand, for any given $\pi = \{u_s(.)\}_s$, the inequality (1.93) gives $J_\pi(0, \hat{x}_0, K) \leq T^{k+1}(0)(\hat{x}_0)$, so $J(0, \hat{x}_0, K) \leq T^{k+1}(\hat{x}_0)$, and taking limits, gives $J(0, \hat{x}, \infty) \leq T^\infty(0)(\hat{x})$. Then, for $\hat{x} = x$, using $T^\infty(0)(x) = J(x)$, gives $J(0, x, \infty) \leq J(x)$. So $\lim_{K \to \infty} J(0, x, K) = J(x)$ even in this case. $\qquad\square$

Finally, note that the case where $P_t(V_{t+1}|.)$ and $g(t, .)$ do depend on v_t can be taken care of by using an auxiliary state $\hat{x}_t$ governed by $\hat{x}_{t+1} = V_{t+1}$. Replacing the x_t above by $(x_t, \hat{x}_t)$ means that this case is also covered.

Remark 1.43 (Proof of certain assertions in Sections 1.2 and 1.6). In Sections 1.2 and 1.6, the assertion claimed in case (P), namely that if $J_{u(.)}(x) = T(J_{u(.)})(x)$, then $J_{u(.)}(x) = J(x)$, follows immediately from $J_{u(.)} \geq T^\infty(0) = J$, see Lemma 1.41 and Theorem 1.40.

Let us next prove in case (N) that $u(.)$ is optimal if it satisfies $T_{u(.)}(J)(x) = T(J)(x), x \in \mathcal{X}(x_0)$. As $J = T(J) = T_{u(.)}$, using a result from Lemma 1.40, namely $\tilde{J} = T(\tilde{J}) \Rightarrow \tilde{J} \geq T^\infty(0)$, for the case (P), with g replaced by $-g$, and with no possible choice of controls other than $u(.)$, (so $T = T_{u(.)}$), for $\tilde{J} = -J$, we get that $-J \geq -T_{u(.)}^\infty(0)$. Furthermore, $-T_{u(.)}^\infty(0) = -J_{u(.)}$, (by equality in (1.94)). So $J_{u(.)} \geq J$. $\qquad\square$

Remark 1.44 (Well-behavior of u_*). In the case of a nondiscrete $\mathcal{V}$, we did not state any well-behavior property of u_*, but at least it can be assumed to be what is called "Borel-measurable," see the Appendix. A so-called measurable selection result is needed to obtain $u_*(.)$, (note that $U^k(x)$ is closed, and $x \to \text{distance}(q, U^k(x))$ is Borel-measurable for each $q \in \mathbb{R}^r$). $\qquad\square$

Remark 1.45 (Comment on $T^k(h)$). In cases (B), (P), (N+), $\lim_{k \to \infty} T^k(\hat{h})(x) = J(x)$, $x \in \mathcal{X}(x_0)$, for any bounded function $\hat{h} = \hat{h}(x)$. $\qquad\square$

This can be shown by an easy modification of the above proofs.

Other Methods

Assume again (1.70), with subsequent modifications. Recall that, for any policy $u(x)$, the following equality holds

$$J_{u(.)}(x) = g(x, u(x)) + \alpha E_{u(x)}[J_{u(.)}(f(x, u(x), V))|x], \tag{1.95}$$

or $J_{u(.)} = T_{u(.)}(J_{u(.)})$. One (complicated!) way to see this is that this is the Bellman equation in case of no possible choice of control other than $u(.)$, in which case $J_{u(.)}$ is the optimal value function.

Consider what we shall call the finite case, where it is assumed that U, V are finite sets. This has as consequence that, given a fixed start point x_0, for some countable set S, for any $t, X_t \in S$, (so $\mathcal{X}(x_0) \subset S$). We shall even assume S to be finite in this "finite" case. Then, given $u(.)$, (1.95) represents a set of k linear equations, one for each of the k elements in S. The construction of $J_{u(.)}(x), x \in S$, means solving these linear equation where $J_{u(.)}(x), x \in S$, are unknowns. In this finite case, condition (B) is satisfied in case $\alpha < 1$.

Policy Iteration

Policy iteration is a stepwise approximate solution method and works as follows. Choose an arbitrary initial policy $u_0(x)$, calculate $J_{u_0(.)}(x)$, and then the control $u_1(x)$ yielding maximum when forming $T(J_{u_0})(x)$, then calculate $J_{u_1(.)}(x)$, and next calculate the control $u_2(x)$ yielding maximum when forming $T(J_{u_1})(x)$ and so on. (We assume that such maxima are attained.) Suppose that for any $u = u(x)$, any $\hat{x}_0 \in \mathcal{X}(x_0)$, for $\pi = (u(.), u(.), \ldots)$,

$$\liminf_{k \to \infty} \alpha^k E_\pi[J_u(X_k) | \hat{x}_0] = 0, \tag{1.96}$$

that $\alpha < 1$ and that $g \geq M$ for some constant M (call this case (P+)). (When (B) holds, (1.96) automatically holds.) In case (P+), it turns out that $J_{u_i}(x) \to J(x)$ when $i \to \infty$, for any $x \in \mathcal{X}(x_0)$, so for i large, u_i is approximately optimal.

To see this, we need only consider the case $g \geq 0$. Because for any i, $T_{u_i}(J_{u_i})(x) = J_{u_i}(x)$, then $T_{u_{i+1}}(J_{u_i})(x) = T(J_{u_i})(x) \geq T_{u_i}(J_{u_i})(x) = J_{u_i}(x)$, and then $T^2_{u_{i+1}}(J_{u_i})(x) \geq T_{u_{i+1}}(J_{u_i})(x) \geq J_{u_i}(x)$, and in general $T^k_{u_{i+1}}(J_{u_i})(x) \geq J_{u_i}(x)$. Now, it can be shown that the nondecreasing sequence $\{T^k_{u_{i+1}}(J_{u_i})(x)\}_k$ has a limit $T^\infty_{u_{i+1}}(J_{u_i})(x)$ equal to $J_{u_{i+1}}(x)$, at least when (P+) holds. To see that $T^\infty_{u_{i+1}}(J_{u_i})(x) = J_{u_{i+1}}(x)$, observe that the equality version of (1.93), yields

$$E_\pi \left[\sum_{0 \leq s \leq k} \alpha^s g(X_s, u_{i+1}(X_s)) + \alpha^{k+1} J_{u_i}(X_{k+1}) | \hat{x}_0 \right] = T^{k+1}_{u_{i+1}}(J_{u_i})(\hat{x}_0), \tag{1.97}$$

where X_s is the solution of the difference equation given $\hat{x}_0 \in \mathcal{X}(x_0)$ and $u_s(x) = u_{i+1}(x)$ for all s. (In the current case, (1.93) holds for any non-negative $\hat{h}$.)

Using (1.96), taking limits as $k \to \infty$ in (1.97), we get $J_{u_{i+1}}(\hat{x}_0) = T^\infty_{u_{i+1}}(J_{u_i})(\hat{x}_0) \geq J_{u_i}(\hat{x}_0)$, (which of course implies $J_{u_{i+1}}(\hat{x}_0) \geq T_{u_{i+1}}(J_{u_i})(\hat{x}_0)$).

When (P+) holds, it can further be shown that $\lim_{i \to \infty} J_{u_i}(x) = J(x)$, the optimal value function: Note that, for any u, as $T_{u_{i+1}}(J_{u_i}) = T(J_{u_i})$, by the last inequality, for $x \in \mathcal{X}(x_0)$,

$$g(x, u) + \alpha E_u[J_{u_i}(f(x, u, V)) | x] = T_u(J_{u_i})(x) \leq T(J_{u_i})(x) \leq J_{u_{i+1}}(x).$$

Letting $i \to \infty$ here yields

$$\lim_i J_{u_{i+1}}(x) \geq g(x,u) + \alpha E_u[\lim_i J_{u_i}(f(x,u,V))|x],$$

hence $\lim_i J_{u_i}(x) \geq T(\lim_i J_{u_i})(x)$.

Thus, by Lemma 1.41 and Theorem 1.40, $\lim_i J_{u_i} \geq T^\infty(0) = J$. Evidently, all $J_{u_i} \leq J$, so $\lim_i J_{u_i}(x) = J(x)$, the optimal value function.

In the finite case, as S is finite, the collection of control functions is finite, and so in this case we cannot have $J_{u_{i+1}}(x) > J_{u_i}(x)$ for some $x \in S$ for an infinite number of i's, so in this case an optimal policy will be found after a finite number of steps.

In the finite case, $J_{u_i(.)}(x)$ can be calculated by using (1.95), and, in general, $T^k_{u_i}(J_{u_{i-1}}) \approx J_{u_i}$, when k is large.

Linear Programming

We here confine our attention to the finite case where $\alpha < 1$, $(\mathcal{X}(x_0) =: S$, with $S, \mathcal{V}, U$ finite). For any function $\tilde{J}$, $\tilde{J} \geq T(\tilde{J}) \Rightarrow \tilde{J} \geq T^2(\tilde{J}) \geq \ldots \geq T^k(\tilde{J}) \geq \lim T^k(\tilde{J}) = J = T(J)$, so also $\sum_x \tilde{J}(x) \geq \sum_x J(x)$, and $J(x), x \in S$, solves the problem $\min \sum_x \tilde{J}(x)$ subject to $\tilde{J}(x) \geq T(\tilde{J})(x), x \in S$. Now, the last inequalities can also be written

$$\tilde{J}(x) \geq g(x,u) + \alpha E_u[\tilde{J}(f(x,u,V))|x] \quad \text{for all } x,u.$$

This is a finite set of linear inequalities in $\tilde{J}(x)$. Finding the solution vector of this linear programming problem, i.e., the minimizing vector with components $J^*(x)$, $x \in S$, means finding the optimal value function in the dynamic programming problem: $J^*(x) = J(x)$.

1.10 Semi-Markov Decision Models

We now modify the basic setup slightly. The time lengths between the stages shall now be stochastic. Thus transitions (changes in the state) occur at stochastic time points $\tau_1, \tau_2 = \tau^2 + \tau_1, \tau_3 = \tau^3 + \tau_2, \ldots$, $\tau_1 \geq 0$, $\tau^i > 0$. Let $\tau^0 = \tau_0 = 0$. The change in the state at time τ_j is influenced by a stochastic variable V_j (see (1.99) below). Given $V_j = v_j$, the next V_{j+1} in the process is determined by the probabilities $P(v_{j+1}|x,u,v_j,\tau^{j+1})$. (We can also allow continuous variable v_{t+1}, in which case densities $v_{j+1} \to p(v_{j+1}|x,u,v_j,\tau^{j+1})$ are given.) Given that a transition occurred at $\tau_j \geq 0$, the next time $\tau_{j+1} := \tau^{j+1} + \tau_j \geq \tau_j$ at which a transition occurs is a random variable, and τ^{j+1} has the cumulative distribution $\tau^{j+1} \to F(\tau^{j+1}; x_j, u_j, v_j, \tau^j), \tau^{j+1} \in [0, \infty)$. We have to ensure that an infinite number of transitions do not occur in a finite interval, so we require: There exist numbers $\delta > 0$ and $\delta' \in (0,1)$ such that, for all x, v_j, u, τ^j,

$$Pr[\tau^{j+1} \leq \delta|x,u,v_j,\tau^j] = F(\delta; x, v_j, u, \tau^j) \leq 1 - \delta'. \tag{1.98}$$

(To see this, note that $\Pr[\tau_0,\ldots,\tau_N \in [0,\delta)] \le (1-\delta')^{N+1}$, by rules of conditional probabilities, so the probability for an infinite number of transition points τ_i in $[0,\delta)$ is zero. Next, the event that an infinite number of transition points occur in $[0,2\delta)$ has also probability zero, because then an infinite number of points belong to $[0,\delta)$ or, if not to this set, then to $[\delta,2\delta)$, both these events having probability zero. An infinite number of transitions cannot occur in $[0,3\delta)$, by a similar argument, neither in $[0,n\delta)$ for any n, by continuing this argument.)

The state equation is now written

$$x_{j+1} = f(x_j, u_j, v_{j+1}, \tau^{j+1}), \quad x_0 \text{ (and } V_0 = v_0) \text{ fixed.} \tag{1.99}$$

In the interval (τ_j, τ_{j+1}) the state $x(t)$ equals x_j. Thus, $x(t)$ is defined for all t in $(0,\infty)$, and with probability 1, $x(t)$ is a piecewise constant function on any interval $[0,a)$ (on $[0,a)$ it has, with probability 1, a finite number of jumps).

The criterion to be maximized is

$$E\left[\sum_{j=0}^{\infty} \{e^{-\alpha\tau_j} f^0(X_j, u_j(X_j, V_j, \tau_j), V_j, \tau^j) \right.$$

$$\left. + \int_{\tau_j}^{\tau_{j+1}} e^{-\alpha s} h(X_j, u_j(X_j, V_j, \tau_j), V_j, \tau^j) ds\} \right]$$

$$= E\left[\sum_{j=0}^{\infty} e^{-\alpha\tau_j} \{f^0(X_j, u_j(X_j, V_j, \tau_j), V_j, \tau^j) \right.$$

$$\left. + h(X_j, u_j(X_j, V_j, \tau_j), V_j, \tau^j)(1 - e^{-\alpha\tau^{j+1}})/\alpha\} \right], \tag{1.100}$$

where the τ_j's $(j \ge 1)$ can take values anywhere in $[0,\infty)$, so the horizon is infinite. The functions f, f^0, and h, as well as the number $\alpha > 0$ are given entities, f, f^0, and h continuous. A control u_j is chosen each time a transition has occurred, i.e. at each time point τ_j, $u_j \in U$, U a given set in $\mathbb{R}^r$. We wish to maximize the criterion over all policy sequences $\{u_i(X_i, V_i, \tau_i)\}_{i=0}^{\infty}$.

We shall treat the current topic in a slightly nonrigorous manner, let us nevertheless mention that, of course, certain assumptions have to be satisfied for the above expectations to be defined. (Sufficient conditions are for example that f^0, f, and h are all bounded functions, then even the supremum of the criterion is then finite. Weaker conditions exist, allowing Lipschitz continuity in x of some rank K, provided $f(0, u, v, \tau^j), f^0(0, u, v, \tau^j), h(0, u, v, \tau^j)$ are bounded functions and K is small enough.)

Let $f_0(x, u, v, \tau^j) :=$

$$f^0(x, u, v, \tau^j) + \int_0^{\infty} \{h(x, u, v, \tau^j)(1 - e^{-\alpha t})/\alpha\} dF(t; x, u, v, \tau^j),$$

(dF calculated with respect to t). Define the infinite horizon current optimal value function $J(j,x_j,v_j,\tau_j,\tau^j)$ to be equal to

$$\max E[\Sigma_{i=j}^{\infty} e^{-\alpha(\tau_i-\tau_j)} f_0(X_i,u_i(X_i,V_i,\tau_i),V_i,\tau^i)|x_j,v_j,\tau^j] \qquad (1.101)$$

where (τ_i,X_i) is the process determined by the u_i's, starting at the given point (τ_j,x_j) (v_j also given). (The word "current" is used to indicate that we discount back to τ_j, not to $t = 0$. Note that $\tau_i - \tau_j = \Sigma_{k=j+1}^{i}\tau^k$.) The maximum is taken over all policy sequences $\{u_i(X_i,V_i,\tau_i)\}_{i=j}^{\infty}$. Observe that, by the autonomy in the problem and the infinite horizon, $J(j,x_j,v_j,\tau_j,\tau^j)$ is independent of j and τ_j and can be written $J(x,v,\tau)$, where $x = x_t, v = v_t, \tau = \tau^j$. The following Bellman equation then holds:

$$J(x,v,\tau) = \max_u \{f^0(x,u,v,\tau)$$
$$+E[h(x,u,v,\tau)(1-e^{-\alpha\hat{t}})/\alpha + e^{-\alpha\hat{t}}J(\hat{X},\hat{V},\hat{t})|x,v,\tau]\} \quad (1.102)$$

(expectation with respect to the stochastic variables $\hat{X},\hat{V},\hat{t}$), i.e.,

$$J(x,v,\tau) = \max_u\{f_0(x,u,v,\tau) + \int_0^{\infty} E[J(\hat{X},\hat{V},t)|x,u,v,t]e^{-\alpha t}dF(t;x,u,v,\tau)\},$$

$$(1.103)$$

where $\hat{X} = f(x,u,\hat{V},t)$, and where

$$E[J(\hat{X},V,t)|x,u,v,t] = \sum_{\hat{v}} P(\hat{v}|x,u,v,t)J(f(x,u,\hat{v},t),\hat{v},t)$$

in the discrete probability case. To see that this is correct, one can imagine that the time between stages are all equal to one, and that τ^j is "included" in the stochastic variable V_j. Then the Bellman equation (1.68) yields (1.102). Again it can be expected that optimal controls are of the form $u_j(x_j,v_j,\tau^j) = u(x_j,v_j,\tau^j)$, in particular, the controls do not depend on j.

As said above, the current topic will be treated somewhat informally. The infinite horizon should require of us to state properties holding in cases similar to (B), (P), and (N) in Section 1.6 above, however, we refrain from doing that.

In pure stopping problems, where $U = \{u^s,u^c\}$ and $h(x,u^s,v,\tau) = 0$, we confine our interest to problems where we can stop only at jump points (the "moment after" such jumps), so let us have this as a requirement. In this case, the Bellman equation is

$$J(x,v,\tau) = \max\{f^0(x,u^s,v,\tau),$$
$$f_0(x,u^c,v,\tau) + \int_0^{\infty} E[J(f(x,u^c,\hat{V},t),\hat{V},t)|x,u^c,v,t]e^{-\alpha t}dF(t;x,u^c,v,\tau)\},$$

(the expectation E is with respect to $\hat{V}$).

Example 1.46. At a workplace for garbage collection, used PCs are received in accordance with a Poisson process with intensity λ. At any point in time, the persons working there can decide to take their truck and transport all the PCs to a reprocessing factory, which is reckoned to cost K. As long as $x = i$ PCs are in the store, then, per unit of time, a cost $c(i)$ is incurred, $c(0) = 0$, $c(i)$ increasing to infinity with i. It is easily seen that the decision to ship away the PCs will be taken at the moments in time PCs arrive. At which point in time should PCs be trucked away?

The control u takes values in $\{0, 1\}$, $u = 0$ means the PCs are trucked away, and $u = 1$ means the opposite. The state equation is $x_{j+1} = 1 + x_j u_j$. The criterion is

$$E\left[\sum_j -K(1 - u_j)e^{-\alpha \tau_j} + \int_{\tau_j}^{\tau_{j+1}} -c(X_j)u_j e^{-\alpha t} dt\right],$$

$\alpha > 0$.

Solution. For $x = i$, the Bellman equation for this problem is:

$$J(i) = \max\{-K + J(1)\gamma, -c(i)/\alpha + (c(i)/\alpha + J(i+1))\gamma\}, \qquad (*)$$

where $\gamma = \lambda/(\lambda + \alpha)$. To see this, first note that we shall now find the maximum of two terms, one term stemming from $u = 0$ and one from $u = 1$. The latter case, which corresponds to keeping the PCs in store, yields the term $E[-c(i)((1 - e^{-\alpha \tau})/\alpha) + J(i+1)e^{-\alpha \tau}] = \int_0^\infty \lambda e^{-\lambda t}[-c(i)((1 - e^{-\alpha t})/\alpha) + J(i+1)e^{-\alpha t}]dt = -c(i)/\alpha + (c(i)/\alpha + J(i+1))\gamma$. Next, $u = 0$ gives the term $-K + E[J(1)e^{-\alpha \tau}] = -K + J(1)\gamma$.

Let us guess that the optimal policy is as follows: There is a natural number i^*, such that the PCs are trucked away, if and only if their number exceeds or equals i^*. Then $J(i^*) = -K + J(1)\gamma$. Using

$$J(i) = -c(i)/\alpha + (c(i)/\alpha + J(i+1))\gamma = -c(i)(1 - \gamma)/\alpha + \gamma J(i+1)$$

for $i < i^*$, by backwards recursion $J(i^* - 1), \ldots, J(1)$ can be found. We get

$$J(1) = (-K + J(1)\gamma)\gamma^{i^*-1} - (1 - \gamma)[c(i^* - 1)\gamma^{i^*-2} + c(i^* - 2)\gamma^{i^*-3} + \ldots + c(1)]/\alpha.$$

Hence, $J(1)(1 - \gamma^{i^*}) = -K\gamma^{i^*-1} - (1 - \gamma)\phi(i^*)/\alpha$ where

$$\phi(i^*) = c(i^* - 1)\gamma^{i^*-2} + c(i^* - 2)\gamma^{i^*-3} + \ldots + c(1),$$

so $J(1) = (1 - \gamma^{i^*})^{-1}[-K\gamma^{i^*-1} - (1 - \gamma)\phi(i^*)/\alpha]$. Now, by $(*)$, i^* is the largest i^* such that

$$-c(i^* - 1)(1 - \gamma)/\alpha + J(i^*)\gamma$$
$$= -c(i^* - 1)(1 - \gamma)/\alpha + (-K + J(1)\gamma)\gamma > -K + J(1)\gamma,$$

or

$$K > \gamma J(1) + c(i^* - 1)/\alpha$$

Using the expression for $J(1)$, we get

$$K > \gamma(1-\gamma^*)^{-1}[-K\gamma^{*-1} - (1-\gamma)\phi(i^*)/\alpha] + c(i^*-1)/\alpha$$

which is equivalent to

$$\alpha K > c(i^*-1)(1-\gamma^*) - (1-\gamma)\gamma\phi(i^*). \qquad (**)$$

So, i^* is the largest i^* such that this inequality holds. To see that such a i^* exists, note first that $\alpha K > c(0)(1-\gamma) - (1-\gamma)\gamma\phi(1) = -(1-\gamma)\gamma\phi(1)$, and next that $\phi(i^*) \le c(i^*-1)/(1-\gamma)$, as can be seen by replacing $c(i^*-2),\ldots,c(1)$ by $c(i^*-1)$ in the formula for $\phi(i^*)$, and summing the arising series. The right-hand side of (**) is hence $\ge c(i^*-1)(1-\gamma^*-\gamma)$ and for large i^*, αK (the left-hand side) is even $\le c(i^*-1)(1-\gamma^*-\gamma)$.

(A precise argument for the fact that the above policy *is* optimal is omitted. Similar to the arguments used in the infinite horizon job search case, it is however easy to show that the above policy – and $J(i)$-function – is the only one satisfying the equilibrium optimality equation. And we are in a situation similar to the (N) case, where we then can conclude that the policy is indeed optimal.)   □

Further reading. Among a great number of books on dynamic programming, we mention only Ross (1983), (1992), Bertzekas (1976), and, with measure theory, Bertsekas and Shreve (1978), Hernandez-Lerma and Lasserre (1996), Stokey et al. (1989), and Puterman (1994).

1.11 Exercises

Exercises for Section 1

1.1. (Bertsekas) A farmer annually produces x_k units of a certain crop and stores $(1-u_k)x_k$ units of his production, where $0 \le u_k \le 1$, and invests the remaining $u_k x_k$ units, thus increasing the next year's production to a level x_{k+1} given by

$$x_{k+1} = x_k + W_k u_k x_k, \qquad k = 0, 1, \ldots, N-1.$$

The scalars W_k are bounded independent random variables with identical probability distributions that depend neither on x_k nor on u_k. Furthermore, $E\{W_k\} = \bar{w} > 0$. The problem is to find the optimal policy that maximizes the total expected product stored over N years:

$$E\left\{ X_N + \sum_{k=0}^{N-1} (1-u_k)X_k \right\}.$$

Show that an optimal control law is given by:
(1) If $\bar{w} > 1$, then $u_0(x_0) = \cdots = u_{N-1}(x_{N-1}) = 1$.
(2) If $0 < \bar{w} < 1/N$, then $u_0(x_0) = \cdots = u_{N-1}(x_{N-1}) = 0$.
(3) If $1/N \le \bar{w} \le 1$, then

$$u_0(x_0) = \cdots = u_{N-\bar{k}-1}(x_{N-\bar{k}-1}) = 1, u_{N-\bar{k}}(x_{N-\bar{k}}) = \cdots = u_{N-1}(x_{N-1}) = 0$$

where $\bar{k}$ is such that $1/(\bar{k}+1) \le \bar{w} < 1/\bar{k}$.

1.2. In the examples in the text in Section 1.1, and also in the subsequent exercises for Section 1.1, the situation is so simple that the $J(t, x_t, v_t)$-functions have the same structure for all t, e.g., in one problem, all are quadratic in x_t. In most problems this is not so. Consider for example the problem

$$\max E\left[\sum_{s=0}^{s=T-1} \ln u_s - x_T\right],$$

s.t. $x_{t+1} = 1 + V_{t+1} x_t u_t, x_0 = 1, u_t \in (0, 1], V_t \in \{1, 2\}, V_t$ i.i.d.,
$\Pr[V_t = 1] = 1/2$.

Then $J(T, x_T) = -x_T$, $J(T-1, x_{T-1}) = -\ln x + \ln(2/3) - 2$, and $J(T-2, x_{T-2}) = -2 + \ln(2/3) - (1/2)\ln(1+x) - (1/2)\ln(1+2x)$. Show these formulas.

1.3. Consider the problem:

$$\max E\left[-\delta \exp(-\gamma X_T) + \sum_{0 \le t \le T-1} -\exp(-\gamma u_t)\right],$$

where u_t are controls taking values anywhere in $\mathbb{R}$, δ and γ are given positive numbers, and where

$$X_{t+1} = 2X_t - u_t + V_{t+1}, \qquad x_0 \text{ given.}$$

Here $V_{t+1}, t = 0, 1, 2, \ldots, T-1$, are identically and independently distributed. Moreover, $K := E\left[\exp(-\gamma V_{t+1})\right] < \infty$. Show that the optimal value function $J(t, x)$ can be written $J(t, x) = -\alpha_t \exp(-\gamma x)$, and find a backwards difference equation for α_t. What is α_T?

1.4. (Blanchard & Fischer) Solve the problem

$$\max E\left[\sum_{t=0}^{T-1}(1+\theta)^{-t}\ln(C_t) + k(1+\theta)^{-T}\ln A_T\right],$$

where w_t and C_t are controls, k and θ given positive numbers, and

$$A_{t+1} = (A_t - C_t)[(1+r_t)w_t + (1+V_{t+1})(1-w_t)],$$

where r_t is a given sequence of positive numbers. The stochastic variables V_t are bounded and independently and identically distributed.

1.5. Solve the problem

$$\max E\left[\sum_{t<T} 2(u_t)^{1/2} + aX_T\right], \quad u_t \ge 0, \quad a > 0, \quad x_0 > 0, \quad T \text{ fixed,}$$

where

$$X_{t+1} = X_t - u_t \text{ with probability } 1/2, X_{t+1} = 0 \text{ with probability } 1/2.$$

1.6. Solve the problem

$$\max E\left[\sum_{0\le t\le T}(1-u_t)X_t\right], \qquad x_0 = 1$$

subject to $u_t \in [0,1]$ and

$$X_{t+1} = X_t + u_t X_t + V_{t+1},$$

where $V_{t+1} \ge 0$ is exponentially distributed with parameter λ, (i.e., the density of $V_{t+1} \ge 0$ is $\phi(v) := \lambda e^{-\lambda v}$), ($V_t$ i.i.d.).

1.7. (Hakanson) Let x_t denote capital, y_t income, c_t consumption (a control), and z_t investment with uncertain return (another control). The balance $x_t - c_t - z_t$ is placed in a bank and earns a return r equal to 1 plus the interest rate. Let the gross rate of return on the uncertain investment (i.e., z) be β_t (so β_t equals 1 plus an uncertain net rate of return). Assume that the random variables β_t are bounded and independently and identically distributed. Then

$$X_{t+1} = (\beta_{t+1} - r)z_t + r(X_t - c_t) + y_t.$$

Assume that $E\beta_t > r$. Let $K > 0$, $\gamma \in (0,1)$ be given numbers. The maximization problem is

$$\max E\left\{\sum_{0\le t\le T-1}[\alpha^t(c_t)^\gamma]/\gamma + \alpha^T(K/\gamma)(X_T)^\gamma\right\},$$

where $c_t \ge 0$, $z_t \ge 0$.
(a) Solve the problem, i.e., find the optimal controls. (Assume that $c_t > 0$ and $x_{t+1} > 0$.) *Hint*: When maximizing w.r.t. (c,z), first maximize w.r.t. z. When maximizing w.r.t. z, use the fact that for an arbitrarily given number $b > 0$, one has that

$$\max_{z\ge 0} E\left[\{rb + (\beta - r)z\}^\gamma\right] = b^\gamma a \quad \text{where } a = \max_{s\ge 0} E\left[\{r + (\beta - r)s\}^\gamma\right]. \quad (*)$$

Don't try to find a, use it as a known parameter in the solution of the problem. Formally, we let the w^γ–function be $-\infty$ when $w < 0$. Write expressions of the form $\{y + r(x-c) + (\beta - r)z\}^\gamma$ as $\{r(\frac{1}{r}y + x - c) + (\beta - r)z\}^\gamma$ when using $(*)$.
(b) Discuss dependence on parameters in the problem, including the distribution of β.
(c) Show $(*)$.

1.8. Consider the problem

$$\max E\left[\sum_{0\le t\le T-1}\{(1-u_t)X_t^2 - u_t\} + 2X_T^2\right]$$

subject to (T fixed and)

$$X_{t+1} = u_t X_t V_{t+1}, u \in U := [0,1],$$

where $V_{t+1} = 2$ with probability $1/4$, $V_{t+1} = 0$ with probability $3/4$, (V_t i.i.d.).
(a) Find $J(T,x)$, $J(T-1,x)$, and $J(T-2,x)$, and corresponding controls. Be aware
that the maximands will be convex in the control u, so maxima will be situated at
corners (ends of U).
(b) Find $J(t,x)$ for general t, and the corresponding controls.

1.9. Solve the problem

$$\max E\left[\sum_{0 \leq t \leq T-1} (u_t)^{1/2} + a(X_T)^{1/2} \right], \quad u_t \geq 0,$$

a, T given positive numbers, subject to

$$X_{t+1} = (X_t - u_t)V_{t+1}, \quad x_0 = 1,$$

where $V_{t+1} = 0$ with probability $1/2$, $V_{t+1} = 1$ with probability $1/2$, (V_t i.i.d.). *Hint:*
Try $J(t,x) = 2a_t x^{1/2}, a_t > 0$.

1.10. Solve the problem

$$\max E[(x_T)^{1-\alpha}/(1-\alpha)], \quad u_t \in (0,1),$$

$\alpha > 0$, $\alpha \neq 1$, subject to

$$X_{t+1} = X_t + u_t V_{t+1} X_t, \quad x_0 > 0,$$

where the V_{t+1}'s are i.i.d., $V_{t+1} \in \{-1,1\}$, $\Pr[V_{t+1} = 1] = p > 1/2$.

1.11. Solve the problem

$$\max E\left[\sum_{t=0}^{T-1} -u_t^2 - X_T^2 \right], \quad u_t \in \mathbb{R}$$

when $X_{t+1} = X_t V_{t+1} + u_t, x_0$ a given number, $V_{t+1} \in \{0,1\}$, $\Pr[V_{t+1} = 1|V_t = 1] = 3/4$, $\Pr[V_{t+1} = 1|V_t = 0] = 1/4$, ($V_t$ a Markov chain). *Hint:* Try $J(t,x_t,1) = -a_t x_t^2, J(t,x_t,0) = -b_t x_t^2$.

Exercises for Section 1.2

1.12. Consider the problem

$$\max E\left[\sum_{t=0}^{\infty} (-u_t^2 - X_t^2)\alpha^t \right], \quad \alpha \in (0,1), \quad u_t \in \mathbb{R},$$

$$X_{t+1} = X_t + u_t + V_{t+1}, \quad E(V_{t+1}) = 0, \quad E(V_{t+1})^2 = d.$$

(V_t bounded and i.i.d.) (a) Guess that $J(x)$ is of the form $ax^2 + c$, and insert it in (1.10) to determine a, b, and c.

(b) Solve the finite horizon problem by assuming a quadratic form of the value function. (We now sum only up to time T.) Find $J(0, x_0, T)$, let $T \to \infty$ and prove that the solution in (a) is optimal (we are in case N).

1.13. Solve the problem

$$\max E \left[\sum_{t=0}^{\infty} \beta^t (\ln u_t + \ln X_t) \right], X_{t+1} = (X_t - u_t)V_{t+1}, x_0 > 0, u_t \in (0, x_t),$$

where $\beta \in (0, 1)$, $V_t > 0$, and all V_t i.i.distributed, with $|E \ln V_t| < \infty$. Drop final proof of optimality of the control (i.e., discussion of satisfaction of conditions like P. or N.)

1.14. In Example 1.8, by calculating explicitly J^{u_*}, show that $J(x) = J^{u_*}$.

Exercises for Section 1.3

1.15. Solve the dynamic programming problem

$$\max E \left[\sum_{0 \le t \le T} (1 - u_t)X_t \right], \qquad x_0 = 1$$

subject to $u_t \in [0, 1]$ and

$$X_{t+1} = X_t + u_t X_t + V_{t+1},$$

where V_{t+1} is exponentially distributed with parameter $1/x_t$ (i.e., the density of V_{t+1} is $\phi(v) := \left\{ e^{-v/x_t} \right\} / x_t$).

1.16. Use the stochastic Euler equation to solve the problem

$$\max E \sum_{t=0}^{2} [1 - (V_{t+1} + X_{t+1} - X_t)^2 + 1 + V_3 + X_3], X_0 = x_0,$$

$X_1, X_2, X_3 \in \mathbb{R}$, all V_t bounded and i.i.d. with $EV_t = 1/2$, x_0 given.

1.17. Let $\{u_t^*(z_t)\}_t$ be a finite sequence satisfying the optimality equation, together with a finite sequence of functions $\{J(t, z_t)\}_t$, and let $\{X_t^*\}_t$ correspond to $\{u_t^*(z_t)\}_t$. Prove the optimality of $\pi^* := \{u_t^*(z_t)\}_t$, by elaborating the following argument (the setup is as in the proof of the optimality equation): Let $\pi := u_t(z_t)_t$ be an arbitrarily given control sequence, with corresponding sequence $\{X_t\}_t$. Define $J(t, z_t, \pi) = E^\pi [\sum_{s=t}^T f_0(s, X_s, u_s(Z_s)) | z_t]$, where the conditional probabilities used to calculate the expectation are $P_t(V_t | z_t, u_t(Z_t))$ (which explains the superscript π on the expectation). Trivially, $J(T, z_T, \pi) \le J(T, z_T, \pi^*) = J(T, z_T)$. By backwards induction, let us prove that $J(t, z_t, \pi) \le J(t, z_t, \pi^*) = J(t, z_t)$. Assume that the inequality is correct for t. Now,

$$J(t-1,z_{t-1},\pi) = E^{\pi}\left[\sum_{s=t-1}^{T} f_0(s,X_s,u_s(Z_s))|z_{t-1}\right]$$

$$= f_0(t-1,x_{t-1},u_{t-1}(z_{t-1})) + E^{\pi}\left[E^{\pi}\left[\sum_{s=t}^{T} f_0(s,X_s,u_s(Z_s))|z_{t-1},Z_t\right]|z_{t-1}\right]$$

$$= f_0(t-1,x_{t-1},u_{t-1}(z_{t-1})) + E^{\pi}\left[E^{\pi}\left[\sum_{s=t}^{T} f_0(s,X_s,u_s(Z_s))|Z_t\right]|z_{t-1}\right]$$

$$= f_0(t-1,x_{t-1},u_{t-1}(z_{t-1})) + E^{\pi}[J(t,Z_t,\pi)|z_{t-1}]$$

$$\leq f_0(t-1,x_{t-1},u_{t-1}(z_{t-1})) + E^{\pi}[J(t,Z_t)|z_{t-1}]$$

$$\leq \max_{u}\{f_0(t-1,x_{t-1},u) + E^{u}[J(t,Z_t)|z_{t-1}]\} = J(t-1,z_{t-1}).$$

If $\pi = \pi^*$, the inequalities are equalities, hence $J(t-1,z_{t-1}) = J(t-1,z_{t-1},\pi^*)$.

Exercises for Section 1.4

1.18. Solve Example 1.19 in Section 1.3 (a hard end constraint), by using the stochastic maximum principle.

1.19. In Exercise 1.5, let $a = 0$, and add $EX_T \geq 0$ as a soft constraint. Solve this problem by means of the stochastic maximum principle as well as by means of Remark 1.30.

1.20. In Exercise 1.10, replace $x_T^{1-\alpha}/(1-\alpha)$ by $\ln x_T$, and solve the problem by means of the stochastic maximum principle.

1.21. Solve Exercise 1.3 by means of the stochastic maximum principle.

1.22. Consider the deterministic problem $\max_{u_0 \in \mathbb{R}}[\max\{-u_0^2,-2\}]$ subject to $X_1 = u_0$, $X_0 = 0$, $X(1) \geq 1$. Show that $u_0 = 1$ satisfies the necessary first-order conditions (at this point we have differentiability) but that the procedure in Remark 1.30 does not work (essentially, the Lagrangian in this nonlinear programming problem is not maximized).

1.23. Explain the following proof of sufficiency of Theorem 1.24, even for the end constraints in Remark 1.27, and for $p_T(x_T^*,v_T)$ satisfying the transversality conditions in that remark. Show also that essentially the same proof works for history-dependent controls $u_t(v_{\to t})$. Below, $H(s,x,u,\psi) = f_0(s,x,u) + \psi f, f(T,.,.,.)) = 0$, $\psi_{T+1} = p_T$, and $\psi_{s+1} = \psi(s+1,f(s,x_s^*,u_s^*(x_s^*,v_s),v_{s+1}),v_{s+1})$.

$$E\left\{ \sum_{s=0}^{T} f_0(s, X^*, u_s^*) - f_0(s, X_s, u_s) \right\}$$

$$= E\left\{ \sum_s H(s, X_s^*, u_s^*, \psi_{s+1}) - H(s, X_s, u_s, \psi_{s+1}) \right.$$

$$\left. - \psi_{s+1}(f(s, X_s^*, u_s^*, V_{s+1}) - f(s, X_s, u_s, V_{s+1})) \right\}$$

$$\geq E\left\{ \sum_s H_x(s, X_s^*, u_s^*, \psi_{s+1})(X_s^* - X_s) + H_u(s, X_s^*, u_s^*, \psi_{s+1})(u_s^* - u_s) \right.$$

$$\left. - \psi_{s+1}(f(s, X_s^*, u_s^*, V_{s+1}) - f(s, X_s, u_s, V_{s+1})) \right\}$$

$$\geq E\left\{ \sum_s E[H_x(s, X_s^*, u_s^*, \psi_{s+1})(X_s^* - X_s) + H_u(s, X_s^*, u_s^*, \psi_{s+1})(u_s^* - u_s) \right.$$

$$\left. - \psi_{s+1}(f(s, X_s^*, u_s^*, V_{s+1}) - f(s, X_s, u_s, V_{s+1})|v_{\to s}] \right\}$$

$$\geq E\left\{ \sum_s E[H_x(s, X_s^*, u_s^*, \psi_{s+1})(X_s^* - X_s) - \psi_{s+1}(X_{s+1}^* - X_{s+1})|v_{\to s}] \right\}$$

$$\geq E\left[\sum_{s=0}^{T} \psi_s(X_s^* - X_s) - \psi_{s+1}(X_{s+1}^* - X_{s+1}) \right]$$

$$\geq E[\psi_0(x_0^* - x_0) - \psi_{T+1}(X_{T+1}^* - X_{T+1})] = 0.$$

1.24. (Nonexistence of admissible Markov controls)

(a) Consider the following system. Let $x_0 = 0, X_1 = V_1 \in \{0,1\}, X_2 = X_1 V_2, X_3 = X_2 u_2 + u_2, U = \{-1,1\}, V_2 \in \{0,2\}, \Pr[V_1 = 1] = \Pr[V_2 = 2] = 1/2$, and let us require that $EX_3 = 1$. Show that no admissible Markov controls $u_t(x_t)$ exist, but that history dependent controls exist.

(b) Introduce the following changes in the system in (a): $X_2 = V_2 \in \{0,2\}, \Pr[V_2 = 2|v_1 = 1] = 1/2, \Pr[V_2 = 0|v_1 = 0] = 1$, so $E[V_2|v_1] = v_1$. Show that no admissible Markov controls $u_t(x_t, v_t)$ (but history dependent controls) exist.

Exercises for Section 1.5

1.25. Solve the optimal stopping problem where we shall choose τ to maximize

$$E\left[\sum_{0 \leq t \leq \tau - 1} c_t + Y_\tau^2 \right], \quad c_t = -1, \quad \tau \leq T.$$

Here Y_t is a stochastic process that develops according to

$$Y_{t+1} = Y_t + V_{t+1}, \quad \text{where}$$

$$Pr\big[V_{t+1} = 1\big] = \frac{1}{4t}, \quad \text{and} \quad Pr\big[V_{t+1} = 0\big] = 1 - \frac{1}{4t},$$

V_{t+1} taking only the values 1 and 0, the V_t's being independent. (If we stop at $t = \tau$, we get the reward Y_τ^2 dollars, at all times $t < \tau$, we have had to pay 1 dollar.)

1.26. Solve the optimal stopping problem $\max_\tau EX_\tau$, $X_{t+1} = (X_t - 1/2)V_{t+1}$, $V_{t+1} = 0$ with probability $2/3$, $V_{t+1} = 1$ with probability $1/3$ ($\tau \leq T$) (V_t i.i.d.).

1.27. A surgeon, Jane, is going to carry out a certain operation, the success of which has a probability $1/2$ (the successes are i.i.d.). During the course of maximum three days, Jane is going to perform the operation twice each day. She may stop at the end of the first or second day and definitely stops at the end of the third day. She wants to have a flattering record of operations, using as success criterion 48 times the fraction $s_t/2t$ of successful operations, where s_t is the number of successful operations up to and including day t, $t = 1, 2, 3$. Find an optimal stopping policy that maximizes the expected value of the success criterion.

1.28. (Ross) John drives to his job. He can either park at his workplace, which costs him $K > 0$, or else he can park for free in the street leading up to his workplace, at any parking place numbered $s = 1, 2, 3, \ldots$, starting the numbering from the entrance to his workplace. (He passes place s before passing place $s - 1$.) The probability that any parking place in the street is available is p. John wants to minimize the cost, which equals s if he parks in the street (the effort of walking) and equals K otherwise. *Hint*: You will need the number n^* being the greatest number n such that $(q^{n+1} - q)/p + q^{n-1}K \geq 1$, where $q = 1 - p$, and the formula

$$np + (n-1)qp + \cdots + 2q^{n-2}p + q^{n-1}p = n + (q^{n+1} - q)/p.$$

1.29. Solve the problem $\max_\tau EX_\tau$, where τ is subject to choice in $[0, 3]$,

$$X_{t+1} = (X_t + 1)V_{t+1}, \quad x_0 > 0 \text{ given},$$

and where $V_{t+1} = 0$ with probability $1/2$ and $V_{t+1} = 1$ with probability $1/2$ (V_t i.i.d.).

1.30. Solve the stopping problem:

$$\max_\tau EX_\tau$$

subject to $\tau \in [0, T]$, T fixed, and

$$X_{t+1} = X_t/2 + y_t V_{t+1}, \quad t = 0, 1, 2, \ldots T, \ X_0 = 1$$

$$y_{t+1} = 3y_t V_{t+1}/2,$$

$y_0 = 1$, where $V_{t+1} = 1$ with probability $1/2$ and $V_{t+1} = 0$ with probability $1/2$, (V_t i.i.d.). Does a one-stage look-ahead policy apply? (*Hint*: The sets B_t can be restricted to $\mathcal{X}_t(x_0) \times \mathcal{V}$.) (*Hint*: The sets B_t can be restricted to $\mathcal{X}_t(x_0) \times \mathcal{V}$.)

1.31. Solve the problem $\max_\tau EX_\tau^2$ subject to $\tau \in [0, T]$, T fixed, and $X_{t+1} = V_{t+1}$, $t = 0, 1, 2, \ldots, T$, where V_{t+1} is uniformly distributed in $[0, 1]$ (V_t i.i.d.).

Exercises for Section 1.6

1.32. Mary owns a piece of land on which new sites for silver mines all the time are found with a certain probability. The total amount of silver in all deposits known at time t is x_t, and $X_{t+1} = x_t + a$ with probability $\mu \in (0, 1)$ (a new bed has been found), $X_{t+1} = x_t$ with probability $1 - \mu$. The real price of silver is e^{-bt}, declining all the time. Mary wants to sell all deposits at a certain time. Which time is optimal? (The horizon is infinite.)

1.33. Solve the job search problem in the case where earlier job offers can be recalled. The horizon is infinite.

1.34. In Example 1.38 (job search with infinite horizon), prove that $J^{u*} = \max\{v, v'\}$.

1.35. Consider the infinite horizon problem

$$\max E[X_\tau \alpha^\tau] \quad \text{when} \quad X_{t+1} = (X_t + 1)V_{t+1}, \; X_0 > 0,$$

where X_0 is given and $V_t \in \{0, 1\}$, $\Pr[V_t = 1] = 1/2$, all V_t i.i.d., $\alpha \in (0, 1)$.

 The stopping rule is: Stop the first time $x \geq k$ (Why such a rule ?). Find an unsolvable equation for k. *Hint*: Given this rule, show that $J(0) = \sum_{j=k}^\infty (\alpha/2)^j k = k(\alpha/2)^k/(1 - \alpha/2)$.

1.36. Consider an infinite horizon optimal stopping problem where all rewards are positive when u^s is used, and no rewards occur when u^c is used. One might believe than it cannot be optimal to operate with a policy for which there is a positive probability that we never stop (use u^s); here is a counterexample. Consider the following Markov chain with states $k = 1, 2, \ldots$, where $\Pr[1|1] = 1$, $\Pr[1|2] = 1/2$, $\Pr[3|2] = 1/2$, $\Pr[1|3] = 1/4$, $\Pr[4|3] = 3/4$, and generally, $\Pr[1|k] = 1/4^{k-2}$, $\Pr[k+1|k] = 1 - 1/4^{k-2}$, $k > 2$. No reward is obtained when we do not stop (use u^c), and a reward $g(k) > 0$ is obtained when we stop when the state is k, where $g(1) = 1$, and $g(k) = \beta(k)/2 > 0$, $k \geq 2$, where $\beta(k)$ equals the expected discounted reward obtained from following the policy of stopping the first time the state equals 1, given that we start at k at time 0. (A discount factor $\alpha \in (0, 1]$ is given.) Trivially, it is then never optimal to stop at $k \geq 2$, so when starting in any $k_0 \geq 1$, it is optimal to stop the first time $k = 1$. Being in case (P), show that for this policy $u(.)$, $T(J^{u(.)}) = J^{u(.)}$. Show also that when starting in $k_0 \geq 2$, the probability of stopping sometime is $< 2/3$.

1.37. (a) Show in case (N), with $g \leq 0$, that Markov policies are as good as history-dependent ones. *Hint*: With $(u_t^+)_s$ being the s-th control in the sequence u_t^+, from Remark 1.17, we know that $\sum_{s=0}^\infty E[f_0((u_t^+)_s)] \leq \sum_{s=0}^\infty E[f_0((u_{t+1}^+)_s)]$, (shorthand notation). Show that u_∞^+ is better than any u_t^+, by using the preceding

inequality and the following property (actually a special case of the general Fatou's lemma in disguise). Let $a_n^m \leq 0, n, m = 0, 1, 2, \ldots$. Then $\sum_{n=0}^{\infty} \limsup_{m \to \infty} a_n^m \geq \limsup_{m \to \infty} \sum_{n=0}^{\infty} a_n^m$. (Put $a_n^m = E[f_0((u_m^+)_n)]$.)
(b) Show the same result in case (P), (use then $J(0, x, v, T) \to J(x, v)$).

Comment added. Here, the expression "are as good as" implies that the supremum of the criterion is the same for the two types of controls. It does not automatically follow that when the supremum is attained by a history-dependent control, it can even be realized by a Markov control, though frequently, for example if $\mathcal{V}$ is finite, it does follow. This makes no problem for the sufficient conditions related to (B), (P), and (N).

Exercises for Section 1.7

1.38. In Example 1.39. by induction prove that $\beta_{t-1}(q_{t-1})$ is nondecreasing in q_{t-1}. *Hint:* ϕ and ψ are nondecreasing in q_{t-1}, e.g., use

$$\phi(q_{t-1}) = \frac{a}{a + \left(\dfrac{1}{q_{t-1}} - 1\right)b}.$$

1.39. Consider Example 1.32 (job search) again but assume that the density $\phi(v)$ now contains an unknown parameter θ, which is unknown, but takes only one of two known values θ_1 and θ_2 and for which an initial probability mass distribution g_0 is given, giving positive mass to the two values θ_1 and θ_2 of θ. The distribution of θ is updated in a Bayesian manner, i.e., if we at time t believe g_t and observe v_t, then $g_{t+1}(\theta)$ is proportional to $g_t(\theta)\phi(v_t, \theta)$, $(g_t(\theta) = g_t(v_{\to t-1}, \theta))$. Show that again there is a reservation price $v^t, t = 0, 1, \ldots, T - 1$, working in the same manner as before, and describe how it is determined.

1.40. Pete can sit for an examination in a given course an unlimited number of times. Each time his exam papers are corrected by the same teacher. Teachers differ in their willingness to let persons (here Pete) pass exams, and this willingness is expressed by the fraction Q of correct results required. The Q's (teachers, in a sense) are sampled from a distribution (density) on $[0, 1]$ initially believed to be $\psi_0(q)$. Pete has a probability of $F(x)$ of producing a fraction of at most x of correct answers, F continuous, increasing. Pete gets no information about his scores but only whether he has passed or not. Pete gets a reward of 10 dollars from his father in case he passes but has to pay one dollar per exam. Find the optimal stopping rule for Pete (it hardly pays to continue indefinitely if he recurrently fails).

   Hint: Pete updates his beliefs about the teacher. See if a one-step look-ahead policy works.

1.41. Two urns A and B are given. A has two white balls, B has one white and one black ball. An urn is drawn at random (equal probabilities) and given to Adam. (That the urn is drawn in this manner is known to him.) He is presented with the following situation: At times $t = 0, 1, \ldots$, he may draw a ball from the urn, inspect it, and put it back. Each draw costs him $1/32$ dollars. After each draw he can decide if he wants to stop or not. If he stops he has the possibility to buy the urn at the price of 1.50 dollars. In case he buys the urn he can sell the white balls it contains for one dollar a piece. Find an optimal stopping policy.

Hint: Consider a one-stage look-ahead policy. When stopping at time t, the expected reward is $\max[0, x_t + 1 - 1.5]$, $x_t = \Pr[\text{the urn is A}]$, the probability after draw t.

Exercises for Section 1.9

1.42. Consider the infinite horizon case where $U = \{u^c, u^s\}$, and assume that (1.73) holds. Let $E = E[\ |x_0, v_0]$. Prove that if $M' < \infty$, then $J_T \uparrow J$, which implies that J is the supremum of the criterion when confining the stopping times to belong to the set of stopping times that stop a.s. *Hint*: Let $V^\infty = (v_0, V_1, V_2, \ldots)$, let $\tau = \tau(V^\infty)$ be any stopping time, let $A := \{V^\infty : \tau(V^\infty) < \infty\}$ and let $\tau_T = \min\{T, \tau\}$. Define $I_{T,\tau} := I_{T,\tau}(v_0, V_1, \ldots) := \sum_{0 \le s \le \tau_T - 1} \alpha^s g(X_s, u^c, V_s) + \alpha^{\tau_T} g(X_{\tau_T}, u^s, V_{\tau_T})$. Evidently, for $V^\infty \in A$, $0 \ge I'_{T,\tau} := I_{T,\tau} - \alpha^{\tau_T} g(X_{\tau_T}, u^s, V_{\tau_T}) \downarrow I_{\infty,\tau} - \alpha^\tau g(X_\tau, u^s, V_\tau)$, $E[I'_{T,\tau} 1_A] \to E[I'_{\infty,\tau} 1_A]$, and (by (1.73) and $M' < \infty$),

$$\liminf_T E[\alpha^{\tau_T} g(X_{\tau_T}, u^s, V_{\tau_T}) 1_A] \ge E[\alpha^\tau g(X_\tau, u^s, V_\tau) 1_A],$$

hence $\liminf_T E[I_{T,\tau} 1_A] \ge E[I_{\infty,\tau} 1_A]$. Now, for any T, $\liminf_T I_{T,\tau}(1 - 1_A) \ge I_{\infty,\tau}(1 - 1_A)$ (when $\alpha = 1$, the right-hand side is $-\infty$, and when $\alpha < 1$, $I_{T,\tau}(1 - 1_A) \ge I_{\infty,\tau}(1 - 1_A) - \alpha^T M'$). Hence, $\liminf_{T \to \infty} E[I_{T,\tau}] \ge E[I_{\infty,\tau}]$, so $\liminf_{T \to \infty} \sup_{\hat{\tau}} E[I_{T,\hat{\tau}}] \ge \liminf_{T \to \infty} E[I_{T,\tau}] \ge E[I_{\infty,\tau}]$. Thus, $\liminf_{T \to \infty} J_T(0, x_0, v_0) \ge J(x_0, v_0)$. On the other hand, $J_T(0, x_0, v_0) \le J(x_0, v_0)$, the former being a supremum over a smaller set of policies.

1.43. Assume that (1.73) holds, and that $\alpha = 1$. Prove that an optimal policy τ^* stops at some finite time with probability 1. Prove also that the problem can be reduced to one for which (N) holds. See (a) and (b) below.

(a) For $\alpha = 1$, stopping at once gives $g(x_0, v_0) > -\infty$. What is the expected total reward if we don't stop with positive probability?

(b) Denote the original problem by P, and the criterion value (total expected reward) by $J(x_0, v_0, \tau)$ when using the stopping time $\tau = \tau(V_1, V_2, \ldots)$. Define $\tilde{g}(x, u^c, v) = g(x, u^c, v), \tilde{g}(x, u^s, v) = g(x, u^s, v) - M$ (recall $g \le M$), and denote by $\tilde{P}$ the problem where $\tilde{g}$ is the instantaneous reward function, and by $\tilde{J}(x_0, v_0, \tau)$ the corresponding criterion value. Prove that $\tilde{J}(x_0, v_0, \tau) = J(x_0, v_0, \tau) - M$ when τ stops with probability 1, hence $J(x_0, v_0) - M = \tilde{J}(x_0, v_0)$, the optimal values in problem P and $\tilde{P}$ differ by the amount M.

(c) Show that $J(0, x_0, v_0, T) \to J(x_0, v_0)$. *Hint*: Show this in case of $\tilde{P}$. (Note that

(N) holds for problem $\tilde{P}$). Look through the proof of $J(0,x_0,v_0,T) \to J(x_0,v_0)$ in case (N) to see that (1.91) and (1.92) (actually continuity in u) are not needed when $U = \{u^c, u^s\}$.

(d) Show the equilibrium optimality equation, using that it does hold for problem $\tilde{P}$.

(e) Assume that $J(x_0,v_0)$ is inserted in the optimality equation (1.72). Show that maximizing the right-hand side of the equation gives the optimal control. *Hint*: Use problem $\tilde{P}$ and results for case (N).

1.44. In the infinite horizon case in which the "one-stage look-ahead" set B is absorbing, show that the one-stage look-ahead policy is optimal if (1.73) and $M' < \infty$ hold, or if $g(x,u^c,v) \le g(x,u^s,v)$ in case (P) and (N). *Hint*:

(a) For (N) and (1.73) (for (N) we can assume $g \le 0 =: M$). For any horizon T, with $U_T = \{u^s\}$, the policy, π, stop first time $(x_t, v_t) \in B_t$, is optimal if $\{B_t\}_t$ are absorbing. Here $B_t = B, t < T$. Now, the policy π: stop first time $(x_t, v_t) \in B$ gives maximum in $J_T(0,x_0,v_0) = \max\{g(x_0,u^s,v_0), g(x_0,u^c,v_0) + \alpha E[J_T(0,f(x_0,u^c,V_1),V_1)|x_0,v_0]\}$, and when $(x_t, v_t) \notin B$ the second number in $\max\{\}$ is $\ge \varepsilon +$ the first number, ε independent of T, see the finite horizon proof. Let $T \to \infty$, noting that $M \ge J_T(0,x_0,v_0) \ge g(x_0,u^s,v_0)$, so

$$
\begin{aligned}
J(x_0,v_0) \\
&= \max\{g(x_0,u^s,v_0), g(x_0,u^c,v_0) + \alpha \lim_T E[J_T(0,f(x_0,u^c,V_1),V_1)|x_0,v_0]\} \\
&= \max\{g(x_0,u^s,v_0), g(x_0,u^c,v_0) + \alpha E[J(f(x_0,u^c,V_1),V_1)|x_0,v_0]\},
\end{aligned}
$$

with π yielding the maximum. So the Bellman equation is satisfied by $J(x_0,v_0)$, with π yielding maximum.

(b) For (P), $J_T(0,x_0,v_0) = J(0,x_0,v_0,T,\pi) \to J_\pi^0(x_0,v_0)$, so $J_\pi^0(x_0,v_0) = \max\{g(x_0,u^s,v_0), g(x_0,u^c,v_0) + \limsup_T \alpha E[J(0,f(x_0,u^c,V_1),V_1,T,\pi)|x_0,v_0]\} = \psi_\pi\{g(x_0,u^s,v_0), g(x_0,u^c,v_0) + \alpha E[J_\pi^0(f(x_0,u^c,V_1),V_1)|x_0,v_0]\}$, with π actually yielding the first maximum, and where $\psi_\pi\{,\}$ means the choice of the first or second number according to the π-rule (the ε-argument in (a) also works here). Because the last right-hand side equals $J_\pi^0(x_0,v_0)$, the Bellman equation is satisfied by $J_\pi^0(x_0,v_0)$ and π.

1.45. In this problem, assume (B) in (1.70) and the setup in Section 1.9.

(a) Let $M = \max\{|M_1|, |M_2|\}$, $|\check{J}| := \sup_x |\check{J}(x)|$ and let $\check{J}, J^1$, and J^2 be any bounded functions. Prove the inequalities $|T_{u(.)}(J^1) - T_{u(.)}(J^2)| \le \alpha|J^1 - J^2|$, $|T(J^1) - T(J^2)| \le \alpha|J^1 - J^2|$, $|T^k(J^1) - T^k(J^2)| \le \alpha^k|J^1 - J^2|$, $|T_{u(.)}^k(J^1) - T_{u(.)}^k(J^2)| \le \alpha^k|J^1 - J^2|$, $|T^k(\check{J}) - T^{k-1}(\check{J})| \le [M+2|\check{J}|]\alpha^{k-1}$, and $|T_{u(.)}^k(\check{J}) - T_{u(.)}^{k-1}(\check{J})| \le [M+2|\check{J}|]\alpha^{k-1}$. *Hint*: For the second inequality, use $g(x,u) + \alpha E_u[J^i(f(x,u,V))|x] \le g(x,u) + \alpha E_u[J^j(f(x,u,V))|x] + \alpha E_u[|J^i(f(x,u,V)) - J^j(f(x,u,V))||x]$, for the fifth that $|T(\check{J}) - \check{J}| \le |T(\check{J})| + |\check{J}| \le M+2|\check{J}|$.

(b) Prove that the equations $\bar{J} = T(\bar{J})$ and $\bar{J} = T_{u(.)}(\bar{J})$ have unique bounded solutions. (*Hint*: If, say, the first one has two solutions J_1, J_2, then $|J_1 - J_2| = |T(J_1) - T(J_2)| \le \alpha|J_1 - J_2|$.)

(c) Use $|T^i(\hat{J}) - T^k(\hat{J})| \le \Sigma_{j=k+1}^i |T^j(\hat{J}) - T^{j-1}(\hat{J})|$, to prove, for $i > k$, that $|T^i(\hat{J}) - T^k(\hat{J})| \le \alpha^k(M + 2|\hat{J}|)/(1 - \alpha) =: \beta_k(\hat{J})$. So $T^i(\hat{J})$ is a convergent sequence, with limit, say $\hat{J}$, and $|\hat{J} - T^k(\hat{J})| \le \beta_k(\hat{J})$. Similarly, show that $T^i_{u(.)}(\hat{J})$ converges to a limit denoted $\hat{J}_{u(.)}$, and that $|\hat{J}_{u(.)}) - T^m_{u(.)}(\hat{J})| \le \beta_m(\hat{J})$.

(d) Here and in (e) below, let the limit $\hat{J}$ correspond to $\bar{J} = 0$. Define u_k by $T(T^{k-1}(0)) = T_{u_k}(T^{k-1}(0))$, i.e., u_k yields maximum at the k-th iteration. Use (c) and $|T_{u_k}(\hat{J}) - T_{u_k}(T^{k-1}(0))| = |T_{u_k}(\hat{J}) - T^k(0)| \le \alpha\beta_{k-1}(0)$ to show $|T_{u_k}(\hat{J}) - \hat{J}| \le 2\beta_{k-1}(0)$ and, by induction, $|T^m_{u_k}(\hat{J}) - T^{m-1}_{u_k}(\hat{J})| \le \alpha^{m-1}2\beta_{k-1}(0)$, and hence $|T^m_{u_k}(\hat{J}) - \hat{J}| \le \Sigma_{n=1}^m |T^n_{u_k}(\hat{J}) - T^{n-1}_{u_k}(\hat{J})| \le 2\beta_{k-1}(0)/(1 - \alpha)$ and $|\hat{J}_{u_k} - \hat{J}| \le 2\beta_{k-1}(0)/(1 - \alpha)$.

(e) Use (b) to show $\hat{J} = J, \hat{J}_{u_k(.)} = J_{u_k(.)}, J$ and $J_{u_k(.)}$, being, respectively, the optimal value of the criterion, and its value for $u(.) = u_k(.)$. Hint: By (a) and (c), $\hat{J} = T^\infty(0) \leftarrow T^{i+1}(0) = T(T^i(0)) \rightarrow T(T^\infty(0)), \hat{J}_{u_k} = T^\infty_{u_k} \leftarrow T^{i+1}_{u_k}(0) = T_{u_k}(T^i_{u_k}(0)) \rightarrow T_{u_k}(T^\infty_{u_k}(0)), J = T(J), J_{u_k} = T_{u_k}(J_{u_k})$.

1.46. Prove that, for the conclusions related to the cases (B), (P), (N) in (1.70) to hold, it suffices that these three conditions hold for all $x \in \mathcal{X}(x_0) := \cup_t \mathcal{X}_t(x_0)$. Similarly, the suprema $\sup_x$ in Exercise 1.45 can be replaced by $\sup_{x \in \mathcal{X}(x_0)}$, the functions considered being bounded over $\mathcal{X}(x_0)$. Hint: Essentially, the three conditions were only used for states belonging to $\mathcal{X}(x_0)$. (We might redefine g by letting $g(x, u) = 0$, for $x \notin \mathcal{X}(x_0)$. Then g is still continuous in u, if it was continuous before. Note that for $h = 0$, for $x \notin \mathcal{X}(x_0)$, $0 = T_{u(.)}(h)(x) = g(x, u(x)) + \alpha E[h(f(x, u(x), V))|x]$, $0 = T(h)(x) = \sup_u\{g(x, u) + \alpha E[h(f(x, u, V))|x]\}$.)

1.47. Prove that, for positive $M, \beta, a, \gamma > 1$ given, when $\alpha\gamma^\beta \in (0, 1), |g(x_t, u, V)| \le M(1 + |x|^\beta)$, and $|x_t| \le a\gamma^t$ for $x_t \in \mathcal{X}_t(x_0)$, then this case can be reduced to the case (B). Hint: In a new problem, let $\hat{\alpha} = \alpha\gamma^\beta$ be the discount factor, and let the instantaneous reward be $\hat{g}(y, z, u, v) = g(z \cdot y, u, V)/z^\beta$, where $y_t = x_t/z_t, z_{t+1} = \gamma z_t$, $z_0 = a$, so $y_{t+1} = f(z_t \cdot y_t, u, V)/\gamma z_t$. Here, $|\hat{g}|$ is bounded for all $(y, z) \in \mathcal{X}(y_0, z_0) \subset [-1, 1] \times [a, \infty)$.

1.48. Consider the deterministic problem

$$\max \sum_{s=0}^\infty (1/2)(x_t + u_t)2^{-s}, x_{t+1} = x_t - u_t, u_t \in \mathbb{R}.$$

Show that $\bar{J}(x) = x$ satisfies the infinite horizon equilibrium optimality equation, and that, however, $J(x) = \infty$, e.g., consider the criterion value for a sequence such that $x_t + u_t = 2^t$ for all t, $x_0 = x$. (This means $u_t = 2^t - x_t$, x_t the solution of $x_{t+1} = 2x_t - 2^t$, x_0 given.) It was said above that the fact that a function $\bar{J}$ satisfies the Bellman equation is not sufficient for knowing that $\bar{J}$ is the optimal value function. This problem is an example of that phenomenon. In this example, any u gives maximum in the equilibrium optimality equation. (If u_t is replaced by $u_t + u_t^2$ in the state equation, still $\bar{J}(x) = x$ satisfies the Bellman equation, now for $u = 0$, still $u_t = 2^t - x_t$ yields an infinite criterion value.)

1.49. Using Lemma 1.42 in Section 1.9, show the following assertions: For a pair $(\hat{J}, u(x,v))$ satisfying the Bellman equation, note that in case P $\hat{J} = J^u$ automatically holds (so $u(x,v)$ is optimal) if, for any $\hat{x}_0 \in \mathcal{X}(x_0), v \in \mathcal{V}, \lim_{t\to\infty} \alpha^t E[\hat{J}(X_t, V_t)|\hat{x}_0, v] = 0$, for X_s being a sequence of solutions corresponding to $u_s = u(x,v)$ and the V_s's, starting at $(0, \hat{x}_0)$. In case N, $\hat{J}$ is the optimal value function (so $u(x,v)$ is optimal) if this limit condition holds even for arbitrary choices of the controls $u_s = u_s(x,v)$. (*Hint*: Using (1.93), show $J_u = T_u^\infty = \hat{J}$ in case (P) and $J \le T^\infty(\hat{J} = \hat{J}$ in case (N)); note that (1.93) holds even for any $\hat{h} \ge 0$ in case (P) and $\hat{h} \le 0$ in case (N)).

Exercises for Section 1.10

1.50. Mary owns a piece of land that is continuously explored for silver deposits. The total amount of silver in all deposits known at time t is x_t. New deposits of size V_t are found according to a Poisson process with intensity λ, so x_t jumps to $x_t + V_t$ each time a find is made, $V_t \ge 0$ and i.i.d., $EV_t = a$. The price of silver is constant equal to 1, and the discount factor is e^{-bt}. Mary wants to sell all deposits at a certain time. Which time is optimal?

Hint: In pure stopping problems, one-step look-ahead policies are again optimal when the set B is absorbing (i.e., $(x, v, \tau^0) \in B \Rightarrow (f(x, u^c, V, \tau^1), V, \tau^1) \in B$, with conditional probability 1, given $(x, v, \tau^0))$, where $B = \{(x, v, \tau^0):$

$$f^0(x, u^s, v, \tau^0) \ge f_0(x, u^c, v, \tau^0)$$
$$+ E[e^{-\alpha\tau^1} f_0(f(x, u^c, V, \tau^1), u^s, V, \tau^1)|x, v, u^c, \tau^0]\}.$$

1.51. Prove that when f^0 and h are bounded, the supremum of the criterion in (1.100) is finite.

Hint: $\Pr[\tau_0, \ldots, \tau_N \in [0, \delta)] \le (1 - \delta')^{N+1}$, so the expected value of the number of transitions in $[0, \delta)$ is $\le K$ for some constant K. On $[n\delta, (n+1)\delta)$, if one transition has occurred, the expected value of the number of further transitions is also $\le K$ (so the expected value of the number of transitions in $[n\delta, (n+1)\delta)$ is $\le 1+K$).

Chapter 2
The HJB Equation for Deterministic Control

This short chapter digresses from the main subject of the book and discusses how ideas in dynamic programming in continuous time *deterministic* control problems lead to sufficient conditions for optimality. In the next chapters, these ideas will be carried over to a stochastic setting. The main reason for including the current chapter is that here the ideas can be presented in a very simple setting. Also, some proofs presented here form the basis of proofs in the next chapter.

The continuous time analog of the dynamic programming equation is called the HJB equation and will be presented first. A tool for solving this partial differential equation is to construct a "field of extremals," that is, a set of solutions to the so-called characteristic equations corresponding to the HJB equation (actually the equations of the maximum principle).

2.1 The Control Problem and Solution Tools

Let $f : \mathbb{R}^{n+r+1} \mapsto \mathbb{R}^n$ and $S : \mathbb{R}^n \mapsto \mathbb{R}$ be given functions, and consider the following control problem over a fixed time interval $[0, T]$,

$$\max S(x(T)), \tag{2.1}$$

subject to

$$\dot{x} = f(t, x(t), u(t)) \text{ v.e.}, \quad x(0) = x^0 \in \mathbb{R}^n, \tag{2.2}$$

$$u(t) \in U \subset \mathbb{R}^r \text{ for all } t \tag{2.3}$$

$$\begin{aligned} \text{(a)} \quad & x_i(T) = x_i^1, && i = 1, \ldots, l, \\ \text{(b)} \quad & x_i(T) \geq x_i^1, && i = l+1, \ldots, m, \\ \text{(c)} \quad & x_i(T) \text{ free}, && i = m+1, \ldots, n. \end{aligned} \tag{2.4}$$

Here, v.e. (= virtually everywhere) as well as v.a. (= virtually all) mean for all t except a finite or countable number of points. Assume that S is C^1, that f and f_x (the

Jacobian with respect to x of f) are continuous, and that all control functions $u(.)$, $u(t) \in U$, are piecewise continuous. The set of such functions is denoted U'. The problem is to maximize $S(x(T))$ over U', more precisely we can write the problem as $\max_{u(.) \in U'} S(x^{u(.)}(T))$ subject to (2.4), where $x(t) = x^{u(.)}(t)$ is the solution on $[0,T]$ of (2.2) for $u(.)$ inserted in f. The solution $x^{u(.)}(t)$ is called *admissible* if also (2.4) is satisfied. If it also yields a maximum of $S(x^{\hat{u}}(T))$ among all admissible solutions $x^{\hat{u}}(T)$, it is called an optimal solution, and the corresponding control is called an optimal control. The entities T, U, f, x^0, x_i^1, and S are fixed.

This is a standard control problem where the objective is a "scrap value" function, giving the value of the "stock" $x(T)$ that remains at the terminal time. A control problem with an integral criterion can be written as problem (2.1)–(2.4) by adding an extra state variable. This is shown in Remark 2.13 below. Formulating the problem as above makes proofs shorter and more transparent.

Let us first recall the maximum principle for the above problem.

Theorem 2.1 (Necessary condition). *Let $(x^*(t), u^*(t))$ be an optimal pair in problem (2.1)–(2.4). Then for some p_0, $p_0 = 0$ or $p_0 = 1$, and some continuous function $p(t)$, with $(p_0, p(t)) \neq (0,0)$ for all t, the following (necessary) conditions hold:*

$$u^*(t) \quad maximizes \quad p(t)f(t, x^*(t), u) \quad subject \ to \ u \in U \ for \ v.a. \ t \qquad (2.5)$$

$$\dot{p}(t) = -p(t)f_x(t, x^*(t), u^*(t)) \quad for \ v.a. \ t \qquad (2.6)$$

(a') $\quad p_i(T)$ *no conditions,* $\qquad\qquad\qquad\qquad\qquad i = 1, \dots, l$

(b') $\quad p_i(T) \geq p_0 \dfrac{\partial S(x^*(T))}{\partial x_i}$ *(with = if $x_i^*(T) > x_i^1$),* $\quad i = l+1, \dots, m$ $\qquad (2.7)$

(c') $\quad \dfrac{p_i(T) = p_0 \partial S(x^*(T))}{\partial x_i},$ $\qquad\qquad\qquad\qquad i = m+1, \dots, n$

$\qquad\qquad\qquad\qquad\qquad\qquad\qquad\qquad\qquad\qquad\qquad\qquad\qquad\qquad\qquad\qquad\qquad$ □

Written out, the condition (b') in (2.7) means that for $i = l+1, \dots, m$, always $p_i(T) \geq p_0 \frac{\partial S(x^*(T))}{\partial x_i}$ and if $x_i^*(T) > x_i^1$, then in fact $p_i(T) = p_0 \frac{\partial S(x^*(T))}{\partial x_i}$. Moreover, (a') means there are no conditions on $p_i(T)$, $i = 1, \dots, l$.

Suppose that the interval $[0,T]$ is replaced by $[s,T]$ for some $s \in [0,T)$ and that x^0 is replaced by y. Let $u = u(.)$ belong to U'. The corresponding solution $x(t) := x^u(t)$ of (2.2) on $[s,T]$, with initial point (s,y), is called (s,y)-*admissible* if it satisfies (2.4), moreover $J^u(s,y) := S(x^u(T;s,y))$, and we call $(x(.), u(.)) := (x^u(.), u(.))$ a (s,y)-admissible pair. We shall assume that unique solutions to (2.2) exist on all $[s,T]$ for any (s,y), $s \in [0,T)$, $y \in \mathbb{R}^n$, any $u(.) \in U'$. Uniqueness in fact follows from the existence and continuity of f_x. The problem obtained by this replacement, where $J^u(s,y)$ is the objective, is referred to as problem $P(s,y)$.

Define the *value function* $V(s,y)$ as the supremum of $J^u(s,y)$ over all (s,y)-admissible solutions $x^u(t)$:

$$V(s,y) = \sup_{u \in U(s,y)} S(x^u(T;s,y)) \qquad (2.8)$$

$s \in [0,T)$, $y \in \mathbb{R}^n$, where

$$U(s,y) := \{u \in U' : x^u(t) \text{ is } (s,y)\text{-admissible}\}. \tag{2.9}$$

(The assumption of existence on all $[s,T]$ of $x^u(.)$ for any s,y, any $u(.)$ can be relaxed, if we agree that $V(s,y)$ is only defined for (s,y) such that $U(s,y)$ is nonempty, i.e., contains at least one control u for which $x^u(.)$ is defined on all $[s,T]$ and is (s,y)-admissible).

Two simple lemmas can be proved:

Lemma 2.2. *For any (s,y)-admissible solution $x^u(t)$, the value function $V(t,x^u(t))$ is nonincreasing in t for $t \in [s,T]$.* $\square$

Proof. Let 1_I be the indicator function of any given set I (i.e., $1_I(s) = 1$ if $s \in I$, $1_I(s) = 0$ otherwise), let $x^u(.)$ be a given (s,y)-admissible solution, and let $t' < t''$ be two points in $[s,T]$. For any element v in $U(t'', x^u(t''))$ the control $w_v := w_v(t) := v1_{[t'',T]}(t) + u1_{[0,t'')}(t)$ belongs to $U(t', x^u(t'))$ as well as to $U(t'', x^u(t''))$ (the solution $x^{w_v}(t)$ starting at $(t', x^u(t'))$ of course satisfies $(t'', x^{w_v}(t'')) = (t'', x^u(t''))$ and also (2.4) (as $v \in U(t'', x^u(t''))$). Moreover, whether $x^{w_v}(t)$ starts at $(t', x^u(t'))$ or at $(t'', x^u(t''))$, the solutions coincide on $[t'',T]$, and trivially $x^{w_v}(T) = x^v(T)$ ($x^v(t)$ starts at $(t'', x^u(t''))$). Then

$$\sup_{v \in U(t'', x^u(t''))} S(x^v(T)) = \sup_{v \in U(t'', x^u(t''))} S(x^{w_v}(T))$$
$$\leq \sup_{\hat{u} \in U(t', x(t'))} S(x^{\hat{u}}(T)).$$

The inequality follows because $U(t', x(t'))$ contains all $w_v(.)$ and perhaps also other controls. Thus $V(t, x^u(t))$ is nonincreasing in t. $\square$

Lemma 2.3. *If $x^*(t)$ is an optimal solution in the original problem (2.1)–(2.4), then $t \to V(t, x^*(t))$ is a constant function on $[0,T]$.* $\square$

Proof. By Lemma 2.2, for $t \in [0,T]$,

$$V(0,x^0) \geq V(t,x^*(t)) \geq V(T,x^*(T)) = S(x^*(T)).$$

But $S(x^*(T))$ equals $V(0,x^0)$ by the optimality of $x^*(.)$, so the inequalities must be equalities. $\square$

Let the *target set M* be defined by

$$M := \{x \in \mathbb{R}^n : x_i = x_i^1 \text{ for } i = 1,\ldots,l, \ x_i \geq x_i^1 \text{ for } i = l+1,\ldots m\}. \tag{2.10}$$

Let the *controllable set $\tilde{Q}$* be the set of points (s,y) with $0 < s < T$ from which it is possible to reach the target set M at time T. Thus,

$$\tilde{Q} := \{(s,y) : y \in \mathbb{R}^n, \ 0 < s < T, \ \text{an } (s,y)\text{-admissible } x^u(t) \text{ exists}\}. \tag{2.11}$$

Let $u(s+)$ mean a right limit of $u(.)$ at s. Then we have the following theorem.

Theorem 2.4 (The value function satisfies the HJB equation). *Let (s,y) be an interior point in the controllable set $\tilde{Q}$ at which the value function $V(s,y)$ is differentiable. Then*

$$V_s(s,y) + V_y(s,y)f(s,y,v) \leq 0 \quad \text{for all } v \in U \tag{2.12}$$

and if $(x_*(t), u_*(t))$ is an optimal pair in $P(s,y)$, then

$$u_*(s+) \in argmax_{v \in U}\{V_y(s,y)f(s,y,v)\} \tag{2.13}$$

and

$$V_s(s,y) + V_y(s,y)f(s,y,u_*(s+)) = 0. \tag{2.14}$$

$\square$

Proof. Given any $v \in U$, choose the interval $D := [s - \delta, s + \delta]$ so small that the solution $x_v(t)$ of (2.2) for $u(t) = v$ on D, with $x(s) = y$, belongs to $\tilde{Q}$ for $t \in D$. Because $(s + \delta, x_v(s + \delta)) \in \tilde{Q}$, there exists a control $u^+(t)$ that is $(s + \delta, x_v(s + \delta))$-admissible. Define $u(t) = u^+(t)$ for $t > s + \delta$ and $u(t) = v$ on D, and denote the corresponding (s,y)-admissible function by $x^u(t)$. Because $V(t, x^u(t))$ is non-increasing in t, its derivative at $t = s$ is nonpositive, which gives the inequality (2.12). Equation (2.14) follows from noting that the right-hand derivative of the constant function $V(t, x_*(t))$ at $t = s$ equals zero, the derivative being $V_s(s, x_*(s)) + V_y(s, x_*(s))f(s, x_*(s), u_*(s+))$. Hence, (2.13) follows from (2.12) and (2.14). $\square$

Combining (2.13) and (2.14), the following equation holds for $(s,x) = (s, x_*(s))$

$$V_s(s,y) + \sup_{u \in U} V_y(s,y)f(s,y,u) = 0 \tag{2.15}$$

This equation is called the *Hamilton–Jacobi–Bellman equation*, or the HJB equation, or the partial differential equation of dynamic programming.

Remark 2.5 (A proof of the maximum principle in the case V is C^2).* Let $l = m$ (i.e., no inequality terminal conditions). Suppose that $(x^*(t), u^*(t))$ is an optimal pair in $P(0, x^0)$, such that $(s, x^*(s))$ belongs to the interior of $\tilde{Q}$ for all s. Assume that V is C^2 in the interior of $\tilde{Q}$. Then the maximum principle (2.5)–(2.7) is satisfied for $p(s) = V_y(s, x^*(s))$ and $p_0 = 1$. The proof is as follows: Note that $u^*(.)$ is optimal in $P(s, x^*(s))$, $(S(x^*(T)) = V(s, x^*(s)))$. The maximum condition (2.5) is an immediate consequence of (2.13) and the definition of $p(s)$. Next, let $(s,y) \in \tilde{Q}$. Then, by (2.12), $0 \geq V_s(s,y) + V_y(s,y)f(s,y,u^*(s+))$ and by (2.14), the inequality is an equality if y is replaced by $x^*(s)$. Hence $x^*(s)$ maximizes the right-hand side of the last inequality, so if $u^*(s+) = u^*(s)$, then $0 =$

$$V_{sy}(s, x^*(s)) + V_{yy}(s, x^*(s))f(s, x^*(s), u^*(s)) + V_y(s, x^*(s))f_x(s, x^*(s), u^*(s)).$$

The sum of the two first addends equals $(d/dt)(V_y(s, x^*(s)) = \dot{p}(s)$, and so (2.6) holds. Finally, for $s = T$, $V = S$ on the target set M, which implies $p_i(T) := V_{y_i}(T, x^*(T)) = S_{y_i}(x^*(T))$ for $i > m$. (The ad hoc assumption on V that V is C^2 frequently fails; a better proof is found for example in Fleming and Rishel (1975).) $\square$

Define the *admissible set* Q to be the set of points $(s, x(s))$, where $s \in (0, T)$ and $x(.)$ is admissible (i.e., $(0, x^0)$-admissible).

A function defined on a set Q' is said to be C^1 in Q' if there exist both an open set Q'' containing Q' and an extension of the function to Q'' that is C^1 on Q''. Note also that cl$Q$ means the closure of Q.

We present first a sufficient condition for optimality in $P(0, x^0)$ in the free end case ($m = 0$ in (2.4)). To formulate it, two equations are needed.

$$W_s(s, y) + \sup_{v \in U} W_y(s, y) f(s, y, v) = 0 \qquad (2.16)$$

$$W_s(s, x^*(s)) + W_y(s, x^*(s)) f(s, x^*(s), u^*(s)) = 0 \quad \text{v.e.} \qquad (2.17)$$

Theorem 2.6 (Verification theorem, sufficient condition, free end). *Consider problem (2.1)–(2.3) and let $(x^*(t), u^*(t))$ be a $(0, x^0)$-admissible pair. Suppose $W(s, y)$ is a continuous function on clQ, which is C^1 in Q, and satisfies the HJB equation (2.16) for all (s, y) in Q, as well as*

$$W(T, y) \geq S(y) \quad \text{for all} \quad (T, y) \in \text{cl}Q. \qquad (2.18)$$

Assume also that $(x^*(t), u^*(t))$ satisfies

$$W(T, x^*(T)) = S(x^*(T)) \qquad (2.19)$$

and $u^*(s) \in \text{argmax}_{v \in U}\{W_y(s, x^*(s)) f(s, x^*(s), v)\}$ v.e., (i.e., the equality (2.17) holds). Then $u^*(t)$ is optimal and $W(0, x^0) = V(0, x^0)$. □

Proof. Along any $(0, x^0)$-admissible trajectory $x(t)$ corresponding to some $u(t)$, the function $W(s, x(s))$ is nonincreasing in s, because (2.16) implies

$$(d/ds)W(s, x(s)) = W_s(s, x(s)) + W_y(s, x(s)) f(s, x(s), u(s)) \leq 0 \quad \text{v.e.}$$

Moreover, by (2.17) there is equality if $u(s) = u^*(s)$, $x(s) = x^*(s)$, implying constancy of $W(s, x^*(s))$. Thus, using (2.18) and (2.19), $S(x^*(T)) = W(T, x^*(T)) = W(0, x^0) \geq W(T, x(T)) \geq S(x(T))$, so $(x^*(t), u^*(t))$ is optimal. □

This theorem implies that one way of solving a control problem is to solve the HJB equation. Examples show that this can be an effective procedure. Having found such a solution, the theorem then furnishes a sufficient condition for optimality.

Let us explain the method in detail: Consider the case with no terminal conditions. (Terminal conditions are discussed in the next section.) The HJB equation (2.16) is used in the following way. First, the maximizing v on the left-hand side of (2.16) is found (we assume the maximum exists). It normally becomes a function $v(s, y, W_y)$. (At this point, consider W_y simply as a parameter.) This function is inserted on the left-hand side of (2.16), which then becomes a nonlinear first-order partial differential equation in $W(s, y)$, namely

$$W_s(s, y) + W_y(s, y) f(s, y, v(s, y, W_y(s, y))) = 0. \qquad (2.20)$$

Then solve this equation for W, using the boundary condition $W(T,y) = S(y)$ (most often, we need only look for functions W satisfying (2.18) with equality). When W has been found, then we can define $u(s,y) := v(s,y,W_y(s,y)) \in$ $\text{argmax}_{v \in U} W_y(s,y) f(s,x,v)$, and this is the optimal control on so-called feedback form. This means that if, at time s, the state is y, then the optimal choice of control at time s is $u(s,y)$. Next, to find the optimal solution for the given initial condition $(0,x^0)$, solve $\dot{x} = f(t,x,u(t,x))$ with $x(0) = x^0$. Denoting the solution $x^*(t)$, the optimal control is then $u^*(t) = u(t,x^*(t))$. The sufficient condition in Theorem 2.6 implies that $u^*(t)$ is optimal.

Example 2.7. Let us apply Theorem 2.6 to the problem:

$$\max \left[-\gamma x_1(T)^2 + x_2(T) \right] \quad \text{s.t.} \quad \dot{x}_1 = ax_1 + bu, \ \dot{x}_2 = -u^2, \ x_1(0) = 1, \ x_2(0) = 0,$$

$u \in U := \mathbb{R}, \gamma > 0$.

Solution. The scrap-value function is $S(x_1,x_2) = -\gamma x_1^2 + x_2$, so we try to find a function W that satisfies

$$W(T,x_1,x_2) = -\gamma x_1^2 + x_2 \qquad [*]$$

(i.e., (2.18) with equality), and (2.16), i.e.,

$$W_s + \max_{u \in \mathbb{R}} \{ W_{x_1}(ax_1 + bu) - W_{x_2} u^2 \} = 0.$$

Carrying out the maximization yields $u = bW_{x_1}/2W_{x_2}$, which inserted into the HJB equation gives the following partial differential equation in the unknown function $W = W(s,x_1,x_2)$:

$$W_s + ax_1 W_{x_1} + \frac{b^2(W_{x_1})^2}{4W_{x_2}} = 0. \qquad [**]$$

We guess that W is of the form $W = cx_1^2 + x_2$, where $c = c(s)$ is a function of time. Imagine that this W is inserted into $[**]$. This yields the equation $\dot{c}x_1^2 + 2cax_1^2 + b^2c^2x_1^2 = 0$ or

$$\dot{c} + 2ca + b^2c^2 = 0.$$

This Bernoullian differential equation can be solved by introducing the variable $d = 1/c$. Then, $\dot{d} = -\dot{c}/c^2$, so $\dot{d} = 2ad + b^2$, with general solution $d = Ce^{2at} - b^2/2a$. The boundary condition $[*]$ yields $[Ce^{2aT} - b^2/2a]^{-1} = -\gamma$, so $C = (b^2/2a - 1/\gamma)e^{-2aT}$ and $c(t) = [(b^2/2a - 1/\gamma)e^{2a(t-T)} - b^2/2a]^{-1}$. So W equals $x_2 + x_1^2 c(s)$ for this $c(s)$. One may want to check that it really satisfies the HJB equation. The control u yielding maximum in the HJB equation is

$$u(s,x_1,x_2) = b\frac{W_{x_1}}{2W_{x_2}} = \frac{bx_1}{(b^2/2a - 1/\gamma)e^{2a(s-T)} - b^2/2a}.$$

Next, let us continue by constructing $u^*(t)$ for the given initial point $(0,1,0)$. We do that only for a specific set of values of a, b, and γ, namely $a = 1$, $b = \sqrt{2}$, and $\gamma = 1/2$. Then $u = -\sqrt{2}x_1(e^{2(t-T)} + 1)^{-1}$, and

$$\dot{x}_1/x_1 = 1 - 2(e^{2(t-T)} + 1)^{-1}.$$

This is a separable equation. Integrating (by using the substitution $v = e^{2(t-T)}$ on the right-hand side and the formula $-1/v(v+1) = 1/(v+1) - 1/v$), yields

$$\ln x_1 = D + t - 2(t-T) + \ln(e^{2(t-T)} + 1)$$

or

$$x_1^*(t) = D' e^{-t+2T} (e^{2(t-T)} + 1) = D'(e^t + e^{2T-t}),$$

where D and D' are (related) arbitrary constants, D' determined by $1 = D'(1 + e^{2T})$, i.e., $D' = 1/(1 + e^{2T})$. Then

$$u^*(t) = -\frac{\sqrt{2}(e^t + e^{2T-t})}{(1 + e^{2T})(e^{2(t-T)} + 1)}.$$

Knowing $u^*(t)$, then $x_2^*(t)$ can be calculated.   □

In the above example, a guess concerning the form of W led to a proposal for the solution, via an ordinary differential equation for c. This method owed its success to pure luck. A general method of solving first-order partial differential equations is to find the "characteristics" of the equation. That is what is actually done when using the maximum principle. In effect, we shall show that when solutions to the maximum principle have been found, then frequently we automatically get a solution of the nonlinear partial differential equation (2.20).

In some cases, the above procedure does not work. Sometimes the optimal value function $V(s,y)$ is not C^1 in Q, indicating that Theorem 2.6 will not work. (See Exercise 2.3.) In other cases, $V(s,y)$ is not continuous at $s = T$, so even if $V(T,y) = S(T,y)$, it may not work to require $W(T,y) = S(T,y)$. This is the case for the example

$$\max[x_2(1) - (1/2)(x_1(1))^2], \text{ s.t. } \dot{x}_1 = u, x_1(0) = 0, \dot{x}_2 = u, x_2(0) = 0, u \in \mathbb{R}.$$

Here $V(s,y_1,y_2) = W(s,y_1,y_2) = 1/2 - y_1 + y_2$, $s < 1$, so in fact $W(1,y_1,y_2) + y_1^2/2 = (y_1 - 1)^2/2 + y_2 \geq y_2$, and (2.18) holds (with inequality!).

When V is not C^1, a generalization of the HJB equation exists, employing so-called viscosity solutions. This matter cannot be considered here (see Fleming and Soner (1993)), but a slight weakening of the conditions appears in Remark 2.10.

2.2 Terminal Conditions

Frequently, when terminal conditions are imposed (see (2.4)), it is impossible to find a function W defined and continuous on all of cl$Q$, with the properties stated in Theorem 2.6. What frequently fails is continuity of W at points in cl$Q \cap (\{T\} \times \mathbb{R}^n)$, as is shown in the next example. The following modification of Theorem 2.6 then holds.

Theorem 2.8 (Verification theorem, sufficient condition, end constrained case). *In the end constrained case ((2.2),(2.3), $m > 0$ in (2.4)), if $W(s,y)$ is defined and* *continuous on* $\{(0,x^0)\} \cup Q$ *and is* C^1 *on* Q, *but not necessarily defined for* $s = T$, *then Theorem 2.6 holds, provided conditions (2.18), (2.19) are replaced by the conditions*

$$\limsup_{t \to T} W(t,x(t)) \geq S(x(T)) \tag{2.21}$$

for all admissible solutions $x(t)$, *and*

$$S(x^*(T)) \geq \limsup_{t \to T} W(t,x^*(t)). \tag{2.22}$$

$\square$

Proof. To obtain this result, replace T by $T' < T$ in the proof of Theorem 2.6. We then obtain $W(T',x^*(T')) = W(0,x^0)) \geq W(T',x(T'))$. Taking $\limsup$ as $T' \to T$, gives $S(x^*(T)) \geq S(x(T))$, even in this case. $\square$

The two boundary conditions in the theorem usually do not completely determine the solution. That actually works to our advantage. Unknown parameters in W also enter the feedback control $u(s,y)$ obtained from the maximization in the HJB equation, and then also $x(t;s,y)$ (the solution of $\dot{x} = f(t,x,u(t,x)), x(s) = y$). Values of these parameters are sought such that the terminal conditions and boundary conditions (2.21), (2.22) are satisfied by $x(t;0,x^0)$. An example of this procedure is given below.

Remark 2.9 (Restrictions on the solutions*). Assume that all solutions are required to belong to a given set $\hat{Q}$ in $\mathbb{R}^{n+1}$ (i.e., in (t,x)-space). In particular, assume that $(t,x^*(t)) \in \hat{Q}$. Replace Q by $Q \cap \hat{Q}$ in Theorem 2.8 (as well as in the next remark). Then optimality of $(x^*(t),u^*(t))$ holds among all admissible pairs $(x(.),u(.))$ for which $(t,x(t)) \in \hat{Q}$ for all $t \in (0,T)$. $\square$

Remark 2.10 (Weaker differentiability conditions on W*). Sometimes the value function is not C^1 in all Q, as required in Theorem 2.8. However, the C^1-assumption on W can be slightly weakened. Only the following assumptions are needed in Theorem 2.8 for $u^*(t)$ to be optimal: W is defined on some open set Q' containing Q and is locally Lipschitz continuous on Q'. The function W is differentiable and satisfies the HJB equation (2.16) in $Q' \setminus Z^*$ for some set Z^*, with the following property. There are given C^1-functions $\psi_i(s,y)$ on Q', $i = 1,\ldots,i'$, such that, for any i, the vector of derivatives (ψ_{is}, ψ_{ix}) is nonzero for all (s,y) such that $\psi_i(s,y) = 0$, and Z^* is contained in $\{(s,y) \in Q' : \psi_i(s,y) = 0$ for some $i\}$. (If Z^* is such a set, we call Z^* *slim*, or better, slim in Q'.) Finally, W is differentiable at all points $(t,x^*(t))$, except a finite number (or countable number) of points, with (2.17) assumed satisfied. $\square$

Proof of Remark 2.10. For the purpose of this proof, let us define Z^* to be "system slim" if, for any admissible solution $x(.)$, for v.e. $s \in (0,T)$ such that $(s,x(s)) \in Z^*$, for any $\varepsilon > 0$, there exist arbitrary small numbers $h > 0$ and corresponding vectors $(a,b) := (a_h,b_h) \in \mathbb{R} \times \mathbb{R}^n$ such that $(t+a,x(t)+b) \notin Z^*$ for v.e. $t \in (s,s+h)$, and $|(a,b)| \leq \varepsilon h$. Assume first that Z^* is system slim. Let $x(s)$ be an arbitrary admissible solution, and let $c < d$ be any two points in $(0,T)$.

Now, for any real-valued locally Lipschitz continuous function $\alpha(t)$ on $[c,d]$, by standard calculus, $\alpha(t) = \alpha(c) + \int_c^t \dot{\alpha}(s)ds$, $t \in [c,d]$, if $\dot{\alpha}$ exists and is continuous in $[0,T] \setminus A$, A containing a finite number of points. A slight extension of this result allows A to be countable. A more general result says that the last equality always holds, i.e., that $\dot{\alpha}$ automatically exists at sufficiently many points (and is well-behaved enough), for the equality to hold. (Actually, this has brought us into measure theory, A being a null set, but perhaps uncountable. See Exercise 2.9 for another way out.) Define $(d_-^+/ds)\alpha(s) := \liminf_{h\downarrow 0} h^{-1}[\alpha(s+h) - \alpha(s)]$.

Let us prove that $W(s,x(s))$ in nonincreasing in $[c,d]$. Using the remarks on $\alpha(.)$, and recalling the local Lipschitz continuity of W, this monotonicity will follow if we show that $(d_-^+/ds)W(s,x(s)) \leq 0$ for v.e. s in (c,d), (note that whenever $(d/ds)W(s,x(s))$ exists, it equals $(d_-^+/ds)W(s,x(s))$). Given any s in (c,d) for which system slimness holds, for any $\varepsilon > 0$, let h and (a,b) be as in the definition of system slimness. For h small, $(t+a,x(t)+b)$ belongs to Q', when $t \in [s,s+h]$. Now,

$$h^{-1}[W(s+h,x(s+h)) - W(s,x(s))]$$
$$= h^{-1}[W(s+h+a,x(s+h)+b) - W(s+a,x(s)+b)]$$
$$+ h^{-1}[W(s+h,x(s+h)) - W(s+h+a,x(s+h)+b)]$$
$$+ h^{-1}[W(s+a,x(s)+b) - W(s,x(s))].$$

Because W is locally Lipschitz continuous with a rank, say K, near $(s,x(s))$, then for h small enough, the absolute values of the two last terms are smaller than $K\varepsilon$, and because $W(t+a,x(t)+b)$ is differentiable for v.e. t in $(s,s+h)$ with derivative ≤ 0 (by the HJB equation), the first term on the right-hand side is nonpositive. It then follows that $(d_-^+/ds)W(s,x(s)) \leq 0$ for v.e. $s \in [c,d]$.

Next, let us show that slimness implies system slimness. Given any admissible $x(.)$, let $s \in (0,T)$ be such that $\dot{x}(.)$ exists near s, and is continuous here. Define $I := \{i : \psi_i(s,x(s)) = 0\}$ and $I^0 = \{i \in I : (d/ds)\psi_i(s,x(s)) = 0\}$. There exists a unit vector (a',b') such that $\psi_{is}(s,x(s))a' + \psi_{ix}(s,x(s))b' \neq 0$ for all $i \in I$. Let $\varepsilon > 0$, assumed to be so small that, for all $t \in [s,s+\varepsilon]$ and for all $(a'',b'') \in \mathbb{R} \times \mathbb{R}^n$ such that $|(a'',b'')| \leq \varepsilon$, the three following properties hold: $\psi_i(t+a'',x(t)+b'') \neq 0$ for $i \notin I$, $\psi_{is}(t+a'',x(t)+b'') + \psi_{ix}(t+a'',x(t)+b'')\dot{x}(t) \neq 0$ for $i \in I \setminus I^0$ and, for some $\gamma > 0$, for $i \in I$,

$$|\psi_{is}(t+a'',x(t)+b'')a' + \psi_{ix}(t+a'',x(t)+b'')b'| \geq \gamma,$$

(γ independent of (t,a'',b'')). Choose $h \in (0,\min\{\varepsilon,1\})$ so small that, for $i \in I^0$, $|\psi_i(t,x(t))| \leq (\gamma/2)\varepsilon h$ when $t \in [s,s+h]$, (recall that $\psi_i(s,x(s)) = 0$, $(d/ds)\psi_i(s,x(s)) = 0$). Let $(a,b) = \varepsilon h(a',b')$. As $\psi_i(t+a,x(t)+b) =$

$$\psi_i(t,x(t)) + \int_0^{\varepsilon h}[\psi_{is}(t+\theta a',x(t)+\theta b')a' + \psi_{ix}(t+\theta a',x(t)+\theta b')b']d\theta,$$

then $|\psi_i(t+a,x(t)+b)| \geq \gamma\varepsilon h - (\gamma/2)\varepsilon h > 0$ for $t \in [s,s+h]$, $i \in I^0$. In fact, for any i, $\psi_i(t+a,x(t)+b) \neq 0$ for v.e. $t \in (s,s+h)$, and $|(a,b)| \leq \varepsilon h$. Hence, system slimness has been proved.

By continuity and monotonicity on $[c,d]$ for any $c > 0$, $W(d,x(d)) \leq W(0,x(0))$. Again, $W(s,x^*(s))$ is constant on $[c,d]$, and so also on $[0,d]$, which then again entails optimality of $u^*(.)$. $\square$

Remark 2.11 (Changed assumptions in Theorem 2.8 and Remark 2.10*). If it is assumed that $t \to W(t,x^*(t))$ is constant in $[0,T)$, then the conclusions in Theorem 2.8 and Remark 2.10 hold even when (2.17) is deleted, together with (in Remark 2.10) the assumption on differentiability of W along $(t,x^*(t))$. (This constancy, a consequence of (2.17), is what is actually used in the proofs of these results above.)
                                                                                      $\square$

Remark 2.12 (Uniqueness of solutions of the HJB equation*). There exist several uniqueness results connected with the solution of (2.16). Here, a single result is indicated (actually taken from the theory of so-called viscosity solutions, Fleming and Soner (1993), p.86). Assume that $\hat{H}(t,x,p) = \sup_{u \in U} pf(t,x,u)$ is C^1 and satisfies $|H_x| \leq K(1+|p|)$, $|f(t,x,u)| \leq K$, with K a given constant, $(t,x) \in \tilde{Q} := [0,T] \times \mathbb{R}^n$, $u \in U$. Then there is at most one function W, which is C^1 on $(0,T) \times \mathbb{R}^n$, satisfies (2.16) and $W(T,.) = S(.)$ and is bounded and uniformly continuous on $\tilde{Q}$. $\square$

Remark 2.13 (Criterion containing an integral). Assume that the criterion (2.1) is replaced by

$$\max\left\{\int_0^T f_0(t,x(t),u(t))dt + S_0(x(T))\right\} \tag{2.23}$$

(S_0 is C^1 and f_0 and f_{0x} are C^0). We then introduce a new scalar state variable $x_0(t)$ governed by $\dot{x}_0(t) = f_0(t,x,u)$, with $x_0(0) = 0$. Then $x_0(T) = \int_0^T f_0(t,x(t),u(t))\,dt$. Introduce the augmented state vector $\tilde{x} = (x_0,x)$ and define $S(\tilde{x}) = x_0 + S_0(x)$. Then the maximization of the criterion in (2.23) is the same as the maximization of $S(\tilde{x}(T))$ in this "augmented system." Hence, we are back in the setting of (2.1)–(2.4). However, let us write down the HJB equation in the case the criterion is as given in (2.23):

$$V_s(s,y) + \max_{v \in U}\{f_0(s,y,v) + V_y(s,y)f(s,y,v)\} = 0. \tag{2.24}$$

The term $f_0(s,y,v)$ in (2.24) is actually $(\partial\tilde{V}/\partial x_0)f_0(s,y,v)$, where $\tilde{V}(s,\tilde{x})$, is the value function in the augmented system. It has the property that $\partial\tilde{V}/\partial x_0 = 1$, moreover $V(s,y)$ in (2.24) equals $\tilde{V}(s,(0,y))$.

In Theorem 2.8, in this case, replace S by S_0 in (2.21) and (2.22) and in (2.18) and (2.19) in the free end case, moreover (2.24) replaces (2.16) (for V replaced by W), and (2.17) is replaced by: For v.e. t,

$$W_s(t,x^*(t)) + f_0(t,x^*(t),u^*(t)) + W_y(t,x^*(t))f(t,x^*(t),u^*(t)) = 0. \tag{2.25}$$

                                                                                      $\square$

Example 2.14 (End constraints). Let us solve the end constrained problem

$$\max \int_0^1 -u^2 \, dt, \quad \dot{x} = u \in \mathbb{R}, \quad x(0) = 0, \quad x(1) = 1.$$

Solution. According to the theory above, $W(s,y)$ should be the optimal value of the criterion in the following problem:

$$\max \int_s^1 -u^2 \, dt, \quad \dot{x} = u \in \mathbb{R}, \quad x(s) = y, \quad x(1) = 1. \tag{$*$}$$

The HJB equation in this example is $W_s + \max_u[-u^2 + W_y u] = 0$, which after maximization ($u = W_y/2$) becomes $W_s + (W_y)^2/4 = 0$. We guess that the solution is of the form $W = a(s) + b(s)y + c(s)y^2$. Insertion into the HJB equation gives

$$\dot{a} + \dot{b}y + \dot{c}y^2 + b^2/4 + bcy + c^2y^2 = 0.$$

We see that if we choose $\dot{a} = -b^2/4$, $\dot{b} = -bc$, and $\dot{c} = -c^2$, the HJB equation is satisfied. Now, the last equation gives $c(s) = 1/(s+D)$, where D is an integration constant. Then $b = E/(s+D)$ and $a = F + E^2/4(s+D)$, where E and F are integration constants. Then $W = F + [E^2/4 + Ey + y^2]/(s+D) = F + [y + E/2]^2/(s+D)$. Now, $W(s,y)$ is ≤ 0, $s \in (0,1)$, (the criterion is nonpositive). When y is large, the second term in the last formula for W dominates, so $s + D < 0$. For $y = -E/2$, $W(s,y) = F \leq 0$. Both terms must then vanish as $s \to 1$ and $y = 1$, because we guess that $W(1,1) = 0$ (see the limits, or limsup's, in (2.21),(2.22)), so we try $F = 0, 1 + E/2 = 0$, i.e., $E = -2$. (In fact, $W(s,1) = 0$, as $u = 0$ is evidently optimal when starting at $(s,1)$, as can be seen directly from the problem formulation. The function $s \to W(s,1)$ would not be constant unless $E = -2$.) Then $W = (y-1)^2/(s+D)$. Now, the feedback control $u(s,y)$ equals $W_y/2 = (y-1)/(D+s)$. Solving the equation $\dot{x} = (x-1)/(D+t)$, we get $x(t) = 1 + C(D+t)$, where C is an integration constant. Now, $x(s) = y, x(1) = 1$, gives $1 + C(D+s) = y, 1 + C(D+1) = 1$, i.e., $D = -1, C = (y-1)/(s-1)$, and $W = (y-1)^2/(s-1)$.

Note that W is not defined at $s = 1$. For $(s,y) = (0,0)$, denote the solution $x^*(t)$, it equals t. Evidently, $W(t,x^*(t)) = (t-1)^2/(t-1) = -(1-t)$ satisfies $\limsup_{t \to 1} W(t,x^*(t)) \leq 0$. Moreover, for any admissible piecewise continuous function $u(t)$, if $\sup_t |u(t)| \leq M$, then $|x(t) - x(1)| = |x(t) - 1| \leq M(1-t)$ and hence

$$|W(t,x(t))| \leq (x(t) - 1)^2/(1-t) \leq M^2(1-t)^2/(1-t) \leq M^2(1-t),$$

so $W(t,x(t)) \to 0$ as $t \to 1$, hence $\limsup_{t \to 1} W(t,x(t)) \geq 0$. Evidently, Theorem 2.8 gives optimality of of $u(t;0,0) = 1$.

We now solve the problem $(*)$ by means of the maximum principle, using $p(t)$ as adjoint variable: Evidently, $u = p(t)/2$, $\dot{p} = 0$, $p(s) = p(t)$ must be so chosen that $x(T) = 1$, i.e., $p(s)$ is determined by $y + \int_s^1 p(s)/2 \, dt = 1$, i.e., $u = p(s)/2 = p(t)/2 = (1-y)/(1-s)$. Then $W(s,y) = -(1-y)^2/(1-s)$. $\square$

Example 2.15 (End constrained problems again). Consider next the two problems:

$$\text{(a) } \max \int_s^1 (x - \tfrac{1}{2}u^2)\, dt, \quad \dot{x} = u \in \mathbb{R}, \quad x(s) = y, x(1) = 0, y \in \mathbb{R},$$

$$\text{(b) } \max \int_s^1 x\, dt, \quad \dot{x} = u \in [-1,1], \quad x(s) = y, x(1) = 0, y \in [s-1, 1-s],$$

$s \in [0,1)$, (s,y) given. Let the optimal values in the two problems be denoted by $V^*(s,y)$ and $W^*(s,y)$, respectively. In both problems $p(t) = 1/2 - t$ (see below), when $y = s = 0$. Along the optimal path for this initial condition, $p(t) = V_y^*(t, x^*(t))$ in the first problem. As should be well-known, this is a relationship that is frequently expected to hold, see, e.g., Remark 2.5. But the corresponding relationship fails in the second case (for $t > 1/2$, $0 = W_y^*(t, x^*(t)) \neq p(t)$ at least if we let W_y^* be a left partial derivative, we are then on the boundary of the domain of definition of W^*). We might then suspect that the HJB equation fails in problem (b), but essentially it does hold.

Solution. Let us give some details. In the first problem, $u(t) = p(t) = C - t$, $x(t) = -t^2/2 + Ct + D$, $V^*(s,y) = \int_s^1 [-t^2/2 + Ct + D - (t-C)^2/2]\, dt$, where C and D satisfies $y = -s^2/2 + Cs + D$ and $0 = -1/2 + C + D$, i.e., $D = (1-s)^{-1}(s^2/2 - s/2 + y)$, $C = 1/2 - D$. From these formulas, V_2^* is easily calculated by differentiating under the integral sign. We then get $V_y^* = \int_s^1 [2C_y t + D_y - CC_y]\, dt = \int_s^1 [-2t/(1-s) + (1+C)/(1-s)]\, dt = -(1-s^2)/(1-s) + 1 + C = C - s$, so $V_y^*(s,y) = p(s) = C - s$ and it is easily seen that V^* satisfies the HJB equation and the boundary conditions (2.21) and (2.22).

In the second problem, $u = 1$, for $t < \sigma$, $u = -1$, for $t > \sigma$, where the switch point σ is determined by $y + \int_s^\sigma 1\, dt + \int_\sigma^1 -1\, dt = 0$, hence $\sigma = \sigma(s,y) = (s+1-y)/2$, $x(t) = y + t - s$ for $t < \sigma$, $x(t) = y + 2\sigma - s - t = 1 - t$ for $t > \sigma$, with $W^*(s,y) = \int_s^\sigma (y + t - s)\, dt + \int_\sigma^1 (1-t)\, dt = \phi(s, s, y, \sigma(s,y))$, where $\phi(s, \tau, y, \sigma) := \int_s^\sigma (y + t - \tau)\, dt + \int_\sigma^1 (1-t)\, dt$. We here and below consider only points (s,y) such that $s - 1 \leq y \leq 1 - s$, as the end condition can only be satisfied for such initial points. Now, $\phi_2 = -\phi_3$, and $\partial\sigma/\partial s = -\partial\sigma/\partial y = 1/2$, and, for $s = \tau$, $W_y^* = \phi_3 - \phi_4/2 = \sigma - s \geq 0$ and $W_s^* = -y + \phi_2 + \phi_4/2 = -y - (\sigma - s)$, $(\phi_1 = -y)$. This means that the HJB equation is satisfied by W^* for points (s,y) satisfying $s - 1 < y < 1 - s$ and even for points (s,y) satisfying $y = 1 - s$, if we agree that partial derivatives at such points equal left derivatives. It is easily seen that the boundary conditions (2.21) and (2.22) are satisfied. (The picture of the optimal solution for $y = s = 0$ is a "tent" with its base (or floor) being the interval $[0,1]$ on the x-axis, and with top in the point $(1/2, 1/2)$). Geometrically it is easily seen that if the point (s,y) is situated on the line $y = 1 - s$, then changing y to $y + \Delta y$, $\Delta y < 0$, makes a second-order change only in the area representing $W^*(s,y)$, so $W_2^*(s,y) = 0$ for $y = 1 - s$.

(Letting $s = t$, the feedback optimal control $u(t,y)$ is given by $u = 1$ if $t < (t + 1 - y)/2$, i.e., $t < 1 - y$, and $u = -1$ if $t = 1 - y$.)

Imagine that in both problems (a) and (b) originally we wanted to solve the problems for $(s,y) = (0,0)$. In problem (a), the set Q is open (it equals $(0,1) \times \mathbb{R}$), in the second one it is not (it is the closed set Q confined between the four lines $y = \pm t$, $y = \pm(1-t)$). In both problems we checked that the proposed value function (V^*

and W^*) satisfied the HJB equation for (at least) $(s,y) \in Q$, in problem (b) we then agree to use the formula for W^* above, also for (s,y) not in Q, (W^* is C^1-extendable to an open set larger than Q). In problem (a), the satisfaction of the HJB equation also follows from general theory (Theorem 2.16 below), whereas in problem (b) the above explicit test has to be carried out (Theorem 2.16 does not apply). As correct boundary conditions hold both in case (a) and (b), in both cases Theorem 2.8 can be invoked to ensure optimality of the controls proposed.   □

Recall that in Example 2.7, the optimal control came directly out specified as a function of t and x. In the case where an arbitrary initial point (s,y) is given, when using the maximum principle, non-feedback controls $t \to u(t;s,y)$ are obtained, which of course will depend on (s,y). For the moment assume them to be right continuous in t. From these controls, feedback controls $u(t,y)$ are obtained simply by putting $u(t,y) = u(t;t,y)$. It may not be immediately apparent that the equation $\dot{x}(t) = f(t,x,u(t,x))$, $x(s) = y$, has the same solution as $\dot{x}(t) = f(t,x,u(t;s,y))$, $x(s) = y$, and sometimes there lies a problem here—but in cases of *unique* solutions of the two differential equations and *unique* non-feedback optimal controls for any starting point (s,y), this must be so: Solving the last one gives a solution $x(t;s,y)$. Now, $\dot{x}(t';s,y) = f(t',x(t';s,y),u(t';s,y)) = f(t',x(t';s,y),u(t';t',x(t';s,y))) = f(t',x(t';s,y),u(t',x(t';s,y)))$, the next to last equality because $u(t';s,y)$ is also optimal in the problem where we start at $(t',x(t';s,y))$ (the "tail is optimal"), and so by uniqueness $u(t';s,y) = u(t';t',x(t';s,y))$. (If the last equality holds, for any t',s,y, $t' > s$, we say that we have a consistent family of controls, one might even argue that if the family was not consistent at the outset, it might be possible to rearrange it in such a manner that this property holds.)

   Above $W(t,x)$ was the solution of a partial differential equation, and we said that in effect $W(t,x)$ could be found by using the "method of characteristics," which here amounts to solving the maximum principle. In fact, when the maximum principle yields a solution $x(t;s,y)$ for all initial points (s,y) in the admissible set Q, then very often $W(s,y) = S(x(T;s,y))$ satisfies the HJB equation, and $x(t;0,x^0)$ is optimal. Let us state this result precisely. (Below, an open set Q^0 appears, frequently it is chosen to equal or include Q.)

   The following preconditions must be satisfied: Assume that there exists an open set Q^* in (t,x,p)-space such that $\hat{H}(t,x,p) := \max_u H(t,x,u,p)$, as well as its first- and second-order derivatives with respect to x and p (exist and) are C^0 here (the maximum assumed to exist for all $(t,x,p) \in Q^*$). Assume also that there exists an open set Q^0 in $(0,T) \times \mathbb{R}^n$, such that for any $(s,y) \in Q^0 \cup \{(0,x^0)\}$, on $[s,T]$ there exist solutions $p(t;s,y)$ and $x(t;s,y)$, with $x(s;s,y) = y$, and with corresponding control $u(t;s,y)$, of the necessary conditions (maximum principle) (2.5)–(2.6), for which the terminal conditions (2.4) and the following transversality conditions are satisfied.

$$p_i(T;s,y) \geq (\partial/\partial x_i)S(x(T;s,y)), i = l+1,\ldots,m$$
$$\text{with equality holding if } x_i(T;s,y) > x_i^1, \tag{2.26}$$
$$p_i(T;s,y) = (\partial/\partial x_i)S(x(T;s,y)), i = m+1,\ldots,n. \tag{2.27}$$

Assume, finally, that $(t, x(t; 0, x^0)) \in Q^0$ for all $t \in (0, T)$. The solutions $x(t; s, y)$ shall here be called *characteristic solutions* (sometimes the name *extremals* are used.)

Theorem 2.16 (Sufficient condition for characteristic solutions). *In addition to the preconditions just stated, assume that, for each $t \in (s, T]$, $(s, y) \rightarrow (x(t; s, y), p(t; s, y))$ is continuously differentiable in Q^0, with $t \rightarrow (x_s(t; s, y), x_y(t; s, y))$ continuous at $t = T$. Assume furthermore, for any $(s, y) \in Q^0 \cup \{(0, x^0)\}$, that $(t, x(t; s, y), p(t; s, y)) \in Q^*$ for all $t \in [s, T)$. Assume also the "consistency condition" that*

$$(x(t'; 0, x^0), p(t'; 0, x^0)) = (x(t'; t, x(t; 0, x^0)), p(t'; t, x(t; 0, x^0))) \tag{2.28}$$

for all $t', t \in [0, T]$, $t' > t$. Assume, finally, that $S(x(T; s, y))$ is continuous on $\{(0, x^0)\} \cup Q^0$ and that

$$\limsup_{t \to T} S(x(T; t, \hat{x}(t))) \geq S(\hat{x}(T)) \tag{2.29}$$

for all admissible solutions $\hat{x}(.)$ for which $(t, \hat{x}(t))$ belongs to Q^0 for all $t \in (0, T)$. Then $(x(t; 0, x^0), u(t; 0, x^0))$ is optimal in the class of such admissible pairs. $\square$

Remark 2.17 (Open set Q). Frequently, the set Q^0 is taken to be the admissible set Q, in which case Theorem 2.16 requires this set to be open for optimality to hold. $\square$

The consistency condition can be dropped if it is explicitly required that $\lim_{t \to T} S\left(x(T; t, x(t; 0, , x^0))\right) = S(x(T; 0, x^0))$. If (2.29) fails for some admissible solutions, then optimality holds in the subset of admissible solutions for which (2.29) holds.

In free end problems, in which Q^0 is taken to be the admissible set, very often $\lim_{s \to T, y \to \hat{y}} x(T; s, y) = \hat{y}$ for all $(T, \hat{y}) \in \text{cl}Q^0$, $((s, y) \in Q^0)$. In this case, of course, (2.29) holds with equality. In end constrained problems, this property often fails in cases where arbitrary large values of $|\dot{x}(t)|$ can occur. (For example, this property fails in Example 2.20 below.)

A weaker condition that implies (2.29) with equality is of course

$$\lim_{t \to T} x(T; t, \hat{x}(t)) = \hat{x}(T) \tag{2.30}$$

for all admissible $\hat{x}(t)$.

Remark 2.18 (Conditions implying the limit condition (2.30)*). The limit condition (2.30) is satisfied if, for any admissible solution $\hat{x}(.)$, $p_T^{\hat{x}(.)} := \lim_{t \to T} p(t; t, \hat{x}(t))$ exists and $(T, \hat{x}(T), p_T^{\hat{x}(.)})$ belongs to Q^*. Another condition implying the limit condition is $\sup_{s, t \geq s} |u(t; s, \hat{x}(s))| < \infty$ for all admissible $\hat{x}(.)$. $\square$

Note that when Theorem 2.16 holds, then $W(s, y) := S(x(T; s, y))$ satisfies the HJB equation (2.16) in Q^0 and $u(s; 0, x^0)$ yields supremum in this equation for $y = x(s; 0, x^0)$.

Usually, the triples $x(t; s, y), p(t, s, y), u(t; s, y)$ are found by first finding a control $\hat{u}(t, x, p)$ maximizing $H(t, x, u, p)$. (When there is a unique maximum of the Hamiltonian, then usually $u(t, x, p)$ becomes continuous in (t, x, p), at least this holds if U is compact. The method works also in other cases.) Next, one solves the equations

$$\dot{x} = f(t, x, \hat{u}(t, x, p)), \quad \dot{p} = -H_x(t, x, \hat{u}(t, x, p), p)$$

with initial condition $x(s) = y$, and terminal conditions (2.4), (2.26), and (2.27). The two differential equations are one variant of what is called the characteristic equations of the HJB equation, and we shall call $x(t; s, y), p(t; s, y), u(t; s, y)$ a characteristic triple. The consistency condition most often follows automatically from the above construction.

Remark 2.19 (Optimality of u(.;s,y)*). In case of Theorem 2.16, if the consistency condition also holds for $(0, x^0)$ replaced by any $(s, y) \in Q^0$, and (2.29) holds for all (s, y)-admissible solutions $\hat{x}(t)$, then, for all $(s, y) \in Q^0$, $u(.; s, y)$ is optimal. $\quad\square$

Example 2.20. Let us solve the problem

$$\max \int_0^{\pi/2} (x^2 - u^2) \, dt, \quad \dot{x} = u, \quad x(0) = 0, \quad x(\pi/2) = 0, \quad u \in \mathbb{R}.$$

Solution. Let $\dot{x}_0 = x^2 - u^2$, $x_0(0) = 0$, so the criterion equals $S(x_0, x) = x_0(\pi/2)$. Here, we have two states, x_0 and x, the adjoint function $p_0(t)$ corresponding to x_0 equals 1, and, for p corresponding to x, $H = x^2 - u^2 + pu$. Now, $H_u = -2u + p = 0$ yields $(\hat{u} =) u = \frac{1}{2}p$. Moreover, $\ddot{x} = \dot{u} = \frac{1}{2}\dot{p}$. Also, $\dot{p} = -\partial H/\partial x = -2x$. Hence, $\ddot{x} = -x$, which has the solution $x(t) = A\cos t + B\sin t$. With $x(\pi/2) = 0$, $B = 0$, so $x(t) = A\cos t$. For the start point (s, y), $s \in [0, \pi/2)$, $y = A\cos s$, and so $A = y/\cos s$, and thus, $x(t; s, y) = \dfrac{y\cos t}{\cos s}$ and $u(t; s, y) = \dfrac{-y\sin t}{\cos s}$. (Actually, solutions for x_0, p_0, x, and p should be written down for arbitrary start points (s, y_0, y) and shown to be C^1 in (s, y_0, y), which is easy and hence omitted, note that $x(t; s, y_0, y) = x(t; s, y)$.) We want to prove that $x_0(\pi/2; s, \hat{x}_0(s), \hat{x}(s)) \to \hat{x}_0(\pi/2)$ when $s \to \pi/2$, for any admissible $(\hat{x}_0(.), \hat{x}(.))$ with corresponding control $\hat{u}(.)$. Let $K = \sup_t |\hat{u}(t)|$. Now, $\hat{x}(\pi/2) = 0$, hence $|\hat{x}(s)| = |\hat{x}(s) - \hat{x}(\pi/2)| \leq K(\pi/2 - s)$ so $\hat{x}(s)/\cos s$ is bounded. This means that the integrand in the expression $x_0(\pi/2; s, \hat{x}_0(s), \hat{x}(s)) = \hat{x}_0(s) + \int_s^{\pi/2} \dot{x}_0(\sigma; s, \hat{x}_0(s), \hat{x}(s)) d\sigma$ is bounded, hence $x_0(\pi/2; s, \hat{x}_0(s), \hat{x}(s)) \to \hat{x}_0(\pi/2)$ when $s \to \pi/2$. By Theorem 2.16 (and subsequent comments), $u(t; 0, 0) \equiv 0$ is optimal. (Here, we can let $Q^* = (0, \pi/2) \times \mathbb{R} \times \mathbb{R} \times (1/2, 3/2) \times \mathbb{R}$, Q^* consist of points (t, x_0, x, p_0, p), and $Q^0 = (0, \pi/2) \times \mathbb{R}^2$.) $\quad\square$

It would be more difficult to prove optimality in this nonconcave problem if we did not have recourse to Theorem 2.16. Necessary condition gives $u(t) \equiv 0$ as a unique candidate also if $T = \pi/2$ is replaced by, say, $T = 2\pi + \pi/2$, but in the latter case the candidate is not optimal (no optimal solutions exist, no $x(t; s, y)$ exists for $s = \pi/2, y \neq 0$).

In the literature, often a variant of Theorem 2.16 is given, where it is imagined that originally a feedback control function $u(t, x)$ is given, and then unique solutions $x(t; s, y)$ of the equation $\dot{x} = f(t, x, u(t, x))$ satisfying the terminal conditions are postulated to exist for all $(s, y) \in Q$. If Q is open, and $u(t, x)$ is continuously differentiable, at least in the free end case, continuous differentiability of $(s, y) \to (x(t; s, y), p(t; s, y))$ follows from general theory of differential equations.

The continuous differentiability of $(s,y) \rightarrow (x(t;s,y), p(t;s,y))$ throughout Q may fail if $u(t,x)$ is not C^1. Less demanding result can also be found (see, e.g., Fleming and Rishel (1975)), in which $u(t,x)$ is allowed to be less smooth, it may even be discontinuous at certain points).

A question naturally arises as to how well-behaved the function $u(t,x)$ in general will be. A general existence theorem asserts that in the above problem, for conditions occurring in standard Fillipov–Cesari existence theorems, there exists, on the controllable set $\tilde{Q}$, an optimal feedback function $u(t,x)$ that is a so-called Borel measurable function. This type of function can have a lot of discontinuity, nevertheless in some sense values at nearby points are "most often" not too far apart, (see the Appendix for a further discussion). One should note that what can actually be proved is that there exist a Borel measurable function $u(t,x)$ and for each (s,y) in $\tilde{Q}$ an optimal pair $(x(t;s,y), u(t;s,y))$ where the latter function can be expressed as $u(t;s,y) = u(t,x(t;s,y))$. Although solving differential equations $\dot{x} = f(t,x,u(t,x))$, for $u(t,x)$ given, raises difficult questions when $u(t,x)$ is not continuous, nevertheless such results can be proved. (See Haussmann and Lepeltier (1990) for a proof even in a stochastic setting.)

Problems encountered when a given $u(t,x)$-function is discontinuous in x can be indicated as follows: Assume that $u(t,x)$, x real, has a discontinuity as a function of x along a horizontal line $x = x'$ in (t,x)-plane and that $f(t,x,u(t,x'-)) \geq 1$ when $x < x'$ and $f(t,x',u(t,x'+)) \leq -1$ with $u(t,x)$, say, right-continuous in x. In such situations it is difficult to agree on what a solution should be. (Starting from below, the solution would never enter the region $x > x'$, it leaves the region $x < x'$, but having touched the line $x = x'$, it can never reenter the region $x < x'$. Neither can it stay on the line $x = x'$, here $\dot{x} \leq -1$.)

In spite of such problems, in a number of practical applications, lines of discontinuity are crossed without problems. In these applications, when crossing a line of discontinuity, the state luckily locally moves away from the line.

Proof of Theorem 2.16. Define $W(s,y) = S(x(T;s,y))$. Let $\tilde{u} \in \text{argmax}_u \hat{p} f(t,\hat{x},u)$, $(t,\hat{x},\hat{p}) \in Q^*$. From the first-order conditions for a local minimum at $(\hat{x},\hat{p})$ of $(x',p') \rightarrow \hat{H}(t,x',p') - p'f(t,x',\tilde{u})$, it follows that $\hat{H}_x(t,\hat{x},\hat{p}) = \hat{p}\hat{f}_x(t,\hat{x},\tilde{u})$, and $\hat{H}_p(t,\hat{x},\hat{p}) = f(t,\hat{x},\tilde{u})$. This means that $\hat{H}(t,\hat{x},\hat{p}) = \hat{p}\hat{H}_p(t,\hat{x},\hat{p})$ and that $x(.;s,y)$ and $p(.;s,y)$, satisfy

$$(a) \quad \dot{x}(t) = \hat{H}_p(t,x(t;s,y),p(t;s,y)) \quad (b) \quad \dot{p}(t) = -\hat{H}_x(t,x(t;s,y),p(t;s,y)) \quad (2.31)$$

These equations are the characteristic equations of the HJB equation. Write $w(t;s,y) := W_y(s,y) - p(t;s,y)x_y(t;s,y)$ and $\hat{w}(t;s,y) := W_s(s,y) - p(t;s,y)x_s(t;s,y)$. Let (s,y) belong to Q^0. It will be shown that

$$\hat{w}(T;s,y) = 0, w(T;s,y) = 0. \quad (2.32)$$

To see this, let $x^{(j)}$ be either $(\partial/\partial y)x_j(T;s,y)$ or $(\partial/\partial s)x_j(T;s,y)$ and note that

$$\sum_j (p_j(T;s,y) - S_{x_j}(x(T;s,y)))x^{(j)}$$

$$= \sum_{p_j(T;s,y)-S_{x_j}(x(T;s,y))\neq 0} (p_j(T;s,y) - S_{x_j}(x(T;s,y)))x^{(j)}.$$

In the last sum, evidently, by the transversality conditions, the sum runs only over $j = 1,\ldots,m$, at most. If $p_j(T;s,y) - S_{x_j}(x(T;s,y)) > 0$, $j = l+1,\ldots,m$, then, by continuity of $(s',y') \to p_j(T;s',y') - S_{x_j}(x(T;s',y'))$, $p_j(T;s',y') - S_{x_j}(x(T;s',y')) > 0$ for all (s',y') in a neighborhood of (s,y), hence $x_j(T;s',y')) = x_j^1$ here, i.e. $(\partial/\partial y)x_j(T;s,y) = 0$ and $(\partial/\partial s)x_j(T;s,y) = 0$. This evidently also holds for $j \leq l$, so the last sum equals zero. Hence (2.32) holds, as $W_y - px_y = \sum_j[S_{x_j}(\partial/\partial y)x_j - p_j(\partial/\partial y)x_j]$, $W_s - px_s = \sum_j[S_{x_j}(\partial/\partial s)x_j - p_j(\partial/\partial s)x_j]$ (shorthand notation).

Let $(\hat{s},\hat{y})$ be a given point in Q^0, let $T' < T$ be any number close to T, $T' > \hat{s}$, and define $A_{T'} := \{(t,x(t;\hat{s},\hat{y}),p(t;\hat{s},\hat{y})) : t \in [\hat{s},T']\}$. For some $\varepsilon > 0$, $B(A_{T'};\varepsilon) := \{(t,x,q) : \text{dist}((t,x,q),A_{T'}) < \varepsilon\}$ is contained in Q^*. Because $p(\tilde{s};s,y)$ is close to $p(\tilde{s};\hat{s},\hat{y})$ when (s,y) is near $(\hat{s},\hat{y})$, $\tilde{s}$ any given number close to $\hat{s}$, $\tilde{s} > \hat{s}$, then, by standard theory of differential equations, numbers s' and T'' exist, $s' < \hat{s}$, $T'' > T'$, such that, for all (s,y) near $(\hat{s},\hat{y})$, $x(t;s,y)$ and $p(t;s,y)$ are defined on $[s',T'']$, satisfying (2.31), being, uniformly in t, close to $x(t;\hat{s},\hat{y})$ and $p(t;\hat{s},\hat{y})$ for (s,y) near $(\hat{s},\hat{y})$ for any $t \in [s',T'']$ (so $(t,x(t;s,y),p(t;s,y)) \in B(A_{T'};\varepsilon)$, $t \in [s',T'']$), and being C^1 in (s,y) near $(\hat{s},\hat{y})$ for any $t \in [s',T'']$. This holds for any pair $(\hat{s},\hat{y})$ in Q^0. Moreover, the partial derivatives x_t, x_s, x_y, p_t, p_s, and p_y exist and are continuous in (t,s,y) for $t \in (t',T')$ and (s,y) near $(\hat{s},\hat{y})$.

In the calculations to follow, in $\hat{H}(t,x,p)$, $\hat{H}_p(t,x,p)$, and $\hat{H}_x(t,x,p)$, $x(t;s,y)$ and $p(t;s,y)$ are inserted for x and p, and $\partial/\partial y$ and $\partial/\partial s$ means derivatives with respect to y and s, at $(\hat{s},\hat{y})$, after these insertions. (Note that, for (s,y) near $(\hat{s},\hat{y})$, $x_{ty} = x_{yt}$, $x_{ts} = x_{st}$, all exist and are C^0 in (t,y), because $x(t;s,y) = y + \int_s^t \hat{H}_p$.) Let $t \in (s',T')$. Using (2.31)(a) and $\hat{p}\hat{H}_{\hat{p}} = \hat{H}$ (shown above), we get that

$$w_t = -p_t x_y - p x_{yt}$$
$$= (\partial/\partial y)(-px_t) + p_y x_t - p_t x_y$$
$$= (\partial/\partial y)(-\hat{H}) + \hat{H}_x x_y + p_y \hat{H}_p = 0.$$

Similarly, as $x_{ts} = x_{st}$, we get

$$\hat{w}_t = -p_t x_s - p x_{st}$$
$$= (\partial/\partial s)(-px_t) + p_s x_t - p_t x_s$$
$$= (\partial/\partial s)(-\hat{H}) + p_s \hat{H}_p + \hat{H}_x x_s = 0.$$

Because T' was arbitrary, this holds for $t \in (s',T)$.

Now, (2.32) then implies $w(t;\hat{s},\hat{y}) = 0$, $\hat{w}(t;\hat{s},\hat{y}) = 0$ for all t in $[\hat{s},T]$, by continuity in $t \in [\hat{s},T]$. Writing (s,y) instead of $(\hat{s},\hat{y})$, because $w = 0$ and $x_y(s;s,y) = I$, then $p(s;s,y) = W_y(s,y)$. Using $p\dot{x} = p\hat{H}_p = \hat{H}$ (see (2.31) and preceding comments) and $y = x(s;s,y)$ (so $(\partial/\partial s)x(s;s,y) = 0$) yield

$$0 = p(s;s,y)(\partial/\partial s)x(s;s,y)$$
$$= [p(s;s,y)(\partial/\partial t)x(t;s,y) + p(s;s,y)(\partial/\partial s)x(t;s,y)]_{t=s}$$
$$= \hat{H}(s,y,p(s;s,y)) + p(s;s,y)[(\partial/\partial s)x(t;s,y)]_{t=s}$$
$$= \hat{H}(s,y,p(s;s,y)) + W_s(s,y),$$

(the last equality as $\hat{w}$ equals zero). As $p(s;s,y) = W_y(s,y)$ (obtained above), the HJB equation (2.16) holds in Q^0. In particular, the last equality holds for $t = s$, $x(t,0,x^0) = y$, so $p(t;t,x(t;0,x^0)) = W_y(t,x(t;0,x^0))$ and the previous one gives

$$0 = W_s(t,x(t;0,x^0)) + \hat{H}(t,x(t;0,x^0),p(t;t,x(t;(0,x^0)))) \tag{2.33}$$

Thus, (2.17) holds for $(x^*(.),u^*(.)) = (x(.;0,x^0),u(.;0,x^0))$.

Now,

$$W(t,x(t;0,x^0)) = S(x(T;t,x(t;0,x^0))) = S(x(T;0,x^0)),$$

the second equality following from consistency. Then (2.22) holds. As (2.21) is assumed to hold in Theorem 2.16, then all conditions in Theorem 2.8 are satisfied, so $(x(.,0,x^0),u(.,0,x^0))$ is optimal.   □

The proof was formed so that the comment subsequent to Remark 2.17 becomes self-evident. As we saw, consistency automatically implies that

$$W(t,x(t;0,x^0)) = S(x(T;t,x(t;0,x^0))) = S(x(T;0,x^0)),$$

from which $\lim_{t \to T} S(x(T;t,x(t;0,,x^0))) = S(x(T;0,x^0))$ follows; the latter condition can replace consistency in Theorem 2.16.

Proof of Remark 2.18. Note that

$$x(T;t,\hat{x}(t)) = \hat{x}(t) + \int_t^T f(\sigma,x(\sigma;t,\hat{x}(t)),u(\sigma;t,\hat{x}(t)))d\sigma.$$

Consider first the boundedness condition on $u(t;s,\hat{x}(s))$. The boundedness of $u(t;s,\hat{x}(s))$ and $\hat{x}(t)$ gives that $f(\sigma,\hat{x}(s),u(\sigma;s,\hat{x}(s)))$ is bounded, which means, by Gronwall's inequality, that, for s in some interval (s',T), $x(\sigma;s,\hat{x}(s))$ is bounded, $(f_x(\sigma,x,u(\sigma;s,\hat{x}(s))))$ is bounded on a neighborhood of $(s,\sigma,x) = (T,T,\hat{x}(T)))$. Hence, $f(\sigma,x(\sigma;s,\hat{x}(s)),u(\sigma;s,\hat{x}(s)))$ is bounded by some constant C for s in $[s'',T]$, s'' near T, $\sigma \in [s,T]$. Thus,

$$|x(T;t,\hat{x}(t)) - \hat{x}(t)| \leq \int_t^T |f(\sigma,x(\sigma;t,\hat{x}(t)),u(\sigma;t,\hat{x}(t)))|d\sigma \leq C(T-t)$$

for $t \in (s'',T)$. Hence, $x(T;t,\hat{x}(t)) \to \hat{x}(T)$ when $t \to T$.

In case $q := (T,\hat{x}(T),p_T^{\hat{x}(.)})$ belongs to Q^*, then $(\hat{H}_p,\hat{H}_x)$ is bounded by a constant $K \geq 1$ on a ball $B(q,\delta)$ around this point. From the limit condition in Remark 2.18 it follows that for t in some interval $I := [T-\delta',T]$, we have that $(t,\hat{x}(t),p(t;t,\hat{x}(t))) \in B(q,\delta/2)$, for later use δ' is also chosen such that $\delta' < \delta/4K$. For $t \in I$, $\sigma \in [t,T]$,

$$|(x(\sigma;t,\hat{x}(t)),p(\sigma;t,\hat{x}(t)))-(\hat{x}(t),p(t;t,\hat{x}(t)))|\leq K(\sigma-t)\leq K\delta'\leq\delta/4,$$

in fact, $(\sigma,x(\sigma;t,\hat{x}(t)),p(\sigma;t,\hat{x}(t)))$ stays inside $B(q,\delta)$ when σ increase from t to T. From this we get that $\lim_{t\to T}x(T;t,\hat{x}(t))=\hat{x}(T)$. $\square$

Remark 2.21 (Criterion containing an integral). Now, we want to apply the above results to a problem where an integral also appears in the criterion, see (2.23). Similar to what we did in Remark 2.13, the problem is changed to an "augmented" problem where an additional state x_0 appears, with $\dot{x}_0=f_0(t,x,u),x_0(0)=0$, ($f_0$ and f_{0x} are C^0), and with criterion $x_0(T)+S_0(x(T))$. The augmented state vector is now $\tilde{x}=(x_0,x)$. Let p_0 be the adjoint variable corresponding to x_0 and let p correspond to x. The Hamiltonian in the augmented system is written $\tilde{H}(t,x,u,p,p_0):=p_0f_0+pf$.

In this case, we need to modify the assumptions in Theorem 2.16 in the following way. Assume the existence of open sets Q^0 and Q^* and triples $(x(t;s,y),$ $u(t;s,y),\ p(t;s,y))$ for $(s,y)\in Q^0\cup\{(0,x^0)\},t\in[s,T]$, satisfying the earlier terminal and transversality conditions (2.4), (2.26), (2.27) such that $u(t;s,y)$ maximizes $\tilde{H}(t;x(t;s,y),u,p(t;s,y),1)$ v.e., and such that $\dot{x}(t;s,y)=f(t,x(t;s,y),u(t;s,y))$ v.e., $x(s)=y$ and

$$\dot{p}(t;s,y)=-\tilde{H}_x(t,x(t;s,y),u(t,s,y),p(t;s,y),1)\ \text{v.e.}\qquad(2.34)$$

Moreover, assume, for $(s,y)\in\{(0,x^0)\}\cup Q^0$, that $(t,x(t;s,y),p(t;s,y))$ belongs to Q^* for all $t\in[s,T)$, and that consistency (see (2.28)) holds.

Define $x_0(t;s,y)$ by

$$\dot{x}_0(t;s,y)=f_0(t,x(t;s,y),u(t;s,y)),\ x_0(s;s,y)=0.\qquad(2.35)$$

The function $x_0(t;s,y)$ exists on $[s,T]$ for $(s,y)\in\{(0,x^0)\}\cup Q^0$, assume that $x_0(T;s,y)+S_0(x(T;s,y))$ is continuous on $\{(0,x^0)\}\cup Q^0$, and that $(s,y)\to(x_0(t;s,y),$ $x(t;s,y),p(t;s,y))$ is C^1 in $(s,y)\in Q^0$, for each $t\in(s,T]$, with $t\to(x_{0s}(t;s,y),$ $x_{0y}(t;s,y),x_s(t;s,y),x_y(t;s,y))$ continuous at $t=T$.

Define $\hat{H}(t,x,p,p_0)=\max_{u\in U}\tilde{H}(t,x,u,p,p_0)$ and assume, finally, that $\hat{H}(t,x,$ $p,1)$, as well as its first and second derivatives with respect to x and p (exists and) are C^0 in Q^*. Finally, assume that $\limsup_{t\to T}[x_0(T;t,\hat{x}(t))+S_0(x(T;t,\hat{x}(t)))]\geq S_0(\hat{x}(T))$, for any admissible $\hat{x}(t)$. Then again $u(t;0,x^0)$ is optimal. (If $\lim_{t\to T}[x_0(T;t,$ $x(t;0,x^0))+S_0(x(T;t,x(t;0,x^0)))]=S_0(x(T;0,x^0))$, again consistency, (2.28), can be dropped.) $\square$

To see that the assertion in this remark is correct, define $\tilde{Q}^*=\{(t,x_0,x,p_0,p):(t,x,p/p_0)\in Q^*,x_0\in(-\infty,\infty),p_0\in(1/2,3/2)\},\ \tilde{Q}^0=\{(t,x_0,x):(t,x)\in Q^0,x_0\in\mathbb{R}\}$ and for $(s,y_0,y)=(s,\tilde{y})\in\tilde{Q}^0\cup\{(0,0,x^0)\}$, let

$$(x_0(t;s,\tilde{y}),x(t;s,\tilde{y}),p(t;s,\tilde{y}))=(x_0(t;s,y)+y_0,x(t;s,y),p(t;s,y)).$$

As $\hat{H}(t,x,p,p_0)=p_0\hat{H}(t,x,p/p_0,1)$, then $\hat{H}(t,\tilde{x},p,p_0)=\hat{H}(t,x,p,p_0)$ has continuous first- and second-order derivatives with respect to $\tilde{x},p$ and p_0 in $\tilde{Q}^*$. Let $\tilde{p}=(p,p_0)$, $\tilde{x}(t;s,\tilde{y})=(x_0(t;s,\tilde{y}),x(t;s,\tilde{y}))$, $\tilde{p}(t;s,\tilde{y})=(p(t;s,\tilde{y}),1)$. The "augmented" collection $\tilde{Q}^0, \tilde{Q}^*, \hat{H}(t, \tilde{x}, \tilde{p})$, $(\tilde{x}(t; s, \tilde{y}))$, $(\tilde{p}(t; s, \tilde{y}))$ can now play the roles of Q^0, Q^*, $\hat{H}(t, x, p), x(t; s, y), p(t; s, y)$ in Theorem 2.16.

Remark 2.22 (The C^1-property of $(s, y) \rightarrow x_0(t; s, y)*$). Assume the conditions in the preceding remark to be satisfied. The C^1-property of $(s, y) \rightarrow x_0(t; s, y)$ for $(s, y) \in Q^0$, and continuity of $t \rightarrow (x_{0s}(t; s, y), x_{0y}(t; s, y))$ at $t = T$ follow if it is assumed that, for $t \in [s, T)$, $f_0(t, x(t; s, y), u(t; s, y)))$ and its first derivatives with respect to s and y are locally bounded in $(s, y) \in Q^0$, uniformly in $t \in (s, T)$. If the uniform local boundedness of f_0 even holds for $(s, y) \in \{(0, x^0)\} \cup Q^0$, then $x_0(T; s, y)$ is continuous in (s, y) for $(s, y) \in \{(0, x^0)\} \cup Q^0$. (Note that $f_0(t, x(t; s, y), u(t; s, y))) = \hat{H}_{p_0}(t, x(t; s, y), p(t; s, y), 1)$, so the derivatives exist for each $t \in (s, T)$ and are continuous.)

The two properties of $f_0(t, x(t; s, y), u(t; s, y))$ are satisfied if for any $(s, y) \in \{(0, x^0)\}) \cup Q^0$,

$$(t, x(t; s, y), p(t; s, y)) \in Q^* \text{ for } t \in [s, T], \tag{2.36}$$

(i.e., also for $t = T$). □

It is easy to see that the asserted implications of (2.36) are correct, let us however merely show that when (2.36) holds for $(s, y) \in Q^0$, then $x_0(t; s, y)$ is C^1 in Q^0, for $t \in (s, T]$. Note first that, for any $(s, y) = (\hat{s}, \hat{y})$ in Q^0, a set A_T, (see $A_{T'}$ defined in Proof of Theorem 2.16 above), belongs to Q^*, and (using that $\dot{x}_0(t; s, y) = \hat{H}_{p_0}$), similar to what was noted in that proof, standard theory of differential equations says that for all (s, y) near $(\hat{s}, \hat{y})$, solutions $\tilde{x}(t; s, y), p(t; s, y)$ exist, for t in some interval $[s', T'']$, (s', T'') containing $[\hat{s}, T]$, being C^1 in (s, y) and uniformly close to $(\tilde{x}(t; \hat{s}, \hat{y}), p(t; \hat{s}, \hat{y}))$ $t \in (s', T'']$.

Remark 2.23 (Changed preconditions for Theorem 2.16).* Sometimes, the conditions presupposed in Theorem 2.16 are too demanding. Let us change the differentiability assumptions on $\hat{H}(t, x, p)$ as follows: For a moment, call a function of (t, x, p) "1–smooth in an open set" V if the function and its first derivatives with respect to x and p (exist and) are C^0 in V. Assume that there exist C^1-functions $\phi_k, k = 1, \ldots, k^*$ on clQ^*, such that, for any point (t, x, p) in Q^*, $\phi_k(t, x, p) = 0$ for at most one k and such that $\hat{H}$, as well as it first and second derivatives with respect to x and p are C^0 in $Q^* \setminus Z, Z := \{(t, x, p) \in Q^* : \phi_i(t, x, p) = 0 \text{ for some } i\}$. Assume also that $\hat{H}_x(t, x, p)|_A$ and $\hat{H}_p(t, x, p)|_A$, ($|_A$ means restricted to A), have 1-smooth extensions to an open set containing cl$A$ for any set A of the form $\cap_i \Phi^i$, $\Phi^i = \{(t, x, p) \in Q^* : \phi_i(t, x, p) > 0\}$, or $\Phi^i = \{(t, x, p) \in Q^* : \phi_i(t, x, p) < 0\}$ (the direction of the inequality sign may depend on i).

Let us also weaken the differentiability assumptions on $(s, y) \rightarrow (x(t; s, y), p(t; s, y))$: Let $\hat{Z}$ be given subset of Q^0, $\hat{Z}$ closed, such that

$$\hat{Z} \supset \{(s, y) \in Q^0 : \phi_i(s, y, p(s; s, y)) = 0 \text{ for some } i\}. \tag{2.37}$$

Define $Q_{s,y} := \{t \in (s, T) : \phi_k(t, x(t; s, y), p(t; s, y)) = 0 \text{ for some } k\}$. Assume for any $(s, y) \in Q^0 \setminus \hat{Z}, t \in (s, T), t \notin Q_{s,y}$, that $(s', y') \rightarrow (x(t; s', y'), p(t; s', y'))$ is C^1 in a neighborhood of (s, y). Moreover, for $(s, y) \in Q^0 \setminus \hat{Z}, t \in (s, T)$, assume that

$$\phi_k(t, x(t; s, y), p(t; s, y)) = 0 \Rightarrow 0 \neq \phi_{kt}(t, x(t; s, y), p(t; s, y))$$
$$+ \phi_{kx}(t, x(t; s, y), p(t; s, y))\dot{x}(t^{\pm}; s, y) + \phi_{kp}(t, x(t; s, y), p(t; s, y))\dot{p}(t^{\pm}; s, y),$$

$$(2.38)$$

(i.e., the right-hand side is nonvanishing both for the left limit and the right limit). Assume, finally, that $S(x(T; s, y))$ is locally Lipschitz continuous on $\{(0, x^0)\} \cup Q^0$ and that

$$Z' := \{(s, y) \in Q^0 : \phi_i(T, x(T; s, y), p(T; s, y)) = 0 \text{ for some } i\} \qquad (2.39)$$

and $\hat{Z}$ (see (2.37)) are slim in Q^0. Except for the above changes in the differentiability assumptions, assume that the other assumptions in Theorem 2.16 are kept unchanged. Then $u(t, 0, x^0)$ is optimal.

In this remark, it is not necessary explicitly to assume the property that $t \rightarrow (x_s(t; s, y), x_y(t; s, y))$ is defined and continuous at $t = T$.   □

Remark 2.24 (Dropping slimness of $Z'*$). In Remark 2.23, the assumption that Z' is slim can be deleted, provided, for any $(s, y) \in Q^0$, $\phi_i(T, x(T; s, y), p(T; s, y)) = 0$ for at most one i and (2.38) holds also for $t = T$, when $t^{\pm}$ is replaced by $T-$.   □

Remark 2.25 (Criterion containing an integral). In case of Remark 2.21, where an integral appears in the criterion, the conditions assumed in Remark 2.23 should be applied to $\hat{H}(t, x, p, p_0)$, $\tilde{x}(t; s, \tilde{y})$, $\tilde{p}(t; s, \tilde{y})$ instead of $\hat{H}(t, x, p)$, $x(t; s, y)$, $p(t; s, y)$. Assume from now on in this remark that ϕ_k, $k = 1, \ldots, k^*$ do not depend on x_0 and p_0. Then of course the definitions of the entities $\hat{Z}$, and Z' in the augmented system (both required to be slim), reduces to (2.37)(2.39), and the augmented version of (2.38) reduces to (2.38).

It is now assumed that $\hat{H}(t, x, p, 1)$, as well as its first and second derivatives with respect to x and p, are C^0 in $Q^* \setminus Z$, and that $\hat{H}(t, x, p, 1)$ has the extension property of $\hat{H}(t, x, p)$. The nontangentiality condition (2.38) is kept unchanged. In this case, we in addition assume that even $(s', y') \rightarrow x_0(t; s', y')$ is C^1 in a neighborhood of any (s, y) in $Q^0 \setminus \hat{Z}$, $t \notin Q_{s, y}$, $t \in (s, T)$, and that $x_0(T; s, y) + S_0(x(T; s, y))$ is locally Lipschitz continuous on $\{(0, x^0)\} \cup Q^0$. (If $p(s'; s', y')$ is C^1 in a neighborhood of any (s, y) in $Q^0 \setminus \hat{Z}$, and is locally Lipschitz continuous on $\{(0, x^0)\} \cup Q^0$, the two last properties automatically follow. Compare Remark 2.27 below.) Then, again $u(t; 0, x^0)$ is optimal.   □

Remark 2.26 (Changes in the premises of Remarks 2.23, 2.25).* Assume, at any $(s, y) \in Q^0 \setminus \hat{Z}$, continuity of $p(T; s, y)$ and existence and continuity of $t \rightarrow (x_s(t; s, y), x_y(t; s, y))$ at $t = T$ (in case of Remark 2.25, existence and continuity at $t = T$ of $t \rightarrow (x_{0s}(t; s, y), x_{0y}(t; s, y), x_s(t; s, y), x_y(t; s, y))$. Then, in Remarks 2.23 and 2.25, slimness of Z' can be omitted and clA can be replaced by $(clA) \cap ((0, T) \times \mathbb{R}^{2n})$.   □

Proof of Remark 2.23. Define again $W(s, y) := S(x(T; s, y))$. Define $Z^* = Z' \cup \hat{Z}$ and write $v = (s, y)$. Let $\hat{v} := (\hat{s}, \hat{y})$ belong to $Q^0 \setminus Z^*$. By compactness of $[\hat{s}, T]$ and nontangentiality ((2.38)), there is at most a finite number of "crossing point"

$\hat{t} \in (\hat{s}, T)$ in $Q_{\hat{v}} = Q_{\hat{s},\hat{y}}$. Let $\hat{t}$ be any point in $Q_{\hat{v}}$, such that for some (single) i $\phi_i(\hat{t}, x(\hat{t}; \hat{v}), p(\hat{t}; \hat{v})) = 0$. There exists a set A of the type of the remark, such that $(t, x(t; \hat{v}), p(t; \hat{v})) \in A$ for $t < \hat{t}$, t close to $\hat{t}$. By the extension property of $\hat{H}(t, x, p)$ and (2.31), for $t < \hat{t}$, t close to $\hat{t}$, the solution $(x(t; v), p(t; v))$ has an extension $(x^-(t; v), p^-(t; v))$ to a slightly larger interval, in fact to an interval $[t''_-, t'_-]$, $t''_- < \hat{t}, t'_- > \hat{t}$, with $v \to (x^-(t; v), p^-(t; v))$ being continuously differentiable near $\hat{v}$ for all t in $[t''_-, t'_-]$. Similarly, the solution $(x(t; v), p(t; v)), t > \hat{t}, t$ close to $\hat{t}$ has an extension $(x^+(t; v), p^+(t; v))$ to an interval $[t''_+, t'_+]$, $t''_+ < \hat{t} < t'_+$, with $v \to (x^+(t; v), p^+(t; v))$ being continuously differentiable near $\hat{v}$, for all t in $[t''_+, t'_+]$. By the nontangentiality condition (2.38), a C^1-function $\hat{T}(v)$, v near $\hat{v}$, exists such that $\hat{T}(\hat{v}) = \hat{t}$, and $\phi_i(\hat{T}(v), x^-(\hat{T}(v); v), p^-(\hat{T}(v); v)) = 0$. Now, for $t^* > \hat{t}$, t^* close to $\hat{t}$, as $x^+(t^*; v) = x^-(\hat{T}(v), v) + \int_{\hat{T}(v)}^{t^*} \dot{x}^+(s; v) ds$, then, neglecting the small term (integral) arising from differentiating under the integral sign,

$$x_v^+(t^*; \hat{v}) \approx -\dot{x}^+(\hat{t}; \hat{v})\hat{T}'(\hat{v}) + \dot{x}^-(\hat{t}; \hat{v})\hat{T}'(\hat{v}) + x_v^-(\hat{t}; \hat{v}). \tag{2.40}$$

By the fact that $u(\hat{t}-; \hat{v})$ and $u(\hat{t}+; \hat{v})$ maximize the Hamiltonian pf at $\hat{t}$, then $p(\hat{t}; \hat{v})\dot{x}^+(\hat{t}; \hat{v})) = p(\hat{t}; \hat{v})\dot{x}^-(\hat{t}; \hat{v})$, and multiplying by $p(\hat{t}; \hat{v})$ in (2.40), we get $p(\hat{t}; \hat{v})x_v^+(t^*; \hat{v}) \approx p(\hat{t}; \hat{v})x_v^-(\hat{t}; \hat{v})$. Letting $t^* \downarrow \hat{t}$,

$$p(\hat{t}; \hat{v})x_v^+(\hat{t}; \hat{v}) = p(\hat{t}; \hat{v})x_v^-(\hat{t}; \hat{v}). \tag{2.41}$$

Evidently, by the extension property, continuity of $v \to (p(T; v), x(T; v))$ at $\hat{v}$, existence of $x_v(T; \hat{v})$, and continuity of $t \to x_v(t; \hat{v})$ at $t = T$ hold for all $\hat{v} \in Q^0 \setminus Z^*$. Then as in the proof of Theorem 2.16,

$$x_i(T; v) = x_i^1 \text{ for } v \text{ close to } \hat{v} \in Q^0 \setminus Z^* \tag{2.42}$$

if the first inequality in (2.26) is strict for $(s, y) = \hat{v}$. For $(s, y) = \hat{v} \in Q^0 \setminus Z^*$, by this result it follows from (2.26), (2.27) that (2.32) holds, as in the proof of Theorem 2.16.

Now, for $(s, y) = \hat{v} \in Q^0 \setminus Z^*$, $w_t = 0$ and $\hat{w}_t = 0$ for $t \notin Q_{\hat{v}}$, by the same argument as in the proof of Theorem 2.16, and by (2.41) $w(t; s, y)$ and $\hat{w}(t; s, y)$ are constant for all $t \in [s, T]$, the constants being equal to zero, by (2.32). From this, as in the proof of Theorem 2.16, it follows that $W(s, y)$ satisfies the HJB equation in $Q^0 \setminus Z^*$. Furthermore, by definition of W and consistency, $t \to W(t, x(t; 0, x^0))$ is constant on $[0, T)$. Finally, by assumption, $S(x(T; s, y))$ is locally Lipschitz continuous on $\{(0, x^0)\} \cup Q^0$ and Z^* is slim. Hence, Remarks 2.10 and 2.11 yield optimality of $u(t; 0, x^0)$. $\square$

Remark 2.26 has in fact the same proof, existence, and continuity with respect to t, of $x_s(t; \hat{v})$ and $x_y(t; \hat{v})$ at $t = T$ now being postulated. In this case, the extension properties are not needed for $t = T$, hence allowing the replacement of cl$A$ by cl$A \cap ((0, T) \times \mathbb{R}^{2n})$.

Proof of Remark 2.24 (sketch). Define $Z_i = \{v \in Q^0 : \phi_i(T, x(T; v), p(T; v)) = 0\}$. First, note that for any $\hat{v}$ in Z', hence $\hat{v} \in Z_i$ for some i, we have enough differentiability for the arguments in proof of Theorem 2.16 to work near T, if $\hat{T}_i'(\hat{v}) = 0$, where $\hat{T}_i(v)$ is a C^1-function defined in a ball B around $\hat{v}$, such that $\hat{T}_i(\hat{v}) = T$ and, for all $v \in B$, $\phi_i(\hat{T}_i(v), x(\hat{T}_i(v); v), p(\hat{T}_i(v); v)) = 0$. To see this, let $(x^-(.;v), p^-(.;v))$ be an extension of $(x(.;v), p(.;v))$ slightly beyond T, v close to $\hat{v}$. As $x(\hat{T}(v); v) = x^-(\hat{T}(v); v)$, then $x(T; v) - x^-(\hat{T}(v); v)$ is of the second order, hence $x_v(T, \hat{v})$ exists, and equals $[(d/dv)x^-(\hat{T}(v); v)]_{v = \hat{v}} = [(\partial/\partial v')x^-(T; v')]_{v' = \hat{v}}$. Furthermore, $t \to x_v(t, \hat{v}) = x_v^-(t, \hat{v})$, $t \leq T$ is continuous at $t = T$. So it is only for $\hat{v}$ in $\check{Z}_i := \{\hat{v} \in Z_i : \hat{T}_i'(\hat{v}) \neq 0\}$ for some i that differentiability of $v \to x(T; v)$ may fail at $\hat{v}$. But locally, $\check{Z}_i$ is slim as we shall see. (We could have, in an obvious way, "localized" the requirement of slimness in Remark 2.10.)

If $\hat{v} \in Z_i$ (with $\hat{T}_i(v)$ and B as above), for some ball $\hat{B} \subset B$ around $\hat{v}$, we have that $\hat{T}_i'(v) \neq 0$ for $v \in \hat{B}$, and $Z_i \cap \hat{B} \subset \{v \in \hat{B} : \hat{T}_i(v) = T\}$. But the latter set (and so $Z_i \cap \hat{B}$) is slim in $\hat{B}$ (in $\hat{B}$, $v \to \hat{T}_i(v) - T$ behaves as a ψ-function).

Hence, Remark 2.24 follows.   □

Remark 2.27 (The Lipschitz continuity of $S(x(T; s, y))$*). In Remark 2.23, if $p(s; s, y)$ is locally Lipschitz continuous on $\{(0, x^0)\} \cup Q^0$ and "general consistency" holds, i.e., $(x(t'; s, y), p(t'; s, y)) = (x(t'; t, x(t; s, y)), p(t'; t, x(t; s, y)))$, $t \geq t' > s$, then $S(x(T; s, y))$ is locally Lipschitz continuous here. To see this, note that for any $\hat{v} = (\hat{s}, \hat{y}) \in \{(0, x^0)\} \cup Q^0$, by the extension property of $\hat{H}$, $\hat{H}_p$ and $\hat{H}_x$ and their first derivatives with respect to x and p are bounded on a open neighborhood B of $(\hat{v}, p(\hat{s}; \hat{v}))$ (of course, only a finite number of extensions is what needs to be considered). Choose a $\check{s}$ such that $\phi_i(\check{s}, x(\check{s}; \hat{s}, \hat{y}), p(\check{s}; \hat{s}, \hat{y})) \neq 0$ for all i, and such that for all $t \in (\hat{s}, \check{s}]$, all (s, y) near $(\hat{s}, \hat{y})$, $(t, x(t; s, y), p(t; s, y)) \in B$. Because the right-hand side of the equation for $(\dot{x}, \dot{p})$ is bounded and Lipschitz continuous in (x, p) for (t, x, p) in B, then, evidently, by Gronwall's inequality, $(x(\check{s}; s, y), p(\check{s}; s, y))$ is Lipschitz continuous in (s, y) near $(\hat{s}, \hat{y})$. Similarly, an open neighborhood B' of $(T, x(T; \hat{v}), p(T; \hat{v}))$ exists, such that all "needed" extensions of $\hat{H}_p$ and $\hat{H}_x$ and their first derivatives are bounded on B' (for any set $A = \cap_i \Phi^i$ of the type of Remark 2.23, for which $(T, x(T; \hat{v}), p(T; \hat{v})) \in \mathrm{cl}A$, an extension is needed). Choose a $T' < T$ such that $(t, x(t; \hat{v}), p(t; \hat{v})) \in B'$ for $t \in [T', T]$, and $\phi_i(T', x(T'; \hat{v}), p(T'; \hat{v})) \neq 0$ for all i. Because $y'' \to (x(T'; \check{s}, y''), p(T'; \check{s}, y''))$ is C^1 near $x(\check{s}; \hat{v})$, $v \to (x(T'; v), p(T'; v))$ is Lipschitz continuous near $\hat{v}$. But then, by Gronwall's inequality (and, similar to what happened on B, by boundedness and Lipschitz continuity on B'), for $t \in [T', T]$, $(x(t; T', x(T', v)), p(t; T', x(T', v)))$ belongs to B' for v close to $\hat{v}$, and $v \to (x(T; v), p(T; v)) = (x(T; T', x(T', v)), p(T; T', x(T', v)))$ is Lipschitz continuous near $\hat{v}$. The asserted property of $S(x(T; s, y))$ follows.   □

Remark 2.28 (The C^1-property of $(x(t; s, y), p(t; s, y))$*). It suffices to assume in Remark 2.23, for any $(s, y) \in Q^0 \setminus \hat{Z}$, for some $t > s$, t arbitrarily close to s, that $(s', y') \to (x(t; s', y'), p(t; s', y'))$ is C^1 in a neighborhood of (s, y). The last property is evidently equivalent to $p(s'; s', y')$ being C^1 near (s, y).

To sketch how this is proved, consider again the proof of Remark 2.23. Let $\hat{v} = (\hat{s}, \hat{y})$ be a point in $Q^0 \setminus \hat{Z}$. First, note that if $\check{s}$ is the first point in $Q_{\hat{v}}$ to the right of $\hat{s}$, then evidently, $(s, y) \to (x(t; s, y), p(t; s, y))$ is C^1 near $\hat{v}$, for any $t \in (\hat{s}, \check{s})$. Observe also that because $(s, y) \to (x(t; s, y), p(t; s, y))$ is C^1 for (s, y) close to $(\hat{s}, \hat{y})$, for t close to $\check{s}$, then, by (2.31) and the existence of $\hat{T}(v)$, this C^1-property also holds for $t > \check{s}, t$ close to $\check{s}$, for (s,y) close to $(\hat{s},\hat{y})$. By extension of this argument, $(x(t;s,y),p(t;s,y))$ is C^1 in $Q^0 \setminus \hat{Z}$ close to $(\hat{s},\hat{y})$, for any $t \notin Q_{\check{s},\check{s}}, t \in (\hat{s},T)$. From this the assertion follows. $\square$

Remark 2.29 (Changed end conditions*). The conclusions of Theorem 2.16 and Remark 2.23 also hold when the end conditions are changed to read $g^i(x(T)) = 0, i = 1,\ldots,n', g^i(x(T)) \geq 0, i = n'+1,\ldots,n''$, where g^i are C^1. Then the tranversality conditions on $p(t;s,y)$ read $p(T;s,y) = \sum_i \Lambda^i(s,y)g_x^i(x(T;s,y)) + S_x(x(T;s,y))$, where $\Lambda^i(s,y)$ are certain numbers, being, for $i > n', \geq 0$ and even equal to zero if $g^i(x(T;s,y)) > 0$. It is assumed that $\Lambda^i(s,y)$ is continuous in Q^0 for $i > n'$. $\square$

(A new proof of (2.32) is needed in this case: Now $\sum_i \Lambda^i(s,y)g_x^i(x(T;s,y)) = \sum_{\Lambda^i(s,y)\neq 0} \Lambda^i(s,y)g_x^i(x(T;s,y))$. When $\Lambda^i(s,y) > 0$ for some $i > n'$, then $\Lambda^i(s',y') > 0$ for (s',y') near (s,y), so $g^i(x(T;s',y')) = 0$ for all such (s',y'). This also holds for $i \leq n'$. Hence, $(\partial/\partial y)g^i(x(T;s,y)) = 0$ and hence $\sum_i \Lambda^i(s,y)g_x^i(x(T;s,y))x_y(T;s,y) = 0$. A similar argument holds for $(\partial/\partial s)$, so (2.32) follows.)

Remark 2.30 (The domain of definition of f and f_0*). In Theorem 2.16, f needs only be defined on $Q^0 \times U$, provided $Q^* = \{(t,x,p) \in Q^* : (t,x) \in Q^0\}$, and for Remark 2.21, add that even f_0 needs only be defined on $Q^0 \times U$. Similar comments pertain to Remarks 2.23 and 2.25. Moreover, in Theorems 2.8, 2.6, and in Remark 2.13, it suffices that f_0 and f are continuous, so no differentiability assumptions are needed ($f_0 = 0$ in the theorems). Assuming this, in Remarks 2.23, 2.25, it is only needed to assume that f_x and f_{0x} exist and are C^0 for $(t,x) \in Q^0 \setminus \hat{Z}, u \in U$. $\square$

Example 2.31. Consider the problem

$$\max \int_0^2 tu\,dt, \dot{x} = u \in [0,1], x(0) = 0, x(2) \leq 1.$$

Solution. The solution is trivial, let us nevertheless apply Remark 2.25 to solve it. Let $Q^0 = (0,2) \times (-1,1)$. Evidently, p is independent of t, so $p(t;s,y) = p(s,y) \leq 0$. When $y > s-1$, we can put $p(s,y) = -y-1$, $u(t;s,y) = 1$ for $t \in (1+y,2)$, $u(t;s,y) = 0$ for $t \in (s,1+y)$, $x(t;s,y) = \max\{y,t-1\}$. If $y \leq s-1$, $p(s,y) = 0$ and $x = y+t-s$. Evidently, $\hat{H}$ is C^2 in $Q^* := (0,2) \times \mathbb{R} \times \mathbb{R}$, except at points (s,y,p) where $s+p = 0$. Hence, let $\phi(s,y,p) := s+p$. Then $Z' = \emptyset$, and $\hat{Z} = \{(s,s-1) : s \in (0,2)\}$ is slim. Moreover, $(s,y) \rightarrow (x(t;s,y),p(t;s,y))$ is C^1 near any (s,y) such that $y \neq s-1$, for $t \in (s,2), t \notin \{t \in (s,2) : t-1 = y\}$, (the last set contains $Q_{s,y} = \{t \in (s,2) : \phi(t,x(t;s,y),p(t;s,y)) = 0\}$). Finally, $(d/dt)\phi(t,x(t;s,y),p(t;s,y)) = (d/dt)(t+p(s,y)) = 1 \neq 0$, so (2.38) is satisfied. Now, $p(s;s,y)$ is discontinuous, but $x_0(t;s,y) = \int_{\max\{s,y+1\}}^t r\,dr$ is locally Lipschitz continuous in $\{(0,0)\} \cup Q^0$ for $t = 2$ and C^1 in $(s,y) \in Q^0 \setminus \hat{Z}$, for $t \in (s,2), t \neq y+1$. In fact, all conditions in Remark 2.25 hold. $\square$

2.3 Infinite Horizon

In the original problem (2.1)–(2.4), let us replace T by ∞. The terminal conditions must then be modified to read:

$$\lim_{t \to \infty} x_i(t) = x_i^1, i = 1,\ldots,l, \tag{2.43}$$

(the limit required to exist), and

$$\liminf_{t \to \infty} x_i(t) \geq x_i^1, i = l+1,\ldots,m. \tag{2.44}$$

Some of the x_i^1's, $i = l+1,\ldots,m$ may be ∞. We now want to maximize $\lim_{T \to \infty} S(x(T))$. However, we get into trouble if this limit does not exist, so instead here we use catching up optimality: We define the pair $(x^{u^0}(.), u^0(.))$ to be catching up optimal in the infinite horizon problem if

$$\liminf_{s \to \infty} \{S(x^{u^0}(s)) - S(x^u(s))\} \text{ is } \geq 0 \text{ for all admissible pairs } (x^u(.), u(.)). \tag{2.45}$$

In the case where $\lim_{s \to \infty} S(x^u(s))$ exists and is finite for all admissible $x^u(t)$, catching up optimal is the same as "ordinarily optimal", i.e., giving the largest value to $\lim_{T \to \infty} S(x^u(T))$.

The next theorem is the infinite horizon variant of Theorem 2.8:

Theorem 2.32 (Verification theorem, infinite horizon). *Let $W(s,y)$ be a C^1 function on a set $\hat{Q}$ in (t,x)-space that satisfies*

$$\liminf_{T \to \infty} [W(T, x(T)) - S(x(T))] \geq 0 \tag{2.46}$$

for all $(0, x^0)$-admissible solutions $x(t)$ that stays in $\hat{Q}$, (i.e., $(t, x(t)) \in \hat{Q}$ for $t \in [0, T]$). Assume also that W satisfies the HJB equation (2.16) in $\hat{Q}$. Furthermore, let $(x^*(t), u^*(t))$ be an $(0, x^0)$-admissible solution staying in $\hat{Q}$ such that

$$\liminf_{T \to \infty} [S(x^*(T)) - W(T, x^*(T))] \geq 0 \tag{2.47}$$

and such that

$$W_s(t, x^*(t)) + W_y(t, x^*(t)) f(t, x^*(t), u^*(t)) = 0 \quad v.e. \tag{2.48}$$

Then $(x^*(.), u^*(.))$ is catching up optimal in the set of all admissible pairs $(x(.), u(.))$ for which $x(.)$ stays in $\hat{Q}$. $\square$

Proof. As before, $W(t, x^*(t)) = W(0, x^0) \geq W(t, x(t))$ ($x(t)$ admissible). Then $S(x^*(t) - S(x(t)) \geq S(x^*(t)) - W(t, x^*(t)) + W(t, x(t)) - S(x(t))$. Taking liminf on both sides and using (2.46) and (2.47) yields catching up optimality of $(x^*(.,.), u^*(.,.))$. $\square$

Note that when f is independent of t, and a criterion $\int_0^\infty f^0(x(t),u(t))e^{-\alpha t}dt$, $\alpha >$ 0, is going to be maximized (note that f^0 is independent of t), then this is the same as maximizing $\lim_{T\to\infty} x_0(T)$, $\dot{x}_0 = f^0 e^{-\alpha t}$, $x_0(0) = 0$. (If necessary, turn to catching up optimality.) In this case, we can expect $W(s,y)$ to be of the form $h(y)e^{-\alpha s}$, where $h(.)$ satisfies the "current value" HJB equation $0 = -\alpha h + \max_u\{f^0 + h_x f\}$, which in the scalar case (x one-dimensional) is an ordinary differential equation in h. (Note that $h(.)$ satisfies the last equation iff $W = h(x)e^{-\alpha t}$ satisfies the ordinary HJB equation.)

2.4 Free Terminal Time Problems, with Free Ends

Consider the scrap value problem $\max S(T,x(T))$ (where S is C^1), subject to (2.2),(2.3), where we allow T to be subject to choice in a given set $[T_1,T_2]$, $T_2 > T_1 \geq 0$ (the end state is free). Assume for the moment that $[0,T]$ is replaced by $[s,T]$, and, moreover that the maximum of $S(T,x^u(T))$ is found for all triples $(x^u(t),u(t),T)$ where $x(.)$ starts at (s,y) and T belongs to $[\max\{s,T_1\},T_2]$. (Such a triple is called (s,y)-admissible and "admissible" means $(0,x^0)$-admissible.) This problem is denoted $P^{s,y}$, and the maximal value of its criterion is denoted $W^*(s,y)$. Let the set G^* be defined by:

$$G^* = \{(s,y) \in [T_1,T_2] \times \mathbb{R}^n, W^*(s,y) > S(s,y)\}. \tag{2.49}$$

By definition of G^*, $W^*(s,y) \leq S(s,y)$ for $(s,y) \in G_* := ([T_1,T_2] \times \mathbb{R}^n) \setminus G^*$. But, of course, the optimal value is never strictly smaller than $S(s,y)$, because we always have the possibility to stop immediately at (s,y) and get $S(s,y)$. So, $W^*(s,y) = S(s,y)$ for $(s,y) \in G_*$ and we can define $G_* = \{(s,y) : W^*(s,y) = S(s,y), s \in [T_1,T_2]\}$. Moreover, it is natural to call G^* the continuation region: We don't stop immediately at (s,y) when we are in G^*, because by continuing, we can get something better: $W^*(s,y)$ instead of $S(s,y)$.

Let us write down some relationships that can be expected to hold. Assume that $W^*(s,y)$ is continuous in (s,y). Then $W^*(s,y)$ satisfies:

$$W^*(s,y) = S(s,y) \text{ for } (s,y) \in \partial G_* \tag{2.50}$$

where ∂G_* denotes the boundary of G_* (in fact, G_* is a closed set).

At each point $(s,y) \in ([0,T_1] \times \mathbb{R}^n) \cup G^*$ at which $W^*(s,y)$ is differentiable in (s,y), the following equality (HJB equation) holds for $W = W^*$:

$$0 = W_s(s,y) + \sup_{u \in U} W_y(s,y) f(s,y,u). \tag{2.51}$$

(At least it holds if $P^{s,y}$ has an optimal control.) The argument for this relationship is as before.

Let $(x(.;s,y),u(.;s,y),T)$ denote an optimal triple in problem $P^{s,y}$, the functions being defined on $[s,T]$, with $T \in [\max\{s,T_1\},T_2]$, $x(s;s,y) = y$. If we stop as soon as possible in case we are indifferent to continuing or stopping, then, if $T > s$, $(T, x(T; s, y))$ belongs to ∂G_*: The point $(T, x(T; s, y))$ does not belong to G^*, otherwise, it would be strictly better to continue a little longer. But it cannot belong to the interior of G_*, because then we could as well have stopped earlier.

Let us define $u(t; s, y) = u$ for $t > T$, u any given vector in U. Letting $(d^{\pm}/dT)$ denote right (+) and left (-) derivatives, by optimality, it follows that $(\pm 1)(d^{\pm}/dT)$ $S(T, x(T; s, y)) \leq 0$ if $T \in (\max\{s, T_1\}, T_2)$. Moreover, $(d^-/dT)S(T, x(T; s, y)) \geq 0$ if $T = T_2$, and $(d^+/dT)S(T, x(T; s, y)) \leq 0$ if $T = \max\{s, T_1\}$. (If $T = s$, of course $x(T; s, y) = x(s; s, y) = y$.) Now, $(d^{\pm}/dT)S(T, x(T; s, y)) =$

$$S_T(T, x(T; s, y)) + S_x(T, x(T, s, y))f(T, x(T; s, y), u(T\pm; s, y)).$$

Let the admissible set Q now consist of all points $(s, x^u(s))$, $s \in (0, T)$, where $x^u(.)$ belongs an admissible triple $(x^u(.), u(.), T)$, and let D^* be the set of all $(T, y) \in$ cl$Q$, $T \in [T_1, T_2]$ that satisfy the following conditions:

$$S_T(T, y) + \sup_{u \in U} S_x(T, y)f(T, y, u) = 0 \text{ if } T \in (T_1, T_2),$$
$$(\geq 0 \text{ if } T = T_2, \leq 0 \text{ if } T = T_1). \tag{2.52}$$

Not seldom, we can expect that $D^* = \partial G_*$, because, roughly, both sets contain points (T, y) at which it does not pay to continue (if possible, i.e., $T < T_2$), nor stop at an earlier point in time (if possible, i.e., $T > T_1$). The solution method then consists in finding a solution W of the HJB equation on an open subset D containing $Q \cap$ $((0, T_1) \times \mathbb{R}^n)$, such that $D^{**} := \text{cl}Q \cap (\partial D) \cap ([T_1, T_2] \times \mathbb{R}^n) \subset D^*$, and such that W is C^1 in D, continuous on cl$D$ and satisfies $W(s, y) > S(s, y)$ for $(s, y) \in D$, $s \geq T_1$, and $W(s, y) = S(s, y)$ for $(s, y) \in D^{**}$. In fact, a sufficient condition is connected to this method, we then need to put in further conditions, on the other hand, though (2.52) (or $D^{**} \subset D^*$) is useful when constructing a solution W, it need not formally be included in the theorem to follow.

Theorem 2.33 (Sufficient condition for the HJB equation, free end, free terminal time). *Assume that open subsets Q'' and D of $\mathbb{R}^{n+1}$ and a function $W(s, y)$, defined on Q'' have been found, $Q \subset Q''$, $D \subset Q''$, $W(s, y)$ being C^1 in D and in $Q'' \backslash \text{cl}D$, and locally Lipschitz continuous in clQ'', satisfying the HJB equation (2.51) in D, and, for $s \geq T_1$, satisfying $W(s, x) > S(s, x)$ if (s, x) in $Q \cap D$, $W(s, x) \geq S(s, x)$ if (s, x) in cl$Q$. Moreover, assume that $Q \cap ((0, T_1) \times \mathbb{R}^n) \subset D$. Assume, furthermore the "HJB inequality":*

$$0 \geq W_s + \sup_{u \in U} W_x(s, y)f(s, y, u) \tag{2.53}$$

for $(s, y) \in Q \backslash \text{cl}D$, $s \in (T_1, T_2)$. Assume also that $W(t, x) = S(t, x)$ for $(t, x) \in$ $(\text{cl}Q) \cap \partial D$, $t \geq T_1$. Moreover, assume that $\partial^* D := [(T_1, T_2) \times \mathbb{R}^n] \cap (\partial D) \cap Q''$ is slim in Q''. Assume, finally, that there exists a control function $u^*(s)$, with corresponding solution $x^*(s)$ defined on some interval $[0, T^*]$, $T^* > 0$, $T^* \in [T_1, T_2]$, which yields the supremum in the HJB equation (2.51) for $(s, y) = (s, x^*(s)) \in D$, and such that

$T^* = \sup\{t > 0 : (s, x^*(s)) \in D \text{ for } s \in (0, t)\}$. *Then* $(x^*(t), u^*(t), T^*)$ *is an optimal triple in the class of all admissible triples* $(x(t), u(t), T)$, $T \in [T_1, T_2]$. $\square$

Remark 2.34 (Solution method for the HJB equation). As before, a standard method for solving the HJB equation is to obtain solutions of the characteristic equations, for (s, y) in an open set Q^0 containing Q. In this case, imagine that solutions of the characteristic equations, denoted $x(t; s, y, T)$, $p(t; s, y, T)$, with corresponding controls $u(t; s, y, T)$, defined for t in $[s, T]$, have been found for any (s, y) in Q^0, any T in $[\max\{s, T_1\}, T_2]$, satisfying $p(T; s, y, T) = S_x(T, x(T; s, y, T))$. Next, assume that points $T = T(s, y)$ in $[\max\{s, T_1\}, T_2]$ have been found satisfying $\eta(T; s, y) :=$

$$S_T(T, x(T; s, y, T)) + \max_u S_x(T, x(T; s, y, T)) f(T, x(T; s, y, T), u) = 0$$

if $T \in (\max\{s, T_1\}, T_2)$, ≥ 0 if $T = T_2$, ≤ 0 if $T = \max\{s, T_1\}$. (Hopefully, such a point T exists.) For $T = T(s, y)$, write $x(t; s, y) = x(t; s, y, T(s, y))$ $p(t; s, y) = p(t; s, y, T(s, y))$. Note that the function $W(s, y) := S(T(s, y), x(T(s, y); s, y))$ automatically satisfies the HJB equation for $s < T(s, y)$, provided enough differentiability is exhibited by the "characteristic entities" $x(t; s, y), p(t; s, y)$ and $T(s, y)$, $(s, y) \in \{(s, y) \in Q^0 : T(s, y) > s\}$. Defining $D = \{(s, y) \in Q^0 : s < T(s, y)\}$ and $Q'' = Q^0$, one may then test if the conditions in Theorem 2.33 hold, with the reasonable proposal $W(s, y) = S(s, y)$ for (s, y) such that $s = T(s, y)$. A sufficiency result formulated directly in terms of the "characteristic entities" can actually be stated; it is omitted here (but see Chapter 3, Remark 3.54). $\square$

Remark 2.35. Note that if $T_1 = 0$ and a function W has been found, satisfying $W(0, x^0) = S(0, x^0)$ and both (2.53) and $W(s, y) \geq S(s, y)$ in Q, being C^1 in Q, then $T^* = 0$ is optimal. $\square$

Proof of Theorem 2.33 (sketch). Similar to the proof of Theorem 2.6, (d/ds) $W(s, x^*(s)) = 0$ for s in $(0, T^*)$, and for any $(0, x^0)$-admissible triple $(x(.), u(.), T)$, $(d/ds)W(s, x(s)) \leq 0$ for any $(s, x(s)) \notin \partial^* D$, s in $(0, T)$, $s \neq T_1$. If the last inequality holds, except perhaps for a countable number of points s, we can integrate and get $S(T^*, x(T^*)) = W(T^*, x(T^*)) = W(0, x^0) \geq W(T, x(T)) \geq S(T, x(T))$. If the set of exceptional points is uncountable, an argument similar to that in the proof of Remark 2.10 can be used to give the same result. Hence, $(x^*(.), u^*(.))$ is optimal. $\square$

Proof of Remark 2.34 (sketch). The formulations in the remark are somewhat sketchy, yet, let us give some explanations of the assertions. Fix $(\hat{s}, \hat{y})$ in Q^0, such that $\hat{s} < \hat{T} := T(\hat{s}, \hat{y}) \in (T_1, T_2)$, write $\hat{x}(t; s, y) = x(t; s, y, \hat{T})$, and let $\hat{W}(s, y) = S(\hat{T}, \hat{x}(\hat{T}; s, y))$. Evidently, at $(\hat{s}, \hat{y})$, $\hat{W}(s, y)$ satisfies the HJB equation, by the fixed horizon results. Then also the function $W(s, y) := S(T(s, y), x(T(s, y); s, y))$ satisfies the HJB equation at $(\hat{s}, \hat{y})$, because $\hat{W}_s = W_s$ and $\hat{W}_y = W_y$, as we shall see. Extend $t \to \hat{x}(t; s, y)$ a little beyond $\hat{T}$ in such a way that $[(\partial^-/\partial t)\hat{x}(t; s, y)]_{t=\hat{T}} = [(\partial^+/\partial t)\hat{x}(t; s, y)]_{t=\hat{T}}$. Then, note that

$$\eta(\hat{T}, \hat{s}, \hat{y}) := [(\partial/\partial T)S(T, x(T; s, y))]_{T=\hat{T}, s=\hat{s}, y=\hat{y}}$$
$$= S_T(\hat{T}, \hat{x}(\hat{T}; \hat{s}, \hat{y})) + S_x(\hat{T}, \hat{x}(\hat{T}; \hat{s}, \hat{y}))[(\partial/\partial t)\hat{x}(t; \hat{s}, \hat{y})]_{t=\hat{T}}.$$

Hence, if $\eta(T(\hat{s},\hat{y});\hat{s},\hat{y}) = 0$ (which does hold if $T(\hat{s},\hat{y}) \in (T_1,T_2)$), then at $(\hat{s},\hat{y})$,

$$W_s(\hat{s},\hat{y}) := (\partial/\partial s)S(T(\hat{s},\hat{y}),x(T(\hat{s},\hat{y});\hat{s},\hat{y})) = S_2(\hat{T},\hat{x}(\hat{T};\hat{s},\hat{y}))\hat{x}_s(\hat{T};\hat{s},\hat{y})$$
$$=: \hat{W}_s(\hat{s},\hat{y})$$

and

$$W_y(\hat{s},\hat{y}) := (\partial/\partial y)S(T(s,y),x(T(s,y);s,y)) = S_2(\hat{T},\hat{x}(\hat{T};\hat{s},\hat{y}))\hat{x}_y(\hat{T};\hat{s},\hat{y})$$
$$=: \hat{W}_y(\hat{s},\hat{y}).$$

At least, these equalities evidently hold if $T(s,y)$ is differentiable at $(\hat{s},\hat{y})$, and it is no problem to assume this to hold if $T(\hat{s},\hat{y}) \in (T_1,T_2)$. Considering next the case where $T(\hat{s},\hat{y})$ equals T_1 or T_2, the equalities also hold if $T(s,y)$ equals T_1 or T_2 even in a neighborhood of $(\hat{s},\hat{y})$. In this case, it seems more problematic if no such neighborhoods exist. But then, normally, $\eta(T(\hat{s},\hat{y}),\hat{s},\hat{y}) = 0$, and all that is needed for the equalities to hold when $\eta(T(\hat{s},\hat{y}),\hat{s},\hat{y}) = 0$ is that $T(s,y)$ is locally Lipschitz continuous, which can be assumed without problem (i.e., most often this will turn out to be the case in the problem at hand). $\square$

Remark 2.36 (Restrictions on the solutions*). Assume that all solutions are required to belong to a given set $\hat{Q}$ in $\mathbb{R}^{n+1}$ (i.e., in (t,x)-space). Replace Q by $Q \cap \hat{Q}$ in Theorem 2.33 and assume that $(t,x^*(t)) \in \hat{Q}, t \in (0,T^*)$. Then $(x^*(t),u^*(t))$ is optimal among all admissible triples $(x(.),u(.),T)$ for which $(t,x(t)) \in \hat{Q}, t \in (0,T)$. $\square$

Remark 2.37 ($T_2 = \infty$). If $T_2 = \infty$, and $T^* < \infty$, Theorem 2.33 yields optimality among all admissible triples $(x(.),u(.),T), T \in [T_1,\infty)$. $\square$

Example 2.38. Consider the problem

$$\max \int_0^T [-u^2 - x - t^2]dt \text{ s.t. } \dot{x} = u < 0, x(0) = -4, T \geq 0, x(T) \text{ free.}$$

Solution. We rewrite the problem by letting $\dot{x}_0 = -u^2 - x - t^2$, $x_0(0) = 0$, in which case we want to maximize $x_0(T)$. We solve the problem for the initial point $(0,0,-4)$ replaced by (s,y_0,y), $s \geq 0$, $y,y_0 \in \mathbb{R}$. Using the maximum principle, with p and p_0 as adjoint variable corresponding to x and x_0, we get $\dot{p}_0 = 0$, $p_0(T) = 1$, $p_0(t) \equiv 1$, so $H = -u^2 - x - t^2 + pu$. Write $\tilde{y} = (y_0,y)$. Next, $\dot{p} = 1$, $p(t;s,\tilde{y},T) := p(t) = t - T$, and $u = (t-T)/2$ maximizes H. Hence, $x(t;s,\tilde{y},T) = y + (t-T)^2/4 - (s-T)^2/4$, $W(s,\tilde{y}) := x_0(T;s,\tilde{y}) = y_0 + \int_s^T (-y - (t-T)^2/2 + (s-T)^2/4 - t^2)dt = y_0 - y(T-s) - (s-T)^3/12 - T^3/3 + s^3/3 =: \phi(s)$, where $T = T(s,\tilde{y}) = \max\{s,(-y)^{1/2}\}$ for $y < 0$. The last formula follows from the fact that the equality in (2.52) reduces to $-y - T^2 = 0$. For $y \geq 0$, surely $T = T(s,y) = s$ by (2.52). (In general, we have $T = T(s,y) = \max\{s,(\max\{0,-y\})^{1/2}\}$.) By Remark 2.34 we know that the HJB equation is satisfied for (s,y) such that $s > 0$ and $s^2 + y < 0$. Moreover, for $s = T(s,y)$, i.e., $y + s^2 \geq 0$ we have that $0 \geq \sup_u\{-u^2 - y - s^2\}$, i.e., (2.53) is satisfied for $W = y_0$. Evidently, the set $\{(s,y) : s > 0, s^2 + y = 0\}$ is slim. For $(s,y) \in D := \{(s,y_0,y) : s > 0, y < 0, s^2 + y < 0, y_0 \in \mathbb{R}\}$, $W(s,\tilde{y}) > S(\tilde{y}) = y_0$, because $\phi((-y)^{1/2}) = y_0$ and $\phi'(s) < 0$, for $s < (-y)^{1/2}$. Now, for $(s,\tilde{y}) = (0,0,-4)$,

$x(T;0,0,-4) = -4 - T^2/4$, and then $T^2 + x(T;0,0,-4)) = 0$ gives $T = 4/3^{1/2}$. Remark 2.37 gives optimality of $u = (t - 4/3^{1/2})/2$.   □

Further reading. Fields of extremals has quite a long history, we here mention only the treatment of this subject in Bliss (1946), Boltyanskii (1971), and Fleming and Rishel (1975). These books also discuss the HJB equation.

2.5 Exercises

Exercises for Section 2.1

2.1. Solve the problem

$$\max \left[\int_0^T u^{1/2} dt + x(T)^{1/2} \right], \dot{x} = -u < 0, x^0 \text{ given } > 0.$$

Hint: Rewrite the problem to read $\max[x_0(T) + x(T)^{1/2}], \dot{x}_0 = u^{1/2}, x_0(0) = 0$, and try a solution of the HJB equation of the form $W = x_0 + \phi(t)x^{1/2}$.

2.2. Solve the problem

$$\max[x_2(1) - (1/2)(x_1(1))^2], \text{ s.t. } \dot{x}_1 = u, x_1(0) = 0, \dot{x}_2 = u, x_2(0) = 0, u \in \mathbb{R}.$$

Hint: See solution presented at the end of Section 2.1.

2.3. Consider the problem

$$\max y(1), \dot{y} = ux, y(0) = 0, \dot{x} = 0, x(0) = x^0, y(1), x(1) \text{ free, } u \in [0,1].$$

The solution is evidently $u \equiv 1$ if $x^0 \geq 0$, $\equiv 0$ if $x_0 < 0$. Show that $x^0 \mapsto V(0, y^0, x^0)$, $y^0 = 0$ is not differentiable at $x^0 = 0$ (V the optimal value function).

Exercises for Section 2.2

2.4. Solve the following two problems, (a) and (b), by means of characteristic solutions.

$$\max \int_0^1 (-u^2 + x) dt, \dot{x} = u \in \mathbb{R}, x(0) = 0, \text{ (a): } x(1) = 1, \text{ (b): } x(1) \geq 1.$$

Hint for (b): Here Remark 2.23 is needed. The solutions $x(t;s,y)$ will sometimes satisfy $x(1;s,y) > 1$. A single ϕ-function is needed, namely, $\phi(s,y,p) = p^{(a)}(1;s,y)$, where $p^{(a)}(.;.,.)$ is the adjoint function in problem (a). For (s,y) such that $p^{(a)}(1;s,y) > 0$, by consistency in (a), $p^{(a)}(1;t,x(t;s,y))=p^{(a)}(1;s,y) > 0$. It may be explicitly checked (or argued as in (i) below) that this holds also for

$p^{(a)}(1;s,y) < 0$, in which case we in (b) are in the situation where $x(1;s,y) > 1$. Hence, the premise in (2.38) "never happens."

(i) $p^{(a)}(1;t,x(t;s,y)) = 0$ can never happen in case $p^{(a)}(1;s,y) < 0$. By contradiction, assume the last equality. Then there is an (a)-solution $x^{(a)}(.;t,x(t;s,y))$, that by uniqueness equals $x(.;t,x(t;s,y))$ $(= x(.;s,y)$ by consistency), and the single proposal for $x^{(a)}(.;s,y)$ would be $x(.,s,y)$, and by uniqueness again $x^{(a)}(.;s,y) = x(.,s,y)$. By consistency, $0 = p^{(a)}(1;x(t;s,y)) = p^{(a)}(1;t,x^{(a)}(t;s,y)) = p^{(a)}(1;s,y)$, a contradiction.

2.5. Consider the following problem

$$\max \left[\int_0^T -u^2 dt + x(1)^2 \right], \dot{x} = u > 0, x(0) = x^0 > 0, x(T) \text{ free.}$$

(a) Solve the problem by means of characteristic solutions for $T = 1/2$. (b) Let $T = 2$. Show that no optimal solution exists (show that an arbitrary large constant control yields an arbitrary large criterion value). Do characteristic solutions exist?

2.6. Solve the following problem by means of (a) Remark 2.10 and (b) Remark 2.23.

$$\max \int_0^1 (1-u)x dt, \dot{x} = u \in [0,1], x(0) = 1/2, x(1) \text{ free.}$$

Exercises for Section 2.3

2.7. Let $\alpha^2 > 8$, $\alpha > 0$, and solve the following problem by means of Theorem 2.32:

$$\max \int_0^\infty (-u^2/2 + x^2)e^{-\alpha t} dt, \dot{x} = u > 0, x(0) = x^0 > 0, x(\infty) \text{ free.}$$

Hint: Try $h(x) = ax^2$ in the current value HJB equation.

Exercises for Section 2.4

2.8. Solve the following problem by means of Theorem 2.33

$\max\{\int_0^T [-t/2 + u^{1/2}] dt + x(T)/2, T \in [0,8], \dot{x} = -u \in (-\infty,0], x(0) = 1, x(1)$ free.

2.9. (Theoretical problem) Consider again the proof of Remark 2.10. Define $d_-\alpha(s) = \liminf_{h\downarrow 0} h^{-1}[\alpha(s+h) - \alpha(s-h)]$. Show that slimness even implies the following version of system slimness: $(t+a, x(t)+b) = (t+a_h, x(t)+b_h) \notin Z^*$ holds for arbitrary small h, for v.e. $t \in (s-h, s+h)$, s any continuity point of $\dot{x}(s)$. From this, prove that $d_-W(s,x(s)) \leq 0$. Next, prove the monotonicity of $W(s,x(s))$ on any closed interval $[a,b]$ (and hence on $[c,d]$) at which $\dot{x}(s)$ is continuous, by covering $(a,b]$ by a finite number of disjoint intervals $(a_i, a_{i+1}]$ for which $W(a_i, x(a_i)) - W(a_{i+1}, x(a_{i+1})) \geq -\varepsilon(a_{i+1} - a_i)$, ε any given positive number.

Chapter 3
Piecewise Deterministic Optimal Control Problems

Piecewise deterministic control problems involve systems governed by deterministic differential equations but influenced by sporadic stochastic disturbances. In certain situations, in an otherwise deterministic control system, it may happen that the state jumps at certain stochastic points of time. Examples are sudden oil finds, or sudden rain, sudden discoveries of metal deposits, or perhaps a sudden war. Similarly, in seemingly deterministic processes, the dynamics may suddenly change character: At certain stochastic points in time, the right-hand side of the differential equation governing the system changes form, such changes being effected by jumps in a (dummy) state variable. Examples of such phenomena are sudden inventions, sudden ecological disasters, earthquakes, floods, storms, fires, the sudden capture of a criminal, which suddenly change the prospects of the firm, the society, the agriculture, the criminal, ... A systematic method for solving such problems, based on the HJB equation (the Hamilton–Jacobi–Bellman equation) for the problem, is presented in Davis (1993), *Markov Models and Optimization*, and also discussed below. In this chapter, a related method is presented first, which is closely connected to a common solution method in deterministic control theory. For problems with a bound on the number of possible jumps, it yields a recursive solution procedure.

First, stochastic control problems with a fixed horizon are discussed, where solution tools are fields of extremals (characteristic solutions) and a suitable version of the HJB equation. Similar solution tools are used also when we later on turn to optimal stopping problems, i.e., problems where also the horizon is subject to choice.

The chapter contains statements of several theoretical results. Proofs are mainly placed at the end of the chapter or completely omitted.

3.1 Free End, Fixed Terminal Time Problems

Consider the following control system:

$$\dot{x} = f(t, x, u), \quad t \in [0, T], \quad x(0) = x^0 \in \mathbb{R}^n, \quad u \in U \subset \mathbb{R}^r. \tag{3.1}$$

A. Seierstad, *Stochastic Control in Discrete and Continuous Time*,
DOI: 10.1007/978-0-387-76617-1_3,
© Springer Science+Business Media, LLC 2009

Here, the control region U, the initial point x^0, $f : \mathbb{R}^{1+n+r} \to \mathbb{R}^n$ and the terminal time T are fixed. Equation (3.1) is required to hold v.e. (i.e., for all t except a countable number of points). The vector u is a control, subject to choice in U. At certain jump time-points τ_i, $0 < \tau_1 < \tau_2 < \ldots$, the state jumps according to

$$x(\tau_j+) - x(\tau_j-) = g(\tau_j, x(\tau_j-)), \tag{3.2}$$

where g is a fixed function. (Here, $x(\tau_j+) = $ right limit, $x(\tau_j-) = $ left limit.) The points τ_i are random variables taking values in $(0, \infty)$. It will often be assumed that at most $N < \infty$ jump points, τ_j can occur in $[0, T]$. The stochastic assumptions on the jump points τ_i are as follows: Given τ_{i-1}, $i > 1$, τ_i is exponentially distributed in (τ_{i-1}, ∞) with parameter (jump intensity) $\lambda(i) > 0$, so the density of $\tau_i = \tau > \tau_{i-1}$ is $\lambda(i)e^{-\lambda(i)(\tau-\tau_{i-1})}$. Moreover, τ_1 is exponentially distributed in $(0, \infty)$ with parameter $\lambda(1)$. It is assumed that $\sup_i \lambda(i) < \infty$. (In some remarks, these stochastic assumptions will be generalized.) The case where a maximum of N jumps can occur is the case where $\lambda(i) = 0$, $i > N$. The problem to be solved is:

$$\max_{u(.,.)} E \left[\int_0^T f_0(t, x, u)dt + \sum_{\tau_j < T} g_0(\tau_j, x(\tau_j-)) + h_0(x(T-)) \right] \tag{3.3}$$

subject to (3.1), (3.2). There are no terminal conditions (later on such ones will be introduced). In (3.3), f_0, g_0, and h_0 are fixed functions, f_0 measures the running benefit obtained from the process, h_0 is a scrap value (or bequest function), and g_0 measures the benefit obtained at jump times. The functions g, g_0, h_0 are assumed to be C^2, and f_0, f_{0x}, f, and f_x (the Jacobian of f with respect to x) are C^0. Write $\omega = (\tau_1, \tau_2, \ldots)$, $\tau_j < \tau_{j+1}$. The control functions $u(t, \omega) = u(t, \tau_1, \tau_2, \ldots)$ that will be used are, separately, piecewise and right-continuous in t and piecewise continuous in each τ_j. The control functions are assumed to depend (only) on the history up till now, hence for any given t, $u(t, \omega)$ depends only on the τ_i's having occurred (i.e., $\tau_i \leq t$). Such controls are called nonanticipating. (In Remark 3.1 below, a more precise description of these controls are given.) The optimization problem is to maximize the criterion in (3.3) in the class of nonanticipating controls. We then imagine that u equals $u(t, \omega)$ in (3.1) and (3.3), and that x and $x(.)$ in (3.3) equals $x(t, \omega) := x^u(t, \omega)$, the solution of the equations in (3.1), (3.2), for $u = u(t, \omega)$. The function $x^u(t, \omega)$ is constructed by solving (3.1) successively on the intervals $(0, \tau_1), (\tau_1, \tau_2), (\tau_2, \tau_3), \ldots$. First, (3.1) is solved on $(0, \tau_1)$, using $(0, x^0)$ as initial point. Next, (3.1) is solved on (τ_1, τ_2) using $(\tau_1, x^u(\tau_1+, \omega))$ as initial point. Then, (3.1) is solved on (τ_2, τ_3), using $(\tau_2, x^u(\tau_2+, \omega))$ as initial point, and so on. The initial states $x^u(\tau_1+, \omega), x^u(\tau_2+, \omega), \ldots,)$ are, successively, all obtained from (3.2). For convenience, the function $t \to x(t, \omega)$ is taken to be left-continuous (sometimes we write $x(t-, \omega)$ in certain formulas, although we could as well have written $x(t, \omega)$), moreover $t \to x(t, \omega)$ is continuous at $t = 0$ and on each interval $(\tau_j, \tau_{j+1}) \cap (0, T)$. It is assumed that for any bounded $u = u(t, \omega)$, $x^u(t, \omega)$ exists on all $[0, T]$, a.s. (With probability 1, $\omega = (\tau_1, \tau_2, \ldots)$ has the property that $\{j : \tau_j < T\}$ is finite, so a.s. the above construction terminates after a finite number of steps.) The functions $x^u(t, \omega)$ become piecewise continuous in each τ_j.

A solution method will be presented in a moment. The method yields candidates for the optimal controls that come out in a slightly different form than $u(t, \omega)$. After presenting the method, the relationship between the various types of controls is discussed.

Remark 3.1 (Comment on nonanticipating controls*). A more precise definition of nonanticipating controls is as follows: For any given $\omega = (\tau_1, \tau_2, \ldots)$ and t, let $i(t, \omega)$ be the largest index i such that $\tau_i \leq t$. Then $u(t, \omega)$ is called nonanticipating if for any t, any $\omega = (\tau_1, \tau_2, \ldots)$ and $\omega' = (\tau_1', \tau_2', \ldots)$, the equality $u(t, \omega) = u(t, \omega')$ holds whenever $i(t, \omega) = i(t, \omega')$ and $\tau_j = \tau_j'$ for $j \leq i(\omega, t)$. $\square$

3.2 Extremal Method, Free End Case

In this section it is assumed that the maximal number of jumps N is finite. In this case, one standard, elementary, method of solving the problem is as follows.

First, find a control $\hat{u}(t, x, p)$ such that

$$\hat{u}(t,x,p) \text{ maximizes } H(t,x,u,p) := f_0(t,x,u) + pf(t,x,u) \text{ for } u \in U. \qquad (3.4)$$

(In this maximization, t, x, p are just parameters in the problem. H is called the Hamiltonian function for the problem.) Then let us write down the so-called characteristic equations. (Below H_x, g_{0x}, h_{0x} denote gradients with respect to x, g_x is a Jacobian matrix with respect to x, and I is the identity matrix.)

$$\dot{x}(t) = f(t,x,\hat{u}(t,x,p)), \qquad (3.5)$$

$$\dot{p}(t) = -H_x(t,x,\hat{u}(t,x,p),p) + \lambda(j+1)p$$
$$\quad -\lambda(j+1)[g_{0x}(t,x) + p(t;t,x+g(t,x),j+1)(I+g_x(t,x))], \qquad (3.6)$$

with boundary conditions

$$x(s) = y, \quad p(T) = [h_{0x}(x)]_{x=x(T)}. \qquad (3.7)$$

Here, the pair (s, y) is arbitrary, $s \in [0, T]$. The ordinary differential equations with boundary conditions (3.5)–(3.7) are solved by backwards recursion: First (3.5)–(3.7) are solved on $[s, T]$ for $j = N$, in which case $\lambda(N+1) = 0$ and (3.6) reduces to $\dot{p}(t) = -H_x(t,x,\hat{u}(t,x,p),p)$. The solution pair $x(.)$, $p(.)$ obtained is denoted $x(t;s,y,N)$, $p(t;s,y,N)$. Then (3.5)–(3.7) are solved on $[s, T]$ for $j = N - 1$. In this case, the known function $p(.;.,.,N)$ is inserted in (3.6). The pair of solutions obtained is denoted $x(t;s,y,N-1)$, $p(t;s,y,N-1)$. Then (3.5)–(3.7) are solved on $[s, T]$ for $j = N - 2$, with $p(.;.,.,N-1)$ inserted in (3.6), in which case $x(t;s,y,N-2)$, $p(t;s,y,N-2)$ are obtained. In this manner, we work backwards until $j = 0$ is reached and $x(t;s,y,j)$, $p(t;s,y,j)$ have been obtained for $j = 0, 1, \ldots, N$. The controls

$$u(t;s,y,j) := \hat{u}(t,x(t;s,y,j),p(t;s,y,j)), \quad j=0,1,\ldots,N,$$

are our candidates for the optimal controls (they yield nonanticipating candidates for optimality, see Remark 3.2 below). We call the solutions $x(t;s,y,j)$, $j=0,1,\ldots,N$, characteristic solutions (and $u(t;s,y,j)$, $p(t;s,y,j)$, characteristic controls, respectively characteristic adjoint functions). Finally, $(x(t;s,y,j), u(t;s,y,j))$, $j=0,1,\ldots,N$, are called characteristic pairs, and $(x(t;s,y,j), u(t;s,y,j), p(t;s,y,j))$ characteristic triples. Sometimes the word "extremal" is used instead of the word "characteristic solution," and we call the above method the *extremal method*.

One reason why we, at each step, solve (3.5)–(3.7) for arbitrary initial points (s,y) is that we need to know the controls $u(t;s,y,j)$: The stochastic jumps bring the state to points (s,y) that we cannot predict beforehand. The control $u(t;s,y,j)$ tells us how the system ought to be controlled from jump time no. j until the next jump, given that the j-th jump brings the (path of the) state to a point $(s,y) = (\tau_j, y)$.

Remark 3.2 (Characteristic pairs yield nonanticipating candidate controls). Let us show that characteristic pairs give rise to non-anticipating candidate controls, ("candidate" = candidate for optimality). For the given initial point $(0,x^0)$, the characteristic pairs $x(t;s,y,j)$, $u(t;s,y,j)$ give rise to nonanticipating functions $x(t,\omega)$, $u(t,\omega)$ in the following way: Again, $\omega = (\tau_1,\tau_2,\ldots)$. For $t \le \tau_1, x(t,\omega) = x(t;0,x^0,0)$. For $\tau_1 < t \le \tau_2$,

$$x(t,\omega) = x(t;\tau_1,x(\tau_1-,\omega) + g(\tau_1,x(\tau_1-,\omega)),1).$$

Continuing in this manner, in general, by recursion, we have that, for $\tau_j < t \le \tau_{j+1}$,

$$x(t,\omega) = x(t;\tau_j,x(\tau_j-,\omega) + g(\tau_j,x(\tau_j-,\omega)),j).$$

Furthermore, for $\tau_j \le t < \tau_{j+1}$,

$$u(t;\omega) := u(t;\tau_j,x(\tau_j+,\omega),j)$$

Evidently, $u(t;\omega)$ is a nonanticipating control of a special type.   □

If we also let $p(t;\omega) := p(t;\tau_j,x(\tau_j+,\omega),j)$ for $\tau_j \le t < \tau_{j+1}$, a triple $x(t,\omega)$, $u(t,\omega)$, $p(t,\omega)$ is obtained that satisfies the necessary conditions stated below (Theorem 3.8). Hence $u(t,\omega)$ is a candidate for optimality. If the collection $u(t;s,y,j)$ gives rise to an optimal control $u(t,\omega)$ in problem (3.1)–(3.3) (which we may hope that it does!), we call the collection $u(t;s,y,j), j=0,1,\ldots,N$ optimal in the problem.

Example 3.3. Consider the problem:

$$\max E\left[\int_0^T (-u^2/2)dt + ax(T)\right], \qquad \dot{x} = u \in \mathbb{R}, \; x(0) = 0,$$

with a possibility for a single, unit upwards jump in $x(t)$ at $\tau \in [0,\infty)$, with τ being exponentially distributed with intensity λ (i.e., the jump point τ is distributed with density $\lambda e^{-\lambda\tau}$).

Solution. The maximum condition gives $\hat{u} = p$, and the characteristic equations for $j = 1$ (i.e., after a jump) becomes $\dot{p} = 0$, $\dot{x} = p$, with $p(T) = a$. As above, t denotes running time. Hence, $u(t;s,y,1) = p(t;s,y,1) = a$, $x(t;s,y,1) = y + a(t-s)$.

Next, for $j = 0$, the characteristic equations become: $\dot{x} = p$, $\dot{p} = -\lambda a + \lambda p$, with $p(T;s,y,0) = a$. Then, $p(t;s,y,0) = a + Ce^{\lambda t}$ where C is determined by $p(T;s,y,0) = a$, so $C = 0$, and $p(t;s,y,0) = a$. Thus, $u(t;s,y,0) = a$ is the control also before the jump, which should not come as a surprise. Sufficient conditions presented below (Theorem 3.9) secure that the optimal control has been found. $\square$

Next, replace $ax(T)$ in the criterion by $ax(T)^2/2$, with $1 - aT > 0$.

Solution. As above, $\hat{u} = p$, $p(t;s,y,1) = p(T;s,y,1)$ and $x(t;s,y,1) = y + p(T;s,y,1)$ $(t-s)$. Now $p(T;s,y,1)$ is determined by the condition that $ax(T;s,y,1) = a(y + p(T;s,y,1)(T-s)) = p(T;s,y,1)$, which gives $p(T;s,y,1) = ay/(1+a(s-T))$, so $u(t;s,y,1) = ay/(1+a(s-T))$ and $x(t;s,y,1) = y + ay(t-s)/(1-a(T-s))$.

Let us find the characteristic solutions for $j = 0$. The characteristic equations become $\dot{p} = -\lambda a(x+1)/(1+a(t-T)) + \lambda p$, $\dot{x} = p$. From this pair of equations, the following second-order differential equation for x follows:

$$d^2x/dt^2 = -\lambda a(x+1)/(1+a(t-T)) + \lambda dx/dt. \qquad [*]$$

We prove below that this equation has the solution

$$x(t;s,y,0) = -1 + (C\beta(t) + D)(1 + a(t-T)), \qquad [**]$$

where $\beta(t) = \int_T^t e^{\lambda\sigma}/(1+a(\sigma-T))^2 d\sigma$ and where C and D are two arbitrary (integration) constants. The boundary conditions are $x(s;s,y,0) = y$ and $p(T;s,y,0) = ax(T;s,y,0)$, so (recalling $\dot{x} = u = p$), we get $\dot{x}(T;s,y,0) = ax(T;s,y,0)$. Now, $[**]$ yields that $\dot{x}(T,s,y,0)$ equals $C\dot{\beta}(T) + (C\beta(T) + D)a = Ce^{\lambda T} + aD$, so the next to last equality becomes $Ce^{\lambda T} + aD = a(-1+D)$, which gives $C = -ae^{-\lambda T}$. Knowing this, from $[**]$ and $x(s;s,y,0) = y$, the constant D is determined, and the result is $D = (y+1)/(1+a(s-T)) + ae^{-\lambda T}\beta(s)$. The candidate for the optimal policy is hence to use $u(t;0,0,0) = (d/dt)[-ae^{-\lambda T}\beta(t) + 1/\{1-aT\} + ae^{-\lambda T}\beta(0)](1+a(t-T))]$ as long as no jump has occurred. When a jump has occurred at some τ, we use the control $u(t,\omega) := u(t,\tau) := u(t;\tau,x(\tau-;0,0,0)+1,1) = a(x(\tau-;0,0,0)+1)/(1+a(\tau-T))$ from then on (this control is independent of $t, t \geq \tau$). (Note that from the characteristic triples we have obtained nonanticipating controls.)

To prove $[**]$, define $w(t)$ by $w(1+a(t-T)) = x+1$. Then

$$(dw/dt)(1+a(t-T)) + wa = dx/dt$$

and

$$(d^2w/dt^2)(1+a(t-T)) + 2(dw/dt)a = d^2x/dt^2,$$

so

$$(d^2w/dt^2)(1+a(t-T)) + 2(dw/dt)a = -\lambda a(x+1)/(1+a(t-T)) + \lambda dx/dt$$
$$= -\lambda aw + \lambda(dw/dt)(1+a(t-T)) + \lambda aw,$$

and hence
$$d^2w/dt^2 = [\lambda - 2a/(1+a(t-T))]dw/dt.$$

Integrating this separable equation, we get $dw/dt = Ce^{\lambda t}/(1+a(t-T))^2$, so $w(t) = C\beta(t) + D$. Here C and D are two integration constants. By definition of $w(t)$, $x(t;s,y,0) = -1 + w(t)(1+a(t-T))$ and [**] follows.

Do sufficient conditions apply here? Yes, but because the scrap value function may be convex rather than concave (it takes place when $a > 0$), a theorem not using concavity is needed, in this example Theorem 3.11 below is applicable.   □

Heuristic Derivation of the Extremal Method

Let us simplify the problem slightly by letting $g_0 \equiv 0$. Let $J(s,y,j)$ be the optimal value function in the problem, given the initial condition $x(s) = y$, and given that exactly j jumps have occurred in $[0,s]$. Hence,

$$J(s,y,j) = \max_{u=u(.,.)} E\left[\int_s^T f_0(t,x^u(t,\omega),u(t,\omega))dt + h_0(x^u(T))|s,y,j\right]. \qquad (3.8)$$

The maximization is carried out over all pairs $(x^u(t,\omega),u(t,\omega))$, $x^u(t,\omega)$ a solution starting at (s,y), $(x^u(t,\omega),u(t,\omega))$ depending only on $\tau_{j+1},\tau_{j+2},\ldots$
Evidently,

$$J(T,y,j) = h_0(y). \qquad (3.9)$$

We want to be able to use the dynamic programming equation from Chapter 1, so we now consider time to be discrete. Let s be a given time point and let $s+h$ be the next time point, h small. When h is small, we can simplify the situation by assuming that either a single jump occurs at $s+h$ with probability λ^*h, $\lambda^* := \lambda(j+1)$, or no jump occurs at $s+h$ with probability $1 - \lambda^*h$. Still, assume that j jumps have occurred in $[0,s]$. Below, k is a stochastic variable, for which $\Pr[k = j+1)] = \lambda^*h$, and $\Pr[k = j]) = 1 - \lambda^*h$. For any given vector u in U, for the moment let $x^u(t)$ be the continuous solution of (3.1) on $[s,s+h)$, starting at (s,y). The reward obtained over the period $(s,s+h)$ is now $f_0(s,y,u)h$. Using the dynamic programming equation yields

$$\begin{aligned}
J(s,y,j) &= \max_{u\in U}\{f_0(s,y,u)h + E[J(s+h,x^u(s+h),k)|s,y,j]\} \\
&= \max_{u\in U}\{f_0(s,y,u)h + (1-\lambda^*h)J(s+h,x^u(s+h),j) \\
&\qquad + \lambda^*hJ(s+h,x^u(s+h) + g(s+h,x^u(s+h)),j+1)\} \\
&\approx \max_{u\in U}\{f_0(s,y,u)h + (1-\lambda^*h)[J(s,y,j) + J_s(s,y,j)h \\
&\qquad + J_x(s,y,j)f(s,y,u)h] + \lambda^*hJ(s,y+g(s,y),j+1)\}.
\end{aligned}$$

To obtain the second equality, note that when a jump occurs at $s+h$, the state after the jump is $x^u(s+h) + g(s+h,x^u(s+h))$, and to obtain the approximate equality, we have used $J(s+h,x^u(s+h),j) \approx J(s,y,j) + (d/ds)J(s,x^u(s),j)h =$

$J(s,y,j)+J_s(s,x^u(s),j)h+J_x(s,x^u(s),j)f(s,y,u)h$. Moreover a small error, as compared with h, is made by replacing $\lambda^*hJ(s+h,x^u(s+h)+g(s+h,x^u(s+h)),j+1)\}$ by $\lambda^*hJ(s,y+g(s,y),j+1)$. Finally, if we drop all terms containing h^2, because they are "extremely small," we get

$$J(s,y,j) \approx \max_{u\in U}\{f_0(s,y,u)h+J(s,y,j)-\lambda^*hJ(s,y,j)+J_s(s,y,j)h$$
$$+J_x(s,y,j)f(s,y,u)h+\lambda^*hJ(s,y+g(s,y),j+1)\}.$$

The error may be seen to be small compared with h. Canceling the two terms $J(s,y,j)$ and dividing by h, we get

$$0 = \max_{u\in U}\{f_0(s,y,u)+J_x(s,y,j)f(s,y,u)\}$$
$$+J_s(s,y,j)+\lambda^*J(s,y+g(s,y),j+1)-\lambda^*J(s,y,j) \quad (3.10)$$

(where $\lambda^* := \lambda(j+1)$), which is the so-called HJB equation in the problem. If $(x^*(.,\omega),u^*(.,\omega))$ is optimal in the problem, then, for any given ω, after exactly j jumps, $u^*(s,\omega)$ yields the maximum in the above dynamic programming equation, and so also in the last equation, for $y = x^*(s,\omega)$.

The HJB equation will reappear in a sufficient condition for optimality below (Theorem 3.37).

Define $\check{p}(s,j) := J_x(s,x^*(s,\omega),j)$ (still the assumption is that exactly j jumps have occurred before s). Then, from (3.9), evidently,

$$\check{p}(T,j) = h_{0x}(x^*(T,\omega)) \quad (3.11)$$

(when exactly j jumps have occurred before T). For $H(s,x,u,p) = f_0 + pf$, the following maximum condition is easily obtained: After exactly j jumps, $(s > \tau_j)$,

$$H(s,x^*(s,\omega),u^*(s,\omega),\check{p}(s,j)) = \max_u H(s,x^*(s,\omega),u,\check{p}(s,j)). \quad (3.12)$$

In fact, for $y = x^*(s,\omega)$, denoting the sum of terms not under the max–sign in (3.10) by A, we have that $\max_{u\in U}\{f_0(s,y,u) + J_x(s,y,j)f(s,y,u)\} = -A = f_0(s,y,u^*(s,\omega))+J_x(s,y,j)f(s,y,u^*(s,\omega))$, which entails (3.12).

Define $L(s,y,u) =$

$$f_0(s,y,u)+J_x(s,y,j)f(s,y,u)+J_s(s,y,j)$$
$$+\lambda^*J(s,y+g(s,y),j+1)-\lambda^*J(s,y,j).$$

By (3.10), $0 \geq L(s,y,u^*(s,\omega))$ for any (s,y), and $0 = L(s,x^*(s,\omega),u^*(s,\omega))$. Hence as a function of y, $L(s,y,u^*(s,\omega))$ has a maximum at $x^*(s,\omega)$. Thus, $L_y(s,x^*(s,\omega),u^*(s,\omega)) = 0$. Calculating $L_y(s,x^*(s,\omega),u^*(s,\omega))$, using a shorthand notation and that $(d/dt)\check{p}(t,j) = J_{xt}(t,x^*(t),j)+J_{xx}\dot{x}^*(t,\omega) = J_{tx}+J_{xx}f$, we get

$$0 = f_{0x}+J_{tx}+J_{xx}f+J_xf_x+\lambda^*J_x(j+1)(I+g_x)-\lambda^*J_x(j)$$
$$= f_{0x}+d\check{p}/dt+\check{p}f_x+\lambda^*J_x(j+1)(I+g_x)-\lambda^*J_x(j),$$

that is, when exactly j jumps have occurred and $s > \tau_j$,

$$(d/ds)\check{p}(s,j) = -f_{0x}(s,x^*(s,\omega),u^*(s,\omega)) - \check{p}(s,j)f_x(s,x^*(s,\omega),u^*(s,\omega))$$
$$-\lambda(j+1)\hat{p}(s,j+1)(I+g_x(s,x^*(s,\omega))) + \lambda(j+1)J_x(s,x^*(s,\omega),j), \quad (3.13)$$

where $\hat{p}(s,j+1) = J_x(s,x^*(s,\omega)) + g(s,x^*(s,\omega)),j+1))$. Above, we wrote $\check{p}(s,j) = J_x(s,x^*(s,\omega),j)$, we might have written "in more detail" $\check{p}(s,j) := \check{p}(s,x^*(s,\omega),j) := J_x(s,x^*(s,\omega),j)$. Then, of course, $\hat{p}(s,j+1) = \check{p}(s,x^*(s,\omega)) + g(s,x^*(s,\omega)),j+1)$. So, evidently, for $(t,x) = (s,x^*(s,\omega))$, the adjoint equation (3.6) in the extremal method is the same as (3.13) above. $\square$

Later on necessary conditions for optimality will be stated (see Theorem 3.8), and the above arguments intuitively explain how these necessary conditions arise.

In the next example, a maximum of N jumps can occur, $1 \leq N < \infty$.

Example 3.4. Consider the problem

$$\max E \left\{ \int_0^1 (-u^2/2)dt + x(1) \right\},$$

subject to $\dot{x} = u \in \mathbb{R}, x(0) = 0, x(\tau_j+) - x(\tau_j-) = x(\tau_j-)$.

We assume that a maximum of $N < \infty$ jumps can occur, and that τ_j is exponentially distributed in $[\tau_{j-1}, \infty)$ with intensity λ_j, all λ_j different. An interpretation of the problem might be that x is the value of an investor's stock holding in a research firm, which now and then makes an invention. Each time an invention occurs, the value of the stock doubles. Moreover, any other change in the value of the stock (using $u \neq 0$) is costly. (The manner this is taken care of is debatable.) The best opportunities for inventions are exploited first, so we may imagine that λ_j is a decreasing sequence.

Solution. Maximization of the Hamiltonian yields $\hat{u} = p$, hence $u(t;s,y,j) = p(t;s,y,j)$. It remains to find $p(t;s,y,j)$. Now, $(d/dt)p(t;s,y,N) = 0$, so $p(t;s,y,N) = p(1;s,y,N) = 1$. This function is is independent of the starting point (s,y). This will also be seen to hold for $j < N$, so we drop (s,y) in the symbols for the adjoint functions. Next, $(d/dt)p(t;N-1) = -2\lambda_N p(t;N) + \lambda_N p(t;N-1)$ $(1+g_x = 2)$, which has the solution $p(t;N-1) = 2 - e^{\lambda_N(t-1)}$. (To determine this solution also $p(1) = 1$ has been used.) The formula is easily obtained by using $e^{-\lambda_N(t-1)}$ as integrating factor (i.e., by multiplying on both sides of the differential equation by this factor). The method is shown in detail in the next calculation. Let us find $p(t;N-2)$:

$$(d/dt)p(t;N-2) = -2\lambda_{N-1}p(t;N-1) + \lambda_{N-1}p(t;N-2)$$
$$= -2\lambda_{N-1}2 + 2\lambda_{N-1}e^{\lambda_N(t-1)} + \lambda_{N-1}p(t;N-2).$$

Using $e^{-\lambda_{N-1}(t-1)}$ as integrating factor, i.e., multiplying on both sides by this factor, and rearranging, we get

$$e^{-\lambda_{N-1}(t-1)} \cdot (d/dt)p(t;N-2) - \lambda_{N-1}e^{-\lambda_{N-1}(t-1)}p(t;N-2)$$
$$= -4\lambda_{N-1}e^{-\lambda_{N-1}(t-1)} + 2\lambda_{N-1}e^{(\lambda_N - \lambda_{N-1})(t-1)}.$$

The left-hand side of the last equality equals $(d/dt)[e^{-\lambda_{N-1}(t-1)}p(t;N-2)]$, so integrating on both sides yields $e^{-\lambda_{N-1}(t-1)}p(t;N-2) =$

$$4e^{-\lambda_{N-1}(t-1)} + 2\lambda_{N-1}(\lambda_N - \lambda_{N-1})^{-1}e^{(\lambda_N - \lambda_{N-1})(t-1)} + C,$$

where C is an integration constant. Multiplying by $e^{\lambda_{N-1}(t-1)}$ on both sides and using $p(1,N-2) = 1$ yield $p(t;N-2) = 4 + 2\lambda_{N-1}(\lambda_N - \lambda_{N-1})^{-1}e^{\lambda_N(t-1)} + (1 - 4 - 2\lambda_{N-1}(\lambda_N - \lambda_{N-1})^{-1})e^{\lambda_{N-1}(t-1)}$.

In general, we guess that $p(t;N-j)$ has the following form: $p(t;N-j) = 2^j + \sum_{0 \le i \le j-1} a_j^i e^{\lambda_{N-i}(t-1)}$. Let us prove that $p(t;N-(j+1))$ has a similar form. We have

$$(d/dt)p(t;N-(j+1))$$
$$= -2\lambda_{N-j}p(t;N-j) + \lambda_{N-j}p(t;N-(j+1))$$
$$= -2\lambda_{N-j}2^j - 2\lambda_{N-j} \cdot \sum_{0 \le i \le j-1} a_j^i e^{\lambda_{N-i}(t-1)} + (\lambda_{N-j})p(t;N-(j+1)).$$

Using the integrating factor $e^{-\lambda_{N-j}(t-1)}$ in the differential equation for $p(t;N-(j+1))$ yields that $p(t;N-(j+1))$ has the form: $p(t;N-(j+1)) = 2^{j+1} + \sum_{0 \le i \le j} a_{j+1}^i e^{\lambda_{N-i}(t-1)}$, where

$$a_{j+1}^i = 2\lambda_{N-j}a_j^i/(\lambda_{N-j} - \lambda_{N-i}), \quad i < j. \qquad [a]$$

The last coefficient in the expression for $p(t;N-(j+1))$, a_{j+1}^j is an integration constant. It is determined by $p(1;N-(j+1)) = 1$, so

$$1 - 2^{j+1} - \sum_{0 \le i \le j-1} a_j^i = a_{j+1}^j. \qquad [b]$$

Using [a] for $a_{j+1}^i, a_j^i, \ldots, a_{i+2}^i$ yields the following product for a_{j+1}^i:

$$a_{j+1}^i = a_{i+1}^i \Pi_{i+1 \le m \le j} 2\lambda_{N-m}/(\lambda_{N-m} - \lambda_{N-i}), \quad i < j. \qquad [c]$$

The coefficients $a_{k+1}^k, k = 0, 1, 2, \ldots, N-1$, are determined by the following difference equation, obtained by combining [b] and [c]:

$$a_{k+2}^{k+1} = 1 - 2^{k+2} - \sum_{0 \le i \le k} a_{i+1}^i \Pi_{i+1 \le m \le k-1} 2\lambda_{N-m}/(\lambda_{N-m} - \lambda_{N-i}). \qquad [d]$$

First, as $p(t;N-1) = 2 - e^{\lambda_N(t-1)}$, we get $a_1^0 = -1$. Then, [d] gives us all $a_{k+1}^k, k = 1, 2, \ldots, N-1$. By [c], for any $j, j < k-1$ we know all a_k^j. Thus, we know $p(t,N-j) = u(t,N-j)$. A sufficient condition based on concavity (Remark 3.10 below) gives the optimality of $u(t,N-j), j = 0, 1, \ldots, N$ (or of the associated nonanticipating control $u(t,\omega)$). $\qquad\square$

In Remark 3.2, we saw that characteristic pairs give rise to nonanticipating controls. Let us give some comments on how so-called Markov controls arise.

Remark 3.5 (Characteristic controls yield Markov candidate controls). When constructing the characteristic controls $u(t;s,y,j)$ by the method above, we most often
obtain that

$$u(t';s,y,j) = u(t';t,x(t;s,y,j),j), \qquad t' \geq t > s. \tag{3.14}$$

In fact, for any given j, when $\hat{u}(t,x,p)$ is unique and C^1, for any given initial point
$(s,y,\hat{p})$ solutions to (3.5) and (3.6), are unique. Next, the method tells us to seek
values of the parameter $\hat{p}$ such that $p(T) = h_{0x}(x(T;s,y,j))$. Often this condition
gives a unique value of $\hat{p}$. At least under such circumstances, (3.14) will hold.

Let us discuss (3.14) a little more. Suppose that the solution starts in (s,y,j) (as
j is mentioned, it means that exactly j jumps have already occurred in $[0,s)$). The
control $u(t;s,y,j)$ then (hopefully) describes the optimal behavior from s on. At
any time $t' \geq s$, the control value $u(t';s,y,j)$ should be used as long as no further
jumps occur. Let x be a point that is reached by $x(.;s,y,j)$ at time t, and assume
that even at the times t and $t' > t$ only j jumps have occurred. At the time t' we
have then two prescriptions for which control value to use, namely $u(t';s,y,j)$ and
$u(t';t,x(t;s,y,j),j)$. These two values should coincide, and that is what (3.14) require them to do. Define $u(t,x,j) := u(t;t,x,j)$. Assuming (3.14), and then in particular that (3.14) holds for $t = t'$, we see that $u(t,x,j)$ gives, at all points (t,x), the
unique control to use. There is no "doubling" of proposed control values to use,
which would occur if (3.14) failed to hold. A control of the form $u(t,x,j)$ is called a
Markov control. Most often, (3.14) holds, so most often characteristic controls give
rise to unique Markov candidate controls.   □

Remark 3.6 (Nonanticipating, characteristic, and Markov controls*). First, Markov
controls and nonanticipating controls will be compared. Given that j jumps have
occurred at $\tau_1, \ldots, \tau_j$, a nonanticipating control $u(t,\omega)$ makes it possible to calculate
the state x at time t by using (3.1) and (3.2). At such a point (t,x), the control
function prescribes the control value to be $u(t;\omega)$. Of course, even if different ω's
(i.e., different $(\tau_1, \ldots, \tau_j)$) by chance lead us to the same point (t,x), different control
values (depending on ω) may be prescribed. However, the above system has the
property that the future development of the system depends only on t,x and j and not
when the jumps $\tau_1, \ldots, \tau_j$ have occurred. We can therefore expect optimal behavior
at time s to depend only on (t,x,j). Thus, an optimal control $u^*(t,\omega)$ will prescribe
the same control value at (t,x), irrespective of how the system came to x at time t,
given that j jumps have occurred. Hence, by this loose argument, optimal controls
can frequently be expected to be of the type $u(t,x,j)$, i.e., Markov controls. We
saw in Remark 3.3 that characteristic controls most often yield Markov controls, as
the consistency condition (3.14) most often holds. They also yield nonanticipating
controls, as we saw in Remark 3.2.

If Markov controls $u(t,x,j)$ are given, let us imagine that these are inserted into
(3.1), which then are solved on $[s,T]$ for initial condition (s,y). The solutions obtained are denoted $x(t;s,y,j)$ (they satisfy $x(s;s,y,j) = y$), and corresponding controls are $u(t;s,y,j) = u(t,x(t;s,y,j),j)$. (Normally they automatically satisfy the
consistency condition (3.14).) As in Remark 3.2, solutions $x(t;s,y,j)$ give rise to

nonanticipating solutions $x(t, \omega)$, and then also to nonanticipating controls $u(t, \omega)$ $(= u(t, x(t, \omega), j), t \in (t_j, \tau_{j+1}]$.

The solution method above yields characteristic controls $u(t; s, y, j)$, which also give rise to nonanticipating candidate controls $u(t, \omega)$ with corresponding solutions $x(t, \omega)$, as we have seen. $\qquad \square$

Remark 3.7 (Explicit dependence on j). The functions f_0, f, g_0, g can be allowed to depend explicitly on j. In f_0 and f, j has value 0 as long as no jump has occurred (i.e., in $(0, \tau_1)$), the value 1 when exactly one jump has occurred (i.e., in (τ_1, τ_2)), the value 2 when exactly two jumps have occurred (i.e., in (τ_2, τ_3)), and so on. To simplify some statements, let $\tau_0 = 0, \omega^0 = 0, \omega = (\tau_0, \tau_1, \ldots)$, and let $\omega^j :=$ $(\tau_0, \ldots, \tau_j)$. The differential equation is thus

$$\dot{x} = f(t, x, u, j), \ t \in (\tau_j, \tau_{j+1}), \ x(0) = x^0, \ u \in U, \tag{3.15}$$

the jump condition is, for $j > 0$,

$$x(\tau_j +) - x(\tau_j -) = g(\tau_j, x(\tau_j -), j), \tag{3.16}$$

(no jump at τ_0), and the criterion is

$$E \left[\sum_{j \geq 0} \int_{\min\{T, \tau_j\}}^{\min\{T, \tau_{j+1}\}} f_0(t, x, u, j) dt + \sum_{0 < \tau_j < T} g_0(\tau_j, x(\tau_j -), j) + h_0(x(T-)) \right]. \tag{3.17}$$

The solution procedure now uses the four relationships:

$\hat{u}(t, x, p, j)$ maximizes

$$H(t, x, u, p, j) := f_0(t, x, u, j) + pf(t, x, u, j) \text{ for } u \in U \tag{3.18}$$

$$\dot{x}(t) = f(t, x, \hat{u}(t, x, p, j), j) \tag{3.19}$$

$$\dot{p}(t) = -H_x(t, x, \hat{u}(t, x, p, j), p, j) + \lambda(j+1)p - \lambda(j+1)[g_{0x}(t, x, j+1)$$
$$+ p(t; t, x + g(t, x, j+1), j+1)(I + g_x(t, x, j+1))], \tag{3.20}$$

with boundary conditions

$$x(s) = y, \quad p(T) = [h_{0x}(x)]_{x = x(T; s, y, j)}. \tag{3.21}$$

From now on, this explicit dependence on j will be assumed. For a pair $(x(t, \omega), u(t, \omega))$ to be called admissible, we shall, in addition to the satisfaction of (3.15), (3.16), also require that $u(t, \omega)$ is bounded ($\sup_{t, \omega} |u(t, \omega)| < \infty$).

In many practical problems, it is the dynamics (i.e., the right-hand side of the state equation) that changes abruptly at time points $\tau_1, \tau_2, \ldots$, while there are no jumps in the state, in fact both g and g_0 may equal zero. Such problems form an important case to which the present setup applies. $\qquad \square$

3.3 Precise Necessary Conditions and Sufficient Conditions

As before, let "v.e." (virtually everywhere) mean everywhere except a finite (or countable) number of points, and let "a.s." mean almost surely, i.e., with probability 1. As $\omega^j = (\tau_0, \ldots, \tau_j)$, we write $x^*(t, \omega^j) := x^*(t, \tau_0, \ldots, \tau_j) := x^*(t, \tau_0, \ldots, \tau_j, T, T+1, \ldots)$. Similar definitions hold for the $p(t, \omega^j)$ and $u^*(t, \omega^j)$ functions. The following maximum principle holds.

Theorem 3.8 (Necessary condition, free end). *Assume that $x^*(t, \omega)$, $u^*(t, \omega)$, is an optimal pair in problem (3.15)–(3.17), such that $u^*(t, \omega)$ is bounded and nonanticipating. Then, for each ω, there exists a nonanticipating function $t \mapsto p(t, \omega)$, continuous in t between the points τ_i in ω, differentiable v.e., such that, a.s., for v.e. $t > \tau_j$, $j = 0, 1, 2, \ldots$,*

$$u \mapsto f_0(t, x^*(t, \tau_0, \ldots, \tau_j), u, j) + p(t, \tau_0, \ldots, \tau_j) f(t, x^*(t, \tau_0, \ldots, \tau_j), u, j)$$

$$\text{has a maximum at } u^*(t, \tau_0, \ldots, \tau_j) \text{ in } U. \tag{3.22}$$

Moreover, $t \mapsto p(t, \tau_0, \ldots, \tau_j)$, $t > \tau_j$, satisfies v.e.

$$
\begin{aligned}
\dot{p}(t, \tau_0, \ldots, \tau_j) = &-f_{0x}(t, x^*(t, \tau_0, \ldots, \tau_j), u^*(t, \tau_0, \ldots, \tau_j), j) \\
&- p(t, \tau_0, \ldots, \tau_j) f_x(t, x^*(t, \tau_0, \ldots, \tau_j), u^*(t, \tau_0, \ldots, \tau_j), j) \\
&- \lambda(j+1) g_{0x}(t, x^*(t, \tau_0, \ldots, \tau_j), j+1) \\
&- \lambda(j+1) p(t+, \tau_0, \ldots, \tau_j, t)\{I + g_x(t, x^*(t, \tau_0, \ldots, \tau_j), j+1)\} \\
&+ \lambda(j+1) p(t, \tau_0, \ldots, \tau_j). \tag{3.23}
\end{aligned}
$$

$$p(T, \tau_0, \ldots, \tau_j) = h_{0x}(x^*(T-, \tau_0, \ldots, \tau_j)). \tag{3.24}$$

The function $p(t, \tau_0, \ldots, \tau_j), t > \tau_j$, is bounded, continuous in $t \in (\tau_j, T]$, and for each such t, piecewise continuous in each $\tau_i, i \leq j$, and the limit $p(\tau_j+, \tau_0, \ldots, \tau_j)$ exists. □

Let us state precise conditions for Theorem 3.8 to hold; later on these conditions are called the Standard System Conditions. We shall continuously assume that f_0, f, f_{0x}, and f_x (exist and) are continuous, and that g_0, g, and h_0 are C^2. (For the purpose of Theorem 3.8, it would suffice to assume that f_0, f, g_0, g, h_0, f_{0x}, f_x, g_{0x}, g_x, h_{0x}, exist and are, separately, piecewise continuous in t, and separately continuous in x and in u.)

It is assumed that the following "growth conditions" are satisfied: The five functions f_0, f, g_0, g, h_0 are Lipschitz continuous with respect to x with a common rank $\kappa^n < \infty$, $n = 1, 2, \ldots$, for $u \in U \cap B(0, n)$, independent of (t, u, j). Moreover it is assumed that the five functions satisfy an inequality of the following form when playing the role of ϕ: For some constants α_n, κ_n

$$|\phi(t, x, u, j)| \leq \alpha_n + \kappa_n |x|, \quad \text{for all } (t, x, u, j), u \in U \cap B(0, n). \tag{3.25}$$

Finally, as before, $\sup_j \lambda(j) < \infty$.

If $N < \infty$, the growth conditions can be replaced by the condition that $\sup_{t,\omega} |x^*(t,\omega)| < \infty$.

From now on, these Standard System Conditions are assumed. Exceptions in certain remarks are explicitly stated, let us note that if these growth conditions hold even for $n = \infty$, then in the theorem, $u^*(.,.)$ need not be bounded. (See Remark 5 and Remark 3 in Seierstad (2001).)

Formally, to solve a control problem completely, one has to find all possible controls $u^{**}(t,\omega)$ satisfying the necessary conditions in Theorem 3.8. Now, these conditions *are* satisfied by a candidate $u^{**}(t,\omega)$ obtained from characteristic controls, so if one can prove that no other $u^{**}(t,\omega)$ satisfies (3.22)–(3.24), a unique candidate for optimality has been found. If one knows that the problem must have an optimal control, then it must be this unique candidate.

So far, we have not provided results concerning existence of optimal controls, they are, however, similar to the Fillipov–Cesari results in deterministic control theory. In fact, with the Standard System Conditions, and provided U is compact and $\{(f_0(t,x,u,j) + \gamma, f(t,x,u,j)) : u \in U, \gamma \leq 0\}$ is convex for all (t,x,j), an optimal nonanticipating control exists (being perhaps only "Borel measurable," see the Appendix for this term). (When, later on "hard" terminal constraints are added, we have to add that an admissible pair must exist.) See Seierstad (2008).

Connected with the necessary condition above is a sufficient condition, based on concavity: Define $\hat{H}(t,x,p,j) = \sup_{u \in U} H(t,x,u,p,j)$. Then the following theorem holds. (See Seierstad (2001).)

Theorem 3.9 (Sufficient condition based on concavity). *Suppose the triple* $(x^*(t,\omega), u^*(t,\omega), p(t,\omega))$ *satisfies the necessary conditions (3.22)–(3.24), with* $(x^*(t,\omega), u^*(t,\omega))$ *satisfying (3.15), (3.16), $u^*(.,.)$ bounded and nonanticipating, $p(t,\omega)$ having properties as in Theorem 3.8. Suppose, furthermore, that* $x \to h_0(x)$ *and* $x \to g_0(t,x,j)$ *are concave, and, for each* $j, \omega^j, t > \tau_j$, *that* $x \to \hat{H}(t,x,p(t,\omega^j),j)$, *and* $x \to p(t,\omega^j)g(t,x,j)$ *are concave. Then* $(x^*(t,\omega), u^*(t,\omega))$ *is optimal in the set of admissible pairs.* $\square$

The concavity condition on $\hat{H}(t,x,p(t,\omega^j),j)$ holds in particular when $(x,u) \to H(t,x,u,p(t,\omega^j),j) := f_0(t,x,u,j) + p(t,\omega^j)f(t,x,u,j)$ is concave.

Remark 3.10 (Sufficient concavity conditions for characteristic controls, $N < \infty$). Similar to Theorem 3.9, sufficient for the characteristic triple $x(t;s,y,j), u(t;s,y,j)$, $p(t;s,y,j)$ to yield an optimal nonanticipating control is concavity of $x \to g_0(t,x,j)$, $x \to h_0(x)$, $x \to \hat{H}(t,x,p(t;s,y,j),j)$, and $x \to p(t;s,y,j)g(t,x,j)$ for all (s,y,j). In addition, we assume that the control $u^*(t,\omega)$ corresponding with the characteristic triple is a bounded function.

(Here, we imagine that characteristic triples have been found for any j, any (s,y) in $Q(j)$, even in $\{(0,x^0)\} \cup Q(0)$ in case $j = 0$, $Q(j)$ as defined in (3.26) below.) $\square$

Above, we have dropped stating assumptions concerning the dependence of $x(t;s,y,j)$, $u(t;s,y,j)$, and $p(t;s,y,j)$ on t,s, and y. To a large extent, we can avoid stating explicitly such assumptions, by saying that, implicitly (for example in the preceding remark), it is assumed that the nonanticipating triples

$(x(t,\omega),u(t,\omega),p(t,\omega))$ arising from the characteristic triples satisfy the standard assumptions. However, we do postulate that $t \to (x(t;s,y,j),p(t;s,y,j))$ is continuous in $t \in [s,T]$ and that $t \to u(t;s,y,j)$ is piecewise continuous in $t \in [s,T]$.

When at most N jumps can occur, there is a recursive procedure related to the necessary conditions. Loosely speaking, first $(x^*(t;\tau_0,\dots,\tau_N), u^*(t;\tau_0,\dots,\tau_N), p(t;\tau_0,\dots, \tau_N))$ is constructed, then $(x^*(t;\tau_0,\dots,\tau_{N-1}), u^*(t;\tau_0,\dots,\tau_{N-1}), p(t;\tau_0,\dots, \tau_{N-1}))$, and so on. One way of systematizing this procedure was shown in the solution method connected with (3.4)–(3.7) and (3.18)–(3.21) above (the extremal method).

Now, Theorem 3.8 also holds if there is no bound on the number of jumps that can occur ($N = \infty$). But, in this case no recursive procedure is available. Still, in principle, one seeks triples $(x^*(t;\tau_0,\dots,\tau_j), u^*(t;\tau_0,\dots,\tau_j), p(t;\tau_0,\dots,\tau_j))$ satisfying the necessary conditions for $j = 0,1,2,\dots$. Sometimes, it may pay instead to turn to the so-called HJB equation described later on.

The next theorem gives a sufficient condition not requiring concavity. Roughly the sufficient condition is as follows: Assume that for all initial points $(s,y) \in [0,T] \times \mathbb{R}^n$, characteristic pairs $x(t;s,y,j), p(t;s,y,j)$ exist that are C^1 in (s,y). Then the nonanticipating pair $(x^*(t,\omega),u^*(t,\omega))$ corresponding with this collection of characteristic pairs is optimal.

Frequently, we can restrict the set of initial points (s,y) needed to be considered, and we need some assumptions on the maximized Hamiltonians $\hat{H}(t,x,p,j)$. To formulate the theorem some definitions are needed: Let

$$Q_{x(.,.)}(j) := \{(t,x(t,\omega^j)) : t \in (\tau_j,T), \text{ for some } \omega^j\},$$
$$Q(j) := \cup_{x(.,.)} Q_{x(.,.)}(j), \tag{3.26}$$

the union taken over all admissible solutions $x(.,.)$. Thus $Q(j)$ consist of all points (s,y) that can be reached after j jumps, considering all possible admissible solutions and locations of these j jumps. In the next theorem also, sets $Q^0(j) \subset (0,T) \times \mathbb{R}^n$ appear, being (for each j) larger than $Q(j)$. Sometimes, these sets can be constructed by taking the union of $Q_{x(.,.)}(j)$ for all solutions $x(.,.)$, starting at time 0 at given start points $\hat{x}^0$, where $\hat{x}^0$ runs through a (perhaps small) open ball around x^0. In other cases, $Q^0(j)$ can, for example, be the whole set $(0,T) \times \mathbb{R}^n$.

Theorem 3.11 (Sufficiency based on a field of characteristic solutions, $N < \infty$).
For each $j = 0,1,2,\dots,N$, the following assumptions are made: A continuous function $\hat{u}(t,x,p,j)$ exists that gives maximum of the Hamiltonian in (3.18) for each point (t,x,p) in a given open set $Q^*(j) \subset \mathbb{R}^{1+2n}$. Open sets $Q^0(j) \subset (0,T) \times \mathbb{R}^n$, $Q^0(j)$ containing $Q(j)$, and solutions $x(t;s,y,j), p(t;s,y,j)$ on $[s,T]$ of (3.19)– (3.21) (the characteristic equations with boundary conditions) have been found for $(s,y) \in \tilde{Q}_0(j) := \mathrm{cl}Q^0(j)$, being continuous in (t,s,y) for $(s,y) \in \tilde{Q}_0(j), t \in [s,T]$, and C^1 in $(s,y) \in Q^0(j)$ for any $t \in (s,T]$, and any j. For all $t \in [s,T], (s,y) \in \tilde{Q}_0(j),$

$$(t,x(t;s,y,j),p(t;s,y,j)) \text{ belongs to } Q^*(j) \tag{3.27}$$

and $(t,x(t;s,y))$ belongs to $Q^0(j)$ for all $t \in (s,T)$, for all $(s,y) \in Q^0(j)$, for $j=0$, even for all $(s,y) \in \{(0,x^0)\} \cup Q^0(0)$. Moreover, for all $j < N$,

$$\text{if } (s,y) \in Q^0(j), \text{ then } (s,y+g(s,y,j+1)) \in Q^0(j+1). \tag{3.28}$$

The pair $(x^*(t,\omega),u^*(t,\omega))$ corresponding to the pairs $(x(.,.,j),u(.,.,j))$ is admissible. Finally, $\hat{H}(s,x,p,j)$ and its first and second derivatives with respect to x and p (exist and) are continuous in $Q^*(j)$. Then $u(t;s,y,j) := \hat{u}(t,x(t;s,y,j),p(t;s,y,j))$, $j = 0,1,2,\ldots,N$, are optimal controls (i.e., $(x^*(t,\omega),u^*(t,\omega))$ is optimal in the set of admissible pairs). $\qquad\square$

For $s = T$, of course $x(T;s,y,j) = y$, $p(T;s,y,j) = h_{0x}(y)$. Continuity of $\hat{u}(t,x,p,j)$ is actually not needed, for example, we might require only that $\hat{u}$ is (say) separately piecewise continuous in each real variable it depends upon, and that $t \rightarrow u(t;s,y,j)$ is piecewise continuous.

Lack of continuity may arise when $\hat{u}(t,x,p,j)$ is nonunique, so what needs actually be presupposed in general is that there exists, for each j, a triple $u(t;s,y,j)$, $x(t;s,y,j)$, $p(t,s,y,j)$ for which $\hat{u}(t,x(t;s,y,j),p(t;s,y,j))$ equals $u(t;s,y,j)$, with $u(t;s,y,j)$ assumed to be piecewise continuous in t, and such that the characteristic equations (3.19), (3.20), (3.21) are satisfied by $x(t;s,y,j),u(t;s,y,j),p(t;s,y,j)$ (so to speak: insert and check!), and with no well-behavior conditions on $\hat{u}(t,x,p,j)$.

Although the results in the theorem do follow from the stated premises, note that, hidden in the premises is the fact that, in order for the stated differentiability of $(s,y) \rightarrow (x(t;s,y,j),p(t;s,y,j))$ to hold, the system must, in most cases, satisfy further differentiability properties.

Actually, the condition $N < \infty$ in the theorem can be dropped, provided certain additional conditions are required, including that $(t,x) \rightarrow p(t;t,x,j)$ is C^1 in $Q^0(j)$, see Remark 3.32 below. (There is then no "elementary" extremal method of finding the entities $x(t;s,y,j),p(t;s,y,j)$, but if such entities somehow have been found for $j = 0,1,\ldots$, a modification of the theorem still holds.)

This theorem does not require concavity assumptions. So why use the preceding Theorem 3.9 (or Remark 3.10) at all? The answer is that when concavity holds, these tools are easier to apply, and in particular, there may be situations where concavity holds, but where it is impossible to find open sets $Q^0(j)$ and entities $x(t;s,y,j),p(t;s,y,j)$ satisfying all conditions in Theorem 3.11. (Note than in a latter remark, Remark 3.45, the conditions on the dependence of $x(t;s,y,j),p(t;s,y,j)$ on (s,y) are slightly weakened.)

Example 3.12. Example 3.3 revisited Consider the case of the scrap value function $ax(T)^2/2$. Let $Q^0(j) = (0,T) \times \mathbb{R}$ for $j = 0,1$. In this case, functions $x(t;s,y,j)$, $p(t;s,y,j)$, $j = 0,1$, were found for each (s,y), having the required continuity and C^1-properties. Moreover, $\hat{H}$ and $\hat{u}$ are C^2 in $Q^*(0) = Q^*(1) = \mathbb{R}^3$, and (3.28) holds. Hence, $u(t;s,y,j)$, $j = 0,1$, is an optimal collection. $\qquad\square$

Connected to Theorem 3.11, there is a sufficient condition that involves solutions of a sequence of partial differential equation, the HJB equations of the problem.

In fact, the solutions $(x(t;s,y,j),p(t;s,y,j))$ yield the solutions to these HJB equations. A discussion of this theme is postponed to the next section on end constrained problems.

Remark 3.13 (The C^1-property in Theorem 3.11*). Sometimes, the C^1-property needs not be explicitly required. So Theorem 3.11 even holds when the C^1-property of $(s,y) \rightarrow (x(t;s,y,j),p(t;s,y,j))$ is deleted, provided it is assumed that $(s,y) \rightarrow p(s;s,y,j)$ is locally Lipschitz continuous in $Q^0(j)$. In addition the following consistency condition is then required: For $T \geq t' \geq t > s$, $(s,y) \in \tilde{Q}_0(j)$,

$$(x(t';s,y,j),p(t';s,y,j)) = (x(t';t,x(t;s,y,j),j),p(t';t,x(t;s,y,j),j),j)). \quad (3.29)$$

(Most often, this property automatically holds.)   □

Remark 3.14 (Time- and state-dependent jump intensity). Assume that $\lambda(j)$ is a C^1-function of (t,x). Then the density of $\tau = \tau_j \in [\tau_{j-1},\infty)$ is

$$\lambda(\tau,x(\tau),j)e^{\int_{\tau_{j-1}}^{\tau} -\lambda(\sigma,x(\sigma),j)d\sigma},$$

where $x(s) = x(s,\omega^{j-1})$, $(x(s) = x(T,\omega^{j-1})$ for $s > T)$. Now an auxiliary state variable z is needed, together with a modified adjoint equation:

$$\dot{z} = \lambda(t,x,j+1)\{z - g_0(t,x,j+1) - z(t;t,x+g(t,x,j+1),j+1)\}$$
$$\quad - f_0(t,x,\hat{u}(t,x,p),j), \quad (3.30)$$
$$\dot{p} = -H_x(t,x,\hat{u}(t,x,p),p,j) + \lambda(t,x,j+1)p$$
$$\quad - \lambda(t,x,j+1)g_{0x}(t,x,j+1)$$
$$\quad - \lambda(t,x,j+1)p(t;t,x+g(t,x,j+1),j+1)(I+g_x(t,x,j+1))$$
$$\quad - \lambda_x(t,x,j+1)[g_0(t,x,j+1)+z(t;t,x+g(t,x,j+1),j+1)-z]. \quad (3.31)$$

Instead of finding characteristic pairs satisfying (3.19)–(3.21) for $N, N-1, N-2, \ldots$, one now has to find characteristic triples $x(t;s,y,j)$, $p(t;s,y,j)$, $z(t;s,y,j)$, for $j = N, N-1, \ldots$ that satisfy simultaneously (3.19), (3.30) and (3.31), with $x(s) = y$, $p(T) = h_{0x}(x(T))$, $z(T) = h_0(x(T))$. See Seierstad (2001).   □

Remark 3.15 (Stochastic jump sizes). A stochastic disturbance $V_j \in \mathbb{R}^{r^*}$ may be included in $g_0(t,x,j)$ and $g(t,x,j)$. Thus, we now write $g_0(t,x,V_j,j)$ and $g(t,x,V_j,j)$. Here g, g_0 and their first and second derivatives are C^0 in (t,x,v) (the continuity in (t,x) being uniform in v), and the growth conditions in the Standard System Conditions (recall (3.25)) must hold uniformly in V_j as far as g_0 and g are concerned. The jump equation now reads $x(\tau_j+) - x(\tau_j-) = g(\tau_j,x(\tau-),V_j,j)$. For simplicity, it is assumed that all V_j's, $j = 1,2,\ldots$, belong to a given bounded set. If the stochastic variable V_j has a distribution that depends on τ_j, then the solution method connected with (3.18)–(3.21) again works when $N < \infty$, though in (3.20) we replace

$$g_{0x}(t,x,j+1)+p(t;t,x+g(t,x,j+1),j+1)(I+g_x(t,x,j+1))$$

by

$$E[g_{0x}(t,x,V_{j+1},j+1)$$
$$+p(t;t,x+g(t,x,V_{j+1},j+1),j+1)(I+g_x(t,x,V_{j+1},j+1))|t] \quad (3.32)$$

(The expectation is with respect to V_{j+1}.)

In ω, we now include all outcomes of the V_j's, so $\omega = (\tau_0, v_0, \tau_1, v_1, \tau_2, v_2, \ldots)$, ($v_0$ having no effect), and nonanticipating entities at time t now depend on the outcomes (τ_i, v_i) having occurred before t, e.g., when $\tau_j \le t < \tau_{j+1}$, then, for $\omega^j = (\tau_0, v_0, \ldots, \tau_j, v_j)$, $p(t, \omega)$ equals

$$p(t, \omega^j) := p(t, \tau_0, v_0, \ldots, \tau_j, v_j, T, V_{j+1}, T+1, V_{j+2}, \ldots).$$

The distribution of V_j can be allowed to depend on all τ_i, $i \le j$, and all earlier outcomes v_i, $i < j$. Then the necessary conditions of Theorem 3.8 still apply, provided we in (3.23) replace

$$g_{0x}(t, x^*(t, \tau_0, \ldots, \tau_j), j+1)$$
$$+p(t+, \tau_0, \ldots, \tau_j, t)\{I + g_x(t, x^*(t, \tau_0, \ldots, \tau_j), j+1)\}$$

by

$$E[g_{0x}(t, x^*(t, \tau_0, v_0, \ldots, \tau_j, v_j), V_{j+1}, j+1)$$
$$+p(t+, \tau_0, v_0, \ldots, \tau_j, v_j, t, V_{j+1})$$
$$\times \{I + g_x(t, x^*(t, \tau_0, v_0, \ldots, \tau_j, v_j), V_{j+1}, j+1)\}$$
$$| \tau_0, v_0, \ldots, \tau_j, v_j, t)]. \quad (3.33)$$

The function $p(t, \omega^j)$ is bounded, continuous in t, $t > \tau_j$, with $p(\tau_j+, \omega^j)$ existing. Moreover, for each such t, $p(t, \omega^j)$ is separately piecewise continuous in each τ_i, and in each v_i, $i \le j$, when the expectation $E[\alpha(V_{j+1})|\tau_0, v_0, \ldots, \tau_j, v_j, \tau_{j+1}]$ is continuous in $\tau_1, v_1, \ldots, \tau_j, v_j, \tau_{j+1}$ for any piecewise continuous function $\alpha(V_{j+1})$, which is assumed. □

Remark 3.16 (Reformulation of the necessary conditions*). Define $\mu((t,\infty)|\tau_0, \ldots, \tau_j) := \exp[-\lambda(j+1)(t-\tau_j)], \mu(t|\tau_0, \ldots, \tau_j) := 1 - \mu((t,\infty)|\tau_0, \ldots, \tau_j)$, $j = 0, 1, 2, \ldots, (\tau_0 = 0)$, $\mu(\tau_0, \ldots, \tau_j) := \mu(\tau_1, \ldots, \tau_j)$, the latter being the simultaneous cumulative distribution of $(\tau_1, \ldots, \tau_j)$ corresponding with the $\mu(t|\tau_0, \ldots, \tau_j)$'s. Define also

$$p^*(t, \tau_0, \ldots, \tau_j) := \mu((t,\infty)|\tau_0, \ldots, \tau_j) p(t, \tau_0, \ldots, \tau_j).$$

Let $D\mu$ denote the density of the cumulative distribution $\mu(\tau_1, \ldots, \tau_j)$, $(D\mu(\tau_0) := 1)$, and let, as before, a dot above a function mean a derivative with respect to time t. Then $p^*(t, \tau_0, \ldots, \tau_j)$ satisfies the differential equation (3.35) below. In fact, the necessary conditions (3.22)–(3.24) yield the following necessary conditions: Assume that $x^*(t, \omega)$, $u^*(t, \omega)$ is an optimal pair in problem (3.15)–(3.17). Then, there exists a nonanticipating function $p^*(t,\omega)$, such that for $\Lambda_0 = 1$, a.s., for v.e. $t > \tau_j$,

$$u \mapsto [\Lambda_0 f_0(t,x^*(t,\tau_0,\ldots,\tau_j),u,j)\mu((t,\infty)|\tau_0,\ldots,\tau_j)$$
$$+ p^*(t,\tau_0,\ldots,\tau_j)f(t,x^*(t,\tau_0,\ldots,\tau_j),u,j)]D\mu(\tau_0,\ldots,\tau_j)$$

has a maximum at $u^*(t,\tau_0,\ldots,\tau_j)$ in U. $\qquad$ (3.34)

Moreover, (still for $\Lambda_0 = 1$), for each ω, $t \to p^*(t,\tau_0,\ldots,\tau_j)$, $t > \tau_j$ is differentiable v.e., and satisfies v.e.

$$\dot{p}^*(t,\tau_0,\ldots,\tau_j)$$
$$= -\Lambda_0 f_{0x}(t,x^*(t,\tau_0,\ldots,\tau_j),u^*(t,\tau_0,\ldots,\tau_j),j)\mu((t,\infty)|\tau_0,\ldots,\tau_j)$$
$$- p^*(t,\tau_0,\ldots,\tau_j)f_x(t,x^*(t,\tau_0,\ldots,\tau_j),u^*(t,\tau_0,\ldots,\tau_j),j)$$
$$- \Lambda_0 g_{0x}(t,x^*(t,\tau_0,\ldots,\tau_j),j+1)\mu(t|\tau_0,\ldots,\tau_j)$$
$$- p^*(t+,\tau_0,\ldots,\tau_j,t)\{I + g_x(t,x^*(t,\tau_0,\ldots,\tau_j),j+1)\}\mu(t|\tau_0,\ldots,\tau_j). \quad (3.35)$$

Finally,

$$p^*(T,\tau_0,\ldots,\tau_j) = h_{0x}(x^*(T-,\tau_0,\ldots,\tau_j))\mu((T,\infty)|\tau_0,\ldots,\tau_j). \quad (3.36)$$

Here $p^*(t,\omega^j)$ is bounded, continuous in t, $t > \tau_j$, and is separately piecewise continuous in each τ_i, $i \le j$, and the limit $p^*(\tau_j+,\omega^j)$ exists. $\qquad \square$

The term $D\mu(\tau_0,\ldots,\tau_j)$ in (3.34) is entered only for the purpose of the next remark, here it is always positive and can be deleted. Moreover, a value $\Lambda_0 \ne 1$ is needed later on.

Remark 3.17 (General distribution of jump times*). The necessary conditions in Remark 3.16 also hold, if the probability distribution of the τ's are not of the type above. In fact, the above necessary conditions (3.34)–(3.36) hold for any given sequence of bounded conditional densities $\dot{\mu}(t|\tau_0,\ldots,\tau_j)$, $j = 0,1,\ldots$, for $t = \tau_{j+1} \in [\tau_j,\infty)$, $\dot{\mu}(t|\tau_0,\ldots,\tau_j)$ separately piecewise continuous in t and in each τ_i, $i > 0$ (for $j = 0$, the density is simply $\dot{\mu}(t) = \dot{\mu}(t|\tau_0)$). If $N = \infty$, for the necessary conditions to hold, a condition, (3.37) below, is needed. See Seierstad (2001). Define $\Theta_m(t,\omega^j)$ to be the conditional probability that exactly m more jumps occur in $[0,t]$, given ω^j, $\tau_j < t$. Assume the following inequality

$$\Theta_m(t,\omega^j) \le \Phi(t,j)v(t,j)^{m-j}, \quad (3.37)$$

for some positive numbers $\kappa, \Phi(t,j), v(t,j), v(t,j) \in (0,1/(1+\kappa))$, where $|g(t,x,j)| \le \alpha + \kappa|x|$ for all t,x,j, α some given positive number, κ being also a Lipschitz rank of $x \to g(t,x,j)$, $j = 1,2,\ldots$. Again a stochastic disturbance V can be allowed, with modifications as described in Remark 3.15.

The necessary conditions, with $\Lambda_0 = 1$ are sufficient if

$$x \to \max_{u \in U}\{f_0(t,x,u,j)\mu((t,\infty)|\tau_0,\ldots,\tau_j) + p^*(t,\omega)f(t,x,u,j)\}D\mu(\tau_0,\ldots,\tau_j),$$

$x \to p^*(t,\omega)g(t,x,j)$, $x \to g_0(t,x,j)$, and $x \to h_0(x)$ are concave for $t \in (\tau_j, \tau_{j+1})$.

The theory above applies also to the case where f and f_0 depend on the τ_j's, provided, for any given (x,u), these functions are nonanticipating in (t,ω). In Theorem 3.8, as well as in (3.34), (3.35), in the functions f_0, f, f_{0x}, f_x, we simply replace j by $\tau_0, \ldots, \tau_j$. Now, we must assume that the Lipschitz continuity is uniform in ω, (i.e., the Lipschitz ranks κ^n are independent of ω), and that inequality (3.25) holds for f_{0x} and f_x for all ω. Moreover, it is assumed that f_0, f, f_{0x} and f_x are separately piecewise continuous in each τ_i.

The same solution procedure applies as before. Let us relate it to the necessary conditions (3.34)–(3.36), (with $\Lambda_0 = 1$):

First, let $\hat{u}(t,x,p^*,\tau_0,\ldots,\tau_j)$ maximize $f_0(t,x,u,\tau_0,\ldots,\tau_j)\mu((t,\infty)|\tau_0,\ldots,\tau_j) + p^*f(t,x,u,\tau_0,\ldots,\tau_j)$. Then, in the case $N < \infty$, consider the equations

$$\dot{x}(t) = f(t,x,\hat{u}(t,x,p^*,\tau_0,\ldots,\tau_j),\tau_0,\ldots,\tau_j) \qquad (3.38)$$

and

$$\begin{aligned}
\dot{p}^*(t) = &-f_{0x}(t,x,\hat{u}(t,x,p^*,\tau_0,\ldots,\tau_j),\tau_0,\ldots,\tau_j)\mu((t,\infty)|\tau_0,\ldots,\tau_j) \\
&-p^*f_x(t,x,\hat{u}(t,x,p^*,\tau_0,\ldots,\tau_j),\tau_0,\ldots,\tau_j) \\
&-g_{0x}(t,x,j+1)\dot{\mu}(t|\tau_0,\ldots,\tau_j) \\
&-p^*(t+,\tau_0,\ldots,\tau_j,t)[I+g_x(t,x,j+1)]\dot{\mu}(t|\tau_0,\ldots,\tau_j). \qquad (3.39)
\end{aligned}$$

Boundary conditions for $(x(t),p(t))$ are again, for $s = \tau_j$, the two equalities $x(\tau_j) = y$ and

$$p^*(T) = h_{0x}(x(T))\mu((t,\infty)|\tau_0,\ldots,\tau_j). \qquad (3.40)$$

In case $N < \infty$, again these equations can be solved backwards. First for $j = N$, solutions of the equations are found, which we denote $x(t;y,\tau_0,\ldots,\tau_N)$, $p^*(t;y,\tau_0,\ldots,\tau_N)$. Next, for $j = N - 1$, solutions of the equations are found, which we denote $x(t;y,\tau_0,\ldots,\tau_{N-1}), p^*(t;y,\tau_0,\ldots,\tau_{N-1})$, for $p^*(t+,\tau_0,\ldots,\tau_{N-1},t)$ replaced by $p^*(t,x+g(t,x,N),\tau_0,\ldots,\tau_{N-1},t)$ (so in this case a knowledge of $p^*(t;y,\tau_0,\ldots,\tau_{N-1},t)$ is needed), and so on.

Further comments are given in Remark 3.44 below, where it is shown how this new type of problem can be rewritten to be of a type earlier considered, at least in a special case. $\qquad \square$

3.4 Problems with Terminal Constraints

Let us next consider problems with terminal constraints. The tools available in the general case are a little complicated, so we first consider a very special situation, where x is a scalar, the jump sizes are constant, the scrap value function is zero, and where the terminal inequality condition $x(T-,\omega) \geq \hat{x}$ a.s., is imposed, $\hat{x}$ a given number. Such a terminal condition is sometimes called hard, as opposed to a soft one, which takes the form $Ex(T-,\omega) \geq \hat{x}$, $\hat{x}$ a given number. Below, mainly hard end constraints are treated. Soft end constraints are briefly treated in Remark 3.35 below. Thus, we now have a system that, in addition to (3.15)–(3.17), is required to satisfy the conditions:

$$x \in \mathbb{R}, \ h_0 \equiv 0, \ g \equiv c \neq 0, \ x(T-,\omega) \geq \hat{x} \text{ a.s.}, \ N < \infty, \tag{3.41}$$

and that f_0 and f are independent of j. Now, if j jumps have occurred, then as long as no further jumps occur, $x(t;s,y,j)$ has to be steered in such a manner that if none, or one or more jumps occur in the future (after t), we are able to satisfy the end constraint. At most $N - j$ jumps can occur after s. Even if t is very close to T, such a number of jumps can occur with positive probability. If $c > 0$, we often need only be concerned with satisfying the inequality $x(T-,s,y,j) \geq \hat{x}$: If this inequality is satisfied, and given any t close to T, and also given that j jumps have occurred in $[0,t]$, then if one or more jumps occur after t, it is easier to obtain $x(T-;s,y,j') \geq \hat{x}$, $j' > j$ (sometimes these inequalities will even be automatically satisfied). If $c < 0$, · a more detailed discussion is needed. We will only consider two situations.

$$\sup_{u \in U} f(t,x,u) = +\infty \ \text{ for all } \ (t,x), \tag{3.42}$$

or, for some constant K,

$$\sup_{u \in U} f(t,x,u) \leq K \ \text{ for all } \ (t,x). \tag{3.43}$$

Let $\hat{x}^j = \hat{x}$ both in case c is positive, and in case (3.42) (even if c is negative). In the case where c is negative and (3.43) holds, let $\hat{x}^j = \hat{x} - (N - j)c$. In this case, we have to arrange it so that if j jumps have occurred, then we must make sure that $x(T-;s,y,j) + kc \geq \hat{x}$ for $k = 0,\ldots,N - j$ (k jumps downwards can occur arbitrarily close to T). The most demanding equality is obtained for $k = N - j$. So, in general we have to seek solutions $x(t;s,y,j)$ satisfying the end condition

$$x(T;s,y,j) \geq \hat{x}^j. \tag{3.44}$$

As usual, there is a slackness condition that the adjoint function must satisfy at time T. The functions $p(T;s,y,j)$ must satisfy the transversality conditions

$$p(T;s,y,j) \geq 0, = 0 \text{ if } x(T;s,y,j) > \hat{x}^j, j = 0,\ldots,N. \tag{3.45}$$

The extremal method for problem (3.15)–(3.17), (3.41), is the same as before, except that the boundary conditions are:

$$x(s;s,y,j) = y, \ (3.44) \text{ and } (3.45) \tag{3.46}$$

Thus, the extremal method now consist of (3.18)–(3.20), (3.46).

Example 3.18. Let us solve the problem $\max E\{\int_0^3 (3-s)(1-u)ds\}$, subject to $\dot{x} = u \in [0,1]$, $x(0) = 0$, $x(3) \geq 1$ with probability 1.

There is a probability with intensity $\lambda > 0$ that a single unit jump downwards occurs.

Solution. The conditions (3.18)–(3.20), (3.46) will be applied. After a jump, the solution is $u(t;s,y,1) = 0$ if $y \geq 1$, and if $y < 1$, $u(t;s,y,1) = 0$ for $t < y+2$, while $u(t;s,y,1) = 1$ for $t > y+2$. To see this, note that $p(t;s,y,1) = p(s;s,y,1)$. From the maximum condition, evidently $u = 0$ to begin with, where $3-t > p$ (if at all) and $u = 1$ at the end (if at all). Consider the case where we do have a point $\sigma \in (0,1)$ at which the control switches from 0 to 1. Then, $x(t;s,y,1) = y$ for $t < \sigma$, $x(t;s,y,1) = y+t-\sigma$ for $t > \sigma$, with σ determined by $x(3;s,y,1) = 1$ i.e. $y+3-\sigma = 1$, or $\sigma = y+2$. At $s = \sigma$, by the maximum condition, $3-\sigma = p(\sigma;s,y,1)$ (both $u = 0$ and $u = 1$ yield maximum in the maximum condition at σ), so $p(t;s,y,1) = 1-y$ (p is independent of s and t). The formula for $p(t;s,y,1)$ is valid if $\sigma \in (s,3)$, i.e., $y \in (s-2,1)$. It works also for $\sigma = s$, ($y = s-2$) and we use it, though then any $p(t;s,y,1) \geq 1-y$ would also work. The general formula for $p(t;s,y,1)$ is then $p(t;s,y,1) = \max\{0, 1-y\}$, $y \geq s-2$ (from points (s,y), $y < s-2$, it is not possible to reach $(3,1)$ or above).

For $j = 0$ we need only consider $(s,y) = (0,0)$, below $p(t)$ is a shorthand for $p(t;0,0,0)$, similarly $x(t) := x(t;0,0,0)$.

We guess, or try the possibility, that $x(t) - 1 \leq 1$ for all t, in which case $p(t;t,x(t)-1,1) = 2-x(t)$. Now, $dp/dt = \lambda p(t) - \lambda p(t;t,x-1,1) = \lambda p - \lambda 2 + \lambda x(t)$, while $dx/dt = 0$ if $3-t > p$ and $dx/dt = 1$ if $3-t < p$. We guess that the two last inequalities occur on intervals $[0,\alpha)$, and $(\alpha,3]$, respectively (this will be confirmed). Thus on the first interval, $x(t) = 0$, and so $p(t) = 2 + Ce^{\lambda t}$ here, for some integration constant C. On the last interval, $x(t) = D+t$, D an integration constant, and we guess that $x(3) = 2$ (a jump may occur close to $s = 3$, i.e. $x(T-)$ has to equal at least 2). Thus, $D = -1$, which, by $x(0) = 0$, means $\alpha = 1$, and which gives that $dp/dt = \lambda p(t) - \lambda 2 + \lambda(t-1)$ for $t \in (1,3)$.

The last equation has the solution $p(t) = 3 - t - 1/\lambda + e^{\lambda(t-1)}/\lambda$ for $t \in (1,3)$ (the integration constant is determined by $3 - \alpha = p(\alpha)$, i.e., $2 = p(1)$). For $t \in (0,1)$, $p(t) = 2$, (here $p(t) = 2 + Ce^{\lambda t}$, and $2 = p(1)$ gives $C = 0$).

Let us confirm a couple of properties: The transversality condition of (3.45) is satisfied ($p(3) > 0$). Moreover, as $3-t$ is decreasing and $p(t)$ is nondecreasing, the maximum condition is satisfied by $u = 0$ before $\alpha = 1$, and by $u = 1$ afterwards. Note that a solution has to be found such that the end constraint is met with probability 1. This requires $x(T-;0,0,0) \geq 2$. The solution $x(t;0,0,0) = \max\{0, t-1\}$ satisfies the latter inequality and it describes the optimal behavior before a jump. After a jump, the optimal behavior is as follows: If the jump occurs in $[0,1)$, we use $u = 0$ from then on until $t = 1$, and then $u = 1$, if the jump occurs in $[1,3)$, we use $u = 1$ all the time after the jump.

Note that the solution is the same for all $\lambda > 0$. Sufficient conditions, see Remark 3.26 below, apply here. $\square$

Example 3.19. Consider the problem

$$\max E\left[\int_0^T -u^2/2dt\right], \quad \dot{x}=u,\ x(0)=-2,\ x(T)\geq 0 \text{ a.s.},\ u\in\mathbb{R},$$

with a possibility for a single, unit upwards jump in $x(t)$ at $\tau \in [0,\infty)$, τ exponentially distributed with intensity λ.

Solution. When $y \geq 0$, evidently $u(t;s,y,1)=u(t;s,y,0)=0$ are optimal (with the p-function vanishing), so we consider only y's satisfying $y < 0$. To find $u(t;s,y,1)$ and $u(t;s,y,0)$, the solution method (3.18)–(3.20), (3.46) is used. We seek a solution $x(t;s,y,1)$ satisfying $x(T;s,y,1) \geq 0$. In fact, we guess that, after a jump has occurred, we steer x in such a way that $x(T;s,y,1) = 0$. (The guess is confirmed below.) As $u = p(t;s,y,1) \equiv \hat{p}(s,y)$ by the maximum condition, we choose $\hat{p}(s,y)$ such that if we start at (s,y), we end at $(T,0)$, i.e., $x(T;s,y,1) = y + \int_s^T \hat{p}(s,y)dt = 0$, or $\hat{p}(s,y) = -y/(T-s)$. Moreover, $u(t;s,y,1) = y/(s-T), x(t;s,y,1) = y(t-T)/(s-T)$ for $y < 0$.

To find $u(t;s,y,0)$, $y < 0$, consider the characteristic equations: $\dot{x}=p$, $\dot{p} = -\lambda p(t;t,x+1,1)+\lambda p = \lambda(x+1)/(T-t)+\lambda p$, (drop the first term if $x+1 \geq 0$). Differentiating the first equation and inserting from the second and first equations give $d^2x/dt^2 = \lambda dx/dt - \lambda(x+1)/(t-T)$ for $x \leq -1$ (drop the last term in the opposite case), with boundary conditions $x(s;s,y,0)=y$, $x(T;s,y,0)=0$. Defining $w(t)$ by $w(t)(t-T) = x+1$, we get $dx/dt = (dw/dt)(t-T)+w$ and $d^2x/dt^2 = (d^2w/dt^2)(t-T)+2dw/dt = \lambda dx/dt - \lambda(x+1)/(t-T) = \lambda(dw/dt)(t-T) + \lambda w - \lambda w$, so $d^2w/dt^2 = (\lambda - 2/(t-T))dw/dt$. Integrating this separable equation, we get $\dot{w}(t) = Ce^{\lambda t}/(t-T)^2$, and $w(t) = D + \int_s^t Ce^{\lambda\sigma}/(\sigma-T)^2d\sigma$ (C and D are integration constants). So $w(t)(t-T)-1 = x(t) =: x(t;s,y,0) = \beta(t;s,y) := (t-T)[D + \int_s^t Ce^{\lambda\sigma}/(\sigma-T)^2d\sigma]-1 = (T-t)[(y+1)/(T-s) - \int_s^t Ce^{\lambda\sigma}/(T-\sigma)^2d\sigma] - 1$, as long as $x(t) \leq -1$ (we have used $x(s) = y$ to determine D). The equation for x is simpler when $y > -1$, then $x(t;s,y,0) = D' + C'e^{\lambda(t-s)}$, D' and C' integration constants. With $x(T) = 0$, $x(s) = y$, this gives $x(t;s,y,0) = \beta^*(t;s,y) := -ye^{\lambda T}/(e^{\lambda s} - e^{\lambda T})+ye^{\lambda t}/(e^{\lambda s} - e^{\lambda T})$, for $y \in (-1,0)$.

Ultimately, before any jump, we need to know the solution $x(t;s,y,0)$ only for $(s,y) = (0,-2)$. The solution $x(t;0,-2,0)$ is "smoothly pasted" together at the point $t'' < T$ at which the line $x = -1$ is crossed, by putting $\beta(t'',0,-2) = -1$ $(= -\beta^*(t'',t'',-1))$, as well as by requiring the first derivatives of $t \to \beta$ and $t \to \beta^*$ at t'' to be equal, which yields $\lambda e^{\lambda t''}/(e^{\lambda T} - e^{\lambda t''}) = [\partial\beta^*(t;t'',-1)/\partial t]_{t=t''} = [\partial\beta(t;0,-2)/\partial t]_{t=t''} = (-1)[-1/T - \int_0^{t''} Ce^{\lambda\sigma}/(\sigma-T)^2d\sigma]-Ce^{\lambda t''}/(T-t'')$. (The first characteristic equation does tell that $\dot{x}$ is continuous.) The two equations determine C and t'': The first equation reduces to $-1/T - C\int_0^{t''} e^{\lambda\sigma}/(T-\sigma)^2d\sigma = 0$, using this in the second one, it reduces to $\lambda e^{\lambda t''}/(e^{\lambda T} - e^{\lambda t''}) = -Ce^{\lambda t''}/(T-t'')$, or $\lambda/(e^{\lambda T} - e^{\lambda t''}) = -C/(T-t'')$, which gives $C = (t''-T)\lambda/(e^{\lambda T} - e^{\lambda t''})$. Using this, the first equality becomes

$$-1/T = -\{\lambda(T-t'')/(e^{\lambda T} - e^{\lambda t''})\}\int_0^{t''} e^{\lambda\sigma}/(T-\sigma)^2d\sigma.$$

For $t'' = 0$, the right-hand side is 0, while for $t'' = T$ it is $-\infty$ (when $t'' \to T$, the curly bracket has a positive limit, while the integral diverges). So a solution in $(0, T)$ for t'' exists, which then determines C.

Having found C and t'', then $x(t; 0, -2, 0) = \beta(t; 0, -2)$ for t in $(0, t'')$ and $x(t; 0, -2, 0) = \beta^*(t; t'', -1)$ for $t \geq t''$. Using sufficient conditions based on concavity (Remark 3.26 below), it can be shown that the solutions $x(t; s, y, 1)$ and $x(t; 0, -2, 0)$ are optimal. Here the solution depends on λ.   □

3.5 The General End Constrained Case

The assumption (3.41) will now be dropped, and f_0 and f now depend also on j. In the standard problem (3.15)–(3.17), the following "hard" end conditions are imposed. With probability 1,

$$x_i(T-, \omega) = \hat{x}_i, \qquad i = 1, \ldots, n', \tag{3.47}$$

$$x_i(T-, \omega) \geq \hat{x}_i, \qquad i \geq n' + 1, \ldots, n''. \tag{3.48}$$

Define $U^m := U \cap B(0, m)$. In addition to the Standard System Conditions on the system presented subsequent to Theorem 3.8, it is now assumed, for any m, that f_{0x} and f_x are uniformly continuous in x, uniformly in (t, u, j), $u \in U^m$, that $g_{0x}(t, x, j)$, and $g_x(t, x, j)$ are uniformly continuous in x, uniformly in t, j, and, finally, that, for some constants M'_j, for all t, x,

$$|g_x(t, x, j)| \leq M'_j, \quad |g(t, 0, j)| \leq M'_j. \tag{3.49}$$

$$\sum_{j=1}^{N} M'_j < \infty. \tag{3.50}$$

where N is the bound on the number of jumps that can occur. The last condition is of course trivially satisfied unless $N = \infty$. A slight variant of condition (3.49) is mentioned in Remark 3.23 below. When $N = \infty$, the condition (3.50) is of course very demanding (and rather artificial), but let us mention that it covers the important special case where there are no jumps in the state x, but where the right-hand side of the equation changes form at each time point τ_i. However, sometimes we can do without (3.50) even in the case where an unbounded number of jumps can occur.

We call a nonanticipating pair $(x(t, \omega), u(t, \omega))$ semiadmissible if, a.s., $x(t, \omega)$ exists as a solution of (3.15), (3.16) on all $[0, T]$, $\sup_{t \leq T', \omega} |u(t, \omega)| < \infty$ for any $T' < T$, and $E[\int_J |\dot{x}(t, \omega)| dt < \infty$ and $E \int_J |f_0(t, x(t, \omega), u(t, \omega))| dt < \infty$. Occasionally, full boundedness of controls may be a too restrictive requirement; an example appears in Exercise 3.10.

Let Π be the map $(x_1, \ldots, x_n) \to (x_1, \ldots, x_{n''})$, and let

$$\hat{B} = \{x \in \mathbb{R}^n : x_i = 0, i \leq n', x_i \geq 0, n' + 1 \leq i \leq n''\}. \tag{3.51}$$

The possibility of being able to satisfy (3.47), (3.48) depends on the behavior of the process (i.e., on the "controllability" properties of the process). For example, if f is a bounded function (implying case (3.53) below), we cannot hope to have (3.47) satisfied, when large jumps in $x_i(.)$, $i \leq n'$ occur close to T. We shall only discuss systems satisfying one of the two condition: For some $T' \in [0,T)$, for all x, j,

$$\mathbb{R}^{n''} \subset \Pi\{f(t,x,u,j) - b : u \in U, b \in \hat{B}\} \quad \text{for all} \quad t \in [T',T], \tag{3.52}$$

or for all bounded sets $B^* \subset \mathbb{R}^n$,

$$\sup_{j,x \in B^*, t \in [T',T], u \in U} |\Pi f(t,x,u,j)| < \infty \ , \ g_i \equiv 0 \text{ for all } i \leq n'. \tag{3.53}$$

The problem now consists of (3.15)–(3.17), (3.47), (3.48), (3.49), and either (3.52) or (3.53). For solutions to be admissible, (3.47) and (3.48) have to hold and in case (3.52), we simply steer the system such that (3.47) and (3.48) hold. In case (3.53), the situation is more complicated. Define $g^{0,j}(x) = x$, $g^{1,j}(x) = x + g(T,x,j+1)$, $g^{2,j}(x) = g^{1,j}(x) + g(T,g^{1,j}(x),j+2)$, and recursively $g^{k,j}(x) = g^{k-1,j}(x) + g(T,g^{k-1,j}(x),j+k)$. As long as only j jumps have occurred, we have to steer the system in such a manner that the equalities $x_i = \hat{x}_i$, $i \leq n'$ and the inequalities $(g^{k,j}(x))_i \geq \hat{x}_i$, $k = 0,\ldots,N-j$, $i = n'+1,\ldots,n''$, hold for $x = x(T-;s,y,j)$. The reason is that however close t is to T, further jumps can occur. Let us give some further explanation:

Note that in case (3.53), an admissible solution must not only satisfy (3.48) (and (3.47)), but for $T > \tau_j$, even $g^{k,j}(x(T-,\omega^j)) \geq \hat{x}_i$, $k = 1,2,\ldots,N-j$. To see this, note that when t is close to T, (by (3.53)), $x(t,\omega^j)) \approx x(T-,\omega^j)$, and in (t,T), for any $k = 1,\ldots,N-j$, there is a positive probability that exactly k further jumps occur. Assuming only slightly erroneously that all these jumps occur at t, then $x(T-,\omega^{j+k}) \approx g^{k,j}(x(t,\omega^j)) \approx g^{k,j}(x(T-,\omega^j))$ and the assertion follows.

Remark 3.20 (Extremal method, end constrained case, $N < \infty$). The extremal method for finding solutions described in connection with (3.18)–(3.20) above also works here, the only change needed is a change of the terminal condition on $p(T)$: We will need numbers (multipliers) $\Lambda_i(k,s,y,j), k = 0,\ldots,N-j, i = 1,\ldots,n'', j = 0,1,2\ldots$, with $\Lambda(k,s,y,j) := (\Lambda_1(k,s,y,j),\ldots,\Lambda_{n''}(k,s,y,j),0,\ldots,0) \in \mathbb{R}^n$, to formulate the condition.

$$p(T;s,y,j) = h_{0x}(x(T;s,y,j)) + \sum_{0 \leq k \leq N-j} \Lambda(k,s,y,j)g_x^{k,j}(x(T;s,y,j)),$$

$$\Lambda_i(k,s,y,j) \geq 0, \ i = n'+1,\ldots,n'',$$

$$\Lambda_i(k,s,y,j) = 0 \text{ if } (g^{k,j}(x(T;s,y,j)))_i > \hat{x}_i, \ i = n'+1,\ldots,n'',$$

$$\Lambda_i(k,s,y,j) = 0, \text{for } k > 0, \ i \leq n', \text{ and also for } i > n' \text{ in case of (3.52)},$$

$$\tag{3.54}$$

so $p_i(T;s,y,j) = (h_{0x}(x(T;s,y,j)))_i, \ i > n''.$

To solve the end constrained problem (3.15)–(3.17), (3.47)–(3.49) (i.e., to find characteristic pairs) one applies (3.18)–(3.20), with boundary conditions

$$(3.47), \ (3.48), \ (3.54) \ \text{and} \ x(s;s,y,j) = y. \tag{3.55}$$

(Formally, there may be a need to find also "abnormal" pairs; this is commented upon subsequent to Theorem 3.22 below.)   □

Remark 3.2 above describes how characteristic triples $x(t;s,y,j)$, $u(t;s,y,j)$, $p(t;s,y,j)$ give rise to corresponding non-anticipating entities $x(t,\omega), u(t,\omega), p(t,\omega)$. Similarly, to $\Lambda(k,s,y,j)$ corresponds $\Lambda(k,\tau_0,\ldots,\tau_j) := \Lambda(k,\tau_j,x(\tau_j+,\tau_0,\ldots,\tau_j),j)$.

Remark 3.21 (Which start-points should enter characteristic triples). Let us discuss a little more the start-points (s,y) used to construct the characteristic solutions.

At the outset, we must assume that admissible solutions exist. Now admissibility implies that the terminal conditions are satisfied, and this sense of admissibility should also be understood when reading the definition of $Q(j)$ in (3.26). The just mentioned assumption means that $Q(j)$ is nonempty. As before, solutions $x(t;s,y,j)$ must be constructed for all points (s,y) in $Q(j)$, (as well as for $(0,x^0)$ when $j = 0$).

□

In a sufficient condition below larger open sets $Q^0(j)$ appear. Sometimes they can be constructed again by considering all solutions $x(t,\omega)$, starting at $t = 0$ at all $\hat{x}^0$ in an open ball around x^0 as mentioned in the discussion subsequent to (3.26). Occasionally, even $Q^0(j) = (0,T) \times \mathbb{R}^n$ works. But note that now it can be more difficult (or in some cases impossible) to find such larger *open* sets $Q^0(j)$ that have the property that characteristic solutions can start anywhere in these sets and still satisfy the terminal conditions.

To obtain a maximum principle (necessary condition), a condition is needed that secures that "enough choices" of values of $\dot{x}$ are available: Let $x^*(t,\omega)$ be an optimal solution, and let Π as before be the map $(x_1,\ldots,x_n) \to (x_1,\ldots,x_{n''})$. For some $\eta > 0, c > 0$, some bounded nonanticipating function $z(t,\omega)$, piecewise and left-continuous in t, piecewise continuous in each τ_i, for all ω, for all $t \in [T - \eta, T]$ such that $t \in (\tau_j, \tau_{j+1})$,

$$B(\dot{x}^*(t,\omega) + z(t,\omega), \eta) \subset$$
$$\Pi\{f(t,x^*(t,\omega),u,j) - b : u \in U \cap B(0,c), b \in \hat{B} \cap B(0,c)\}. \tag{3.56}$$

In case of (3.52), this condition often follows more or less automatically.

The following maximum principle will *normally* hold as a necessary condition for optimality. The assumptions on f_0, f, g_0, g, h_0 subsequent to (3.48) and to Theorem 3.8 are again postulated.

Theorem 3.22 (Necessary conditions, end constrained case, $N \leq \infty$). *Let $(x^*(t,\omega), u^*(t,\omega))$ be an optimal pair in the end constrained problem (3.15)–(3.17), (3.47), (3.48), (3.49), (3.50), (3.53), $u^*(t,\omega)$ bounded. Assume (3.56). Then Theorem 3.8 holds with two modification: The transversality condition (3.24) on*

$p(T, \omega)$ *is changed to the following one: There exist a number* $\Lambda_0 \in \{0, 1\}$ *and multipliers* $\Lambda_i(k, \tau_0, \ldots, \tau_j)$, $k = 0, \ldots, N - j$, $i = 1, \ldots, n''$ *such that (3.54) holds for* $(\Lambda_i(k, s, y, j), p(T; s, y, j), x(T; s, y, j), h_0(x(T; s, y, j)))$ *replaced by* $(\Lambda_i(k, \tau_0, \ldots, \tau_j), p(T, \tau_0, \ldots, \tau_j), x^*(T, \tau_0, \ldots, \tau_j), \Lambda_0 h_0(x^*(T, \tau_0, \ldots, \tau_j)))$, $\sup_k |\Lambda_i(k, \tau_0, \ldots, \tau_j)| < \infty$. *Moreover, in (3.22)–(3.23),* f_0, f_{0x}, *and* g_{0x} *are replaced by* $\Lambda_0 f_0$, $\Lambda_0 f_{0x}$, *and* $\Lambda_0 g_{0x}$, *respectively. Furthermore,*

$$\sup_{|z(t, \omega)| < 1} E\left[\int_0^T p(t, \omega) z(t, \omega) dt\right] < \infty, \tag{3.57}$$

(the supremum taken over functions $z(t, \omega)$ *having properties as in (3.56)). Finally, if* $\Lambda_0 = 0$, *then not all functions* $\Lambda_i(k, \tau_0, \ldots, \tau_j)$, $i = 1, \ldots, n''$, $j = 0, 1, \ldots, k = 0, \ldots, N - j$ *vanish.* □

Of course, if $N = \infty$, then $k \in \{0, 1, 2, \ldots\}$. If $N < \infty$, j belongs to $\{0, 1, \ldots, N\}$, otherwise $j = 0, 1, \ldots$ The functions $\Lambda(k, \tau_0, \ldots, \tau_j)$ can normally be assumed to be bounded functions, piecewise continuous in each variable τ_i separately. The condition (3.56) will automatically ensure $\Lambda_0 = 1$ if $z(t, \omega) = 0$.

Unless sufficient conditions (Theorems 3.24 and 3.27 below) are used, to ensure that all candidates for optimality have been found, it is necessary to apply the solution procedure connected with Theorem 3.22 both for $\Lambda_0 = 1$ and $\Lambda_0 = 0$.

Formally, (3.56) is required to hold for $x^*(t, \omega)$. But, as for other constraints qualifications, in practice, we need that (3.56) holds for all admissible pairs, not only for $x^*(t, \omega)$. (Or, more carefully worded: If (3.56) fails for some admissible solution, it must automatically be considered a candidate for optimality.)

The word "normally" was used a couple of times above to indicate that there are cases, rather exceptional ones, in which necessary conditions require multipliers less well-behaved than the above ones. A proof of such conditions appears in Seierstad (2002).

Remark 3.23 (Change in assumptions on g*). Theorem 3.22 holds also in the case where $f_j(t, x, u, j)$, $j \le n''$, is independent of $x_i, i > n''$, and (3.49) holds for g_x replaced by Πg_x. □

The sufficient conditions in Theorem 3.9 now takes the following form (formally we then need not assume (3.56)):

Theorem 3.24 (Sufficient conditions, end constrained problems). *If we substitute the transversality condition of Theorem 3.8 by the one in Theorem 3.22 for* $\Lambda_0 = 1$, *then Theorem 3.9 still holds in the end constrained case, provided all* $\Lambda(k, \tau_0, \ldots, \tau_j) g^{k,j}$ *are concave in x.* □

Remark 3.25 (Sufficient conditions for semiadmissible pairs). In the end constrained problem (3.15)–(3.17), (3.47), (3.48), (3.49), (3.52), or (3.53), the sufficient conditions of Theorem 3.24 yield optimality even for semiadmissible pairs, provided the transversality condition is replaced by $\limsup_{t \to T} E[p(t, \omega) (x(t, \omega) - x^*(t, \omega))] \ge$

$E[h_{0x}(x^*(T,\omega))(x(T;\omega)-x^*(T,\omega))]$ for all semiadmissible solutions. Concavity of $\Lambda(k,\tau_0,..,\tau_j)g^{k,j}$ can then be dropped.

This also holds if $T=\infty$, the concept of optimality, if neccessary, being replaced by sporadic catching up optimality, see the definition subsequent to (3.70) below, in case $h_0 \equiv 0$.   □

Remark 3.26 (Sufficient concavity condition for characteristic triples, $N < \infty$). Similarly, in problem (3.15)–(3.17), (3.47), (3.48), (3.49), (3.52), or (3.53), the characteristic triples $x(t;s,y,j), u(t;s,y,j)$, $p(t;s,y,j)$ yield an optimal control among admissible pairs, if the concavity conditions in Remark 3.10 hold, if the pair $(x^*(.,.),u^*(.,.))$ arising from the characteristic triples is admissible, and if, in case (3.53), all $\Lambda(k,s,y,j)g^{k,j}$ are concave in x.   □

Define $t \to z(t;s,y,j)$ to be the solution of the following differential equation (linear in z), when $(x(t;s,y,j),p(t;s,y,j))$ is inserted for (x,p):

$$\dot{z}(t) = \lambda(j+1)\{z - g_0(t,x,j+1) - z(t;t,x+g(t,x,j+1),j+1)\}$$
$$-f_0(t,x,\hat{u}(t,x,p,j),j), \quad z(T) = h_0(x(T;s,y,j)) \tag{3.58}$$

Assume that all $x(t;s,y,j),p(t;s,y,j)$ are known. Solving (3.58) by using backwards recursion yields solutions $z(t;s,y,j)$. (With $\lambda(N+1)=0$, $z(t;s,y,N)$ can first be constructed from (3.58).) Define

$$J(s,y,j) := z(s;s,y,j) \tag{3.59}$$

Then, normally (for example when the conditions in Theorem 3.27 to follow are satisfied), $J(s,x,j)$ is the optimal value in the problem where already j jumps have occurred, and where the process starts at (s,y).

Theorem 3.27 (Sufficient conditions for a field of characteristic triples, with $N < \infty$ and terminal constraints). *Consider the problem (3.15)–(3.17), (3.47), (3.48), (3.49), (3.52), or (3.53). For each $j = 0,1,2,\ldots,N$, the following assumptions are made: A function $\hat{u}(t,x,p,j)$ exists that gives maximum in (3.18) for each point (t,x,p) in a given open set $Q^*(j) \subset \mathbb{R}^{1+2n}$. Open sets $Q^0(j) \subset (0,T) \times \mathbb{R}^n$, $Q^0(j)$ containing $Q(j)$, and solutions $x(t;s,y,j)$, $p(t;s,y,j)$, $z(t;s,y,j)$ on $[s,T]$ have been found, satisfying, for $u(t;s,y,j) = \hat{u}(t,x(t;s,y,j),p(t;s,y,j))$, $x(s;s,y,j) = y$, (3.19)–(3.20), and (3.58) (the first ones being the characteristic equations of $x(.;.,.,.)$ and $p(.;.,.,.)$), for $(s,y) \in Q_0(0) := \{(0,x^0)\} \cup Q^0(0)$ in case $j = 0$, and for $(s,y) \in Q_0(j) := Q^0(j)$ for any $j > 0$. (The functions $x(t;s,y,j)$, $p(t;s,y,j)$, and $z(t;s,y,j)$, $(s,y) \in Q_0(j)$, are continuous in $t \in [s,T]$ and $u(t;s,y,j)$ is piecewise continuous.) The characteristic pairs now satisfy the present terminal and transversality conditions (3.47), (3.48), and (3.54), with $\Lambda_i(k,s,y,j)$ continuous in $(s,y) \in Q_0(j)$, for $i = n'+1,\ldots,n''$. For all $t \in [s,T)$, $(s,y) \in Q_0(j)$, $(t,x(t;s,y,j), p(t;s,y,j))$ belongs to $Q^*(j)$, $(t,x(t;s,y,j))$ belongs to $Q_0(j)$, and if $(s,y) \in Q^0(j)$, then $(s,y+g(s,y,j+1)) \in Q^0(j+1)$.*

Moreover, $\hat{H}(s,x,p,j)$ and its first and second derivatives with respect to x and p (exist and) are continuous in $Q^*(j)$.

Furthermore, $(s,y) \to (z(t;s,y,j),x(t;s,y,j),p(t;s,y,j))$ is C^1 near any $(\hat{s},\hat{y})$ in $Q^0(j)$, $j = 0,1,2,\ldots,N$, for $t \in (\hat{s},T]$, and, for $j = 0$, $x(T;s,y,0)$ and $z(s;s,y,0)$ are continuous on $\{(0,x^0)\} \cup Q^0(0)$, and, for any j, $t \to (z_s(t;s,y,j),z_y(t;s,y,j), x_s(t;s,y,j),x_y(t;s,y,j))$ is continuous at $t = T$, $(s,y) \in Q^0(j)$.

Finally, the pair $(x^*(t,\omega), u^*(t,\omega))$ corresponding with the pairs $(x(.,.,j), u(.,.,j))$ is semiadmissible.

Then $(x^*(t,\omega),u^*(t,\omega))$ is optimal in the set of all strongly semiadmissible pairs, i.e., in the set S of semiadmissible pairs $(\hat{x}(t,\omega),\hat{u}(t,\omega))$ for which $E[|J(\tau_j, \hat{x}(\tau_j+,\omega),j)1_{[0,T]}(\tau_j)|] < \infty$ for all j and $\lim_{t \to T} z(t;t,\hat{x}(t,\omega^j),j) = h_0(x(T,\omega^j))$ for all $\omega^j, \tau_j < T$, provided $(x^*(t,\omega),u^*(t,\omega))$ belongs to S. □

Remark 3.28 (Strong semiadmissibility). When U is bounded, S coincides with the set of admissible pairs.

Consider next the case of an unbounded U. Usually, it does not restrict the usefulness of the above theorem, if in the theorem it is also assumed that for all $T' \in (0,T)$, $z(s;s,y,j)$ is bounded on bounded subsets of $Q^0(j) \cap ([0,T'] \times \mathbb{R}^n)$. In this case, the last inequality in the theorem (strong semiadmissibility) holds for semiadmissible solutions if, for such solutions, $\lim_{t \to T} z(t;t,\hat{x}(t,\omega^j),j) = h_0(\hat{x}(T,\omega^j))$ and $\lim_{t \to T}\{z(T;t,\hat{x}(t+,\omega^j,t),j+1) - z(t;t,\hat{x}(t+,\omega^j,t),j+1)\} = 0$, for any j, both limits uniform in ω^j. □

In the next example, there are no jumps in the state variable, but the dynamics change spontaneously after a stochastic time point.

Example 3.29. Consider the problem:

$$\max E\left[\int_0^T (-u^2/2)dt\right], \quad \dot{x} = u + 1_{[0,\tau]}(t), \ u \in \mathbb{R}, \ x(0) = 0, \ x(T) = 0 \text{ a.s.}$$

with no jumps in x, τ being exponentially distributed with intensity λ (i.e., the jump point τ is distributed with density $\lambda e^{-\lambda \tau}$).

Solution. The maximum condition gives $\hat{u} = p$. The characteristic equations for $j = 1$ (i.e., after a jump) becomes $\dot{p} = 0$, $\dot{x} = u = \hat{p}(s,y)$, $\hat{p}(s,y)$ an unknown to be determined by $x(T;s,y,1) = y + \int_s^T \hat{p}(s,y)dt = 0$, i.e. $\hat{p}(s,y) = -y/(T-s)$, so $x(t;s,y,1) = y - y(t-s)/(T-s)$.

Note that for $j = 0$, the characteristic equations are $\dot{x} = u+1 = p+1, \dot{p} = \lambda x/(T-t) + \lambda p$, so $\ddot{x} = \dot{p} = \lambda x/(T-t) + \lambda p = \lambda x/(T-t) + \lambda(\dot{x}-1)$. Letting $w(t-T) = x$, we get $\dot{w}(t-T) + w = \dot{x}, \ddot{w}(t-T) + 2\dot{w} = \ddot{x} = \lambda x/(T-t) + \lambda(\dot{x}-1) = -\lambda w + \lambda(\dot{w}(t-T)+w-1) = \lambda\dot{w}(t-T) - \lambda$. So $\ddot{w} = (\lambda - 2/(t-T))\dot{w} - \lambda/(t-T)$. Using the integrating factor $e^{-\lambda t}(t-T)^2$, we get $(d/dt)[\dot{w}e^{-\lambda t}(t-T)^2] = -\lambda e^{-\lambda t}(t-T)$, and integrating yields $\dot{w}e^{-\lambda t}(t-T)^2 = C + e^{-\lambda t}(t-T) + (1/\lambda)e^{-\lambda t}$; C an integration constant. Hence, $\dot{w} = Ce^{\lambda t}/(t-T)^2 + 1/(t-T) + (1/\lambda)/(t-T)^2$. As $(w(s;s,y) :=)w(s) = y/(s-T)$, $w(t) = y/(s-T) + C\beta(s,t) + \ln|t-T| - \ln|s-T| - (1/\lambda)/(t-T) + (1/\lambda)/(s-T)$. where $\beta(s,t) = \int_s^t e^{\lambda r}/(r-T)^2 dr$. Thus, $x(t;s,y,0) = (t-T)y/(s-T) + (t-T)C\beta(s,t) + (t-T)\ln|t-T| - (t-T)\ln|s-T| - (1/\lambda) + (t-T)(1/\lambda)/(s-T)$.

It is easily seen that when $t \to T$, $(t-T)\beta(s,t) \to -e^{\lambda T}$, so $x(T) = 0$ gives $-Ce^{\lambda T} - 1/\lambda = 0$, or $C = -e^{-\lambda T}/\lambda$. Hence, $\dot{w} = (t-T)^{-2}[(t-T) + 1/\lambda - e^{\lambda(t-T)}/\lambda]$.

Now, $z(t;s,y,1) = -y^2(T-t)/2(T-s)^2$, and for any admissible solution $\hat{x}(t,\omega)$ with a bound K on its time derivative $z(s;s,\hat{x}(s+,s),1) = -(\hat{x}(s+,s))^2/2(T-s) \to 0$ when $s \to T$, since $|\hat{x}(s+,s)| = |\hat{x}(s+,s) - \hat{x}(T,s)| \leq K(T-s)$. Moreover, $z(t;s,y,0) =$

$$\int_T^t \left[e^{\lambda(t-r)}(\dot{x}(r;s,y,0) - 1)^2/2 \right] dr + \int_T^t \left[e^{\lambda(t-r)}\lambda(x(r;s,y,0))^2/2(T-r) \right] dr.$$

Now, $\dot{w}$ is bounded (use l'Hopital's rule twice to find $\dot{w}(T-)$), hence $|\dot{x}| = |\dot{w}(t-T) + w| \leq |w(s)| +$ a constant, so, using the rule $(a+b)^2 \leq 2a^2 + 2b^2$, the absolute value of the two last integrands are both $\leq A(w(s))^2 + B$, for some constants A and B. Furthermore, as $w(s;s,y) = y/(s-T)$, then, $|z(s;s,y,0))| \leq \int_s^T [A(w(s))^2 + B]dt = y^2/(T-s) + B(T-s)$. For $y = \hat{x}(t,\omega^0)$ this entity is $\leq \{K(T-s)\}^2/(T-s) + B(T-s)$, which converges to zero, when $s \to T$.

Evidently, $(s,y) \to (z(t;s,y,j), x(t;s,y,j), p(t;s,y,j))$ is C^1 for (s,y) near any $(\hat{s},\hat{y}) \in [0,T) \times \mathbb{R}, t \in (\hat{s},T], j = 0,1$, moreover the derivatives of the three functions with respect to s and y are continuous in t at $t = T$, and $z(s;s,y,j)$ is continuous in $[0,T) \times \mathbb{R}$.

Finally, note that $\dot{x}^*(t,\omega)$, $t > \tau$, has the bound $|x(\tau;0,0,0)|/(T-\tau) = |x(\tau;0,0,0) - x(T;0,0,0)|/(T-\tau) \leq K(T-\tau)/(T-\tau) = K$, where K is a bound on $\dot{x}(t;0,0,0)$, so $(x^*(.,.,), u^*(.,.,.))$ is admissible. Hence Theorem 3.27, with subsequent Remark 3.28, applies for $Q^*(j) = (0,T) \times \mathbb{R}^2$, $Q^0(j) = (0,T) \times \mathbb{R}$, and yields optimality of $u(t;s,y,j)$, $j = 0,1$ in the set admissible pairs, or optimality in this set of the corresponding function $u^*(t,\omega)$. □

Remark 3.30 (Restrictions on admissible functions).* If the assumption $Q(j) \subset Q^0(j)$ is dropped, but $x^*(t,\omega^j) \in Q^0(j)$ for $t \in (\tau_j, T)$ and still $(x^*(t,\omega), u^*(t,\omega)) \in S$, then optimality of $u(t;s,y,j)$ holds in the subset of strongly semiadmissible solutions $\hat{x}(.)$, for which $(t,\hat{x}(t,\omega^j)) \in Q^0(j)$, for $t \in (\tau_j, T)$, for all ω^j. □

Remark 3.31 (Further properties of $z(s;s,y,j)$).* An additional property holds in Theorem 3.27: $J(s,y,j) := z(s;s,y,j)$ is C^1 in $Q^0(j)$ and satisfies the HJB equation (3.64) below, with $u(t;\hat{s},\hat{y},j)$ yielding the supremum in the HJB equation (3.64) for $(s,y) = (t,x(t;\hat{s},\hat{y},j))$ for any $(\hat{s},\hat{y}) \in Q^0(j)$. □

Remark 3.32 (Unbounded number of jump points ($N = \infty$)).* The conclusion in Remark 3.31 even holds for $N = \infty$ provided continuity of $(s,y) \to (z(s;s,y,j), z_y(s;s,y,j), p(s;s,y,j), p_y(s;s,y,j))$ in $Q^0(j)$, $j = 0,1,\ldots$ and the equality $z_y(s;s,y,j) = p(s,s,y,j)$ are postulated, $((s,y) \in Q^0(j))$, and the conditions on J in Remark 3.39 (b) below hold. Moreover, optimality of $(x^*(.,.), u^*(.,.))$ holds in the same way as in Theorem 3.27. □

Sometimes, Theorem 3.27 does not apply, because solutions defined for (s,y) in *open* sets $Q^0(j)$ containing $Q(j)$ do not exist. Consider the following example:

Example 3.33. Consider the problem

$$\max E\left[\int_0^3 ut\,dt\right], \quad \dot{x} = -u, \quad u \in [0,1], \ x(0) = 1, \ x(3-) \geq 0 \text{ a.s.},$$

with a possibility for a single, unit upwards jump in $x(t)$, with probability given by the intensity λ. We shall only consider a "medium sized" value of λ, one for which $1/(e^\lambda - 1) \in (1,2)$.

Solution. (a) Consider first the situation after a jump with initial condition (s,y). Let us construct the solution in this case. To get some "sizable" positive value of the criterion, we have to use $u = 1$ for some time. The constraint $x(3) \geq 0$ limits the time interval on which we can use $u = 1$. It evidently pays to postpone the use of $u = 1$ as long as possible, due to the factor t in the integrand. Hence, the optimal path is evidently $x = y$ until time t'' given by $y = 3 - t''$. Afterwards, $u = 1$ and this yields $x(3) = 0$. Here we assumed $s < 3 - y$, and in this case then, $x(t; s, y, 1) = \min\{y, 3-t\}$. Now, $p(t; s, y, 1)$ is constant, and to have a switch point at t'', we must have $p(t''; s, y) = 3 - y$ by the maximum condition. If on the other hand $s > 3 - y$, we can use $u = 1$ all the time, and still the state ends above $(3,0)$, so $0 = p(3; s, y, 1) = p(t; s, y, 1)$ and $x(t; s, y, 1) = y + s - t$. If $s = 3 - y$, still $u = 1$ all the time, and we can put $p(t; s, y, 1) = 0$. It is easily seen that the maximum condition holds in the two cases $s \geq 3 - y$, $s < 3 - y$.

(b) Before a jump, we consider only the case $(s, y) = (0, 1)$. Our intuition tells us that an optimal policy is as follows. The factor t in the integrand in the criterion makes it optimal to postpone using $u = 1$. When we surely have no jump, $u = 1$ is optimal on a unit interval I at the very end. However, with a positive probability for a jump, it may pay to have $I = (a, a+1)$ at which $u = 1$ somewhat earlier than at the very end. It may be advantageous to make room for having $u = 1$ on an additional set in case of a jump (a jump makes this possible). At least for λ small, presumably we will not exploit the latter possibility fully (or at all), for λ large, we would expect a to be close to (or perhaps equal to) 1. It can never pay to have $a < 1$, for if we experience a jump at any point, x is ≤ 2 here, and we cannot make use of a larger interval than one of length 2 for x to decrease down to zero.

(c) Write $x^*(t) := x(t; 0, 1, 0)$, $u^*(t) := u(t; 0, 1, 0)$, $p(t) := p(t; 0, 1, 0)$. The following maximum condition holds:

$$(u - u^*(t))t + p(t)(u^*(t) - u) \leq 0 \text{ for all } u \in [0, 1] \qquad [*]$$

and $p(t)$ satisfies the differential equation:

$$\dot{p}(t) = -\lambda p(t; t, x^*(t) + 1, 1) + \lambda p(t) \qquad [**]$$

with end condition $p(3) \geq 0, = 0$ if $x^*(3-) > 0$.

(d) We now take for granted that $u^*(t) = 0$ for $t \in [0, a]$, $a \geq 1$, $u^*(t) = 1$ in $(a, a+1)$ and $u^*(t) = 0$ for $t > a + 1$, and $x^*(t) = \min\{1, 1 - (t - a)\}$, for $t \leq a + 1$, a an unknown to be determined. Observe that if a jump occurs at $s \in (a,3]$, then we jump to a point (s,y) satisfying $s \geq 3 - y$. To see this, note that if $s \in (a,3]$, $x^*(s) \geq 1 - (s-a)$, so $y \geq 2 - (s-a) \geq 2 - s + 1 = 3 - s$. From (a), recall that $p(t;s,y,1) = 3 - y$, if $s < 3 - y$, whereas $p(t;s,y,1) = 0$ for $s \geq 3 - y$. Thus, for $s > a$, we jump to a point (s,y) for which $p(s;s,y,1) = 0$.

(e) Define $h^*(t) := \int_t^3 \lambda e^{-\lambda \tau} p(\tau; \tau, x^*(\tau) + 1, 1) d\tau$. Evidently, $p(t) = e^{\lambda t}(h^*(t) + h)$ for some constant $h \geq 0$. Write $\psi(t) := e^{-\lambda t} t - h^*(t) > 0$, then $t - p(t) = e^{\lambda t}(\psi(t) - h)$. Assume for the moment that there exists an interval $[a,b] \subset (0,3)$, $b := a + 1$, in the interior of which $\psi(t) - h > 0$, with the opposite (strict) inequality holding outside of $[a,b]$. Then, by (*), $u^*(t) = 1_{(a,b)}$ (the indicator function of the set (a,b)). If no jump occurs in $[0,3]$, we have guessed $x^*(3-) = 0$, which is consistent with the above choice of control. (There is a positive probability that there is no jump in $[0,3]$, so we must have $x^*(3-) \geq 0$, in the case of no jump in $[0,3]$.) Consider first $t > a$. Then $h^*(t) = 0$ (if we jump at $\tau > t > a$, we jump to a point where $p(\tau; \tau, x(\tau) + 1, 1) = 0$, see (d)) Thus, for $t \geq a$, $\psi(t) = e^{-\lambda t} t =: \phi(t)$. Now, as a and b are switch points, so by [*], $\psi(a) - h = \psi(b) - h = 0$, so $\phi(a) = \phi(b)$, which implies $\alpha(a) := (a+1) e^{-\lambda(a+1)} - a e^{-\lambda a} = e^{-\lambda a}(a(e^{-\lambda} - 1) + e^{-\lambda}) = 0$. The equation has a solution $a = 1/(e^\lambda - 1)$ that by assumption belongs to $(0,2)$, and for this a, $h = e^{-\lambda a} a$. Note that the function $t e^{-\lambda t}$ increases until $t = 1/\lambda$ and then decreases. Because $b e^{-\lambda b} = a e^{-\lambda a}$, $1/\lambda$ belongs to (a,b) and $t e^{-\lambda t} > a e^{-\lambda a}$ in this interval, so $\phi(t) - h > 0$ in (a,b), $\phi(t) - h < 0$ outside of $[a,b]$. Hence, $u = 1$ yields maximum of the Hamiltonian for $t \in (a,b)$, and $u = 0$ yields maximum for $t > b$, see [*], as $t - p(t) = e^{\lambda t}(\psi(t) - h) = e^{\lambda t}(\phi(t) - h)$ here. Consider next $t < a$. For such t, $\psi(t) \leq \phi(t) \leq \phi(a)$, so $t - p(t) = e^{\lambda t}(\psi(t) - h) \leq e^{\lambda t}(\phi(t) - h) < 0$, for such t, hence $u = 0$ maximizes the Hamiltonian.

It is easily seen that the candidates $x(t;s,y,1)$ and $x^*(t)$ satisfy sufficient conditions (concavity holds). Note that here, for $Q^0(0)$ to contain all admissible solutions, it must contain $(t,0) = (t,x^*(t))$ for $t \in (b,3)$, but no (s,y) can belong to $Q^0(0)$ for $s \in (b,3)$, $y < 0$, as from such a point (s,y) no solution exists that satisfies the end condition. So no open set $Q^0(0)$ with the required properties in Theorem 3.27 can be found.   □

Remark 3.34 (Restriction on the time path).* Consider the end constrained optimization problem (3.15)–(3.17), (3.47), (3.48), (3.49), (3.52), or (3.53), with the added restriction that there are given open sets $A(t,j) \subset \mathbb{R}^n$ in which admissible solutions $x(t; \tau_0, \ldots \tau_j)$ are required to stay, $j = 0, 1, \ldots$, (i.e., $(t, x(t; \omega^j)) \in A(t,j)$ for all t, all j). Then Theorem 3.24 holds if the sets $A(t,j)$ are convex. Moreover, the concavity required of the functions occurring in that theorem need only hold in $A(t,j)$ (except for h_0).   □

Remark 3.35 (Soft terminal constraints). Consider the problem where, for simplicity, $h_0 = 0$, and the hard end conditions (3.47), (3.48) are replaced by the following soft ones:

$$E[x_i(T-,\omega)] = \hat{x}_i, \qquad i = 1, \ldots, n' \tag{3.60}$$

$$E[x_i(T-,\omega)] \geq \hat{x}_i, \qquad i \geq n' + 1, \ldots, n''. \tag{3.61}$$

Consider the solution procedure for problem (3.15)–(3.17), (3.47), (3.48), (3.49), (the extremal method). When (3.47), (3.48) are replaced by (3.60), (3.61), the condition (3.54) in this procedure has to be replaced by the following one. Define an auxiliary scrap value function $h_0 = \sum_{i=1}^{n} \tilde{\lambda}_i x_i$, where the multipliers $\tilde{\lambda}_i$ so far are unknowns. Apply the solution procedure for this h_0-function, and at the end, determine the $\tilde{\lambda}_i$'s by using the conditions (3.60), (3.61), and (3.62) below:

$$\tilde{\lambda}_i \geq 0, i = n'+1, \ldots, n'', \tilde{\lambda}_i = 0 \text{ if } E[x_i^*(T;\omega)] > \hat{x}_i, i = n'+1, \ldots, n'',$$
$$\hat{\lambda}_i = 0, i = n''+1, \ldots, n. \tag{3.62}$$

In (3.60), (3.61), and (3.62), $x^*(.,\omega)$ is the nonanticipating solution starting at $(0, x^0)$, constructed by the method described in Remark 3.2. Note that in the current case, the condition (3.50) and no controllability condition like that in (3.56) are needed (for example in the necessary conditions to follow).

In fact, the free end necessary conditions (Theorem 3.8) also hold here, for the auxiliary scrap value function $h_0 = \sum \tilde{\lambda}_i x_i$, with (3.62) satisfied, and with the modification of the introduction of a $\Lambda_0 \in \{0,1\}$ as in Theorem 3.22. (See Seierstad (2001).)   □

Only a very simple example of the use of the preceding remark will now be given.

Example 3.36. Consider the problem:

$$\max E\left[\int_0^T (-u^2/2)dt\right], \qquad \dot{x} = u \in \mathbb{R}, \ x(0) = 0, \ Ex(T) = 1,$$

with a possibility for a single, unit upwards jump in $x(t)$ at $\tau \in [0, \infty)$, with τ being exponentially distributed with intensity λ (i.e., the jump point τ is distributed with density $\lambda e^{-\lambda\tau}$).

Solution. The maximum condition gives $\hat{u} = p$, and the characteristic equations for $j = 1$ (i.e., after a jump) becomes $\dot{p} = 0$, $\dot{x} = p$, with $p(T) = a$, a an unknown to be determined. As always, t denotes running time. Hence, $u(t;s,y,1) = p(t;s,y,1) = a$, $x(t;s,y,1) = y + a(t-s)$.

Next, for $j = 0$, the characteristic equations become: $\dot{x} = p$, $\dot{p} = -\lambda a + \lambda p$, with $p(T;s,y,0) = a$. Then, $p(t;s,y,0) = a + Ce^{\lambda t}$ where C is determined by $p(T;s,y,0) = a$, so $C = 0$, and $p(t;s,y,0) = a$. Thus, $u(t;s,y,0) = a$ is the control also before the jump. It remains to determine a. Now, if $T > \tau$ $x(T;\tau) = x(\tau-;0,0,0) + 1 + a(T-\tau) = a\tau + 1 + a(T-\tau) = 1 + aT$, otherwise $x(T;\tau) = aT$, so all in all $x(T;\tau) = aT + 1_{[0,T]}(\tau)$. Hence $1 = Ex(T;\tau) = aT + 1 - e^{-\lambda T}$, so $a = e^{-\lambda T}/T$.   □

Often the functions $J(s,y,j)$ of (3.59) satisfy partial differential equations called the HJB equations (a fact already known from Chapter 2, see also (3.10)). Related to Theorems 3.27 and 3.11, there is a sufficient condition based on these equations of the problem. To introduce this approach, let us make some introductory comments. We need the following convention: A function defined on a possibly nonopen set A

is C^1 on A if it has an extension to a larger open set, which is C^1. Define the optimal value function by

$$J^*(s,y,j) = \max_{u(.,.)} \left\{ E\left[\sum_{i\geq j} \int_{\tau_i}^{\tau_{i+1}} 1_{[s,T]}(t) f_0(t,x(t,\omega),u(t,\omega),i) dt \right.\right.$$

$$\left.\left. + \sum_{s\leq \tau_i<T, i>j} g_0(\tau_i,x(\tau_i-,\omega),i) + h_0(x(T-,\omega)) | s,y,j \right] \right\}. \quad (3.63)$$

where the maximization is over all so-called (s,y,j)-*admissible* controls $u(.,\omega)$, and where $x(t,\omega) := x^{u(\cdot,\cdot)}(t,\omega)$. An (s,y,j)-admissible control is a nonanticipating control for which $x(t,\omega) = x^{u(\cdot,\cdot)}(t,\omega)$ satisfies the terminal conditions and starts at (s,y,j), exactly j jumps having already occurred, $\tau_j \leq s$, $y = x(s+,\omega)$, $u(.,.)$ independent of τ_k, $k \leq j$. Without being too precise, let it be noted that very often, especially in the free end case, for each j, $J^*(s,y,j)$ is C^1 in (an open set containing) $\mathrm{cl}Q(j)$ and satisfies the j-th HJB equation:

$$0 = J_s(s,y,j) + \sup_{u\in U}\{f_0(s,y,u,j) + J_y(s,y,j) f(s,y,u,j)\}$$

$$+\lambda(j+1)\{g_0(s,y,j+1) + J(s,y+g(s,y,j+1),j+1) - J(s,y,j)\}. \quad (3.64)$$

In the free end case, $J^*(s,y,j)$ satisfies the boundary condition

$$J(T,x,j) = h_0(x). \quad (3.65)$$

In the case where a maximum of N jumps can occur, there is a recursive procedure available for solving the HJB equations. First (3.64), (3.65) are solved for $j = N$, then for $j = N-1$ (then we need the knowledge of $J(t,x,N)$), and next for $j = N-2$ (then we need $J(t,x,N-1)$), and so on backwards. In end constrained cases, (3.66), (3.67) below are used instead of (3.65) (then $J(s,y,j)$ is perhaps not meaningfully defined for $s = T$).

Related to the HJB equations is a sufficient condition, or verification theorem. We need the following boundary conditions.

For all semiadmissible solutions $\hat{x}(t,\omega)$, $\liminf_{s\to T} J(s,\hat{x}(s,\omega),j) \geq h_0(\hat{x}(T,\omega))$,

$$\text{for all } \omega = (\tau_0,\tau_2,\ldots) \text{ for which } \tau_j < T \leq \tau_{j+1}. \quad (3.66)$$

$$\lim_{s\to T} J(s,x^*(s,\omega),j) = h_0(x^*(T,\omega))$$

$$\text{for all } \omega = (\tau_0,\tau_2,\ldots) \text{ for which } \tau_j < T \leq \tau_{j+1}. \quad (3.67)$$

For the next theorem, let us still postulate that the system satisfies the properties stated in connection with (and including) (3.49), and consider again the problem (3.15)–(3.17), (3.47)–(3.49), (3.52), or (3.53).

Theorem 3.37 (Sufficient condition for the HJB equations, $N \leq \infty$). *Assume that $J(s,y,j), j = 0,1,\ldots,$ are C^1 functions on $Q(j)$, continuous on $\{(0,x^0)\} \cup Q(0)$ for*

$j = 0$, satisfying the HJB equation (3.64) for all $(s,y) \in Q(j)$, and the boundary condition (3.66). Assume that for any semiadmissible solution $x(t,\omega)$, for some constants $\alpha_{x(.,.)}$, $\kappa_{x(.,.)}$, for all j, $|J(t,x(t,\omega),j)| \leq \alpha_{x(.,.)} + \kappa_{x(.,.)}|x(t,\omega)|$ when $t \in [\tau_j, \tau_{j+1})$, $t \leq T$. Finally, assume that there exists a semiadmissible pair $(x^*(s,\omega), u^*(s,\omega))$ satisfying (3.67) and, a.s., for any ω, for $\Lambda_0 = 1$, the equality

$$\max_{u \in U} H(t, x^*(t,\omega), u, J_x^*(t, x^*(t,\omega), j), j)$$
$$= H(t, x^*(t,\omega), u^*(t,\omega), J_x^*(t, x^*(t,\omega), j), j) \tag{3.68}$$

for v.e. t such that $\tau_j < t \leq \tau_{j+1}$. Then $u^*(t,\omega)$ is optimal among all semiadmissible controls. □

Remark 3.38 (Stronger conditions*). The conditions (3.66), (3.67) are automatically satisfied if $J(s,y,j)$ can be continuously extended to $Q(j) \cup (\{T\} \times \mathbb{R}^n) \cap \text{cl} Q(j))$, $j = 0, 1, \ldots$, and (3.65) holds for (T,x) in this set. This extension property frequently holds in free end problems. □

Remark 3.39 (Modification of requirements*). **(a)** In Theorem 3.37, note that (3.49) can be replaced by the modification stated in Remark 3.23.
(b) In Theorem 3.37, weaken the growth conditions as follows: For any semiadmissible solution $x(t,\omega)$ for each $T' < T$, assume that for some constants $\alpha_{T',x(.,.)}$, $\kappa_{T',x(.,.)}$, $|J(t,x(t,\omega),j)| \leq \alpha_{T',x(.,.)} + \kappa_{T',x(.,.)}|x(t,\omega)|$ when $t \in [\tau_j, \tau_{j+1})$, $t \leq T'$. Finally, assume that (3.67) holds, and that $E \sup_t |J(t, x^*(t,\omega), \omega)| < \infty$, where $J(t,x,\omega) = \sum_j J(t,x,j) 1_{[\tau_j,\tau_{j+1})}(t)$. Then $(x^*(t,\omega), u^*(t,\omega))$ is optimal in the set of semiadmissible pairs $(x(t,\omega), u(t,\omega))$ for which both (3.66) holds and $E[\inf_t \min\{0, J(t, x(t,\omega), \omega)\}] > -\infty$.
(c) In Theorem 3.37, assume only the weakened growth condition of (b) above, drop (3.66) and and replace (3.67) by the assumption that

$$\lim_{t \to T} E[J(t, x^*(t,\omega), \omega)] = E[h_0(x^*(T,\omega))].$$

Then $(x^*(t,\omega), u^*(t,\omega))$ is optimal among all semiadmissible pairs for which $\limsup_{t \to T} E[J(t, x(t,\omega), \omega)] \geq E[h_0(x(T;\omega))]$. This result even holds if the hard constraints are replaced by soft ones, (3.60), (3.61). □

The relationship between Theorems 3.27 and 3.37 is as follows. A standard manner of solving the HJB equations is using the so-called characteristic equations. For $\hat{H}(s,x,p,j) := \max_u\{f_0(s,x,u,j) + pf(s,x,u,j)\}$, the so-called characteristic equations for the solution of the j-th first-order partial differential equation are

$$\dot{x}(s) = \hat{H}_p(s,x,p,j),$$
$$\dot{p}(s) = -\hat{H}_x(s,x,p,j)$$
$$\qquad - \lambda(j+1)(\partial/\partial x)\{g_0(s,x,j+1) + J(s,x+g(s,x,j+1),j+1) - J(s,x,j)\}$$

It can be shown that these two equations imply that $p(s;s,y,j) = J_y(s,y,j)$, so the equation for p can alternatively be written

$$\dot{p}(s) = -\hat{H}_x(s,x,p,j) + \lambda(j+1)p$$
$$-\lambda(j+1)[g_{0x}(s,x,j+1) + p(s;s,x+g(s,x,j+1),j+1)(I+g_x(s,x,j+1))].$$

If $\hat{H}$ is C^1 in (x,p), the equation for x and the last equation for p can be shown to be the same as (3.19)–(3.20) above. When all $x(t;s,y,j)$ and $p(t;s,y,j)$ have been found, from (3.58), (3.59) functions $J(s,y,j)$ are obtained, and they will satisfy the HJB equation (3.64). A recursive method is available iff $N < \infty$.

When using the HJB equations, we always get candidate controls on Markov form, namely as any control function that yields the maximum in the HJB equation.

Sometimes, it is possible to use Theorem 3.37 but not Theorem 3.27, because *open* sets $Q^0(j)$ cannot be found, for which the conditions in Theorem 3.27 are satisfied. Of course, as just explained, a method for producing the functions $J(s,y,j)$ of Theorem 3.37, especially in case $N < \infty$, is to use characteristic solutions, together with the functions $z(t;s,y,j)$ and put $J(s,y,j) = z(s;s,y,j)$, see (3.58), (3.59). Hopefully, these solutions can be defined for $(s,y) \in Q(j)$. Finally, we can check if the $J(s,y,j)$'s have C^1-extensions to some larger open sets.

Remark 3.40 (Infinite number of jumps, all $\lambda(i)$ equal). When $N = \infty$, there is an infinite family of HJB equations (3.64). So, assume not only that $N = \infty$, but also that all $\lambda(i) = \lambda$ and that g_0, g, f_0, f are independent of j. Then evidently, $J^*(s,y,j)$ is independent of j, and the family of HJB equations reduces to one, namely the one obtained from (3.64) by deleting the arguments j and $j+1$.

In the current case, the adjoint equations also reduce to a single equation. A single HJB equation still holds, even if λ depends on t, x and u, which can be allowed. (See Remark 3.50 below.)   □

Allowing dependence on x in λ, then, introducing an auxiliary state variable, say y, jumping one unit each time a jump time occurs, makes it possible to rewrite a function $\lambda(i)$ as $\lambda(y)$, and hence, from the single HJB equation, to recover the family of HJB equations in Theorem 3.37, and even the possibility that only $N < \infty$ jumps can occur.

Whether all the λ's are equal or not, it is only in rare cases that explicit formulas for solutions of the HJB equation(s) can be found.

In the case considered in the above remark, Davis (1993) discusses the usefulness of the HJB equation for numerical solutions of the problem. He also considers HJB equations in problems with additional features, for example problems where the state jumps each time it reaches a boundary.

Example 3.41. The following problem is essentially Example 3.3, with $N = \infty$. Consider the problem:

$$\max E\left[\int_0^T (-u^2/2)dt + ax(T)\right], \qquad \dot{x} = u \in \mathbb{R}, \ x(0) = 0,$$

with a possibility for an unbounded number of jumps in $x(t)$ with the same intensity λ for all jumps τ_j, and with the same size b, (i.e. $x(\tau_j+) - x(\tau_j-) = b$).

Solution. The single HJB equation is $0 = J_s + \max_u[-u^2/2 + J_y u] + \lambda[J(s,y+b) - J(s,y)]$. The maximum is calculated by putting the first derivative with respect to u equal to zero, it yields the maximum point $u = J_y$. Inserting this in the HJB equation gives $0 = J_s + (J_y)^2/2 + \lambda[J(s,y+b) - J(s,y)]$. We guess that $J(s,y)$ is of the form $\phi(s) + ky$, k a constant. The HJB equation yields that $0 = \phi' + k^2/2 + \lambda kb$, so $\phi(s) = cs + C$, where $c := -k^2/2 - \lambda kb$, and C is an integration constant. Now, $ax = \phi(T) + kx = cT + C + kx$, so $k = a$, $C = -cT$, hence $\phi(s) = c(s - T)$. The control is $u = J_y = a$.

Remark 3.42 (Infinite horizon, unbounded number of possible jumps). In this remark, we shall briefly comment upon a sufficient condition for optimality in the case where $T = \infty$, $h_0 \equiv 0$, $N = \infty$ (an unbounded number of possible jumps), all $\lambda(j) = \lambda$, and where g_0, g, f_0, and f are independent of j. We then look for an optimal pair in the set of admissible pairs $(x(t, \omega), u(t, \omega))$, where admissible now means that $u(t, \omega)$ is bounded on any bounded interval. Then, for all j, $J(s, x, j) = J(s, x, 0)$ and we assume that $J(s, x, 0)$ satisfies the conditions in Theorem 3.37 (the growth condition for all $T' > 0$ as in (b) in Remark 3.39). (Again, there is a single HJB equation.) Then the conclusion of Theorem 3.37 still holds (at least in the sense of catching up optimality or sporadically catching up optimality, see below), provided the boundary conditions (3.66), (3.67) are replaced by the following condition. For all admissible solutions $\hat{x}(.,.)$,

$$\liminf_{s \to \infty} E[J(s, \hat{x}(s; \omega), 0) - J(s, x^*(s, \omega), 0)] \geq 0. \tag{3.69}$$

Here we can allow restrictions at infinity, say of the form $\liminf_{t \to \infty} \hat{x}_i(t) \geq \check{x}_i$ a.s., or $\liminf_{t \to \infty} E\hat{x}_i(t) \geq \check{x}_i$, for a solution $\hat{x}(.,.)$ to be admissible.

We cannot always exclude the possibility that the criterion has infinite values. Then sometimes the following criterion works: $(x^*(.,.), u^*(.,.))$ is called catching up optimal if, for any admissible $x(t, \omega)$,

$$\liminf_{T \to \infty} E\left[\int_0^T \{f_0(t, x^*(t, \omega), u^*(t, \omega)) - f_0(t, x(t, \omega), u(t, \omega))\}dt\right] \geq 0. \tag{3.70}$$

The above sufficient conditions yield catching up optimality in cases where infinite values of the criterion appears.

If liminf in (3.70) is replaced by limsup, we get sporadic catching up optimality, and to obtain this weaker optimality property, one can even replace liminf by limsup in (3.69).   □

A special case of the one in Remark 3.42 is the case where we add the assumptions that $g_0(t, x) = e^{-\beta t}g^0(x)$, $f_0(t, x, u) = e^{-\beta t}f^0(x, u)$, $\beta > 0$, and that g and f are independent of t. Then (normally) $J^*(s, x, 0) = e^{-\beta s}\tilde{J}(x)$, where $\tilde{J}(x)$ satisfies the "current value HJB equation"

$$0 = -\beta\tilde{J} + \max_u\{f^0 + \tilde{J}_y f\} + \lambda\{g^0 + \tilde{J}(y + g(y)) - \tilde{J}(y)\}. \tag{3.71}$$

Turning this around, if $\tilde{J}$ is C^1 and satisfies (3.71), then $J(s,x,0) = e^{-\beta t}\tilde{J}(x)$ is C^1 and satisfies (3.64).

Example 3.43. Consider the problem

$$\max \int_0^\infty (xu)^\alpha e^{-\beta t}\,dt,$$

subject to

$$dx/dt = ax(1-u), x(0) = x^0 > 0, u \in (0,\infty) \text{ and } x(\tau+) = kx(\tau-),$$

where an unbounded number of jumps with intensity λ can occur. It is assumed that $\alpha \in (0,1), \beta > 0, a > 0, k > 0$, and that $\kappa := \beta - \lambda(k^\alpha - 1) - \alpha a > 0$. Write $\kappa' := \kappa/(1-\alpha) > 0$.

Solution. We guess that $\tilde{J} = Ay^\alpha$, $A > 0$. Then the current value HJB equation is $0 = -\beta Ay^\alpha + \max_{u>0}[y^\alpha u^\alpha + \alpha Ay^\alpha a(1-u)] + \lambda(k^\alpha - 1)Ay^\alpha$. Here, y^α can be cancelled, and the maximization gives $u = B$, where $B = (aA)^{1/(\alpha-1)}$, so the HJB equation reduces to $0 = -\beta + B^\alpha/A + \alpha a(1-B) + \lambda(k^\alpha - 1) = -\kappa + B^\alpha/A - \alpha aB$. Expressing B by means of A and using the last equation yield $\kappa = (aA)^{\alpha/(\alpha-1)}/A - \alpha a(aA)^{1/(\alpha-1)} = (1-\alpha)a^{\alpha/(\alpha-1)}A^{1/(\alpha-1)}$. Hence, $A = a^{-\alpha}(\kappa/(1-\alpha))^{\alpha-1}$, and $B = \kappa/a(1-\alpha)$. All admissible $\hat{x}(t,\omega)$ are nonnegative, so $\liminf_{s\to\infty} EJ(s,\hat{x}(s,\omega)) \geq 0$. Moreover, $dx/dt = cx$, for $c = a(1-B)$. Note that for each t, such that $\tau_j < t \leq \tau_{j+1}$, for $t \in (\tau_j, \tau_{j+1})$, $x(t,\tau_0,\ldots,\tau_j) = x(\tau_j+)e^{c(t-\tau_j)} = kx(\tau_j-)e^{c(t-\tau_j)}$. Using this equality for j replaced by $j-1, j-2,\ldots,1$ $(x(\tau_0+) = x^0)$, we get $x(t,\tau_0,\ldots,\tau_j) = k^j x^0 e^{ct}$. From the theory of Poisson processes, we know that the probability of j jumps in an interval $[0,t]$ is $(\lambda t)^j e^{-\lambda t}/j!$. Hence, we get that the expected value $EJ(t,k^j x^0 e^{ct})$, for $m = k^\alpha$, equals $AE[m^j(x^0)^\alpha e^{\alpha ct - \beta t}] = A(x^0)^\alpha e^{(\alpha c - \beta)t}\sum_0^\infty (m\lambda t)^j e^{-\lambda t}/j! = A(x^0)^\alpha e^{(\alpha c - \beta)t}e^{m\lambda t}e^{-\lambda t} = A(x^0)^\alpha e^{(\alpha c + (m-1)\lambda - \beta)t} = A(x^0)^\alpha e^{-(\kappa + \alpha aB)t} = A(x^0)^\alpha e^{-\kappa' t}$. The limit of the last expression, as t goes to infinity, is zero, so (3.69) holds. The optimal (Markov) control is $u(s,y) = B$, thus independent of (s,y). (By a similar calculation, one can show that the value of the criterion is finite for this control, so in effect $u = B$ is not only catching up optimal, but "ordinary" optimal.) □

Remark 3.44 (Generalizations (dependence on ω)*). Also in the hard end constrained case, the functions f_0, f, g_0, and g can be allowed to depend on ω, provided they are nonanticipating in (t,ω) for each x,u, compare with a similar remark (Remark 3.17) in the free end case. (See Seierstad (2001).) Let us restrict the attention to the case where the only change is that $f(t,x,u,j)$ depends also on τ_j, $f(t,x,u,j,\tau_j)$. This can be taken care of by assuming that $f = f(t,x,u,j,w)$, where w is an auxiliary state, for which $\dot{w} = 0$, and for which $w(\tau_j+) - w(\tau_j-) = \tilde{g}(\tau_j, w(\tau_j-))$, $\tilde{g}(t,w) := t - w$. Then f and f_x are assumed to be continuous, (3.25) is assumed to hold uniformly in τ_j as far as f is concerned, and the Lipschitz rank κ^n is independent of τ_j (see the Standard System Conditions connected with (3.25)). Finally, f_w need not exist, and no condition (like (3.49)) needs to be put on $\tilde{g}$. □

Remark 3.45 (Weakened differentiability assumptions on the characteristic solutions, $N < \infty^*$). Assume that (3.29) (i.e., consistency) holds for all $(s,y) \in Q_0(j)$, and that, in these sets, $(s,y) \to z(s;s,y,j)$, $j = 0,1,\ldots,N$, are locally Lipschitz continuous. Then the differentiability conditions on $z(t;s,y,j), x(t;s,y,j), p(t;s,y,j)$, and $\hat{H}(t,x,p,j)$ in Theorem 3.27 can be weakened as follows.

Let $\phi_k(t,x,p,j), k = 1,\ldots,k_j^*$, be C^1-functions in (t,x,p), defined on $\mathrm{cl}Q^*(j)$ and assume that for any $(t,x,p) \in \mathrm{cl}Q^*(j), \phi_k(t,s,p,j) = 0$ for at most one k. Let $Q^1(j)$ be given closed sets containing $\hat{Q}^1(j) := \{(s,y) \in Q^0(j) : \phi_k(s,y,p(s;s,y,j),j) = 0$ for some $k\}$, and define $Q_{s,y,j} := \{t \in (s,T) : \phi_k(t,x(t;s,y,j),p(t;s,y,j),j) = 0$ for some $k\}$, and assume that for any $(s,y) \in Q^0(j) \setminus Q^1(j)$, for any $t \in (s,T), t \notin Q_{s,y,j}$, $(s',y') \to (z(t;s',y',j),x(t;s',y',j),p(t;s',y',j))$ is C^1 for all (s',y') in a neighborhood $N_{t,s,y,j}$ of (s,y). Assume also that $\hat{H}(t,x,p,j)$ and its first- and second-order derivatives with respect to x and p (exist and) are continuous in $\hat{Q}_*(j) := \{(t,x,p) \in Q^*(j) : \phi_i(t,x,p,j) \neq 0$ for all $i\}, j = 0,\ldots,N$. Assume furthermore that $z(t;t,x + g(t,x,j+1),j+1), z_y(t;t,x+g(t,x,j+1),j+1), p(t;t,x+g(t,x,j+1),j+1)$, and $p_y(t;t,x+g(t,x,j+1),j+1)$ (exist and) are continuous in $Q^0(j) \setminus Q^1(j)$. Assume that, for each j, the functions $\hat{H}_x(t,x,p,j)|_A$ and $\hat{H}_p(t,x,p,j)|_A$ have extensions to an open set containing $\mathrm{cl}A$, such that these functions as well as their partial derivatives with respect to x and p (exist and) are continuous on this open set, for any set A of the form $\cap_i \Phi^i$, $\Phi^i = \{(t,x,p) \in Q^*(j) : \phi_i(t,x,p,j) > 0\}$, or $\Phi^i = \{(t,x,p) \in Q^*(j) : \phi_i(t,x,p,j) < 0\}$ (the direction of the inequality sign may depend on i). Assume that, for each $j < N$, the functions $z(t;t,x + g(t,x,j+1),j+1)|_B$ and $p(t;t,x + g(t,x,j+1),j+1)|_B$ have extensions to open sets containing $\mathrm{cl}B$, such that $z_y(t;t,x+g(t,x,j+1),j+1)$ and $p_y(t;t,x+g(t,x,j+1),j+1)$ exist on these open sets, all four functions being continuous here, for any set B of the form $\cap_i \Phi_i$, $\Phi_i = \{(t,x) \in Q^0(j) : \phi_i(t,x,p(t;t,x,j),j) > 0\}$, or $\Phi_i = \{(t,x) \in Q^0(j) : \phi_i(t,x,p(t;t,x,j),j) < 0\}$, (the direction of the inequality sign may depend on i). Assume also that for any given j, for any $(s,y) \in Q^0(j) \setminus Q^1(j)$, and any $t \in (s,T]$,

$$\phi_k(t,x(t;s,y,j),p(t;s,y,j),j) = 0 \Rightarrow$$
$$\phi_{kt} + \phi_{kx}(\partial/\partial t)x(t^{\pm};s,y,j) + \phi_{kp}(\partial/\partial t)p(t^{\pm};s,y,j) \neq 0, \tag{3.72}$$

the partial derivatives ϕ_{kt}, ϕ_{kx} and ϕ_{kp} being evaluated at $(t,x(t;s,y,j),p(t;s,y,j),j)$. (The right-hand side is required to be non-vanishing both for the left limits and the right limits. For $t = T$, let $t^{\pm} = T-$.) Finally, assume that $Q^1(j)$ is slim, $j = 0,1,2,\ldots,N$. $\square$

Remark 3.46. Consider the property:

$$(s,y) \in Q^0(j) \setminus Q^1(j) \Rightarrow (s,y+g(s,y,j+1)) \in Q^0(j+1) \setminus Q^1(j+1). \tag{3.73}$$

If (3.73) holds, and if $z(t;t,x,j+1), z_y(t;t,x,j+1),j+1), p(t;t,x,j+1)$, and $p_y(t;t,x,j+1),j+1)$ are continuous in $Q^0(j+1) \setminus Q^1(j+1)$, then $z(t;t,x + g(t,x,j+1),j+1), z_y(t;t,x+g(t,x,j+1),j+1), p(t;t,x+g(t,x,j+1),j+1)$, and $p_y(t;t,x+g(t,x,j+1),j+1)$ are continuous in $Q^0(j) \setminus Q^1(j)$. The last continuity property is among the assumptions in Remark 3.45.

Moreover, the implications in Remark 3.28 still hold in the setting of Remark 3.45. $\qquad\square$

Remark 3.47 (Changes in the premises of Remark 3.45). Assume, for each j, at any $(s,y) \in Q^0(j) \setminus Q^1(j)$, continuity of $p(T;s,y,j)$ as well as existence of the derivatives in, and continuity at $t = T$ of, $t \to (z_s(t;s,y,j), z_y(t;s,y,j), x_s(t;s,y,j), x_y(t;s,y,j))$. Then cl$A$ can be replaced by $(\text{cl}A) \cap ((0,T) \times \mathbb{R}^{2n})$ and cl$B$ can be replaced by $(\text{cl}B) \cap ((0,T) \times \mathbb{R}^n)$. $\qquad\square$

Example 3.48. Example 3.19 revisited. Theorem 3.27, with Remark 3.45, can be applied to obtain optimality of the proposed candidates in Example 3.19. Let $Q^0(0) = Q^0(1) = (0,T) \times \mathbb{R}$. Earlier we constructed solutions $x(t;s,y,1)$ for all $(s,y) \in Q^0(1)$. Now, we need solutions $x(t;s,y,0)$ for all $(s,y) \in Q^0(0)$. For $y \in [-1,0)$, $x(t;s,y,0) = \beta^*(t;s,y)$, for $y < -1$, $x(t;s,y,0) = \beta(t;s,y)$ for $t \leq t''(s,y)$, $x(t;s,y,0) = \beta^*(t;t''(s,y),-1)$ for $t > t''(s,y)$, where $t''(s,y)$ is obtained by "smooth pasting" again, i.e., by solving the equations $-1 = \beta(t'';s,y) = (T - t'')[(y + 1)/(T-s) - \int_s^{t''} Ce^{\lambda\sigma}/(T - \sigma)^2 d\sigma] - 1$, and $\lambda e^{\lambda t''}/(e^{\lambda T} - e^{\lambda t''}) = [\partial\beta^*(t;t'',-1)/\partial t]_{t=t''} = [\partial\beta(t;s,y)/\partial t]_{t=t''} = (y+1)/(s-T) + \int_s^{t''} Ce^{\lambda\sigma}/(\sigma - T)^2 d\sigma - Ce^{\lambda t''}/(T - t'')$. The two equations determine C and t'': The first equation reduces to $(y+1)/(T - s) - C \int_s^{t''} e^{\lambda\sigma}/(T - \sigma)^2 d\sigma = 0$, using this in the second one, it reduces to $\lambda e^{\lambda t''}/(e^{\lambda T} - e^{\lambda t''}) = -Ce^{\lambda t''}/(T - t'')$, or $\lambda/(e^{\lambda T} - e^{\lambda t''}) = -C/(T - t'')$, which gives $C = (t'' - T)\lambda/(e^{\lambda T} - e^{\lambda t''})$. Using this, the first equality becomes

$$(y+1)/(T - s) = -\{\lambda(T - t'')/(e^{\lambda T} - e^{\lambda t''})\} \int_s^{t''} e^{\lambda\sigma}/(T - \sigma)^2 d\sigma.$$

For $t'' = s$, the right-hand side is zero, for $t'' = T$, it is $-\infty$. So a solution in $(0,T)$ for t'' exists, which then determines C. It is easily seen that C and $t''(s,y)$ are C^1 in (s,y), $y < -1$. Hence, $x(t;s,y,0)$, as well as $\dot{x}(t,s,y,0) = u(t;s,y,0) = p(t;s,y,0)$ are C^1 in (s,y), for $y < -1$ and for $0 > y > -1$, for any $t \in (s,T]$, at least when $x(t;s,y,0) \neq -1$ (more precisely, for any $(\hat{s},\hat{y})$, for t for which $(x(t;\hat{s},\hat{y},0) \neq -1$, then, close to $(\hat{s},\hat{y})$, the C^1 - property holds). For $y \geq 0$, $x(t;s,y,0) = x(t;s,y,1) = y$.

Now, $z(t;s,y,1) = -(\min\{0,y\})^2(T - t)/(s - T)^2$ and $z(t;s,y,0) =$

$$e^{\lambda t} \int_T^t e^{-\lambda\sigma}[\min\{0,y\}\lambda/(e^{\lambda s} - e^{\lambda T})]^2 d\sigma$$

$$= -e^{\lambda t}(e^{-\lambda t} - e^{-\lambda T})[\min\{0,y\}/(e^{\lambda s} - e^{\lambda T})]^2,$$

when $y \in [-1, \infty)$, (for such a y, a jump bring us to a state $y \geq 0$, at which $z(t;t,y,1)$ vanishes, so the f_0–term is what remains in (3.58) for $j = 0$).

Evidently, for $t \in (s,T]$, $(s,y) \to (x(t;s,y,1), p(t;s,y,1), z(t;s,y,1))$ is C^1 in $(s,y) \in (0,T) \times (-\infty,0)$ (and also for $y > 0$). For $(s,y) \in (0,T) \times (-1,0)$ (and also for $y > 0$), $(s,y) \to (x(t;s,y,0), p(t;s,y,0), z(t;s,y,0))$ is continuous, see the explicit formulas for these entities. It is also easily seen that $(s,y) \to (x(t;s,y,0), p(t;s,y,0), z(t;s,y,0))$ is C^1 in $(s,y) \in (0,T) \times (-\infty, -1)$ at least for $t = T$ and for $t \in (s,T)$ such that such that $x(t;s,y,0) \neq -1$ (explicit expressions have been given for $x(t;s,y,0)$

and in effect also for $p(t;s,y,0)$). By using Remarks 3.45 and 3.47, sufficient differentiability properties are then seen to hold (to use these remarks, let $\phi_1 = y+1$, $\phi_2 = y$). In the remarks mentioned, certain further properties are required, including consistency, local Lipschitz continuity of $z(s;s,y,j)$, extension properties, and the "nontangentiality" condition (3.72), all of which are easily checked.

Finally, using the limit conditions in Remark 3.28, let us show that an admissible pair $\hat{x}(t,\omega), \hat{u}(t,\omega)$ is strongly semiadmissible. Here it suffices to show that, uniformly in ω^j, $\lim_{t \to T} z(t;t,\hat{x}(t,\omega^j),j) = 0$, $j = 0,1$. For any semiadmissible pair $(\hat{x}(.,.),\hat{u}(.,.)), \hat{x}(T,\omega) \geq 0$ and $|\hat{x}(s,\omega) - \hat{x}(T,\omega)| \leq K_\tau(T-s)$, where K_τ is a bound on $\hat{u}(t,\omega)$, $s,t > \tau$. Now, $|\min\{0,y\} - \min\{0,y'\}| \leq |y - y'|$, so

$$|\min\{0,\hat{x}(s,\omega)\}| = |\min\{0,\hat{x}(s,\omega)\} - \min\{0,\hat{x}(T,\omega)\}| \leq K_\tau(T-s). \qquad (*)$$

Hence, the above limit condition holds for $j = 1$ for any admissible solution, because then K can be taken to be independent of τ. Thus, strong semiadmissibility follows for admissible solutions, when the below argument in case $j = 0$, is also taken into consideration. (For any semiadmissible solution for which $EK_\tau < \infty$, a similar limit condition yields again strong semiadmissibility.) So, let $j = 0$. For t close to T, $\hat{x}(t,\omega^0) > -1$, and $(*)$ again holds now for some K, so the formula of $z(t;s,y,0)$ for $y \geq -1$ can be used, and it yields that $z(t;t,\hat{x}(t,\omega^0),0) \to 0$ when $t \to T$. Note that $u^*(t,\omega)$ is admissible, as $\dot{x}^*(t,\tau) = \min\{0,x(\tau;0,-2,0)\}/(\tau-T)$, where $|x(\tau;0,-2,0)| = |x(\tau;0,-2,0) - x(T;0,-2,0)| \leq K(T-\tau)$, for some K, by boundedness of $\dot{x}(t;0,-2,0)$. Thus, Remarks 3.45, 3.47, and 3.28 yield that $(u(t,s,y,0), u(t,s,y,1))$ are optimal (i.e., the corresponding nonanticipating pair $(x^*(t),u^*(t))$ with $x^*(0) = -2$), is optimal among all strongly semiadmissible pairs, so in particular among all admissible pairs). $\qquad \square$

Remark 3.49 (Weakened differentiability assumptions in Theorem 3.37*). The conclusion in Theorem 3.37 even holds for the following weakening of its assumptions: For each j, $J(s,y,j)$ is defined and locally Lipschitz continuous in some open set Q'_j containing $Q(j)$. For given C^1-functions $\psi_i(s,y,j)$ on Q'_j, $i = 1,\ldots,i^*_j$, $J(s,y,j)$ is differentiable and satisfies the HJB equation in $Q'_j \setminus Z^*_j$, where Z^*_j are given sets contained in $\{(s,y) \in Q'_j : \psi_i(s,y,j) = 0$ for some $i\}$. (Still $J(s,y,0)$ is continuous on $\{(0,x^0)\} \cup Q(0)$.) The functions $\psi_i(s,y,j)$ have the property that $(s,y) \in Q'_j$, $\psi_i(s,y,j) = 0 \Rightarrow (\psi_{is}(s,y,j), \psi_{iy}(s,y,j)) \neq 0$. Finally, for any ω^j, $J(s,y,j)$ is differentiable at all points $(t,x^*(t,\omega^j))$, $\tau_j < t$, except a finite number (or countable number) of points, with (3.68) assumed satisfied as written. $\qquad \square$

Remark 3.50 ($\lambda = \lambda(t,x,u,j), g_0 = g_0(t,x,V_j,w_j,j)$, $g = g(t,x,V_j,w_j,j)*$). In this remark the intensities $\lambda(j)$ are allowed to depend on (t,x,u), so $\lambda(j) = \lambda(t,x,u,j)$, with $\lambda(.,.,.,.) \leq K$, for some K. Moreover, g_0 and g are assumed to depend on two additional variables V_j, w_j, so $g_0 = g_0(t,x,V_j,w_j,j)$, $g = g(t,x,V_j,w_j,j)$. Here, V_j is a stochastic variable in $\mathbb{R}^{\hat{n}}$, with cumulative distribution $v \to \pi(v;t,x,j)$. If jump number j occurs at $t = \tau_j$ and the state before the jump is $x = x(\tau_j-,\omega)$, then the cumulative distribution of V_j is $\pi(v;\tau_j,x(\tau_j-,\omega),j)$. Furthermore, w_j is a control variable taking values in a given set W_j. It is assumed that at each jump point τ_j

it is possible to choose w_j after having observed the "size" v_j of the jump, and w_j can also depend on all earlier $\tau_i, v_i, \tau_i < t$. Now $\omega = (\tau_0, v_0, \tau_1, v_1, \ldots)$, ($v_0$ having no effect), and the controls $u(t, \omega)$ depend only on τ_i, V_i for $\tau_i \leq t$. The controls $w_j(\omega)$ depend on $\tau_i, V_i, i \leq j$. The functions $g_0, g, g_{0x}, g_x, g_{0xx}, g_{xx}$ are continuous in (x, V, w, t), the continuity being uniform in (t, x, w), and λ and λ_y are continuous in (t, x, u).

Optimal nonanticipating controls are denoted $u^*(t, \omega), w_j^*(\omega)$. The HJB equations are now

$$0 = J_s(s, y, j) + \sup_{u \in U} \left\{ f_0(s, y, u, j) + J_y(s, y, j) f(s, y, u, j) + \lambda(s, y, u, j+1) \right.$$

$$\times E \left[\sup_{w \in W_{j+1}} \{ g_0(s, y, V, w, j+1) + J(s, y + g(s, y, V, w, j+1), j+1) \right.$$

$$\left. \left. -J(s, y, j) \} | s, y, j \right] \right\}. \tag{3.74}$$

The expectation is calculated by means of $\pi(v; s, y, j+1)$.

We are not going to present detailed precise conditions for sufficient conditions to hold, but let us at least assume that for each (s, y), g_0 and g are bounded functions of V, w.

Provided λ depends only on (t, x, j), and π not on x, the equations in the extremal method must be modified as follows: First, in addition to $\hat{u}$, we also need the function $\hat{w}(t, x, v, j+1)$ yielding, by assumption, the maximum in

$$\max_{w \in W_{j+1}} [g_0(t, x, v, w, j+1) + z(t; t, x + g(t, x, v, w, j+1), j+1)].$$

There are now three "states" x, p, z, whose differential equations must be solved simultaneously:

$$\dot{x} = f(t, x, \hat{u}(t, x, p, j), j), \tag{3.75}$$

$$\dot{p} = -H_x(t, x, \hat{u}(t, x, p, j), p, j) + \lambda(t, x, j+1)p$$
$$-\lambda(t, x, j+1)\{E[p(t; t, x + g(t, x, V_{j+1}, \hat{w}(t, x, V_{j+1}, j+1), j+1), j+1)$$
$$\times (I + g_x(t, x, V_{j+1}, \hat{w}(t, x, V_{j+1}, j+1), j+1))$$
$$+ g_{0x}(t, x, V_{j+1}, \hat{w}(t, x, V_{j+1}, j+1), j+1) | t, j]\}$$
$$+ \{z - E[g_0(t, x, V_{j+1}, \hat{w}(t, x, V_{j+1}, j+1), j+1)$$
$$+ z(t; t, x + g(t, x, V_{j+1}, \hat{w}(t, x, V_{j+1}, j+1), j+1), j+1)$$
$$| t, j]\} \lambda_x(t, x, j+1), \tag{3.76}$$

$$\dot{z} = -f_0(t, x, \hat{u}(t, x, p, j), j)$$
$$+ \lambda(t, x, j+1)\{z - E[g_0(t, x, V_{j+1}, \hat{w}(t, x, V_{j+1}, j+1), j+1)$$
$$+ z(t; t, x + g(t, x, V_{j+1}, \hat{w}(t, x, V_{j+1}, j+1), j+1), j+1) | t, j]\}. \tag{3.77}$$

The expectations are calculated by means of $\pi(v;t,j+1)$. For each j, these equations are solved simultaneously in (x,p,z), with side conditions in the free end case $x(s) = y, p(T) = h_{0x}(x(T)), z(T) = h_0(x(T))$ and in the hard end constrained case, (3.55) and $z(T) = h_0(x(T))$ (in which case g is assumed to be independent of w). $\square$

For necessary conditions in the case $\lambda = \lambda(t,x,\omega)$, see Seierstad (2001).

Remark 3.51 (Several stochastic processes*). It is possible to generalize the above theory to the case where there are several stochastic processes at work. So assume now that there are given i^* jump functions $g^i(t,x)$, $i = 1,\ldots,i^*$, and i^* stochastic processes furnishing, for each i, jump points τ_j^i, $j = 1,2,\ldots$, according to intensities $\lambda^i(t,x,u)$. So the state jumps according to $x(\tau_j^i+) - x(\tau_j^i-)) = g^i(\tau_j^i, x(\tau_j^i-))$ at τ_j^i. For each stochastic process, it is assumed that an unbounded number of jumps can occur. Moreover, $g_0 \equiv 0$ (for simplicity), and f_0 and f are independent of j. There is then again a single J-function, and it satisfies

$$0 = \max_{u \in U} \left\{ f_0(s,y,u) + J_x(s,y)f(s,y,u) + J_s(s,y) \right.$$
$$\left. + \sum_i [\lambda^i(s,y,u)J(s,y+g^i(s,y)) - \lambda^i(s,y,u)J(s,y)] \right\}. \quad (3.78)$$

Sufficient conditions related to this equation are similar to those in Theorem 3.37; the precise statement is dropped. $\square$

Again, auxiliary state variables can be introduced to take care of a situation where the λ^i's do depend on the number of jumps of various types having already occurred. In a soccer match, one might imagine that two different processes of the above type determine the two teams' goals, and one might wonder if the "differential game" corresponding with the match could be solved by using the tools presented!

3.6 Optimal Stopping Problems

In this section, control problems are studied in which (also) the terminal time is subject to choice. First, we describe the necessary changes in the extremal method and state a corresponding sufficient condition, based on the field of characteristic solutions obtained. Next, necessary conditions and the approach connected with the HJB equations will be discussed.

The problem studied is now the maximization problem:

$$\max_{u(\cdot,\cdot),T \in [T_1,T_2]} E\left[\sum_{j \geq 0} \int_{\min\{\tau_j,T\}}^{\min\{\tau_{j+1},T\}} f_0(t,x(t,\omega),u(t,\omega),j)dt \right.$$
$$\left. + \sum_{j \in \{j \geq 1 : \tau_j < T\}} g_0(\tau_j, x(\tau_j-,\omega), j) + h_0(T, x(T+,\omega)) \right], \quad (3.79)$$

where the maximization is subject to the standard restrictions (3.15), (3.16) (the differential equation, the given start point, the jump condition, and the restriction $u \in U$) and terminal conditions (if any). A fixed interval $[T_1, T_2]$ is specified, within which T is subject to choice. The maximization in (3.79) is carried out over all *admissible triples* $x(t, \omega), u(t, \omega), T(\omega)$, i.e., triples $x(t, \omega), u(t, \omega), T(\omega)$ where $u(t, \omega)$ is bounded, $x(t, \omega)$ is a solution on $[0, T(\omega)]$ satisfying (3.15), (3.16) and terminal conditions (if any), with $x(0, \omega) = x^0$, corresponding with the nonanticipating $u(t, \omega) \in U$, and where $T(\omega)$ is a *stopping time*, i.e., $1_{[0,T(\omega)]}(t)$ is nonanticipating. Again jump intensities $\lambda(i)$ are specified, and the extremal method will only work if, for some $N < \infty$, $\lambda(i) = 0$ for $i > N$. We mainly consider the case of a free end (no terminal conditions).

(The entity $T+$ in (3.79) means that if we stop immediately when a jump has occurred at some time T, the instantaneous reward we then get is $h_0(T, x(T+))$, where $x(T+) := x(T-) + g(T, x(T-), j)$ if the jump has number j, $x(t) = x(t, \omega)$.)

Write $\omega^j := (\tau_0, \ldots, \tau_j, T_2, T_2 + 1, \ldots)$. Note that $\tau_j = T(\omega) \, (= T(\omega^j))$ may hold for a range of values of ω^j, i.e., with positive probability.

The extremal method will now be changed to allow for the optimal choice of T. It is assumed that f_0, f, g_0, g, and h_0 are C^2. Because T is subject to choice, we expect T to become a function of (s, y, j), $T = T(s, y, j) \in [T_1, T_2]$. For each starting point (s, y, j), we shall also choose T. As we have one more unknown, we need to add further conditions to the basic conditions (3.18)–(3.21) of the extremal method. Let us write down all equations needed for the extremal method. In the method, we need an auxiliary state variable z that we have met before.

Extremal method for free terminal time, free end, $N < \infty$

The extremal method makes use of the following relationships.

$$\hat{u}(t, x, p, j) \text{ maximizes } H(t, x, u, p, j) := f_0(t, x, u, j) + pf(t, x, u, j) \text{ for } u \in U \quad (3.80)$$

$$\dot{x}(t) = f(t, x, \hat{u}(t, x, p, j), j) \quad (3.81)$$

$$\dot{p}(t) = -H_x(t, x, \hat{u}(t, x, p, j), p, j) + \lambda(j+1)p$$
$$-\lambda(j+1)[g_{0x}(t, x, j+1) + p(t; t, x + g(t, x, j+1), j+1)(I + g_x(t, x, j+1))] \quad (3.82)$$

with boundary conditions

$$x(s) = y, \quad p(T) = [h_{0x}(T, x)]_{x = x(T)} \quad (3.83)$$

$$\dot{z}(t) = \lambda(j+1)[z - g_0(t, x, j+1) - J(t, x + g(t, x, j+1), j+1)]$$

$$-f_0(t, x, \hat{u}(t, x, p, j), j), \quad z(T) = h_0(T, x(T)) \quad (3.84)$$

$$\eta(T, x(T; s, y, T, j), p(T; s, y, T, j), j) = 0 \text{ if } T \in (\max\{s, T_1\}, T_2),$$

$$\eta(T, x(T; s, y, T, j), p(T; s, y, T, j), j) \geq 0 \text{ if } T = T_2,$$

$$\eta(T, x(T; s, y, T, j), p(T; s, y, T, j), j) \leq 0 \text{ if } T = \max\{s, T_1\}$$

$$\text{where } \eta(T, x, p, j) := \max_u H(T, x, u, p, j) + h_{0t}(T, x)$$

$$+\lambda(j+1)[g_0(T,x,j+1)+J(T,x+g(T,x,j+1),j+1)-h_0(T,x)] \quad (3.85)$$

$$J(s,y,T,j):=z(s;s,y,T,j), \quad (3.86)$$

$$J(s,y,j)=J(s,y,T(s,y,j),j), \text{ for } s<T(s,y,j), \quad (3.87)$$

$$p(t;s,y,j)=p(t;s,y,T(s,y,j),j), \text{ for } s<T(s,y,j). \quad (3.88)$$

Note that $J(s,y,j)=h_0(s,y)$ and $p(s;s,y,j)=h_{0x}(s,y)$ when $s=T(s,y,j)$. Conditions (3.86), (3.87), (3.88) contain only definitions: we need the entities $p(s;s,y,j+1)$ and $J(s,y,j+1)$ in (3.82), (3.84), and (3.85). Note that s is an arbitrary point in $[0,T_2]$, and y is an arbitrary state. Now T is a parameter that is determined in the solution procedure. Before knowing the correct value of T, we need to solve (3.81)–(3.83), for arbitrary T (as well as arbitrary (s,y,j)), so we write the solution pair as $(x(t;s,y,T,j),p(t;s,y,T,j))$. Moreover, there is not a full simultaneity between the differential equations. In fact, when $x(t;s,y,T,j)$ and $p(t;s,y,T,j)$ are known, we can insert these functions for x and p in (3.84), and solve this linear equation in z, to find the solution $z=z(t;s,y,T,j)$. Once $x(t;s,y,T,j)$ and $p(t;s,y,T,j)$ are known, we can use (3.85) to determine $T=T(s,y,j)$. A full explanation of the use of these equations is thus as follows (again a recursive solution procedure is available). First ("step N"), $x(t;s,y,T,N)$, $p(t;s,y,T,N)$ are found using (3.81)–(3.83), where T is a given parameter. At this step $\lambda(N+1)=0$. Condition (3.84) is used to determine $z(t;s,y,T,N)$ and condition (3.85) is used to determine the (hopefully) optimal $T=T(s,y,N)$, Then, $J(s,y,T,N)$, $J(s,y,N)$, and $p(t;s,y,N)$ are written down, using (3.86), (3.87), (3.88). Next ("step $N-1$"), from (3.81)–(3.83), $x(t;s,y,T,N-1)$, $p(t;s,y,T,N-1)$ are found (in (3.82) the function $p(s;s,y,N)$ just constructed enters). Using the known $J(s,y,N)$, (3.84) yields $z(t;s,y,T,N-1)$, Moreover, using (3.85) (and the known $J(s,y,N)$), the (hopefully) optimal $T=T(s,y,N-1)$ is determined. Then, $J(s,y,T,N-1)$, $J(s,y,N-1)$, and $p(t;s,y,N-1)$ are written down, using (3.86), (3.87), (3.88). Next ("step $N-2$"), from (3.81)–(3.83), $x(t;s,y,T,N-2)$, $p(t;s,y,T,N-2)$ are found (in (3.82) the known function $p(s;s,y,N-1)$ enters). Then $z(t;s,y,T,N-2)$ is found by means of (3.84), using the known $J(s,y,N-1)$. Moreover, inserting the known $J(s,y,N-1)$ when using (3.85), the (hopefully) optimal $T=T(s,y,N-2)$ is determined. Then, $J(s,y,T,N-2)$, $p(t;s,y,N-2)$ and $J(s,y,N-2)$ are written down, using (3.86), (3.87), (3.88). And so on. At each step, quintuples $x(t;s,y,T,j),p(t;s,y,T,j),z(t;s,y,T,j),T(s,y,j),J(s,y,T,j)$, $j=N,N-1,\ldots$, are obtained. (We presuppose that the function $t\rightarrow u(t;s,y,T,j):=\hat{u}(t,x(t;s,y,T,j),p(t;s,y,T,j),j)$ becomes piecewise continuous.)

We call the collection $x(t;s,y,j):=x(t;s,y,T(s,y,j),j)$, $p(t;s,y,j):=p(t;s,y,T(s,y,j),j)$, $J(s,y,j)$, $T(s,y,j)$ a characteristic quadruple, and if we add $u(t;s,y,j):=\hat{u}(t;x(t;s,y,j),p(t;s,y,j),j)$, we speak of a characteristic quintuple. Recall that $s\in[0,T_2]$, and that $s\leq T(s,y,j)\leq T_2$. The function $u(t;s,y,j)$ is defined for $t\in[s,T(s,y,j)]$, and the same goes for $x(t;s,y,j)$, $p(t;s,y,j)$.

Also in the current situation, using the characteristic solutions $x(t;s,y,j)$, we can write down the corresponding nonanticipating solution $x(t,\omega)$, and then the corresponding candidate control $u(t,\omega)$ (and also $p(t,\omega)$). So $x(t;\omega^0):=x(t;0,x^0,0)$ for $t\in[0,\min\{\tau_1,T(0,x^0,0)\}]$, with $T(\omega):=T(\omega^0):=T(0,x^0,0)$ if $T(0,x^0,0)\leq\tau_1$. If $T(0,x^0,0)>\tau_1$, let $x(t;\omega^1):=x(t;\tau_1,x(\tau_1-,\omega^0)+g(\tau_1,x(\tau_1-,\omega^0),1),1)$ for t in

$$(\tau_1, \min\{\tau_2, T(\tau_1, x(\tau_1-, \omega^0) + g(\tau_1, x(\tau_1-, \omega^0), 1), 1)\}].$$

If $T(\tau_1, x(\tau_1-, \omega^0) + g(\tau_1, x(\tau_1-, \omega^0), 1), 1) \le \tau_2$, then $T(\omega) := T(\omega^1) := T(\tau_1, x(\tau_1-, \omega^0) + g(\tau_1, x(\tau_1-, \omega^0), 1), 1)$. If

$$T(\tau_1, x(\tau_1-, \omega^0) + g(\tau_1, x(\tau_1-, \omega^0), 1), 1) > \tau_2,$$

define $x(t; \omega^2) := x(t; \tau_2, x(\tau_2-, \omega^1) + g(\tau_2, x(\tau_2-, \omega^1), 2), 2)$ for t in $(\tau_2, \min\{\tau_3, T(\tau_2, x(\tau_2-, \omega^1) + g(\tau_2, x(\tau_2-, \omega^1), 2), 2)\}]$. If

$$T(\tau_2, x(\tau_2-, \omega^1) + g(\tau_2, x(\tau_2-, \omega^1), 2), 2) \le \tau_3,$$

let $T(\omega) := T(\omega^3) := T(\tau_2, x(\tau_2-, \omega^1) + g(\tau_2, x(\tau_2-, \omega^1), 2), 2)$. If

$$T(\tau_2, x(\tau_2-, \omega^1) + g(\tau_2, x(\tau_2-, \omega^1), 2), 2) > \tau_3,$$

continue by defining $x(t, \omega^3)$ for t in

$$(\tau_3, \min\{\tau_4, T(\tau_3, x(\tau_3-, \omega^2) + g(\tau_3, x(\tau_3-, \omega^2), 3), 3)\}]$$

by $x(t; \omega^3) := x(t; \tau_3, x(\tau_3-, \omega^2) + g(\tau_3, x(\tau_3-, \omega^2), 3), 3)$. And so on.

We see that as long as $T(\tau_k, x(\tau_k-, \omega^{k-1}) + g(\tau_k, x(\tau_k-, \omega^{k-1}), k), k) > \tau_{k+1}$, we continue at least as long as t belongs to (τ_k, τ_{k+1}), while we stop at $t = T(\tau_k, x(\tau_k-, \omega^{k-1}) + g(\tau_k, x(\tau_k-, \omega^{k-1}), k), k)$ the first time $T(\tau_k, x(\tau_k-, \omega^{k-1}) + g(\tau_k, x(\tau_k-, \omega^{k-1}), k), k) \le \tau_{k+1}$.

Note that the function $1_{[0, T(\omega)]}(t)$ is nonanticipating. To see this, observe that if $t > T(\omega)$, and k is the smallest k such that $\tau_{k+1} \ge T(\omega)$, then $t > T(\omega) > \tau_k$ and $T(\omega) = T(\omega^k)$ depends only on $\tau_i, i \le k$.

Remark 3.52 (Free end). Define $\eta^*(T, s, y, j) : \eta(T, x(T; s, y, T, j), p(T; s, y, T, j), j)$. The formulation of (3.85) is chosen such that it also works in case of end constraints, briefly treated in Remark 3.61 below. In the free end case

$$\eta(T, x(T; s, y, T, j), p(T; s, y, T, j), j) =: \eta^*(T, s, y, j)$$
$$= \eta^{**}(T, x(T; s, y, T, j), j)$$

where $\eta^{**}(T, x, j)$

$$= f_0(T, x, \hat{u}(T, x, h_{0x}(T, x), j), j)$$
$$+ h_{0x}(T, x)f(T, x, \hat{u}(T, x, h_{0x}(T, x), j), j) + h_{0t}(T, x)$$
$$+ \lambda(j+1)[g_0(T, x, j+1) + J(T, x + g(T, x, j+1), j+1) - h_0(T, x)].$$

$\square$

It may be shown that $e^{\lambda(j+1)(s-T)}\eta^*(T, s, y, j) = J_T(s, y, T, j)$. Thus, assuming that $J(s, y, T, j)$ gives the optimal value when stopping at T, then (3.85) is a necessary condition for choosing T optimally (e.g., if $T(s, y, j) \in (\max\{s, T_1\}, T_2)$, by necessity, $J_T = 0$ at this point).

Below, two examples are presented showing the use of this procedure. As we in particular want to describe the use of (3.85) in determining the optimal stopping time, we have chosen examples that are quite trivial, in particular no control u appears.

Example 3.53. Consider the pure stopping problem (no control u):

$$\max Ex(T), \text{ when } T \in [0,b], \quad \dot{x} = e^{-t}, x(0) = 0,$$

where $x(t)$ can have two downwards unit jumps with intensity λ, $e^{-b} < \lambda < e^b$, b a fixed positive number.

Solution. The procedure (3.80)–(3.88) will be used. Evidently, $x(t;s,y,T,2) = y + e^{-s} - e^{-t}$, $p(t;s,y,T,2) = 1$. Furthermore, $\eta(T,x,p,2)$ equals $e^{-T} p$ so $\eta^*(T,s,y,2) = e^{-T}$. Moreover, $z(t;s,y,T,2) = z(T;s,y,T,2) = y + e^{-s} - e^{-T}$, $J(s,y,T,2) = y + e^{-s} - e^{-T}$. In this case, we always choose to wait until $t = b$ before stopping, a trivial result $((3.85) \Rightarrow T = b)$. Note that $J(s,y,2) = y + e^{-s} - e^{-b}$.

Next, let us consider $j = 1$. In this case, $\dot{p} = \lambda p - \lambda$. With the end condition $p(T;s,y,T,1) = 1$, this gives $p(t;s,y,T,1) = 1$. Furthermore, $\eta(T,x,p,1) := e^{-T}p + \lambda(x - 1 + e^{-T} - e^{-b} - x)$. Hence, for $\eta^*(T,s,y,1) = e^{-T} + \lambda(-1 + e^{-T} - e^{-b})$, we have that $\eta^*(T,s,y,1) = 0$ if $e^{-T}(1 + \lambda) = \lambda(1 + e^{-b})$, i.e., $T = T^* = -\ln[\lambda(1 + e^{-b})(1 + \lambda)^{-1}] \in (0,b)$. Now, if $T < T^*$, $\eta^*(T,s,y,1) > 0$, and if $T > T^*$, $\eta(T,s,y,1) < 0$. Hence, for $s < T^*$, we cannot have $T(s,y,1) = s$ or $T(s,y,1) = b$, the only possibility is $T(s,y,1) = T^*$. For $s \geq T^*$, $T(s,y,1) = s$. Now, also $x(t;s,y,T,1) = y + e^{-s} - e^{-t}$, and

$$\dot{z}(t) = \lambda[z - (x - 1 + e^{-t} - e^{-b})] = \lambda[z - (x(t;s,y,T,1) - 1 + e^{-t} - e^{-b})] =$$

$\lambda[z - (y - 1 + e^{-s} - e^{-b})]$, so $z(t;s,y,T,1) = y + e^{-s} - 1 - e^{-b} + Ce^{\lambda t}$ where C is determined by $z(T;s,y,T,1) = x(T;s,y,T,1)$. The last equality gives $y + e^{-s} - 1 - e^{-b} + Ce^{\lambda T} = y + e^{-s} - e^{-T}$, which gives $C = e^{-\lambda T}(1 + e^{-b} - e^{-T})$, so $z(t;s,y,T,1) = y - 1 + e^{-s} - e^{-b} + (1 + e^{-b} - e^{-T})e^{\lambda(t-T)}$. Hence,

$$J(s,y,T,1) = y - 1 + e^{-s} - e^{-b} + (1 + e^{-b} - e^{-T})e^{\lambda(s-T)},$$

and for $s < T^*$,

$$J(s,y,1) := J(s,y,T(s,y,1),1) = y - 1 + e^{-s} - e^{-b} + (1 + e^{-b} - e^{-T^*})e^{\lambda(s-T^*)},$$

while for $s \geq T^*$, $J(s,y,1) = y$.

Finally, let us consider $j = 0$. As before, $p(T,0,0,0) = 1$ (we now need only consider $(s,y) = (0,0)$. Moreover, for $T \leq T^*$, $\eta(T,x,1,0) = e^{-T} + \lambda\{x - 2 + e^{-T} - e^{-b} + (1 + e^{-b} - e^{-T^*})e^{\lambda(T-T^*)} - x\} = e^{-T} + \lambda\{-2 + e^{-T} - e^{-b} + (1 + e^{-b} - e^{-T^*})e^{\lambda(T-T^*)}\}$. Hence, $\eta^*(T,0,0,0) = e^{-T} + \lambda\{-2 + e^{-T} - e^{-b} + (1 + e^{-b} - e^{-T^*})e^{\lambda(T-T^*)}\}$ for $T \leq T^*$. For $T = T^*$, (using $e^{-T^*} = \lambda(1 + e^{-b})/(1 + \lambda)$), $\eta^*(T^*,0,0,0) =$

$$e^{-T^*} - \lambda = \lambda[(1 + e^{-b})(1 + \lambda)^{-1} - 1] < \lambda[(1 + \lambda)(1 + \lambda)^{-1} - 1] = 0.$$

Moreover, for $T < T^*$, $(d/dT)\eta^* =$

$$-e^{-T}(1+\lambda) + \lambda^2(1+e^{-b} - e^{-T^*})e^{\lambda(T-T^*)} \leq$$

$-e^{-T^*}(1+\lambda) + \lambda^2(1+e^{-b} - e^{-T^*}) = -\lambda(1+e^{-b})/(1+\lambda) < 0$, to obtain the last equality, $e^{-T^*} = \lambda(1+e^{-b})/(1+\lambda)$ was used. Thus, if T^{**} is the solution of $\eta^*(T,0,0,0) = 0$, then $T^{**} < T^*$ and the optimal stopping time is $\max\{0, T^{**}\}$. (In $(-\infty, T^*)$, $(d/dT)\eta^*(T,0,0,0) < 0$, so if $T^{**} > 0$, then $\eta^*(0,0,0,0) > 0$.) Moreover, in $(T^{**}, T^*]$, $\eta^*(T,0,0,0) < 0$. Finally, in $(T^*, b]$, $\eta^*(T,0,0,0) = e^{-T} - \lambda < e^{-T^*} - \lambda \leq 0$.

The sufficient condition below, (Remark 3.55) can be used to show optimality of this solution. □

Remark 3.54 (Sufficiency based on a field of characteristic solutions, $N < \infty$). Define $Q_{x(.,.)}(j) = \{(t, x(t+, \omega^j)) : t \in [\tau_j, T(\omega^j)]$ for some $\omega^j\}$, and $Q(j) = \cup_{x(.,.)} Q_{x(.,.)}(j)$, the union taken over all admissible pairs $(x(.,\omega), T(\omega))$. Similar to Theorem 3.11, assume the existence of a H-maximizing $\hat{u}(t, x, p, j)$ defined on open sets $Q^*(j) \subset (0, T_2) \times \mathbb{R}^{2n}$, and open sets $Q^0(j)$ satisfying (3.28) and containing $Q(j)$. Assume that, for any $T \in [T_1, T_2]$, solutions $x(t; s, y, T, j)$, $p(t; s, y, T, j)$, $z(t; s, y, T, j)$, $J(s, y, T, j)$, $t \in [s, T]$, have been found, satisfying (3.81)–(3.84), and (3.86) for $\hat{u}$ replaced by $u(t; s, y, T, j) = \hat{u}(t, x(t; s, y, T, j), p(t; s, y, T, j), j)$, together with a function $T(s, y, j)$ satisfying (3.89) below, for all $(s, y) \in clQ^0(j)$. Assume that $x(t; s, y, T, j)$ and $p(t, s, y, T, j)$ are continuous in (t, s, y), $t \in [s, T]$, $(s, y) \in clQ^0(j)$, and that, for any $(\hat{s}, \hat{y})$ in $Q^0(j)$, $x(t; s, y, T, j)$ and $p(t, s, y, T, j)$ are C^1 in $(s, y) \in Q^0(j)$, (s, y) close $(\hat{s}, \hat{y})$, $t \in [s, T]$. Moreover, for any $(s, y) \in Q^0(j)$ $((s, y) \in \{(0, x^0)\} \cup Q^0(0)$ in case $j = 0)$, assume that $(t, x(t; s, y, T, j), p(t; s, y, T, j))$ belongs to $Q^*(j)$ for all T, all $t \in [s, T]$, and that $(t, x(t; s, y, T, j))$ belongs to $Q^0(j)$ for all T, all $t \in [s, T] \cap [0, T_2)$. Furthermore, assume that $T \to J(s, y, T, j)$ is Lipschitz continuous in $[T_1, T_2]$. Finally, assume that $\hat{H}$ and the first and second derivatives of it with respect to x and p exist and are continuous in $Q^*(j)$. Then

$$u(t; s, y, j) = \hat{u}(t, x(t; s, y, T(s, y, j), j), p(t; s, y, T(s, y, j), j), j), j), \quad j = 0, 1, \ldots, N,$$

are optimal, more precisely, the triple $(x^*(t, \omega), u^*(t, \omega), T^*(\omega))$ corresponding to these entities is optimal among all admissible triples $(x(t, \omega), u(t, \omega), T(\omega))$, provided $u^*(.,.)$ is bounded. □

Remark 3.55 (A sufficient condition based on concavity, $N < \infty$). A sufficient condition based on concavity can also be stated. We state only the most essential conditions: Assume that, for any j, any $T \in [T_1, T_2]$, any (s, y) in $Q(j)$, (in $\{(0, x^0)\} \cup Q(0)$ in case $j = 0$), solutions $(x(t; s, y, T, j)$, $p(t; s, y, T, j)$, $z(t, s, y, T, j)$, $J(s, y, T, j)$ of (3.81)-(3.84), and (3.86) have been found, to which there correspond functions $T(s, y, j)$ satisfying (3.85). Moreover, assume that the concavity conditions in Remark 3.10 are satisfied for each T. Finally, for each j, assume that

$$\eta(T', x(T'; s, y, T', j), p(T'; s, y, T', j), j) \geq 0 \text{ if } T' \in [\max\{T_1, s\}, T(s, y, j)),$$
$$\eta(T', x(T'; s, y, T', j), p(T'; s, y, T', j), j) \leq 0 \text{ if } T' \in (T(s, y, j), T_2] \quad (3.89)$$

Then the characteristic solutions are optimal. (See Seierstad and Sydsaeter (1987), Theorem 13, p. 145, for a more carefully worded condition in the deterministic case.)   □

Example 3.56. Consider the pure stopping problem (no control u):

$$\max Ex(T), \text{ when } T \in [0,b], \quad \dot{x} = -x, \quad x(0) = x^0 > 0$$

and where $x(t)$ can have a single unit upwards jumps with intensity $\lambda > x^0$.

Solution. All start values y are chosen to be > 0. After one jump, it surely pays to stop at once, $J(s,y,1) = y$, and $p(t;s,y,1) = 1$.

Let us find $x(t;0,x^0,T,0), p(t;0,x^0,T,0)$. Evidently, $x(t;0,x^0,T,0) = x^0 e^{-t}$ and $\dot{p} = p + \lambda p - \lambda$. With $p(T) = 1$, this gives $p(t,0,x^0,T,0) = [1 - \lambda/(1 + \lambda)]e^{(1+\lambda)(t-T)} + \lambda/(1+\lambda)$. Next, $\eta(T,x,p,0)$ in (3.85) reduces to $\eta(T,x,p,0) := -xp + \lambda(x+1-x) = -xp + \lambda$, so $\eta^*(T,0,x^0,0) = -x^0 e^{-T} + \lambda$. Now, for any T, $\eta^*(T,0,x^0,0) > 0$, so $T(0,x^0,0) = b$.

Remark 3.55 gives directly that this proposal $T(0,x^0,0) = b$ is optimal, because $\eta^*(T,0,x^0,0) \geq 0$ for all T.   □

Let us state necessary conditions for the current problem. Let $x^*(t,\omega), u^*(t,\omega)$, $T^*(\omega)$ be an optimal triple in problem (3.79). Recall that when $\tau_j < T_2$, then $T^*(\omega^j) := T^*(\omega^j, T_2, T_2 + 1, \ldots)$. Moreover, for $t > \tau_j$, $x^*(t,\omega^j) := x^*(t,\omega^j, T_2, T_2 + 1, \ldots)$, and the same type of definition is used for $p(t,\omega^j)$ and $u^*(t,\omega^j)$. The three functions are defined on $[\tau_j, T^*(\omega^j)]$. We need to introduce the function $z^*(t,\omega^j)$ defined on $[\tau_j, T^*(\omega^j)]$, satisfying the two equations: For $t > \tau_j$, v.e.,

$$\dot{z}^*(t,\omega^j) = -f_0(t,x^*(t,\omega^j),u^*(t,\omega^j),j) + \lambda(j+1)z^*$$
$$-\lambda(j+1)[g_0(t,x^*(t,\omega^j),j+1) + z^*(t+,\omega^j,t)] \quad (3.90)$$
$$z^*(T^*(\omega^j),\omega^j) = h_0(T^*(\omega^j),x(T^*(\omega^j)+,\omega^j))), \quad (3.91)$$

($t \to z^*(t,\omega^j)$ continuous). Assume that the Standard System Conditions connected with (3.25) are satisfied (h_0 still C^2).

Theorem 3.57 (Necessary condition for free terminal time, free end, $N \leq \infty$). *Assume that $x^*(t,\omega), u^*(t,\omega), T^*(\omega)$ is an optimal triple in problem (3.79). Then, functions $p(t,\omega)$, defined for $t \in [0, T(\omega))$, exist, with properties as in Theorem 3.8, such that, for any given j, the maximum condition (3.22) and the adjoint equation (3.23) hold for T replaced by $T^*(\omega^j)$ (hence, (3.22) and (3.23) are satisfied for v.e. $t \in (\tau_j, T^*(\omega^j))$). Furthermore, $p(T^*(\omega^j),\omega^j) = h_{0x}(T^*(\omega^j),x^*(T^*(\omega^j)+,\omega^j))$. Moreover, a.s., for $\tau_j < T^*(\omega^j)$,*

$$\eta_*(T^*(\omega^j),\omega^j) = 0 \text{ if } T^*(\omega^j) \in (T_1,T_2),$$
$$\eta_*(T^*(\omega^j),\omega^j) \geq 0 \text{ if } T^*(\omega^j) = T_2, \ \eta_*(T^*(\omega^j),\omega^j) \leq 0 \text{ if } T^*(\omega^j) = T_1, \quad (3.92)$$

where $\eta_*(T,\omega^j) :=$

$$\max_u H(T, x^*(T, \omega^j), u, p(T, \omega^j), j) + h_{0t}(T, x^*(T, \omega^j))$$

$$+ \lambda(j+1)[g_0(T, x^*(T, \omega^j), j+1) + z^*(T, \omega^j, T) - h_0(T, x^*(T, \omega^j))].$$

Finally, a.s., $\eta_*(T^*(\omega^j)+, \omega^j) \leq 0$ *if* $T^*(\omega^j) = \tau_j$.   □

See Seierstad (2001).

The HJB Equation in Optimal Stopping

An alternative solution tool is the HJB equation. Define

$$J^*(s, y, j) := \sup_{u(.,.), T} E\left[\sum_{k \geq j} \int_{\tau_k}^{\tau_{k+1}} 1_{[s,T]}(t) f_0(t, x(t, \omega), u(t, \omega), j) dt\right.$$

$$+ \sum_{k > j, \tau_k < T} g_0(\tau_k, x(\tau_k-, \omega), j)$$

$$\left. + h_0(T, x(T+, \omega)) | s, y, j\right], \tag{3.93}$$

where the maximum is found for all triples $(x(t, \omega), u(t, \omega), T(\omega))$ satisfying the differential equation and the jump condition, with $x(.,.)$ starting at (s, y, j), (i.e., j jumps before s, $\tau_j \leq s$, $y = x(s+, \omega)$), and where we allow $T = T(\omega)$ to be subject to choice in $[\max\{s, T_1\}, T_2]$, $(x(t, \omega), u(t, \omega), T(\omega))$ independent of $\tau_k, k \leq j$ $((x(.,.), u(.,.))$ nonanticipating, $T(\omega)$ a stopping time). Except for Remark 3.61 below, we assume here a free end. When the solution procedure above is used, in well-behaved cases, $J(s, y, j)$ coincides $J^*(s, y, j)$. Let the set $G^*(j)$ be defined by:

$$G^*(j) = \{(s, y) \in [T_1, T_2] \times \mathbb{R}^n, J^*(s, y, j) > h_0(s, y)\}. \tag{3.94}$$

By definition of $G^*(j)$, $J^*(s, y, j) \leq h_0(s, y)$ for $(s, y) \in G_*(j) := ([T_1, T_2] \times \mathbb{R}^n) \setminus G^*(j)$. But, of course, the optimal value is never strictly smaller than $h_0(s, y)$, because we always have the possibility to stop immediately at (s, y) and get $h_0(s, y)$. So, $J^*(s, y, j) = h_0(s, y)$ for $(s, y) \in G_*(j)$. Moreover, $G^*(j)$ is "the continuation region": We don't stop immediately at (s, y) when we are in $G^*(j)$, because by continuing, we get $J^*(s, y, j)$ instead of $h_0(s, y)$.

Assume that $J^*(s, y, j)$ is continuous in (s, y). Then, for each j, $J^*(s, y, j)$ satisfies the following condition: Let $\partial G_*(j)$ denote the boundary of $G_*(j)$. Then

$$J^*(s, y, j) = h_0(s, y) \text{ for } (s, y) \in \partial G_*(j), s \in [T_1, T_2] \tag{3.95}$$

Note that $\partial G_*(j) \cap ((T_1, T_2) \times \mathbb{R}^n) = \partial G^*(j) \cap ((T_1, T_2) \times \mathbb{R}^n)$.

Similar to the fixed horizon case, at each point $(s, y) \in ([0, T_1] \times \mathbb{R}^n) \cup G^*(j)$ at which $J^*(s, y, j)$ is differentiable in (s, y), the following equality (HJB equation) usually holds, for $J = J^*$.

$$0 = J_s(s,y,j) + \sup_{u \in U}\{f_0(s,y,u,j) + J_y(s,y,j)f(s,y,u,j)\}$$

$$+\lambda(j+1)\{g_0(s,y,j+1) + J(s,y+g(s,y,j+1),j+1) - J(s,y,j)\} \qquad (3.96)$$

Finally, the following condition usually holds: For all $(T,y) \in \partial G_*(j)$,

$$h_{0t}(T,y) + \sup_{u \in U}\{f_0(T,y,u,j) + h_{0x}(T,y)f(T,y,u,j)\}$$

$$+\lambda(j+1)\{g_0(T,y,j+1) + J(T,y+g(T,y,j+1),j+1) - h_0(T,y)\} = 0$$

$$\text{if } T \in (T_1,T_2) \ (\geq 0 \text{ if } T = T_2, \ \leq 0 \text{ if } T = T_1). \qquad (3.97)$$

In Chapter 2, in order to derive (2.52) (the analog of (3.97), the left and right derivatives $(d^{\pm}/dT)S(T,x(T;s,y))$ (S the scrap value function) were used to decide whether it pays to continue a little longer or shorter. A similar discussion of whether it pays to continue a little longer or shorter can be carried out in the current situation (it should not pay, in optimum), and this leads to the condition (3.97).

Equality (3.96) is the HJB equation of the problem. The solution of the problem now consists in finding solutions $J(s,y,j)$ of the HJB equation (3.96) on open sets $G(j)$, such that $\partial G(j) \cap ([T_1,T_2] \times \mathbb{R}^n)$ consists of points (T,y) satisfying (3.97) and such that $J(T,y,j) = h_0(T,y)$ at such points.

Also in the optimal stopping case, a sufficient condition is connected with this procedure; some further conditions are then needed.

Theorem 3.58 (Sufficient condition for the HJB equation, free end, free horizon, $N \leq \infty$). *Assume that open subsets $Q''(j)$ and $\hat{G}(j)$ of $\mathbb{R}^{n+1}$ and functions $J(s,y,j)$ defined on $clQ''(j)$ have been found, $\hat{G}(j) \subset Q''(j)$, $Q(j) \subset Q''(j)$, $J(s,y,j)$ being C^1 in $(Q''(j) \cap \hat{G}(j)) \cup (Q''(j) \backslash cl\hat{G}(j))$, satisfying the HJB equation (3.96) in $\hat{G}(j)$ and satisfying*

$$J(t,x,j) > h_0(t,x) \quad \text{for} \quad (t,x) \in Q(j) \cap \hat{G}(j), \ t \geq T_1. \qquad (3.98)$$

Moreover, assume that $Q(j) \cap ([0,T_1) \times \mathbb{R}^n) \subset \hat{G}(j)$ and that $[(T_1,T_2) \times \mathbb{R}^n] \cap \partial \hat{G}(j)$ is slim. Furthermore, assume, for $(s,y) \in clQ(j)$, that $J(s,y,j) \geq h_0(s,y)$, and, for $(s,y) \in [Q(j) \cap ((T_1,T_2) \times \mathbb{R}^n)] \backslash cl\hat{G}(j)$, that

$$(3.96) \text{ is satisfied with the equality replaced by } \geq. \qquad (3.99)$$

Assume also that $J(s,y,j)$ is locally Lipschitz continuous on $clQ''(j)$. Assume that $J(s,y,j) = h_0(s,y)$ for all $(s,y) \in (clQ(j)) \cap \partial \hat{G}(j)$, $s \geq T_1$. Assume, furthermore that, for any admissible pair $x(t,\omega), T(\omega)$, for some positive constants $\alpha_{x(.,.)}, \kappa_{x(.,.)}$, for any j and any ω, for all t, $|J(t,x(t,\omega),j)| \leq \alpha_{x(.,.)} + \kappa_{x(.,.)}|x(t,\omega)|$ if $t \in [\tau_j,\tau_{j+1})$. Assume, moreover, that there exists a control function $u^*(t,\omega)$, with corresponding solution $x^*(t,\omega)$ defined on some interval $[0,T^*(\omega)]$ that yields the supremum in the HJB equation (3.96), for v.e. $s \in (\tau_j,\tau_{j+1})$, $s < T^*(\omega)$ at $(s,y) = (s,x^*(s,\omega^j))$, and satisfies $J(T^*(\omega^j),x^*(T^*(\omega^j)+,\omega^j),j) = h_0(T^*(\omega^j), x^*(T^*(\omega^j)+,\omega^j))$ if $T^*(\omega^j) = \tau_j < T_2$, where $T^*(\omega^j) := T(\tau_1,\ldots,\tau_j,T_2,T_2 + 1,\ldots)$. Finally assume that $1_{[0,T^*(\omega)]}(t)$ is nonanticipating, that $T^*(\omega^j) = \sup\{t :$

$(s, x^*(s, \omega^j)) \in \hat{G}(j)$ for $s \in (\tau_j, t)\}$ $(T^*(\omega^j) = \tau_j$ if the last set is empty), and that $(x^*(.,.), u^*(.,.))$ is admissible. Then $(x^*(t, \omega), u^*(t, \omega), T^*(\omega))$ is an optimal triple in the problem (3.93). □

Remark 3.59 (Method of solving the HJB equation). As before, a standard method for solving the HJB equation is to use the characteristic equations. We then imagine that, for each j, characteristic quintuples have been found for all (s, y) in some open set $Q''(j)$ containing $Q(j) \cap ([0, T_1) \times \mathbb{R}^n)$, and that the open set $\hat{G}(j)$ equals $\{(s, y) \in Q''(j) : s < T(s, y, j)\}$. Note that the functions $J(s, y, j)$ defined in (3.87) automatically satisfy (3.96) for $(s, y) \in \hat{G}(j)$, provided enough differentiability is exhibited by the functions $x(t; s, y, T, j), p(t; s, y, T, j)$, and $T(s, y, j)$, at least when $T(s, y, j) \in (T_1, T_2)$. (Then $J(s, y, j) = h_0(s, y)$ for $(s, y) \notin \hat{G}(j)$, so the HJB inequality needs to be tested for this function $J(s, y, j)$ outside of $\hat{G}(j)$.) □

Example 3.60. Assume in Example 3.56 the possibility of an unbounded number of jumps, all with intensity λ.

Solution. The HJB equation $0 = J_s + J_x(-x) - \lambda J + \lambda J(x+1)$ (as well as the adjoint equation) is now satisfied for a single function J independent of j.

Below, let $y > 0$. Let us do some preliminary calculations. The expected value $z(\sigma)$ of $x(\sigma)$, $\sigma \in (s, b]$ can be found as follows: Given $x(\sigma)$, the conditional expected value of $x(\sigma + d\sigma)$ is approximately equal to $(x(\sigma) + \dot{x}(\sigma)d\sigma)(1 - \lambda d\sigma) + (x(\sigma) + 1)\lambda d\sigma$, the two terms arising, respectively, from the two events: no jump in $(\sigma, \sigma + d\sigma)$, and one jump in $(\sigma, \sigma + d\sigma)$. (For $d\sigma$ small, the possibility of more than one jump may be discarded.) Then $Ex(\sigma + d\sigma) \simeq x(\sigma)(1 - \lambda d\sigma) - x(\sigma)d\sigma(1 - \lambda d\sigma) + (x(\sigma) + 1)\lambda d\sigma \simeq x(\sigma) - x(\sigma)d\sigma + \lambda d\sigma$. Hence, $E[x(\sigma + d\sigma) - x(\sigma)]/d\sigma \simeq -x(\sigma) + \lambda$ (still conditional expectation). Hence, taking unconditional expectation on both sides gives $\dot{z} = -z(\sigma) + \lambda$. This differential equation has the solution $z(t) = \lambda + (y - \lambda)e^{-(t-s)}$ (given $z(s) = y$). Does it pay to stop at b rather than stopping at once? This requires $y < z(b)$, which yields $y(1 - e^{-(b-s)}) < \lambda(1 - e^{-(b-s)})$, or $y < \lambda$. So if this inequality holds, it pays to continue. (In fact in Example 3.56 above, we saw that even if only one jump may occur, it pays to continue.) What is the optimal strategy if $y > \lambda$? We suspect that if y is not too large, it may still pay to continue.

It is difficult to solve the HJB equation in this case. So a guess is needed. Let us calculate the expected reward when starting in (s, y) and waiting until the first jump occurs and then stopping at once, or if no jump occurs stopping at b. (Denote this policy R^*.) The expected reward is evidently

$$\int_s^b (ye^{s-\rho} + 1)\lambda e^{\lambda(s-\rho)}d\rho + \int_b^\infty ye^{s-b}\lambda e^{\lambda(s-\rho)}d\rho$$
$$= \kappa(s, y) := y\lambda(1 + \lambda)^{-1}(1 - e^{(1+\lambda)(s-b)}) + 1 - e^{\lambda(s-b)} + ye^{(1+\lambda)(s-b)}.$$

Now, $\kappa(b, y) = y$. Note also that $\partial\kappa(s, y)/\partial s = ye^{(1+\lambda)(s-b)} - \lambda e^{\lambda(s-b)}$ is negative when $y = \lambda$, $s < b$. Thus, the function $t \to \kappa(t, \lambda)$ stays above λ when going backwards. Hence, the solution $x_*(s)$ of $y = \kappa(s, y)$ satisfies $x_*(s) > \lambda$. The formula is

$$x_*(s) = (1+\lambda)\left(1 - e^{\lambda(s-b)}\right) / \left(1 - e^{(1+\lambda)(s-b)}\right)$$

which evidently is $< 1 + \lambda$ when $s < b$.

We have seen that $1 + \lambda \geq x_*(s) \geq \lambda$. Compared with the above strategy, (R^*), it pays to stop at once if $y > x_*(s)$ but to continue if $y < x_*(s)$. Note also that if $y < \lambda$, then it pays to continue even if only one jump can occur as seen above. (And perhaps we don't stop immediately after that jump.) At such y, we surely know what to do. So from now on, we only consider $y \geq \lambda$.

Note that then $y + 1 \geq \lambda + 1 \geq x_*(s)$. If we propose the set $\{(s,y) : y \geq x_*(s)\}$ to be the "stopping region," then we see that if $\lambda \leq y \leq x_*(s)$, a jump at (s,y) brings $y + 1$ into this stopping set. Thus if we now put $J(s,y) = \kappa(s,y)$ in the set $G_\lambda := \{(s,y) : s < b, \lambda \leq y < x_*(t)\}$, while $J(s,y) = y$ elsewhere in $[\lambda, \infty)$, then theoretical results tell us that $J(s,y)$ satisfies the HJB equation, for (s,y) in G_λ. To explain intuitively the satisfaction of the HJB equation, simply note that $J(s,y)$ as here defined is the optimal value function in a very restricted problem: That one in which we are given no choices, and in which we get the reward $x(\tau) + 1 = ye^{-(\tau-s)} + 1$, when we jump from $(\tau, x(\tau)) \in G_\lambda$ to $(\tau, x(\tau+1))$ and $x(b) = ye^{-(b-s)}$ in case of no jump. However, that $J(s,y)$ satisfies the HJB equation in G_λ can be directly tested by insertion into the equation. To see this, let $x(s) = x(0)e^{-s}$ be an arbitrary solution (i.e., $x(0)$ arbitrary). Insert $x(0)e^{-s}$ for y in (say) the integral formula for $\kappa(s,y)$. When differentiating the ensuing expression with respect to s, note that after differentiating the two integrands, the integrals sum to $\lambda \kappa(s, x(s))$. Differentiating with respect to the lower integration bound gives $-\lambda(x(0)e^{-s} + 1)$. So $\kappa_s(s, x(s)) + \kappa_x(s, x(s))\dot{x}(s) = (d/ds)\kappa(s, x(s)) = -\lambda(x(0)e^{-s} + 1) + \lambda \kappa(s, x(s))$. This evidently yields the satisfaction of the HJB equation at (s,y), for $y = x(s) = x(0)e^{-s}$. We also check that the HJB inequality is satisfied by the function $h_0(s,y) = y$ for $y \in [\lambda, \infty) \setminus G_\lambda$ (which follows from $y > \lambda$). Using sufficiency results (Theorem 3.58), we can conclude that the optimal policy has been found: Also at points $(s,y) \in G_\lambda$, we continue.

A central point in this sufficiency argument is that $J(s,y)$ is (Lipschitz) continuously pieced together along the curve $x_*(t)$.

Above we noted that for $y < \lambda$, we knew that we continued. So the continuation region in fact is $G := \{(s,y) : s < b, y < x_*(s)\}$. We can in principle construct solutions of the HJB equation also for $(s,y) \in G, (s,y) \notin G_\lambda$, by successively constructing solutions in $\{(s,y) : s < b, x_*(s) - 2 \leq y < x_*(s) - 1\}, \{(s,y) : s < b, x_*(s) - 3 \leq y < x_*(s) - 2\}, \ldots$, as we always jump from a given set in this sequence to the preceding one, on which we have already constructed the solution. But we don't need these solutions.   □

Remark 3.61 (End constraints). Assume in problem (3.15), (3.16), (3.79) , with T subject to choice in $[T_1, T_2]$, that the following terminal conditions are introduced: a.s.,

$$x_i(T(\omega)+, \omega) = \hat{x}_i, i = 1, \ldots, n', \tag{3.100}$$

$$x_i(T(\omega)+, \omega) \geq \hat{x}_i, i = n' + 1, \ldots, n''. \tag{3.101}$$

Then Theorem 3.58 also holds in the case of end constraints, provided the following changes are made: Allow triples $(x(t,\omega),u(t,\omega),T(\omega))$ that are *semi-admissible* in the sense that $\sup_{\omega,t\leq T'}|u(t,\omega)1_{[0,T(\omega)]}| < \infty$, for all $T' < T_2$ $E\{|\int_0^{T_2}|\dot{x}(t,\omega)1_{[0,T(\omega)]}|dt\} < \infty$, $E\{\int_0^{T_2}|f_0(t,x(t,\omega),u(t,\omega))1_{[0,T(\omega)]}|dt\} < \infty$, and assume that $(x^*(t,\omega),u^*(t,\omega),T^*(\omega))$ is semiadmissible. Weaken the assumptions on $J(s,y,j)$ to be that it is defined and locally Lipschitz continuous on $(\mathrm{cl}Q''(j))\cap([0,T_2)\times\mathbb{R}^n)$, that $J(s,y,j) = h_0(s,y)$ is required to holds for all $(s,y)\in Q(j)\cap\partial\hat{G}(j)$, $s\in[T_1,T_2)$, and that $J(s,y,j)\geq h_0(s,y)$ is required to hold for all $(s,y)\in Q(j)$, $s\in[T_1,T_2)$. Furthermore, add the condition that, a.s., if $T^*(\omega^j) = T_2$ for some ω^j, then,

$$\lim_{t\uparrow T^*(\omega^j)} J(t,x^*(t,\omega^j),j) = h_0(T^*(\omega^j),x^*(T^*(\omega^j))), \qquad (3.102)$$

and that, for any semiadmissible pair $(x(t,\omega),T(\omega))$, a.s. for any ω^j such that $T(\omega^j) = T_2$,

$$\lim_{t\to T_2} J(t,x(t,\omega^j),j) \geq h_0(T_2,x(T_2,\omega^j)). \qquad (3.103)$$

Finally, assume that for any semiadmissible pair $(x(t,\omega),T(\omega))$, any $T' < T$, for some constants $\alpha_{T',x(.,.)}$, $\kappa_{T',x(.,.)}$, $|J(t,x(t,\omega),j)1_{[0,T(\omega)]}(t)| \leq \alpha_{T',x(.,.)} + \kappa_{T',x(.,.)}|x(t,\omega)|$ when $t\in[\tau_j,\tau_{j+1}), t\leq T'$, and that

$$E\left[\sup_{t<T_2}|J(t,x(t,\omega),\omega)1_{[0,T(\omega)]}(t)|\right] < \infty.$$

In case of the elementary solution procedure, replace the transversality condition $p(T) = [h_{0x}(T,x)]_{x=x(T)}$ in (3.83) on $p(T;s,y,T,j)$ by (3.54) for $\Lambda_i(k,s,y,j) = \Lambda_i(k,s,y,T,j)$, $\Lambda_i(k,s,y,T,j)$ (hopefully) continuous in (s,y) for $i = n' + 1,\ldots,n''$. $\qquad\square$

3.7 A Selection of Proofs*

Proofs are given for sufficient conditions (verification results) based on the HJB equation and characteristic solutions, as well as for the necessity of the HJB equation (only in the fixed terminal horizon case). Certain details are treated with less than full rigor. To simplify the notation, we shall consider the special case where $f_0\equiv 0$ and $g_0\equiv 0$. The general case is treated by applying the "special case" to a redefined problem where an auxiliary additional state x_0 is introduced governed by

$$\dot{x}_0 = f_0(t,x,u,j), x_0(0) = 0, \qquad (3.104)$$

$$x_0(\tau_j+) - x_0(\tau_j-) = g_0(\tau_j,x(\tau_j-),V_j,j), \qquad (3.105)$$

and where the following criterion is maximized

$$E[x_0(T-) + h_0(T, x(T-))].\tag{3.106}$$

Proof of Theorem 3.37

Let $u(t, \omega)$ be a bounded function, with corresponding solution $x(t, \omega)$ of (3.15), (3.16). Assume that the following inequality has been shown:

$$J(\tau_j, x(\tau_j+, \omega^j), j) \geq J(T, x(T, \omega^j), j) e^{\lambda(j+1)(\tau_j - T)}$$
$$+ \int_{\tau_j}^{T} J(t, x(t, \omega^j) + g(t, x(t, \omega^j), j+1), j+1) \lambda(j+1) e^{\lambda(j+1)(\tau_j - t)} dt \tag{3.107}$$

with equality holding if $x(.,.) = x^*(.,.)$ and $u^*(.,.)$ is bounded. The right-hand side equals

$$J(T, x(T, \omega^j), j) e^{\lambda(j+1)(\tau_j - T)}$$
$$+ \int_{\tau_j}^{T} J(\tau_{j+1}, x(\tau_{j+1}+, \omega), j+1) \lambda(j+1) e^{\lambda(j+1)(\tau_j - \tau_{j+1})} d\tau_{j+1}.$$

Hence, for $\tau_j < T$,

$$J(\tau_j, x(\tau_j+, \omega), j) \geq \int_{T}^{\infty} J(T, x(T, \omega^j), j) \lambda(j+1) e^{\lambda(j+1)(\tau_j - \tau_{j+1})} d\tau_{j+1}$$
$$+ \int_{\tau_j}^{T} J(\tau_{j+1}, x(\tau_{j+1}+, \omega), j+1)$$
$$\times \lambda(j+1) e^{\lambda(j+1)(\tau_j - \tau_{j+1})} d\tau_{j+1}$$
$$= E[J(T, x(T, \omega), j) 1_{(T,\infty)}(\tau_{j+1}) | \omega^j]$$
$$+ E[J(\tau_{j+1}, x(\tau_{j+1}+, \omega), j+1) 1_{[0,T]}(\tau_{j+1}) | \omega^j]. \tag{3.108}$$

Then, $J(0, x^0, 0) \geq$

$$E[J(T, x(T, \omega), 0)) 1_{(T,\infty)}(\tau_1)]$$
$$+ E[J(\tau_1, x(\tau_1+, \omega) 1) 1_{[0,T]}(\tau_1)]$$
$$\geq E[J(T, x(T, \omega), 0)) 1_{(T,\infty)}(\tau_1)] + E[E[1_{[0,T]}(\tau_1) J(T, x(T, \omega), 1) 1_{(T,\infty)}(\tau_2) | \omega^1]]$$
$$+ E[E[J(\tau_2, x(\tau_2+, \omega), 2) 1_{[0,T]}(\tau_2) 1_{[0,T]}(\tau_1) | \omega^1]]$$
$$= \sum_{i=0}^{1} E[J(T, x(T, \omega), i) 1_{[\tau_i, \tau_{i+1})}(T)] + E[J(\tau_2, x(\tau_2+, \omega), 2) 1_{[0,T]}(\tau_2)],$$

the first (second) inequality following from using (3.108) for $j = 0$, ($j = 1$). Continuing in this manner, using (3.108) for $j = 2, 3, \ldots$, we get $J(0, x^0, 0) \geq$

$$\sum_{i=0}^{j} E[J(T, x(T, \omega), i) 1_{[\tau_i, \tau_{i+1})}(T)] + E[J(\tau_{j+1}, x(\tau_{j+1}+, \omega), j+1) 1_{[0,T]}(\tau_{j+1})].$$

Now, $\Pr[\tau_{j+1} < T] \to 0$, when $j \to \infty$. Using the growth conditions stated subsequent to Theorem 3.8, it can easily be proved that when $j \to \infty$, then $E[J(\tau_{j+1}, x(\tau_{j+1}+, \omega), j+1)1_{[0,T]}(\tau_{j+1})] \to 0$ and also that the infinite sum to follow exists. (Here boundedness of $u(.,.)$ is needed.) Hence, letting $j \to \infty$, we get

$$J(0, x^0, 0) \geq \sum_{i=0}^{\infty} EJ(T, x(T, \omega), i)1_{[\tau_i, \tau_{i+1})}(T), \tag{3.109}$$

with equality if $x(.,.) = x^*(.,.)$ and $u^*(.,.)$ is bounded.

Let us prove (3.107). Multiplying by $e^{\lambda(j+1)(\tau_j - s)}$ in the "HJB-inequality" corresponding with (3.64), with $y = x(s, \omega^j)$, gives

$$\begin{aligned}
0 &\geq J_s(s, x(s, \omega^j), j)e^{\lambda(j+1)(\tau_j - s)} \\
&\quad + J_y(s, x(s, \omega^j), j)f(s, x(s, \omega^j), u(s, \omega^j), j)e^{\lambda(j+1)(\tau_j - s)} \\
&\quad + \lambda(j+1)J(s, x(s, \omega^j) + g(s, x(s, \omega^j), j+1), j+1)e^{\lambda(j+1)(\tau_j - s)} \\
&\quad - \lambda(j+1)J(s, x(s, \omega^j), j)e^{\lambda(j+1)(\tau_j - s)} \\
&= (d/ds)[J(s, x(s, \omega^j), j)e^{\lambda(j+1)(\tau_j - s)}] \\
&\quad + \lambda(j+1)J(s, x(s, \omega^j)) + g(s, x(s, \omega^j), j+1), j+1)e^{\lambda(j+1)(\tau_j - s)}.
\end{aligned}$$

Equality holds if $(x(.,.), u(.,.)) = (x^*(.,.), u^*(.,.))$, provided $u^*(.,.)$ is bounded. In case $\tau_j < T$, integrating between $s = \tau_j$ and T yields (3.107). (For $j = 0$, $\tau_0 = 0$, even continuity of $J(s, y, 0)$ on $\{(0, x^0)\} \cup Q^0(0)$ is used.)

To prove Theorem 3.37, replace T by T', $T' < T$, in the arguments leading to (3.109). Then (3.109) means that $J(0, x^0, 0) \geq E[J(T', x(T', \omega), \omega)]$, with equality if $x(.,.) = x^*(.,.)$. Here $x(.,.), u(.,.)$ and $x^*(.,.), u^*(.,.)$ are allowed to be semiadmissible. By semiadmissibility and the growth condition on J, letting $T' \to T$ and using (3.66), we get $J(0, x^0, 0) \geq \limsup_{T' \to T} E[J(T', x(T', \omega), \omega)] \geq E[h_0(x(T, \omega)]$, and, by (3.67), equality holds if $x(.,.) = x^*(.,.)$. This means that $(x^*(.,.), u^*(.,.))$ is optimal.

Essentially the same arguments also work in cases (a), (b), and (c) in Remark 3.39. □

Remark 3.62 (Changed premises). Let $x(.,.), u(.,.)$ and $x^*(.,.), u^*(.,.)$ be strongly semiadmissible. If (3.107) holds for $J(T, x(T, \omega^j), j)$ replaced by $h_0(x(T, \omega^j))$ (equality if $x(.,.) = x^*(.,.)$), then $J(0, x^0, 0) \geq Eh_0(x(T, \omega))$ (equality if $x(.,.) = x^*(.,.)$). □

Proof of Theorem 3.27

For simplicity, assume that all semiadmissible pairs are strongly semiadmissible, and that the end conditions (3.47) and (3.48) are required to hold for all ω. Let us first note that from the conditions assumed in the theorem, by backwards induction, it easily follows that

$$(s,y) \rightarrow (z(s;s,y,j), z_y(s;s,y,j), p(s;s,y,j), p_y(s;s,y,j)) \text{ is } C^0 \text{ in } Q^0(j). \quad (3.110)$$

To see this, assume by induction that (3.110) holds for $j+1$ (no such assumption needed when $j = N$). Let us prove that it then holds for j. Essentially, the argument is simply that the functions $z_y(t;s,y,j)$ and $p_y(t;s,y,j)$ (as well as $z(s;s,y,j)$ and $p(s;s,y,j)$) satisfy a set of differential equations with continuous right-hand sides that we don't need to write down, and from this (3.110) follows even for j. But perhaps the following additional arguments are helpful: For any j, for any $(\hat{s}, \hat{y}) \in Q^0(j)$, and for any $T' \in (\hat{s}, T)$, let $A(T', \hat{s}, \hat{y}, j) = \{(t, x(t; \hat{s}, \hat{y}, j), p(t; \hat{s}, \hat{y}, j)) : t \in [\hat{s}, T']\}$, let $B(T', \hat{s}, \hat{y}, j) = \{(t, x(t; \hat{s}, \hat{y}, j) + g(t, x(t; \hat{s}, \hat{y}, j), j+1) : t \in [\hat{s}, T']\}$ and choose an $\varepsilon > 0$ such that $C(T', \hat{s}, \hat{y}, j, \varepsilon) = \{(t, x, q) : \text{dist}((t, x, q), A(T', \hat{s}, \hat{y}, j)) \leq \varepsilon\} \subset Q^*(j)$, and such that $D(T', \hat{s}, \hat{y}, j, \varepsilon) = \{(t, x) : \text{dist}((t, x), B(T', \hat{s}, \hat{y}, j)) \leq \varepsilon\} \subset Q^0(j+1)$. Now, $\hat{H}$ and its first and second derivatives with respect to x and p are bounded on $C(T', \hat{s}, \hat{y}, j, \varepsilon)$ and by the induction hypothesis, $p(s;s,y,j+1)$ and $p_y(s;s,y,j+1)$ are bounded on $D(T', \hat{s}, \hat{y}, j, \varepsilon)$. There exists a $\check{s} < \hat{s}$, such that for (s,y) close to $(\hat{s}, \hat{y})$, the solutions $x(t;s,y,j), p(t;s,y,j)), z(t;s,y,j)$ exist even for t in $[\check{s}, T']$, still being C^1 in (s,y) near $(\hat{s}, \hat{y})$ for such t, moreover $(t, x(t;s,y,j), p(t;s,y,j))$, and $(t, x(t;s,y,j) + g(t, x(t;s,y,j), j+1))$ stay, respectively, in $C(T', \hat{s}, \hat{y}, j, \varepsilon)$ and in $D(T', \hat{s}, \hat{y}, j, \varepsilon)$ when $t \in [\check{s}, T']$. Evidently, $\dot{p}(t;s,y,j)$, $\dot{z}(t;s,y,j)$ as well as $\dot{p}_y(t;s,y,j), \dot{z}_y(t;s,y,j)$ are bounded for $t \in [\check{s}, T']$ for (s,y) close to $(\hat{s}, \hat{y})$. From this (3.110) follows even for j.

By backwards induction, we shall prove that $z_y(s;s,y,j) = p(s;s,y,j)$ for $(s,y) \in Q^0(j)$ (from which it follows that $z_{yy}(s;s,y,j)$ is C^0 at such points). Let $(\tilde{s}, \tilde{y})$ be any given point in $Q(j)$ when $j > 0$, $(\tilde{s}, \tilde{y}) \in \{(0, x^0)\} \cup Q(0)$ in case $j = 0$. Let $z^*(t;s,y,j) := z(t;s,y,j)e^{\lambda(j+1)(\tilde{s}-t)}$, (so $J(\tilde{s}, \tilde{y}, j) = z^*(\tilde{s}; \tilde{s}, \tilde{y}, j)$). Using (3.58), an easy calculation shows that

$$\dot{z}^*(t;s,y,j) = -\lambda(j+1)J(t, x(t;s,y,j)) + g(t, x(t;s,y,j), j+1), j+1)e^{\lambda(j+1)(\tilde{s}-t)},$$

so, by the boundary condition in (3.58),

$$z^*(t;s,y,j) = h_0(x(T;s,y,j))e^{\lambda(j+1)(\tilde{s}-T)}$$
$$+ \int_t^T \lambda(j+1)J(\sigma, x(\sigma;s,y,j)) + g(\sigma, x(\sigma;s,y,j), j+1), j+1)e^{\lambda(j+1)(\tilde{s}-\sigma)}d\sigma.$$
$$(3.111)$$

For a given number j, consider the following deterministic problem, denoted DP below.

$$\max\left\{h_0(x(T))e^{\lambda(j+1)(\tilde{s}-T)} + \int_{\tilde{s}}^T [J(t, x+g(t, x, j+1), j+1)]\lambda(j+1)e^{\lambda(j+1)(\tilde{s}-t)}dt\right\}$$

subject to

$$\dot{x} = f(t, x, u(t), j), x(\tilde{s}) = \tilde{y}, u \in U,$$

with end conditions $x_i(T) = \hat{x}_i, i = 1, \ldots, n', x_i(T) \geq \hat{x}_i, i = n' + 1, \ldots, n''$ in case (3.52) holds, in case of (3.53) add $(g^{k,j}(x))_i \geq \hat{x}_i, k = 1, \ldots, N - j, i = n' + 1, \ldots, n''$. The integrand is defined on $Q^0(j)$. We assume $(\tilde{s}, \tilde{y}) \in Q_0(j)$.

Let us apply the theory in Chapter 2 (Remarks 2.21, 2.30) with $f_0 = J(t, x + g(t, x, j + 1), j + 1)\lambda(j + 1)e^{\lambda(j+1)(\tilde{s}-t)}$ to the above deterministic problem (DP), in which the Hamiltonian is denoted H^* and the maximized Hamiltonian $\hat{H}^*$ and in which $x_0(t; s, y)$ is defined by $x_0(s; s, y) = 0$, $\dot{x}_0(t) = J(t, x + g(t, x, j + 1), j + 1)]\lambda(j + 1)e^{\lambda(j+1)(\tilde{s}-t)}$, where $x = x(t; s, y, j)$. We propose $(x(t; s, y), u(t; s, y),$ $p^*(t; s, y))$ as a characteristic triple in DP, where $(x(t; s, y), u(t; s, y)) = (x(t; s, y, j),$ $u(t; s, y, j))$, and $p^*(t; s, y) = p(t; s, y, j)\exp(\lambda(j + 1)(\tilde{s} - t))$. Using the induction hypothesis that $z_y(s; s, y, j + 1) = p(s; s, y, j + 1)$ (not needed when $j = N$), an easy calculation shows that $p^*(t; s, y)$ is a characteristic adjoint function in DP (i.e., the proposal works, as the maximum condition in DP holds). Now, $\hat{H}^*$ has first- and second-order partial derivative with respect to x and p, at all $(t, x, p) \in Q^* :=$ $\{(t, x, p) \in Q^*(j) : (t, x) \in Q^0(j), t > \tilde{s}\}$ (by the induction hypothesis, $z_{yy}(t; t, x, j + 1)$ is C^0 for $(t, x) \in Q^0(j + 1)$, so second-order partial derivatives of $\hat{H}^*$ do exist and are C^0). Now, $z(t; s, y, j) =$

$$h_0(x(T; s, y, j))e^{\lambda(t-T)} + x_0(T; s, y)e^{\lambda(t-\tilde{s})} - x_0(t; s, y)e^{\lambda(t-\tilde{s})}.$$

The C^1-property of $z(t; s, y, j)$ and $x(t; s, y, j)$ postulated means that $x_0(t; s, y, j), t \in (s, T]$, is C^1 in $Q^0(j)$, use the last equality for $t = s$, and then for arbitrary t. This equality also gives that $x_{0s}(t, s, y, j)$ and $x_{0y}(t, s, y, j)$ are continuous at $t = T$. It is assumed that $h_0(x(T, \omega^j)) := J(T, x(T, \omega^j), j) = \lim_{t \to T} J(t, x(t, \omega^j), j)$, so for $\tilde{s} = \tau_j, \tilde{y} = x(\tau_j +, \omega^j)$, the Remarks 2.21, 2.30 in Chapter 2 yield optimality in DP of $u(t; \tilde{s}, \tilde{y})$ in the set $\{u(t; \omega^j) : (x^{u(\cdots)}(., .), u(., .))$ is semiadmissible$\}$. Moreover, from these results, $z_y^*(s; s, y) = p^*(s; s, y)$, which yields $z_y(s; s, y, j) = p(s; s, y, j)$ for any $s > \tilde{s}$, in fact for any $s > 0$, as $\tilde{s}$ was arbitrary. In particular, the C^0-property of $(z(s; s, y, 0), x(T; s, y, 0))$ on $\{(0, x^0)\} \cup Q^0(0)$ means that $x_0(t; s, y)$ is C^0 on this set, so optimality even of $u(t; 0, x^0, 0)$ in DP follows.

Evidently, for any $x(.) = x(t, \omega^j), x(., .)$ semiadmissible, by $J(\tilde{s}, \tilde{y}, j) = z^*(\tilde{s}; \tilde{s}, \tilde{y}, j)$ and optimality of $u(t; \tilde{s}, \tilde{y})$, for $\tilde{s} = \tau_j, \tilde{y} = x(\tau_j +, \omega^j)$,

$$J(\tilde{s}, \tilde{y}, j) \geq h_0(x(T))e^{\lambda(j+1)(\tilde{s}-T)}$$
$$+ \int_{\tilde{s}}^T J(t, x(t) + g(t, x(t), j + 1), j + 1)\lambda(j + 1)e^{\lambda(j+1)(\tilde{s}-t)}dt \quad (3.112)$$

with equality holding for $x(t) = x(t; \tilde{s}, \tilde{y})$. So, for any semiadmissible $x(t, \omega)$, for $\tilde{s} = \tau_j, \tilde{y} = x(\tau_j +, \omega^j)$,

$$J(\tau_j, x(\tau_j +, \omega^j), j) \geq h_0(x(T, \omega^j))e^{\lambda(j+1)(\tau_j-T)}$$
$$+ \int_{\tau_j}^T J(t, x(t, \omega^j) + g(t, x(t, \omega^j), j + 1), j + 1)\lambda(j + 1)e^{\lambda(j+1)(\tau_j-t)}dt. \quad (3.113)$$

Equality holds for $x(t, w) = x^*(t, \omega)$, $x^*(t, \omega)$ corresponding with the $x(t, s, y, j)$'s. Then Remark 3.62 yields optimality of $u^*(t, \omega)$. $\qquad \square$

Proof of Remark 3.45 (sketch)

Note that, by consistency, $(t, x(t; s, y, j)) \notin \hat{Q}^1(j)$

$$\Leftrightarrow \phi_k(t, x(t; s, y, j), p(t; t, x(t; s, y, j), j)) \neq 0 \text{ for all } k$$
$$\Leftrightarrow \phi_k(t, x(t; s, y, j), p(t; s, y, j), j)) \neq 0 \text{ for all } k$$
$$\Leftrightarrow (t, x(t; s, y, j), p(t; s, y, j)) \notin \hat{Q}_*(j).$$

Assume by backwards induction that the equality $z_y(t; s, y, k) = p(s; s, y, k)$ holds for $k = j + 1$, and let us prove it for $k = j$. Let $(\hat{s}, \hat{y}) \in Q^0(j) \setminus Q^1(j)$ be given. Define $\hat{H}^*(t, x, p, j) := \hat{H}(t, x, p, j) + z(t; t, x + g(t, x, j + 1), j + 1)$. Let $\hat{t}$ be any point such that $(\hat{t}, x(\hat{t}; \hat{s}, \hat{y}, j), p(\hat{t}; \hat{s}, \hat{y}, j)) \notin \hat{Q}_*(j)$, let A be a set of the type of the remark, such that $(t, x(t; \hat{s}, \hat{y}, j), p(t; \hat{s}, \hat{y}, j)) \in A$ for t close to $\hat{t}$, $t < \hat{t}$, and, using the above equivalences, let B be a set of the type of the remark, such that $(t, x(t; \hat{s}, \hat{y}, j)) \in B, t$ close to $\hat{t}$, $t < \hat{t}$. There exist extensions of $\hat{H}(t, x, p, j)$ and of $z(t; t, x + g(t, x, j + 1), j + 1)$, $p(t; t, x + g(t, x, j + 1), j + 1)$ of the type described in the remark to open sets larger than cl$A$, cl$B$, respectively. A similar property holds for $t > \hat{t}$, t close to $\hat{t}$, so extensions $x^-(t; v, j)$, $p^-(t; v, j)$, $x^+(t; v, j)$, $p^+(t; v, j)$, $v = (s, y)$ exist with the same properties as $x^-(t; v)$, $p^-(t; v)$, $x^+(t; v)$, $p^+(t; v)$ in the proof of Remark 2.23 in Chapter 2, (see also Remark 2.30). From the arguments in that proof it is seen, as in the proof of Theorem 3.27, that $u(t; \tilde{s}, \tilde{y}, j)$ is optimal in DP for $(\tilde{s}, \tilde{y}) \in Q_0(j)$, moreover $z_y(s; s, y, j) = p(s; s, y, j)$. So, again, optimality of $(x^*(\cdot, \cdot), u^*(\cdot, \cdot))$ follows.   □

Proof of Theorem 3.11 (sketch)

Using the simultaneous continuity of $x(t; s, y, j)$ and $p(t, s, y, j)$, by backwards induction it is easily proved that $x_y(t; s, y, j)$, $x_s(t; s, y, j)$, $p_y(t, s, y, j)$ $x_y(t; s, y, j)$ are continuous in (t, s, y), for t in any interval $[t', T] \subset (0, T]$ and for $(s, y) \in Q^0(j)$ $s < t'$ (by the induction hypothesis, these entities satisfy differential equations with right-hand sides continuous in (t, s, y), for t in any interval $[t', T] \subset (0, T]$ and for $(s, y) \in Q^0(j)$, $s < t'$). We need to show that the conditions in Theorem 3.27 are satisfied. It suffices to show that $z(t; s, y, j)$ exists and is continuous in (t, s, y), for $(s, y) \in \tilde{Q}_0(j), t \in [s, T]$, and that $z_y(t; s, y, j)$ and $z_s(t; s, y, j)$ exist and are continuous in (t, s, y), for t in any interval $[t', T] \subset (0, T]$ and for $(s, y) \in Q^0(j)$, $s < t'$.

Assume by backwards induction that $z(t; s, y, j + 1)$ is continuous in (t, s, y), for $(s, y) \in \tilde{Q}_0(j), t \in [s, T]$. (No such assumption is needed when $j = N$.) This means, by (3.28), that $z(t; t, x(t; s, y, j) + g(t, x(t; s, y, j), j + 1), j + 1)$ is continuous in (t, s, y) for $(s, y) \in \tilde{Q}_0(j)$, $t \in [s, T]$. Then, evidently, $z(t; t, x(t; s, y, j) + g(t, x(t; s, y, j), j + 1), j + 1)$ is uniformly locally bounded in $(s, y) \in \tilde{Q}_0(j)$ (i.e.,

$$\sup_{t \in [s, T]} |z(t; t, x(t; s, y, j) + g(t, x(t; s, y, j), j + 1), j + 1)|$$

is locally bounded). Hence, by (3.58), $z(t; s, y, j)$ is continuous in (t, s, y), for $(s, y) \in \tilde{Q}_0(j), t \in [s, T]$, as (by (3.58) and $f_0 \equiv 0$),

$$z(t;s,y,j) = h_0(x(T;s,y,j))e^{\lambda(j+1)(t-T)}$$

$$+ \int_t^T \lambda(j+1)z(\sigma;\sigma,x(\sigma;s,y,j)+g(\sigma,x(\sigma;s,y,j),j+1),j+1)e^{\lambda(j+1)(t-\sigma)}d\sigma.$$

By backwards induction, let us prove the continuity of $z_y(t;s,y,j)$. Assume it to hold for $j+1$ (no such assumption is needed when $j = N$). This means that $z_y(t;t,x(t;s,y,j)+g(t,x(t;s,y,j),j+1),j+1)$ is continuous in (t,s,y), for t in any interval $[t',T] \subset (0,T]$ and for $(s,y) \in Q^0(j)$, $s < t'$. Then, evidently, $z_y(t;t,x(t;s,y,j)+g(t,x(t;s,y,j),j+1),j+1)$ is uniformly locally bounded. Now, using continuity of $x_y(t;s,y,j)$ (and so uniform local boundedness of this entity), it is then easily seen by differentiating under the integral sign in the above formula for $z(t;s,y,j)$ that $z_y(t;s,y,j)$ exists and that $z_y(t;s,y,j)$ is continuous in (t,s,y) for t in any interval $[t',T] \subset (0,T]$ for $(s,y) \in Q^0(j)$, $s < t'$. A similar argument yields that $z_s(t;s,y,j)$ has these properties.   □

Proof of Remark 3.13**

We need to prove that $(s,y) \rightarrow (x(t;s,y,j),p(t;s,y,j))$ is C^1 in $Q^0(j)$. We want to prove that $(x(t;s,y,j),p(t;s,y,j))$ is defined for (s,y) in an open set $Q_0(j)$ containing $\tilde{Q}^0(j) := Q^0(j) \cup (clQ^0(j) \cap [\{T\} \times \mathbb{R}^n])$, in fact for any (s,y) near any $(\hat{s},\hat{y}) \in \tilde{Q}^0(j)$, for t in an open interval containing $[\hat{s},T]$, and such that for each such t, $(x(t;s,y,j),p(t;s,y,j))$ is C^1 in (s,y), with $(s,y) \rightarrow (p(s;s,y,j),p_y(s;s,y,j))$ continuous near $(\hat{s},\hat{y})$. (If $(\hat{s},\hat{y}) = (T,\hat{y})$ this represent an extension of the x- and p-functions near such a point.) Assume by backwards induction that these properties hold for $j+1$, and let us then prove them for j. (No induction hypothesis is needed when $j = N$.) First note that, by continuity and (3.28), for $j < N$,

$$(s,y) \in cl\tilde{Q}^0(j) \Rightarrow (s,y+g(s,y,j+1)) \in cl\tilde{Q}^0(j+1). \tag{3.114}$$

Let $(\hat{s},\hat{y})$ be any given point in $\tilde{Q}^0(j)$. Now, $\hat{H}_p$ and $\hat{H}_x$ are C^1 in (x,p) for (t,x,p) in $C(T,\hat{s},\hat{y},j,\varepsilon) \subset Q^*(j)$, and by the induction hypothesis $(p(t;t,y,j+1),p_y(t;t,y,j+1))$ is continuous for (t,y) in $D(T,\hat{s},\hat{y},j,\varepsilon) \subset Q_0(j+1)$ (see proof of Theorem 3.27 for the two last sets mentioned). By standard theory of differential equation, for any initial point (s,y,q) in a neighborhood B of $(\hat{s},\hat{y},\hat{q})$, $\hat{q} := p(\hat{s};\hat{s},\hat{y},j)$, a unique solution $(\tilde{x}(t;s,y,q),\tilde{p}(t;s,y,q))$ exists on an interval $[\hat{s}',T']$ slightly larger than $[\hat{s},T]$, and, for any $t \in (\hat{s}',T')$, the solution is C^1 in (s,y,q), the derivatives being continuous in (t,s,y,q), $t \in (\hat{s}',T')$, (s,y,q) near $(\hat{s},\hat{y},\hat{q})$. Because $(s,y) \rightarrow p(s;s,y,j)$ is continuous, for (s,y) close to $(\hat{s},\hat{y})$, $(s,y) \in \tilde{Q}_0(j)$, $(s,y,p(s;s,y,j))$ belongs to B, and by uniqueness of solutions,

$$x(t;s,y,j) = \tilde{x}(t;s,y,p(s;s,y,j)), \quad p(t;s,y,j) = \tilde{p}(t;s,y,p(s;s,y,j)) \tag{3.115}$$

for such (s,y).

Write $(y,q) = Q$, and $X(t;s,Q) = X(t;s,y,q) = (\tilde{x}(t;s,y,q),\tilde{p}(t;s,y,q))$. For $Q = (y,p(s;s,y,j))$, $t \geq s' \geq s$, $(s,y) \in \tilde{Q}_0(j)$, $y' := x(s';s,y,j)$, using consistency and (3.115) yield $X(t;s,Q) = X(t;s',X(s';s,Q)) = X(t;s',x(s';s,y,j), p(s';s,y,j)) =$

$(x(t;s',x(s';s,y,j),j), p(t;s',x(s',s,y,j),j)) = (x(t;s',y',j), p(t;s',y',j)) = X(t;s', Q')$, where $Q' := (y',p(s';s',y',j))$.

In fact, even for $t = s < s'$, we have $Q = X(s;s,Q) = X(s;s',Q')$, because if this did not hold, even $X(s';s,Q) = X(s';s,X(s;s,Q))$ would differ from $X(s';s, X(s;s',Q')) = Q'$ and we already know that $X(s';s,Q) = X(s';s',Q') = Q'$. Replacing (s,s',y,y') by (s',s'',y',y'') so $y'' = x(s'';s',y',j)$, whether s'' is $< s'$ or $> s'$, we have

$$\tilde{x}(s';s'',y'',p(s'',y'',j)) = x(s';s',y',j)$$
$$\tilde{p}(s';s'',y'',p(s'',y'',j)) = p(s';s',y',j). \tag{3.116}$$

In fact, these properties hold for any $(s',y') \in \tilde{Q}_0(j)$ close to any given $(\hat{s},\hat{y})$ in $\tilde{Q}^0(j)$, $s'' \in [\tilde{s}',T]$ for some number $\tilde{s}' < \hat{s}$, close to $\hat{s}$.

Still, let $(\hat{s},\hat{y}) \in \tilde{Q}^0(j)$ and let $\tilde{s} \in (\tilde{s}',\hat{s})$. Choose a K such that $K \geq$

$$\sup |[(\partial/\partial(y,q))\hat{H}_p(t,\tilde{x}(t;s'',y,q),\tilde{p}(t;s'',y,q)]_{y=x(s'';\hat{s},\hat{y},j),q=p(s'';\hat{s},\hat{y},j)}|,$$

where $\sup = \sup_{t\in(\tilde{s},T],s''\in[\hat{s},T]}$. Assume, for any given $\hat{s}'' \geq \hat{s}$, that a C^1-function $\psi(s'',y'')$ exists near $(\hat{s}'',\hat{y}'')$, $\hat{y}'' = x(\hat{s}'';\hat{s},\hat{y},j)$, such that

$$\tilde{p}(T;s'',y'',\psi(s'',y'')) = h_{0x}(\tilde{x}(T;s'',y'',\psi(s'',y'')))$$

and $\psi(s'',y'') = p(s'';s'',y'',j)$ for $(s'',y'') \in \tilde{Q}^0(j)$. Define

$$|(\partial/\partial y'')\psi(\hat{s}'',y'')_{y''=\hat{y}''}| := \alpha(\hat{s}'').$$

In a sense, ψ surely exists for $s'' = T$, namely $\psi(T,y'') = h_{0x}(y'')$, except that here s'' is not varied, but kept equal to $\hat{s}'' = T$. (In the following argument, s'' is also kept fixed equal to $\hat{s}''$.) Choose $\tilde{s}' \in (\tilde{s},\hat{s}'')$, such that $(\hat{s}'' - \tilde{s}')K(1+\alpha(s'')) < 1/2$. Then, we shall prove that for any $\hat{s}' \in (\tilde{s}',\hat{s}'']$, a C^1-function $\phi(s',y')$ exists for (s',y') near $(\hat{s}',\hat{y}')$, $\hat{y}' := x(\hat{s}';\hat{s},\hat{y},j)$, such that

$$\tilde{p}(T;s',y',\phi(s',y')) = h_{0x}(\tilde{x}(T;s',y',\phi(s',y')))$$

and $\phi(s',y') = p(s';s',y',j)$ for $(s',y') \in \tilde{Q}^0(j)$. Note first that $\Psi(s',y'') := \tilde{x}(s';s'',y'',\psi(s'',y'')) =$

$$y'' + \int_{\hat{s}''}^{s'} \hat{H}_p(\sigma,\tilde{x}(\sigma;s'',y'',\psi(\hat{s}'',y'')),\tilde{p}(\sigma;\hat{s}'',y'',\psi(\hat{s}'',y'')))d\sigma.$$

Denote the integral by Y. The derivative at $\hat{y}'' := x(\hat{s}'';\hat{s},\hat{y},j)$ with respect to y'' of the integral, $Y_{y''}$ (respectively, of the integrand) has a bound $|\hat{s}''-s|)K(1+\alpha(\hat{s}'')) \leq 1/2$, (respectively, $K(1+\alpha(\hat{s}''))$). Then, for $s' \in (\tilde{s}',\hat{s}'')$, at $\hat{y}''$, $(\partial/\partial y'')\Psi(s',y'') = I+Y_{y''}$, with $|Y_{y''}| \leq 1/2$, so $(\partial/\partial y'')\Psi(s',\hat{y}'')$ is invertible. By the inverse function theorem, $y'' \to \Psi(s',y'')$ has an inverse $\Phi(s',y')$ for (s',y') near $(\hat{s}',\hat{y}')$ that is C^1 in (s',y') for (s',y') near $(\hat{s}',\hat{y}')$, such that $y' = \tilde{x}(s';s'',\Phi(s',y'),\psi(\hat{s}'',\Phi(s',y')))$. Define

$$\phi(s',y') = \tilde{p}(s';\hat{s}'',\Phi(s',y'),\psi(\hat{s}'',\Phi(s',y'))).$$

Evidently, $\phi(s',y')$ is C^1. Moreover, for $y'' = \Phi(s',y')$,

$$\tilde{p}(t;\hat{s}'',y'',\psi(\hat{s}'',y''))$$
$$= \tilde{p}(t;s',y',\tilde{p}(s';\hat{s}'',y'',\psi(\hat{s}'',y''))) = \tilde{p}(t;s',y',\phi(s',y')),$$

and

$$\tilde{x}(t;s',y',\phi(s',y')) = \tilde{x}(t;s',y',\tilde{p}(s';\hat{s}'',y'',\psi(\hat{s}'',y'')))$$
$$= \tilde{x}(t;\hat{s}'',y'',\tilde{p}(\hat{s}'';\hat{s}'',y'',\psi(\hat{s}'',y''))) = \tilde{x}(t;\hat{s}'',y'',\psi(\hat{s}'',y'')),$$

and so $\tilde{p}(T;s',y',\phi(s',y')) = h_{0x}(\tilde{x}(T;s',y',\phi(s',y')))$. Moreover, from (3.116) $\phi(s',y') = p(s';s',y',j)$ follows. When $\hat{s} = \hat{s}'' = T$, we can let $\hat{s}'$ be equal to $\hat{s}'' = T$, in which case $\phi(s',y')$ is defined and C^1 in some neighborhood B^T of $(T,\hat{y})$, $((T,\hat{y}) \in clQ^0(j))$. This ϕ-function is denoted $\phi^T(s',y')$. (We have $\tilde{p}(T;T,y,h_{0x}(y)) = p(T;T,y,j), \tilde{x}(T;T,y,h_{0x}(y)) = x(T;T,y,j) = y$ for y in some neighborhood of $\hat{y}$.) In particular, for $\hat{s}'' = T$, for any $\hat{s}' \in [\check{s},T]$, there exists a C^1-function $\phi(s',y')$ defined for (s',y') in some neighborhood $\tilde{B}$ of $(\hat{s}',\hat{y}')$, $\hat{y}' = x(\hat{s}';\hat{s},\hat{y},j)$ (the neighborhood can be assumed to be a subset of B^T, and the ϕ-function can be taken to be ϕ^T restricted to this neighborhood), such that $\tilde{p}(T;s',y',\phi(s',y')) = h_{0x}(\tilde{p}(T;s',y',\phi(s',y')))$, and $\phi(s',y') = p(s';s',y',j)$ for $(s',y') \in \tilde{Q}_0(j) \cap \tilde{B}$, the last equality following from the same arguments as before.

Assume that now that $\hat{s} < T$ and let $\check{s} < T$ be the smallest $s \geq \hat{s}$, such that for any $\hat{s}'' > s$, a C^1-function $\psi(s'',y'')$ exists near $(\hat{s}'',\hat{y}'') := (\hat{s}'',x(\hat{s}'';\hat{s},\hat{y},j))$, such that

$$\tilde{p}(T;s'',y'',\psi(s'',y'')) = h_{0x}(\tilde{x}(T;s'',y'',\psi(s'',y'')))$$

and $\psi(s'',y'') = p(s'';s'',y'',j)$. Now, $p(s'';s'',y'',j)$ is Lipschitz continuous in an open neighborhood B' of $(\check{s},\check{y})$, where $\check{y} := x(\check{s};\hat{s},\hat{y},j)$, with some rank $\tilde{K}$. Choose $\hat{s}'' \in [\check{s},\check{s}+1/4K(1+\tilde{K})]$, $\hat{s}' \in [\check{s}-1/4K(1+\tilde{K}),\check{s})$, $\hat{s}' \geq \check{s}$, such that $(\hat{s}'',\hat{y}'') = (\hat{s}'',x(\hat{s}'';\hat{s},\hat{y},j))$ belongs to B'. Now, $\psi(s'',y'') = p(s'';s'',y'',j)$ near $(\hat{s}'',\hat{y}'')$, so

$$|(\partial/\partial y')\psi(s'',y'')_{[s''=\hat{s}'',y''=\hat{y}'']}| =: \alpha(\hat{s}'') \leq \tilde{K}.$$

As $\hat{s}'' - \hat{s}' \leq 1/2K(1+\tilde{K})$, by the above results, near $(\hat{s}',\hat{y}') = (\hat{s}',x(\hat{s}';\hat{s},\hat{y},j))$ a C^1-function ϕ exists such that $\tilde{p}(T;s',y',\phi(s',y')) = h_{0x}(\tilde{x}(T;s',y',\phi(s',y')))$ and $\phi(s',y') = p(s';s',y',j)$. If $\check{s}$ is $> \hat{s}$, a contradiction has arisen. So $\check{s} = \hat{s}$. In fact, a $\phi(s',y')$ with the above properties even exists near $(\hat{s},\hat{y})$ (to see this, use the last argument again with $\check{s} = \hat{s} = \hat{s}'$). So $(x(t;s,y,j),p(t;s,y,j))$ equals $(\tilde{x}(t;s,y,\phi(s,y)), \tilde{p}(t;s,y,\phi(s,y)))$ for (s,y) close to $(\hat{s},\hat{y})$.

To conclude: Near any $(\hat{s},\hat{y}) \in \tilde{Q}^0(j)$ a C^1-function $\phi(s,y)$ exists, such that $(x(t;s,y,j),p(t;s,y,j)) = (\tilde{x}(t;s,y,\phi(s,y)),\tilde{p}(t;s,y,\phi(s,y)))$ when (s,y) is close to $(\hat{s},\hat{y})$, $(s,y) \in \tilde{Q}_0(j)$. For (s,y) near $(\hat{s},\hat{y}) \in \tilde{Q}^0(j)$, the function $(\tilde{x}(t;s,y,\phi(s,y)), \tilde{p}(t;s,y,\phi(s,y)))$ is defined for t in some open interval I larger than $[\hat{s},T]$, it is C^1 in (s,y), and it, as well as its derivatives with respect to s and y, are continuous in (t,s,y). From this the C^1-property of $(x(t;s,y,j),p(t;s,y,j))$, as well as the continuity of $p(s;s,y,j),p_y(s;s,y,j)$ near $(\hat{s},\hat{y})$ follows (the latter property needed to conclude the induction step). $\qquad\square$

Proof of Remark 3.49

As seen in the proof of Remark 2.10 in Chapter 2, the properties used to prove Theorem 3.37 still hold for the conditions in Remark 3.49, so optimality of $(x^*(.,.),u^*(.,.))$ follows in the same way. $\qquad\square$

Proof of Theorem 3.58 and Remark 3.61

Define $J(t,x,\omega) = J(t,x,j)$ if $\tau_j \le t < \tau_{j+1}$. Let $(x(.,\omega),u(.,\omega),T(\omega))$ be an arbitrary admissible triple and let j be any given number. Assume that there is a countable number of exceptional points $(s,x(s,\omega^j))$, at which the "HJB inequality" does not hold. Multiplying by $e^{\lambda(j+1)(\tau_j-s)}$ in the "HJB inequality" corresponding to (3.96), with $y = x(s,\omega^j)$, $u = u(s,\omega)$, gives

$$
\begin{aligned}
0 \ge\ & J_s(s,x(s,\omega^j),j)e^{\lambda(j+1)(\tau_j-s)} \\
&+J_y(s,x(s,\omega^j),j)f(s,x(s,\omega^j),u(s,\omega^j),j)e^{\lambda(j+1)(\tau_j-s)} \\
&+\lambda(j+1)J(s,x(s,\omega^j)+g(s,x(s,\omega^j),j+1),j+1)e^{\lambda(j+1)(\tau_j-s)} \\
&-\lambda(j+1)J(s,x(s,\omega^j),j)e^{\lambda(j+1)(\tau_j-s)} \\
=\ & (d/ds)[J(s,x(s,\omega^j),j)e^{\lambda(j+1)(\tau_j-s)}] \\
&+\lambda(j+1)J(s,x(s,\omega^j)+g(s,x(s,\omega^j),j+1),j+1)e^{\lambda(j+1)(\tau_j-s)}.
\end{aligned}
$$

Equality holds if $(x(.,.),u(.,.)) = (x^*(.,.),u^*(.,.))$. In case $\tau_j < T(\omega^j)$, integrating between $s = \tau_j$ and $T(\omega^j)$ yields

$$
\begin{aligned}
& J(\tau_j,x(\tau_j+,\omega),j) \ge J(T(\omega^j),x(T(\omega^j)+,\omega^j),j)e^{\lambda(j+1)(\tau_j-T(\omega^j))} \\
&\quad +\int_{\tau_j}^{T(\omega^j)} J(t,x(t,\omega^j)+g(t,x(t,\omega^j),j+1),j+1)\lambda(j+1)e^{\lambda(j+1)(\tau_j-t)}dt \\
&= E[J(T(\omega^j),x(T(\omega^j)+,\omega^j),j)1_{(T(\omega^j),\infty)}(\tau_{j+1})|\omega^j] \\
&\quad +E[J(\tau_{j+1},x(\tau_{j+1}+,\omega),j+1)1_{[\tau_j,T(\omega^j)]}(\tau_{j+1})|\omega^j].
\end{aligned}
$$

The same formula also holds if $T_2 > \tau_j = T(\omega^j)$, then the last term drops out, and only then the plus-sign in $x(T(\omega^j)+,\omega^j)$ makes a difference. (When $\tau_j \ne T(\omega^j)$, $x(T(\omega^j)+,\omega^j) = x(T(\omega^j),\omega^j)$.)

Using the last inequality for $j = 0$ and $j = 1$, $J(0,x^0,0) \ge$

$$
\begin{aligned}
& E[J(T(\omega^0),x(T(\omega^0)+,\omega^0),0)1_{(T(\omega^0),\infty)}(\tau_1)] \\
&+ E[J(\tau_1,x(\tau_1+,\omega),1)1_{[0,T(\omega^0)]}(\tau_1)]
\end{aligned}
$$

$$\geq E[J(T(\omega^0), x(T(\omega^0)+, \omega^0), 0) 1_{(T(\omega^0), \infty)}(\tau_1)]$$

$$+ E[E[1_{[0, T(\omega^0)]}(\tau_1) J(T(\omega^1), x(T(\omega^1)+, \omega^1), 1) 1_{(T(\omega^1), \infty)}(\tau_2) | \omega^1]]$$

$$+ E[E[J(\tau_2, x(\tau_2+, \omega), 2) 1_{[\tau_1, T(\omega^1)]}(\tau_2) 1_{[0, T(\omega^0)]}(\tau_1) | \omega^1]]$$

$$= \sum_{i=0}^{1} E[J(T(\omega^i), x(T(\omega^i)+, \omega^i), i) \Pi_{k=0}^{i-1} 1_{[\tau_k, T(\omega^k)]}(\tau_{k+1}) 1_{(T(\omega^i), \infty)}(\tau_{i+1})]$$

$$+ E[J(\tau_2, x(\tau_2+, \omega), 2) \Pi_{k=0}^{1} 1_{[\tau_k, T(\omega^k)]}(\tau_{k+1})].$$

Here, the product $\Pi_{k=0}^{-1} 1_{[0, T(\omega^k)]}(\tau_{k+1})$ by definition equals 1. Continuing in this manner, we get in general $J(0, x^0, 0) \geq$

$$\sum_{i=0}^{j} E[J(T(\omega^i), x(T(\omega^i)+, \omega^i), i) \Pi_{k=0}^{i-1} 1_{[\tau_k, T(\omega^k)]}(\tau_{k+1}) 1_{(T(\omega^i), \infty)}(\tau_{i+1})]$$

$$+ E[J(\tau_{j+1}, x(\tau_{j+1}+, \omega), j+1) \Pi_{k=0}^{j} 1_{[\tau_k, T(\omega^k)]}(\tau_{k+1})]. \tag{3.117}$$

Letting $j \to \infty$, we get $J(0, x^0, 0) \geq$

$$\sum_{i=0}^{\infty} E[J(T(\omega^i), x(T(\omega^i)+, \omega^i), i) \Pi_{k=0}^{i-1} 1_{[\tau_k, T(\omega^k)]}(\tau_{k+1}) 1_{(T(\omega^i), \infty)}(\tau_{i+1})]$$

$$= EJ(T(\omega), x(T(\omega)+, \omega) \geq Eh_0(T(\omega), x(T(\omega)+, \omega)).$$

To obtain this, one can use the properties of $\alpha_{x(.,.)}$ and $\kappa_{x(.,.)}$. It is then easily seen that the last term in (3.117) $\to 0$ when $j \to \infty$ and that we can replace j by ∞ in the first term in (3.117). As equality holds for $(x(., \omega), u(., \omega), T(\omega)) = (x^*(., \omega), u^*(., \omega), T^*(\omega))$, it follows that $(x^*(., \omega), u^*(., \omega), T^*(\omega))$ is optimal. The case of an uncountable number of exceptional points s is treated as in the proof of Remark 2.10 in Chapter 2.

The modification needed in case of Remark 3.61 is as follows: Define $T_k(\omega) = \min\{T(\omega), T_2 - 1/k\}$, $T_k^*(\omega) = \min\{T^*(\omega), T_2 - 1/k\}$. Now we can obtain $J(0, x^0, 0) \geq E[J(T_k(\omega), x(T_k(\omega)+, \omega), \omega)]$. When $k \to \infty$, the last expression converges to $EJ(T(\omega), x(T(\omega)+, \omega))$, with equality holding if $x(t, \omega) = x^*(t, \omega)$. Optimality then follows from (3.102), (3.103). □

Proof of Remark 3.54 (sketch)

For any given $T < T_2$, extend the solution $x(t; s, y, T, j)$ from $[0, T]$ to a slightly larger interval $[0, T']$, by using the control $u = u(T-; s, y, T, j)$ to the right of T. For $x(.)$ equal to this solution, by (3.112), for T'' near T,

$$J(s, y, T'', j) \geq h_0(T'', x(T''))e^{\lambda(j+1)(s-T'')}$$

$$+ \int_s^{T''} J(t, x(t) + g(t, x(t), j+1), j+1) \lambda(j+1)e^{\lambda(j+1)(s-t)} dt =: \kappa(T'') \tag{3.118}$$

where equality holds if $T'' = T$. Hence, a.e., $(d/dT)J(s,y,T,j) = \kappa'(T) = \eta^*(T,s,y,j)e^{\lambda(j+1)(s-T)}$. By the inequalities satisfied by the latter function, see (3.85), $J(s,y,T,j) \leq J(s,y,T(s,y,j),j) = J(s,y,j)$.

Let $(x(t,\omega),T(\omega))$ be an arbitrary admissible pair. The fixed horizon results, (with horizon $T(\omega^j)$), see (3.113), yield the second inequality in what follows:

$$J(\tau_j,x(\tau_j+,\omega),j) \geq J(\tau_j,x(\tau_j+,\omega),T(\omega^j),j)$$
$$\geq E[h_0(T(\omega^j),x(T(\omega^j)+,\omega^j))1_{(T(\omega^j),\infty)}(\tau_{j+1})|\omega^j]$$
$$+E[J(\tau_{j+1},x(\tau_{j+1}+,\omega),j+1)1_{[\tau_j,T(\omega^j)]}(\tau_{j+1})|\omega^j].$$

Repeated use of this result for $j = 0$, $j = 1,\ldots$, $j = N$, yields, in close analogy to (3.117), that $J(0,x^0,0) \geq$

$$\sum_{i=0}^{j} E[h_0(T(\omega^i),x(T(\omega^i)+,\omega^i))\Pi_{k=0}^{i-1}1_{[\tau_k,T(\omega^k)]}(\tau_{k+1})1_{(T(\omega^i),\infty)}(\tau_{i+1})$$
$$+E[J(\tau_{j+1},x(\tau_{j+1}+,\omega),j+1)\Pi_{k=0}^{j}1_{[\tau_k,T(\omega^k)]}(\tau_{k+1})]. \qquad (3.119)$$

Letting $j \to \infty$ (in fact, putting $j = N$), gives $J(0,x^0,0) \geq E[h_0(T(\omega),x(T(\omega)+,\omega)]$. Equality holds for $(x^*(t,\omega),T^*(\omega))$, hence $(x^*(t,\omega),u^*(t,\omega),T^*(\omega))$ is optimal. $\qquad\square$

Proof of Remark 3.28

Let us first prove that U bounded $\Rightarrow \mathcal{S} =$ the set of admissible pairs.
(a) When U is bounded, by (3.15), (3.16), $x(.;s,y,j)$ is bounded for (s,y) in bounded subsets of $Q^0(j)$. For the moment call $x(.;s,y,j)$ "restricted" when this holds. By backwards induction, using (3.58), even $z(.,s,y,j)$ is restricted. By the latter property and (3.58) again, $\lim_{t\to T}\sup_{(s,y)\in B}|z(t;s,y,j) - z(T;s,y,j)| = 0$ for any bounded set B. Now, $\hat{x}(.,.)$ is bounded, so

$$|z(t;t,\hat{x}(t,\omega^j),j) - z(T;t,\hat{x}(t,\omega^j),j)|$$

is small when t is close to T. Moreover, $z(T;t,\hat{x}(t,\omega^j),j) = h_0(x(T;t,\hat{x}(t,\omega^j),j))$ and $|x(T;t,\hat{x}(t,\omega^j),j) - x(t;t,\hat{x}(t,\omega^j),j)| = |x(T;t,\hat{x}(t,\omega^j),j) - \hat{x}(t,\omega^j))|$ is small, uniformly in ω^j, by the restrictedness of $\dot{x}(.;s,y,j)$, ($\hat{x}(.,.)$ is bounded.) Hence, $z(t;t,\hat{x}(t,\omega^j),j)$ is bounded and $\lim_{t\to T}z(t;t,\hat{x}(t,\omega^j),j) = h_0(\hat{x}(T,\omega^j))$.
(b) Let us prove that the limit conditions in the remark imply strong semiad-missibility for semiadmissible solutions. Let $\varepsilon > 0$, and choose $T'' < T$ so near T that $|z(T;t,\hat{x}(t+,\omega^j,t),j+1) - z(t;t,\hat{x}(t+,\omega^j,t),j+1)\}| \leq \varepsilon$ and $|h_0(\hat{x}(T,\omega^j,t)) - z(T;t,\hat{x}(t+,\omega^j,t),j+1)\}| \leq \varepsilon$ uniformly in $\omega^j,t,t \in [T'',T)$. Then, $|h_0(\hat{x}(T,\omega^j,t)) - z(t;t,\hat{x}(t+,\omega^j,t),j+1)\}| \leq 2\varepsilon$, uniformly in ω^j,t. Now, $z(s;s,y,j)$ is bounded on bounded subsets of $\{(s,y) \in Q^0(j) : s \leq T''\}$, so, $E[|h_0(\hat{x}(T,\omega^{j+1})) - z(\tau_{j+1};\tau_{j+1}, \hat{x}(\tau_{j+1}+,\omega^{j+1}),j+1)\}|1_{[0,T'']}(\tau_{j+1})] < \infty$ ($\hat{x}(.,.)1_{[0,T'']}(t)$ is bounded). By the third

inequality above, $[0,T'']$ can be replaced by $[0,T]$. As $E[|h_0(\hat{x}(T,\omega^{j+1}))|] < \infty$ by semiadmissibility, then $E[|z(\tau_{j+1};\tau_{j+1},\hat{x}(\tau_{j+1}+,\omega^{j+1}),j+1)|1_{[0,T]}(\tau_{j+1})] < \infty$. $\square$

Proof of the necessity of the fixed horizon HJB equation (3.64)

We need a differential formula (infinitesimal generator) for piecewise deterministic equations. In the theorem below, we let g contain a stochastic variable V as in Remark 3.15. We allow $\lambda(j+1)$ to depend on t,x as in Remark 3.14, as well as on $\omega^j = (\tau_0,V_0,\ldots,\tau_j,V_j)$. Moreover the density $\pi(j+1)$ of V_{j+1} is allowed to depend on the time t at which the jump number $j+1$ occurs, as well as on the state x at that time, and on ω^j. The system is assumed to satisfy the conditions described in Remark 3.15. Let $x(t,\omega)$, $u(t,\omega)$ be any given nonanticipating pair. Write $\lambda(t,\omega^j) := \lambda(t,x(t+,\omega^j),\omega^j,j+1)$, $\pi(v;t,\omega^j) := \pi(v;t,x(t+,\omega^j),\omega^j,j+1)$. We assume that $t \to (\lambda(t,\omega^j),\pi(v;t,\omega^j))$ is right-continuous, and that $\lambda(.,.)$ and $\pi(.,.,.)$ are bounded functions.

Theorem 3.63 (Infinitesimal generator). *Let $s,y,\omega^j,\tau_j \leq s$, be given entities, $s \in (0,T)$, and let $x(t,\omega)$ be a given nonanticipating solution starting at $(0,x^0)$, corresponding with a given bounded control $u(t,\omega)$. Let $\phi(t,x,\omega)$ be a given function, such that $\phi(t,x,\omega)$, $\phi_t(t,x,\omega)$ and $\phi_x(t,x,\omega)$, for each ω, are continuous in $(t,x),t \in (\tau_k,\tau_{k+1})$ for each k, with a right limit at τ_k and a left limit at τ_{k+1} as a function of t, is nonanticipating in (t,ω), and, for each (t,x), piecewise continuous in each $\tau_i \leq t$, and in each component of each V_i and such that, by assumption, $t \to \phi(t,x(t+,\omega^j)+g(t,x(t+,\omega^j),V,j+1),\omega^j,t)$ is right-continuous for $t \geq \tau_j$ for each j. We assume that, for some positive constants $\alpha_{x(.,.)}$ and $\kappa_{x(.,.)}$, $|\phi(t,x(t,\omega),\omega)| \leq \alpha_{x(.,.)} + \kappa_{x(.,.)}|x(t,\omega)|$ for all t,ω. Define $\eta(t;s,\omega^j) := E[\phi(t,x(t+,\omega),\omega))|s,x(s+,\omega),\omega^j]$, $t > s$. (The expectation is calculated by conditioning on the assumed fact that exactly j jumps have occurred before s, at $\tau_0,\ldots,\tau_j \leq s$.) Then, given that exactly j jumps have occurred before s, at $\tau_0,\ldots,\tau_j,\tau_j \leq s$,*

$$[\partial^+\eta(t;s,\omega^j)/\partial t]_{t=s} = \gamma(s,\omega^j) \tag{3.120}$$

where $(\partial^+/\partial t)$ means a right derivative and where $\gamma(s,\omega^j) =$

$$\phi_t(s,x(s+,\omega^j),\omega^j) + \phi_x(s,x(s+,\omega^j),\omega^j)f(s,x(s+,\omega^j),u(s+,\omega^j),j)$$
$$+\lambda(s,\omega^j)E[\phi(s,x(s+,\omega^j)+g(s,x(s+,\omega^j),V_{j+1},j+1),\omega^j,s)|$$
$$s,x(s+,\omega^j),\omega^j)] - \lambda(s,\omega^j)\phi(s,x(s+,\omega^j),\omega^j),$$

the conditional expectation being calculated by means of $\pi(v;s+,\omega^j)$. $\square$

Proof of Theorem 3.63 (sketch)

Write $X(t) := x(t,\omega)$. Assume that exactly j jumps have occurred in $[0,s]$. For Δs small, only a second-order error is made by assuming that there is at most one jump point τ_{j+1} in $(s,s+\Delta s)$, with a probability approximately equal to $\lambda(s+,\omega)\Delta s$ for a jump. (Actually, the Standard System Conditions connected with

(3.25) and the growth conditions on ϕ are needed to show that the error is so small.) By the assumptions postulated, a first-order error is made by assuming that $\phi(s+\Delta s, X(s+\Delta s), \omega^j, \tau_{j+1})$ has roughly the same values irrespectively of where the jump τ_{j+1} occurs in $(s, s+\Delta s)$, we use the values this expression has at $s+$ when calculating $E[\phi(s+\Delta s, X(s+\Delta s), \omega)|\omega^j$, a jump in $(s, s+\Delta s), X(s+)] = E[\phi(s+\Delta s, X(s+\Delta s), \omega^j, \tau_{j+1})|\omega^j$, a jump at τ_{j+1} in $(s, s+\Delta s), X(s+)]$, and the expectation is calculated by means of $\pi(.; s+, \omega^j)$, an approximation of the true distribution $\pi(.; \tau_{k+1}, \omega^j)$ needed to calculate the expectation given that the jump occurs at $\tau_{k+1} \in (s, s+\Delta s)$. Using all these approximations,

$$E[\phi(s+\Delta s, X(s+\Delta s), \omega)|s, X(s+), \omega^j]$$
$$\approx \phi(s+\Delta s, X(s+\Delta s), \omega^j)\Pr[\text{no jump in } (s, s+\Delta s)|s, X(s+), \omega^j]$$
$$+E[\phi(s+\Delta s, X(s+\Delta s), \omega)| \text{ a jump in } (s, s+\Delta s), s, X(s+), \omega^j]\Pr[\text{a jump in}$$
$$(s, s+\Delta s)|s, X(s+), \omega^j]$$
$$\approx (1-\lambda(s+, \omega^j)\Delta s)\{\phi(s, X(s+), \omega^j) + \phi_t(s, X(s+), \omega^j)\Delta s$$
$$+\phi_x(s, X(s+), \omega^j)f(s, X(s+), u(s+, \omega^j), j)\Delta s\}$$
$$+\lambda(s+, \omega^j)\Delta s E[\phi(s, X(s+) + g(s, X(s+), V, j+1), \omega^j, s)|s, X(s+), \omega^j].$$

the errors being of the second order. Subtracting $\phi(s, X(s+), \omega)$, dividing by Δs, and letting $\Delta s \to 0$, we get (3.120). $\qquad\square$

(Note, by the way, that $\eta(t; s, \omega^j) = E[\phi(t, x(t, \omega), \omega)|s, x(s+, \omega), \omega^j], t > s$.)

Continued proof of the necessity of (3.64)

It can be proved that for $t > s$,

$$J^*(s, x, j) = \sup_{\hat{u}} E[J^*(t, x^{\hat{u}}(t+, \omega), j + \hat{j}(t, \omega))|s, x, j], \qquad (3.121)$$

where $\hat{j}(t, \omega)$ equals the number of jumps that have occurred in $(s, t]$, $\hat{u} := \hat{u}(., \omega)$ and $x^{\hat{u}}(., \omega)$ starts at (s, x), $\tau_j < s$ (exactly j jumps have occurred in $[0, s)$), $\hat{u}$ independent of τ_i, $i \le j$. For the moment, assume that (3.121) has been proved. Then,

$$0 = \sup_{\hat{u}}\{E[J^*(t, x^{\hat{u}}(t+, \omega), j + \hat{j}(t, \omega))|s, x, j] - J^*(s, x, j)\}/(t-s).$$

Arguing heuristically, letting $t \downarrow s$, we obtain that 0 equals the supremum over $\hat{u}$ of the right derivative of $E[J^*(t, x^{\hat{u}}(t+, \omega_j), j + \hat{j}(t, \omega)))|s, x, \omega^j]$ at $t = s$. For $x = x(s+, \omega)$, the formula for such a derivative is given in connection with (3.120) in Theorem 3.63 for $\phi(t, x, \omega) = J^*(t, x, j + \hat{j}(t, \omega))$ (we can drop the expectation with respect to V). This formula is then the same as the expression on the right-hand side of (3.64) (recall that we operate with the assumption that $g_0 = 0, f_0 = 0$).

Define

$$J_*(s, x, \omega^j) = \sup_{\hat{u}} E[h_0(x^{\hat{u}}(T-, \omega))|s, x, \omega^j], \qquad (3.122)$$

the $x^{\hat{u}}(.,.)$'s starting at (s,x) and given j jumps before s, here $\hat{u}$ also depends on τ_i, $i \leq j$.

A proof similar to the one that gave the equality $J(t, z_{\rightarrow t}) = J(t, z_t)$ in Remark 1.17 in Chapter 1 yields that $J_*(s, x, \omega^j) = J^*(s, y, j)$. Let us now give a heuristic proof of (3.121), and let us write max instead of sup. Any control function $\hat{u}$ can be written as (u', u''), where $\hat{u} = u'$ in $[0, t]$, $\hat{u} = u''$ in $(t, T]$. Then, for $\check{j} = j + \hat{j}(t, \omega)$, $J^*(s, x, j) =$

$$\max_{\hat{u}} E[h_0(x^{\hat{u}}(T-, \omega))|s, x, \omega^j]$$

$$= \max_{u', u''} E[h_0(T, x^{u', u''}(T-, \omega))|s, x, \omega^j]$$

$$= \max_{u'}\{\max_{u''} E[h_0(T, x^{u', u''}(T-, \omega))|s, x, \omega^j]\}$$

$$= \max_{u'}\{\max_{u''} E[E[h_0(T, x^{u', u''}(T-, \omega))|t, x^{u'}(t+, \omega), \omega^{\check{j}}, s, x, j|s, x, \omega^j]\}$$

$$= \max_{u'}\{\max_{u''} E[E[h_0(T, x^{u', u''}(T-, \omega))|t, x^{u'}(t+, \omega), \omega^{\check{j}}|s, x, \omega^j]\}$$

$$= \max_{u'}\{E[\max_{u''} E[h_0(T, x^{u', u''}(T-, \omega))|t, x^{u'}(t+, \omega), \omega^{\check{j}}]|s, x, \omega^j]\}$$

$$= \max_{u'} E[J_*(t, x^{u'}(t+, \omega), \omega^{\check{j}}))|s, x, \omega^j]$$

$$= \max_{u'} E[J^*(t, x^{u'}(t+, \omega), j + \hat{j}(t, \omega))|s, x, j].$$

The last equality follows from the fact that, for j given, $J_*(t, x^{u'}(t+, \omega), \omega^{\check{j}}))$ does not depend on ω^j. One of the equalities above involves an interchange of a maximization ($\max_{u''}$) and an expectation and needs the following argument: Write $\omega^{\check{j}} =: (\omega^j, \omega_{\check{j}})$. The outer expectation can be looked upon as a sum over a discrete set of $\omega_{\check{j}}$'s. The maximization of a sum equals the sum of the maximum of each term, provided the control parameters occurring in the various terms can be chosen independently of each other. This is the case here, for each realization of $\omega_{\check{j}}$, a separate u'' can be chosen. □

In the third line from the bottom of the equalities for $J^*(s, x, j)$ above, the outer expectation is actually an integral over a nondiscrete set of $\omega_{\check{j}}$'s. By using a so-called measurable selection theorem it is possible to find a function $u''(\omega_{\check{j}})$ (unfortunately only known to be measurable) yielding the maximum (or nearly the supremum in the general case) in that line a.s., so the argument used at that point carries over to the nondiscrete case. Certain well-behavior properties of the inner expectation are then needed that are easiest to obtain in the free end case.

In the end-constrained case, we may need to operate with values equal to minus infinity of $\max_{u''}$ in cases where no u'' exists for which the end conditions are satisfied, so, whether we consider the discrete or nondiscrete case, we should, at least by now, perhaps indicate that we actually have had the free end case in mind in this proof!

Theorem 3.63 can be used in the same way to prove necessity of a HJB equation in the case of stochastic jumps (g containing V), where $\pi(v; j+1)$, the density of V_{j+1}, depends on (t, x, ω^j), the time t and state x, at which the jump number $j+1$ occurs, as well as on $\omega^j := (\tau_0, V_0, \ldots, \tau_j, V_j)$, and where $\lambda(j+1)$ depends on (t, x), as well as on ω^j. Then, in 3.122 one *must* condition on (s, x, ω^j), so the optimal value function depends on (s, x, ω^j). Briefly, the formulation of such a HJB equation in the case $g_0 = 0$, $f_0 = 0$ and exactly j jumps have occurred in $[0, s)$, reads:

$$J(s, y, \omega^j) = J_s(s, y, \omega^j) + \max_u \{J_y(s, y, \omega^j)f(s, y, u, j)\}$$
$$+ \lambda(j+1, s, y, \omega^j)E[J(s, y+g(s, y, V_{j+1}, j+1), \omega^j, s, V_{j+1}) - J(s, y, \omega^j)|s, y, \omega^j],$$

where the conditional expectation is taken with respect to V_{j+1}, calculated by means of $\pi(v; s, y, \omega^j, j+1)$. This equation essentially covers the cases discussed in Remarks 3.14 and 3.15.

Further reading. The standard reference, at a more advanced level than the current text, is as mentioned Davis (1993). The articles by Costa et al. (2000), Dempster et al. (1995), Davis et al. (1999), and Ye (1996) develop the theory further in various directions.

3.8 Exercises

Exercises for Sections 3.2, 3.3

3.1. (a) Does it change the optimal control in Example 3.3, if instead of a unit jump, we assume that the jump is a stochastic variable V, with $EV = 1$?
(b) Add the possibility of a second jump in Example 3.3, both for scrap value $ax(T)$ and $ax(T)^2/2$. Solve the problems as far as possible.
    (*Hint:* Now $x(t; s, y, 1), x(t; s, y, 0)$ becomes $x(t; s, y, 2), x(t; s, y, 1)$ etc.)

3.2. Consider the problem

$$\max E\left[\int_0^T \sqrt{u}dt + x(T)\right], \quad \dot{x} = x - u, \ u \in [0, \infty), x(\tau+) - x(\tau-) = x(\tau-),$$

$x(T)$ free. Two jumps can occur with intensity λ. Solve the problem.

3.3. Consider the problem

$$\max E\left[\int_0^T -(u^2/2)dt + x(T)\right], \quad \dot{x} = u, \ x(0) = 0, \ x(\tau+) - x(\tau-) = -(x(\tau-))^2,$$

$u \in \mathbb{R}$, $x(T)$ free. A single jump can occur with intensity λ. Solve the problem. (*Hint:* To find $x(t; s, y, 0)$ for $(s, y) = (0, 0)$, construct a second-order equation for this function.)

3.4. Consider the problem

$$\max E\left\{\int_0^1 \ln u\,dt + x(1)\right\}, \dot{x} = -u,\ u \in (0,\infty), x_0 = 2,$$

$x(1)$ free, where $x(t)$ can jump at most N times, all the times with the same intensity λ. At the jump times τ, the state changes according to $x(\tau+) - x(\tau-) = x(\tau-)$.
(a) Find the optimal controls $u(t;s,y,j)$ for $j = N, N-1, N-2$.
(b) Try to find (a description of) the optimal controls $u(t;s,y,N-k)$ for general $k = 1,2,\ldots$.

3.5. Solve the problem

$$\max E\left[\int_0^T -(u^2/2)dt + ax(T)\right], \text{ subject to } \dot{x} = u + x1_{[0,\tau]}(t),$$

$x(0) = 0$, $u \in \mathbb{R}$, where there are no jumps in $x(.)$, and where the dynamics changes at most one time, at τ, from $\dot{x} = u + x$ to $\dot{x} = u$, $\tau \in [0,\infty)$ occurring with intensity $\lambda \neq 1$ ($x(T)$ free).

3.6. Solve the problem

$$\max E\left[\int_0^1 (1-u)dt + (x(1))^2/2\right], \dot{x} = u - 1, u \in [0,1], x(0) = 1/2,$$

$x(1)$ free, where a single upwards jump of size 2 in the state x may occur with intensity $\lambda > 1$.

Exercises for Section 3.4

3.7. Solve the problem

$$\max E\left[\int_0^2 x\,dt\right], \dot{x} = u \in [-1,1], x(0) = 0, x(2) \leq 1,$$

where x can have a single upward jump of one unit, with intensity λ. *Hint:* As long as no jump has occurred, we have to operate with the constraint $x(2) \leq 0$, otherwise $x(2) \leq 1$ may be violated, as a jump can occur arbitrary close to $t = 2$.

3.8. Consider the problem

$$\max E\left[\int_0^T u^\gamma dt + ax(T)\right], \dot{x} = kx - u, x(0) = c, 0 < \gamma < 1, u \in (0,\infty).$$

With intensity λ, x may experience a single upwards jump of size b, a,b,c,k positive constants. (Interpretation: x is wealth, which gives interest earnings kx, u is consumption, we search for, and may find, an oil field increasing our wealth by b.)

(i) Solve the problem ($x(T)$ free).

(ii) Drop the term $ax(T)$ and instead require $x(T) \geq 0$, find $x(t, s, y, 1)$ and derive a second-order equation for $x(t; 0, c, 0)$.

Exercises for Section 3.5

3.9. Use Theorem 3.27 to solve

$$\max E \left[\int_0^1 -u^2/2 dt \right] \text{ s.t. } \dot{x}_1 = x_2 + u, u \in \mathbb{R}, x_1(0) = 0, x_1(1) = 1 \text{ a.s.,}$$

$$\dot{x}_2 = 0, x_2(0) = 0, x_2(1) \text{ free.}$$

where x_2 has a single unit upwards jump with intensity $\lambda > 0$, no jumps in x_1.

  Hint: Show that $p_1(t; s, y_1, y_2, 1) = u(t; s, y, 1) = (1 - y_1)/(1 - s) - y_2$. Then, before a jump, $\dot{p}_1 = +\lambda p_1 - \lambda[(1 - x_1)/(1 - t) - (y_2 + 1)], \dot{x}_1 = y_2 + p_1$. Find a solvable second-order equation for $w = x_1/(1 - t)$. Further discussion is as in the closely related Example 3.29.

3.10. Consider the problem:

$$\max E \left[\int_0^\tau -u^2/2 dt - \int_\tau^T u dt \right], \dot{x} = u \in \mathbb{R}, x(0) = 0, x(T) = 0 \text{ a.s.,}$$

where x has a single unit upwards jump at the stochastic point τ being exponentially distributed with parameter λ. Find the optimal solutions among all admissible pairs $x(t, \omega), u(t, \omega)$.

  Hint: The optimal control is nonunique after a jump, choose it constant.

Exercises for Section 3.6

3.11. Solve the optimal stopping problem

$$\max E \left[\int_0^T (x - t) dt \right], \dot{x} = 0, x(0) = 0, T \geq 0,$$

when x can have a one unit jump upwards with intensity λ, ($x(T)$ free).

3.12. Solve the optimal stopping problem

$$\max_{T \geq 0, u} E \left[\int_0^T (-t/2 - u^2/2) dt + x(T) \right], \dot{x} = u \in \mathbb{R}, x(0) = 0,$$

where $x(T)$ is free, and x can experience a jump downwards of size 1 (i.e., $x(\tau+) = x(\tau-) - 1$), at a jump point τ exponentially distributed in $(0, \infty)$ with intensity $\lambda < 2/3$.

Chapter 4
Control of Diffusions

In many problems, the system under consideration is continuously influenced by stochastic disturbances. Often, these disturbances are modeled by means of a Brownian motion. Brownian motion is a "stochastic process in continuous time," i.e., it is a function B_t of continuous time t, and for each t, the function value B_t is a stochastic variable. In economics, systems that are modeled by means of Brownian motion include the development of stock prices, oil prices, and prices of other commodities. The development of other factors determining the demand of commodities, as well as the supply of goods, are also sometimes described as influenced by stochastic disturbances of the Brownian type.

Brownian motion can be described in various more or less mathematically demanding ways. We shall introduce Brownian motion in a nonrigorous manner by letting it be a limit of a stochastic process in discrete time. We continue by discussing so-called stochastic integration and stochastic differential equations before turning to optimal control problems and optimal stopping problems.

4.1 Brownian Motion

Consider a stochastic process in the form of a "symmetric, random walk" of the following type. Assume that at times $h, 2h, 3h,\ldots$ we take a step of size c up or down with probability $1/2$. More precisely, suppose there are independent random variables Y_t, for which $Pr[Y_t = c] = Pr[Y_t = -c] = 1/2$, for $t = h, 2h, 3h, \ldots$ Let $X_t = Y_h + \cdots + Y_{nh}$, if $t = nh$ for n a natural number, $X_0 = 0$. Because $E[Y_t] = 0$ and $\text{Var}[Y_t] = E[Y_t^2] = c^2$, we get $E[X_t] = 0$ and $\text{Var}[X_t] = c^2(t/h)$ where $t/h = n$, the number of terms in the sum defining X_t. In the interval $[ih, (i+1)h)$, define $X_t = X_{ih}$. For t in this interval, $E[X_t] = E[X_{ih}]$, $\text{Var}[X_t] = \text{Var}[X_{ih}]$. We shall now let h and c tend to 0, and when the convergence of h and c is suitably coordinated, the resulting limit $B_t := \lim_{h\to 0} X_t$ is a "sensible," nontrivial process defined for all t. Let us try $c = ah^\alpha$, $a > 0$, $\alpha > 0$, a and α fixed. Evidently, $\text{Var}(X_t)$ vanishes when $h \to 0$, if $\alpha > 1/2$. If the convergence of c is too slow, i.e., $0 < \alpha < 1/2$, we get

A. Seierstad, *Stochastic Control in Discrete and Continuous Time*,
DOI: 10.1007/978-0-387-76617-1_4,
© Springer Science+Business Media, LLC 2009

infinite variance. However, the choices $\alpha = 1/2$, $c = h^{1/2}$, and (for simplicity) $a = 1$, make $E[X_t] = 0$ and $Var(X_t) = t$. Both equalities hold even in the limit, i.e., both equalities hold for $\lim_{h \to 0} X_t =: B_t$. The so-called central limit theorem in probability theory says that a sum of many small, independent random variables of the type Y_k is approximately normally distributed, and in the limit exactly so, in fact B_t is $N(0, t)$, i.e., normally distributed with mean 0 and variance t. The limit process B_t is called *Brownian motion* or the *standardized Wiener process*. (See Karatzas and Shreve (1988), Theorem 4.20, for a precise version of these arguments.)

Let us exhibit more properties of B_t: Before taking limits, suppose $(s_j, t_j]$, $j = 1, \ldots, j^*$, are disjoint intervals, s_j, t_j multiples of h. Then $X_{t_j} - X_{s_j}$, $j = 1, \ldots, j^*$ are independent, with zero expectation and variance dependent only on $t_i - s_i$, not on s_i. This property is referred to as having *(weakly) stationary independent increments*. In fact, if $t_j = i'h$, $s_j = ih$, $i' > i$, then $Var[X_{t_j} - X_{s_j}] = (c^2/h)(t_j - s_j) = (t_j - s_j)$. These properties are preserved when taking limits. Also $X_{t_j} - X_{s_j}$ becomes normally distributed in the limit (this difference also being the sum of many small, independent random variables Y_k). To sum up, Brownian motion has the following properties:

(a) $B_0 = 0$
(b) The random variables $B_{t_i} - B_{s_i}$, $i = 1, \ldots, i^*$ are independent, whenever all the intervals $(s_i, t_i]$ are disjoint.
(c) B_t is $N(0, t)$-distributed.
(d) $B_t - B_s$ is $N(0, t-s)$-distributed, when $t > s$.

Note that if s is fixed, then $Z_t = B_{s+t} - B_s$ is again a Brownian motion. Brownian motion is a "Markov process" because it satisfies the defining property of such processes, namely for any given positive numbers t, h, and given any numbers a, b, the conditional probability of the event that B_{t+h} belongs to (a, b), given the the history of B_s up to $s = t$, is the same as the conditional probability of the event that B_{t+h} belongs to (a, b), given B_t. (This simply follows from the fact that, given B_t, the event $B_{t+h} - B_t \in (a - B_t, b - B_t)$ is independent of what has happened before.) Note also that (b) and (d) imply that $E[B_t | B_{s_1}, \ldots, B_{s_k}] = B_{s_k}, t > s_k > \ldots > s_1$. More generally, by the Markov property, Brownian motion satisfies the crucial defining property of so-called "martingales," namely that the conditional expectation of B_t, given the history of B_τ for $\tau \leq s$ equals B_s.

Above we talked about B_t as a limit. In which sense? As an indication of that, note that for any a and b, $Pr[a < X_t < b]$ approximately equals $Pr[a < B_t < b]$ when h is small and t is a multiple of h.

Two properties worth noting are the following ones. Define $M_t = \sup_{0 \leq s \leq t} B_s$. For each t, the non-negative stochastic variable M_t has a density on $[0, \infty)$ proportional to $\varphi^t(z)$, $z > 0$, where $z \to \varphi^t(z)$ is the density of the normal distribution with mean zero and variance t. Also, with probability one, if $b(.)$ is an outcome of a Brownian motion, then for any k, the "k-level set" $A_k = \{t \geq 0 : b(t) = k\}$ is nonempty, closed, with no isolated points, and (a.s.) $\int_0^\infty 1_{A_k}(t)dt = 0$, the integral of the indicator function $1_{A_k}(.)$ of A_k is zero. (This function equals 1 for $t \in A_k$, and 0 for $t \notin A_k$. An isolated point means a point with no close neighbors.) To show the last property, see the proof of a similar property at the end of Example 4.34 below.

Further Discussion of Brownian Motion*

To see how a process like Brownian motion might arise in a stylized economy, imagine an economy with zero interest rate charged by the banks. (A constant nonzero interest rate would lead to geometric Brownian motion defined later on, for simplicity we assume a zero interest rate.) Imagine that the stochastic price of a financial security X_t is continuously observable, where t is continuous time, with X_0 a fixed number. Given X_t, in a well-functioning, zero interest rate market, it seems that it cannot be the case that, for $h > 0$, $E[X_{t+h}|X_t] \neq X_t$, for in that case it is possible, in an expectational sense, to earn money by speculation: If, say, the difference $E[X_{t+h}|X_t] - X_t$ is positive, at time t I would borrow an amount Y times X_t dollars in the bank, buy Y securities at time t, sell the securities and repay the amount YX_t at time $t+h$. The expected return of the operation would be $YE[X_{t+h}|X_t] - YX_t > 0$. If the difference is negative, I would choose a negative value of Y, and come out with a positive expected gain again. (If I owned Y securities worth YX_t, I would sell them at time t, buy them back at time $t+h$ at expected cost $YE[X_{t+h}|X_t] < YX_t$.) So at least if there are risk-neutral actors in the market, is seems that we must have $E[X_{t+h}|X_t] = X_t$.

A similar conclusion can also be reached if we only have (finitely) risk-adverse actors in the market, but make some stronger assumptions concerning the process. We shall again be able to rule out the possibility that $E[X_{t+h}|X_t] \neq X_t$, but only if certain assumptions are made concerning covariances of the process (as well as a certain constancy over time). See A in the Addendum to this chapter.

Replacing h by kh, $k = 0, 1, 2, \ldots$, in the last equality, we get that, for any k,

$$E[X_{t+kh}|X_t] = X_t, \text{ or } E[X_{t+kh} - X_t|X_t] = 0. \tag{4.1}$$

Using the double expectation rule (see Appendix, (5.21)), this property implies that, for $k, k' = 0, 1, 2, \ldots$, $E[X_{t+kh} - X_t] = EE[X_{t+kh} - X_t|X_t] = 0$ and that $0 = E[X_t\{E[X_{t+kh} - X_t|X_t] - E[X_{t+k'h} - X_t|X_t]\}] =$

$$E\{E[(X_{t+kh} - X_{t+k'h})X_t|X_t]\} = E[(X_{t+kh} - X_{t+k'h})X_t] = 0. \tag{4.2}$$

In some cases, it is natural to expect that market conditions do not change over time, i.e., the probability distribution of $X_{t+h} - X_t$ is independent of time t (so-called stationary increments). If n is any natural number, $X_{t+nh} - X_t = \sum_{i=1}^{n}(X_{t+ih} - X_{t+(i-1)h})$. Using this expression, (4.1), (4.2), and stationarity, we get that

$$\text{Var}(X_{t+nh} - X_t) = E(X_{t+nh} - X_t)^2 = E\left(\sum_{i=1}^{n}(X_{t+ih} - X_{t+(i-1)h})\right)^2$$

$$= E\sum_{i,j=1}^{n}(X_{t+ih} - X_{t+(i-1)h})(X_{t+jh} - X_{t+(j-1)h})$$

$$= E\sum_{i=1}^{n}(X_{t+ih} - X_{t+(i-1)h})^2 = n\text{Var}(X_{t+h} - X_t).$$

We get the third equality, as for $i \neq j$, terms $E(X_{t+ih} - X_{t+(i-1)h})(X_{t+jh} - X_{t+(j-1)h})$ at the left-hand side of this equality can be written as

$$E(X_{t+ih} - X_{t+(i-1)h})X_{t+jh} - E(X_{t+ih} - X_{t+(i-1)h})X_{t+(j-1)h}$$

in case $i > j$ (in the opposite case interchange i and j). In both cases, (4.2) yields that these terms vanish. So, let us write down that

$$\mathrm{Var}(X_{t+nh} - X_t) = n\mathrm{Var}(X_{t+h} - X_t). \tag{4.3}$$

Denoting for the moment $\mathrm{Var}(X_{t+1} - X_t)$ by σ^2, it can be shown that $\mathrm{Var}(X_{t+s} - X_t) = s\sigma^2$, $s = k/m$, k, m natural numbers. To show this, we use (4.3) for $n = k$, $h = 1/m$ and for $n = m, h = 1/m$, the latter choices yield $\sigma/m = \mathrm{Var}(X_{t+1/m} - X_t)$ and then the former choices give the asserted equality. We can reasonably expect that $\mathrm{Var}(X_{t+s} - X_t) = s\sigma^2$ for any positive s. Similarly, (replacing kh in (4.1) by s), we can also reasonably expect, for any $s > 0$, that $E[X_{t+s} - X_t | X_t] = 0$, and hence, for $r < t$ $s > 0$, $0 = E[X_{t+s} - X_t | X_r]$ $(= E[X_{t+s} - X_r | X_r] - E[X_t - X_r | X_r])$ and so $E[X_{t+s} - X_t] = 0$. Using this yields, for $s > 0, t \geq r$,

$$\begin{aligned}
\mathrm{Covar}(X_{t+s} - X_t, X_r) &= E[(X_{t+s} - X_t)(X_r - EX_r)] \\
&= EE[(X_{t+s} - X_t)(X_r - EX_r)|X_r] \\
&= EE[(X_{t+s} - X_t)X_r]|X_r] - EE[(X_{t+s} - X_t)EX_r)]|X_t] = 0.
\end{aligned}$$

Finally, let us assume that the conditional density of $X_{t+s} - X_t$ for any $t, s \geq 0$ given X_t is a normal one (in particular then $X_s - X_0$ has this property). The vanishing of the above covariance gives, by general results on normal variables, that $X_{t+s} - X_t$ is independent of X_r, and more generally, independent of all differences $X_{t_i} - X_{s_i}$, $i = 1, \ldots, i^*$, whenever all $(s - i, t_i]$ are disjoint, i.e., the above property (b) of Brownian motion. Moreover, $X_{t+s} - X_t \sim N(0, s\sigma^2)$ so $s \to (X_s - X_0)/\sigma$ is Brownian motion.

The Sample Space

In probability theory, if the sample space Ω contains an uncountably number of points (e.g., $\Omega = \mathbb{R}$), then not all subsets can be given a probability. We always have to imagine that a restricted family $\mathcal{F}$ of subsets A is specified, for which $Pr[\omega \in A] := P(A)$ is defined. This family is usually assumed to be a "σ-field" or a "σ-algebra," which means that the family $\mathcal{F}$ has the property that finite (or countable) unions and intersections of sets from the family also belong to $\mathcal{F}$ and so do also complements of sets from $\mathcal{F}$. (In addition, Ω and $\emptyset$, the empty set, belongs to $\mathcal{F}$.)

An outcome of the Brownian motion will be a path in the (t, B)-plane, i.e., a function in a certain class of functions C. It can be proved that with probability 1, the paths (outcomes) of Brownian motion are continuous. So we may assume that the paths belong to $C_0([0, \infty), \mathbb{R})$, the set of real-valued continuous functions on $[0, \infty)$, with 0 as value at $t = 0$, (indicated by the subscript 0). This is our sample space. The paths of Brownian motion are, generally, not differentiable. This is related to the fact that it is a limit of a sequence of random walks, with steeper and steeper steps (the height of each step is $h^{1/2}$, the width is h).

Probabilists frequently speak of the need to prove that B_t "exists." Roughly, this means proving that there exists on the sample space $\Omega := C_0([0,\infty),\mathbb{R})$, a "probability measure," i.e., probabilities $P(A)$ attached to a suitable class (σ-field) B^* of sets A, containing all sets of the form $A_{t,s,r} := \{b(.) \in \Omega : s \le b(t) \le r\}$ and with

$$P(A_{t,s,r}) = \int_s^r (1/(2\pi t)^{1/2}) \exp(-x^2/2t)dx.$$

(Along the road to a proof, another Ω may at first have been introduced and continuity of paths established.)

Frequently, Brownian motion B_t is introduced by using a sample space Ω different from $C_0([0,\infty),\mathbb{R})$. As with other random variables, one frequently writes $B_t(\omega)$, where ω is "drawn" from Ω, resulting in a continuous function $b(t) = b(t,\omega)$ been drawn (at least a.s.). But then the sets $A^{t,s,r} = \{\omega : s < b(t,\omega) \le r\}$ has the property that $P'(A^{t,s,r}) = P(A_{t,s,r})$, where P' is the probability measure on this new set Ω. This equality makes us indifferent as to working with $C_0([0,\infty),\mathbb{R})$ or this new set Ω. The specific Ω introduced above (namely $\Omega = C_0([0,\infty),\mathbb{R})$) is called the *canonical* sample space (then $\omega = b(.)$, and $B_t(\omega) = B_t(b(.)) = b(t)$). We will always work with this sample space.

Below, functions $u(t)$ will occur that depend on "past values of B_t." Slightly more precisely, this means that for each t, $u(t)$ depends on the values of B_s, $s \le t$. Such functions are called history dependent, or adapted to B_t, or simply adapted. A "simple" function that is evidently adapted to B_t is $u(t) = u(t,b(.)) = \Sigma_{\tau_i < t} a_i(t) 1_{M_i}(b(.))$, where $M_i := \{b(.) : s_i \le b(\tau_i) \le r_i\}$, τ_i are given time points, s_i, r_i are given numbers, and $a_i(t)$ are given piecewise continuous functions. (When an outcome $b(.)$ of Brownian motion on any given interval $[0,t]$ has occurred, we can decide for each set M_i for which $\tau_i < t$ if $b(.)$ belongs to the set or not, hence calculate the value of $u(t,b(.))$.) Evidently, it is natural to call $u(t,b(.))$ history-dependent (dependent on the history of the Brownian motion). The most general functions $u(t,b(.))$ that we work with are (pointwise) limits of sequences of simple history-dependent functions $u(t,b(.))$ (or slightly more general, limits of such sequences for any $t,b(.)$, a.e. $\times$ a.s.). The values of the general functions also become history-dependent. Sometimes, in this text, the nonstandard symbol $b(\to t)$ is used as meaning the values of $b(s)$ for $s \le t$, the same meaning has $B_{\to t}$. Thus, when we write $u(t,B_{\to t})$, then, for each t, this is the stochastic variable that is determined by the stochastic variables $B_s, s \le t$, and $u(t,b(\to t))$ is the outcome of the stochastic variable $u(t,B_{\to t})$ corresponding with the outcome $b(.)$ of the Brownian motion.

We do not want to formalize these nonrigorous considerations. But let us note that in the literature, also concerned with economic applications, adapted functions $u(t)$ are described as follows: For each t, $u(t) = u(t,\omega)$ is "measurable with respect to the σ-field $\mathcal{F}_t$ generated by B_t." Even the term "progressively measurable" is used, it has a similar meaning. In this chapter, we shall not define more precisely these

terms, readers wanting a better idea of the meaning of these concepts, are referred
to the Appendix. The essential point to note is that these concepts entail dependence
on past values of B_t (history dependence on B_t).

(For the specially interested reader: The σ-fields $\mathcal{F}_t$ generated by $B(t)$ are defined
as follows. Given numbers r_i, s_i, t_i, $i = 1, \ldots, n$, the t_i's being distinct, $t_i \leq t$, define
the set $H = \{b(.) \in \Omega : s_i < b(t_i) \leq r_i, i = 1, \ldots, n\}$. Here n, s_i, r_i, and t_i can all
vary, and we may even allow $n = \infty$ or some $r_i = \infty$, $s_i = -\infty$. Then, let $\mathcal{F}_t$ be the
family of all such sets H, as well as complements and finite or countable unions and
intersections of such sets. A more formal definition enlarges $\mathcal{F}_t$ somewhat.)

4.2 Stochastic Calculus

By analogy with difference equations in discrete time involving random processes,
we would like to consider differential equations of the type

$$dx/dt = b(t, x_t) + \sigma(t, x_t)W_t, \tag{4.4}$$

where W_t is "white noise," i.e, the W_t's for varying t are independent, with each W_t
being $N(0, 1)$-distributed. However, there is no suitable stochastic process whose
sample paths (evolution through time) are well behaved and reflect these properties
of W_t. In fact, no process gives sample paths that can be modeled by taking an integral on the right-hand side of the equation. Yet, a solution of a differential equation
X_t should preferably be the integral of its own derivative.

We are going to define a stochastic integral and replace (4.4) by an integral
equation for X_t in which this stochastic integral enters. Recall that an ordinary deterministic integral $\int_J f(s)ds$ (J an interval, f piecewise continuous) is calculated
approximately by a sum of the form $\sum_i f(s^i)(s_{i+1} - s_i)$, where $\{s_i\}$ represents a
fine subdivision of J and $s^i \in [s_i, s_{i+1}]$. Let us apply this to an integral of the form
$\int_J f(s)d\alpha(s)$, where $\alpha(s)$ is differentiable: It is calculated as follows. The integral
equals $\int_J f(s)\alpha'(s)ds$, which is approximated by the sum $\sum_i f(s^i)\alpha'(s^i)(s_{i+1} - s_i)$.
Now, $\alpha'(s^i)(s_{i+1} - s_i) \approx \alpha(s_{i+1}) - \alpha(s_i)$, so $\int_J f(s)d\alpha(s) \approx \sum_i f(s^i)(\alpha(s_{i+1}) -
\alpha(s_i))$. This approximate manner of calculating the last integral can also be used
even in the case where $\alpha(s)$ is not differentiable. The corresponding integral is called
a *Stieltjes integral* (not supposed known here), and α is called an integrator. So the
integral $\int_J f(s)d\alpha(s)$ can be defined even for nondifferentiable α. At least, it works
fine if $\alpha(.)$ is a nondecreasing function, or even if $\alpha(.)$ has "bounded variation,"
which here is taken to mean that it is a difference between two nondecreasing functions. (For our limited purpose, we shall think of the integrator $\alpha(.)$ as also being
continuous. However, in the general definition of Stieltjes integral, continuity is not
needed.)

Note that if $f(s)$ is a piecewise constant function $\sum_i a_i 1_{I_i}$ where the I_i are disjoint
intervals $[s_i, s_{i+1})$, then $\int_J f(s)d\alpha(s)$ equals (exactly) $\sum_i a_i(\alpha(s_{i+1}) - \alpha(s_i))$.

One type of stochastic integral is obtained when α is stochastic. Whether all outcomes of α have derivatives or not, we can calculate the integral $\int_j f(s)d\alpha(s)$ one way or the other, as described above, as long as each outcome of α is an integrator of acceptable type. So, if $\alpha(t) = \alpha(t, \omega)$ (i.e., α depends on the outcome $\omega \in \Omega$), the value of the integral is a stochastic number depending on the outcome ω. It is often said that this random variable is obtained by "pathwise integration" (by carrying out the integration for each outcome path of $\alpha(t, \omega)$, i.e., for each ω), also called "ω-wise" integration. Of course, we can also allow f to be stochastic, $f = f(s, \omega)$ (f piecewise continuous in s) and still calculate such pathwise integrals.

We shall now define integration with respect to Brownian motion B_t. One use of this is that it allows us to tackle the white noise equation above: If W_t is somehow replaced by $B_{t+h} - B_t$, for h small, we get something behaving similar to W_t. The question then becomes if $\sigma(s, X_s)dB_s$ can be integrated. Or, in general, for any adapted function $f(s, B_{\to s})$, will $\int_J f(s, B_{\to s})dB_s$ be well defined? The answer is yes, and the integral will be called *the Ito integral*.

If $f(s)$ is a deterministic piecewise constant function $\sum_i a_i 1_{I_i}$, where the I_i are disjoint intervals $[s_i, s_{i+1})$ forming a partition of J, then $\int_J f(s)dB_s$ can be exactly calculated as $\sum_i a_i(B_{s_{i+1}} - B_{s_i})$. If, however $f(s)$ is not simple but, say continuous, or even (say) right-continuous, then we use the approximate formula $\sum_i f(s^i)(B_{s_{i+1}} - B_{s_i})$ to calculate the integral, where s^i can be chosen anywhere in $[s_i, s_{i+1})$. However, below, we always choose $s^i = s_i$. In a moment, we let f be a stochastic variable, and then the end result (the integral) will in general depend on where in $[s_i, s_{i+1})$ s^i is chosen, in contrast with standard Stieltjes integration mentioned above.

Solutions of stochastic differential equations such as (4.4) must involve integrands f that are stochastic variables.

Above, we indicated a particular interest in integrating expressions such as $\sigma(s, X_s)dB_s$. For stochastic piecewise constant function of the form $f(s) = \sum_i a_i 1_{I_i}(s)$, where we now allow the a_i's to be functions $a_i(B_{\to s})$ (but where the I_i's are deterministic), the integral of f by definition equals $\sum_i a_i(B_{\to s_i})(B_{s_{i+1}} - B_{s_i})$.

As always, one then wants to extend the definition of the integral to more general functions. If $F(s) = F(s, B_{\to s})$, is piecewise, and right-continuous in s, then the integral $\int_J F(s, B_{\to s})dB_s$ is calculated approximately by the sum $\sum_i F(s_i, B_{\to s_i})(B_{s_{i+1}} - B_{s_i})$, for a fine subdivision $\{s_i\}$ of the interval J (we have here put $s^i = s_i$). Making the subdivision finer and finer, in the limit we get the Ito integral of F.

Imagine that for each outcome $b(.)$ of B_t, the integral $\int_J F(s, b(\to s))db(s)$ is calculated approximately by the sum $\sum_i F(s_i, b(\to s_i))(b(s_{i+1}) - b(s_i))$. This sum evidently depends on the outcome $b(.)$ and is (approximately) the value of the stochastic integral $\int_J F(s, B_{\to s})dB_s$ when the outcome of B_t is $b(.)$. It is useful to think of the stochastic integral as arising in this manner, obtained by means of a pathwise Stieltjes integration, to use words introduced above. However, when one looks at the fine mathematical details behind the introduction of such an integral, it will be seen that a slightly different route is taken. In general, the outcomes $b(.)$ are not of bounded variation, so technically the pathwise approach must be modified.

Let us describe in more detail the family of functions that can be integrated against Brownian motion. Let N^* be the family of real functions $F(s) = F(s, B_{\to s})$,

$s \in J = [0, T]$, such that, for each s, $F(s)$ is a random variable that depends on the values $B_{s'}$, $s' \leq s$ (so $F(s)$ is adapted or history dependent). Furthermore, for each outcome $b(.)$ of B_s, the corresponding outcome $f(s) := F(s, b_{\to s})$ of $F(s, B_{\to s})$ is required to be piecewise and right-continuous in s, and $F(s) = F(s, B_{\to s})$ is required to satisfy $E[\int_0^T F(s)^2 ds] < \infty$. This defines the family N^*.

To repeat, the integral $\int_J F(s, B_{\to s}) dB_s$ can be approximately calculated by sums of the type $\sum_{i=0}^{i'-1} F(s_i, B_{\to s_i})(B_{s_{i+1}} - B_{s_i})$, where the subdivision $\{s_i\}_{1 \leq i \leq i'-1}$, $s_0 = 0$, $s_{i'} = T$, $s_{i+1} > s_i$, is fine enough. By choosing the subdivision fine enough (i.e., $\sup_i [s_{i+1} - s_i]$ small enough), the approximation will be very good.

Let us discuss this integral a little more, considering only the case where F is a bounded function. It turns out that there are piecewise constant functions $\phi_n(s)$ converging to $F(s) = F(s, B_{\to s})$ for all $s < T$. Indeed, simply put $\phi_n(s) = \sum_{0 \leq k < n} F(Tk/n) 1_{[Tk/n, T(k+1)/n)}(s)$. The value of the integral $\int_J \phi_n dB_s$ exactly equals a sum of the type already presented, namely $\sum_k F(s_k, B_{\to s_k})(B_{s_{k+1}} - B_{s_k})$, where $s_k = Tk/n$. The values of this sum can be shown to converge in an appropriate sense when $n \to \infty$. The limit of this sequence of random numbers is denoted by $\int_J F(s) dB_s$ and is the Ito integral of F, as earlier indicated. Notice then that, as with other types of integrals, also this integral can be calculated approximately by replacing the original function with a simple function, the integral of which is given by a sum that we can write down.

(Technically speaking, the convergence takes place in "$L_2(\Omega)$-norm," i.e., $E(\int_J \phi_n dB_s - \int_J \phi_m dB_s)^2 \to 0$ when $n, m \to \infty$, in fact, $\int_J \phi_n dB_s$ is a "Cauchy-sequence" in this norm.)

Proof of convergence of $\int_J \phi_n dB_s$ (sketch)*

To sketch a proof of the convergence of $\int_J \phi_n dB_s$, we begin by verifying the so-called *Ito isometry*: Let ϕ be any piecewise constant function $\phi = \sum_{0 \leq i \leq k} F_i 1_{[s_i, s_{i+1})}$, where F_i depends on the history of the Brownian motions up to s_i. Then

$$E\left[\left(\int_0^T \phi dB_t\right)^2\right] = E\left(\int_0^T \phi^2 dt\right) \qquad (4.5)$$

This equality is shown as follows: Let $\Delta B_i = B_{s_{i+1}} - B_{s_i}$. Then, if $i \neq j$, $E[F_i F_j \Delta B_i \Delta B_j] = 0$, (say if $i < j$, $E[F_i F_j \Delta B_i \Delta B_j] = E[F_i F_j \Delta B_i] E[\Delta B_j] = 0$), and $E[F_i F_j \Delta B_i \Delta B_j] = E[F_i^2](s_{i+1} - s_i)$ if $i = j$. Hence, as $\int_0^T \phi dB_t = \sum_i F_i \Delta B_i$,

$$E\left(\int_0^T \phi dB_t\right)^2 = \sum_{i,j} E(F_i F_j \Delta B_i \Delta B_j) = \sum_i E(F_i^2)(s_{i+1} - s_i) = E\int_0^T \phi^2 dt.$$

(In fact, (4.5) holds for any bounded adapted function ϕ.)

Let now F and ϕ_n be defined and related as above. Because $\phi_n - \phi_m$ is simple, $E[(\int_0^T (\phi_n - \phi_m) dB_t)^2] = E(\int_0^T (\phi_n - \phi_m)^2 dt)$. Now, as ϕ_n converges to F for each s, each ω, then $\phi_n - \phi_m \to 0$ for each s, each ω, as $m, n \to \infty$. A theorem on "dominated convergence" in integration theory (see the Appendix) then says that because F and so also all $\phi_n - \phi_m$ are bounded by a constant, it follows that $\int_0^T (\phi_n - \phi_m)^2 dt \to 0$ for each ω. The same theorem says that then even $E(\int_0^T (\phi_n - \phi_m)^2 dt) \to 0$. The Ito isometry finally implies that $E[\int_0^T \phi_n dB_t - \int_0^T \phi_m dB_t]^2 \to 0$, when $n, m \to \infty$. This is the convergence we wanted to establish. It means that when n and m are large, there is "very little difference" between $\int_0^T \phi_n dB_t$ and $\int_0^T \phi_m dB_t$ (at least if this difference is measured by taking the expectation of the square of it). That is, the sequence of integrals is convergent in this sense. (The sequence is a "Cauchy sequence in $L^2(\Omega)$-norm," see the Appendix). We get the same limit for all sequences of piecewise constant functions ϕ_n converging to F for each s, each ω, the limit does not depend on the choice of sequence. □

Above, the functions $F(s)$ that we integrated were history dependent on B_s. Let M_s^j, $j = 1, \ldots, j^*$ be a given collection of stochastic processes that includes B_s, and assume instead that $F(s)$ is history dependent on the M_s^j's. Provided $B_t - B_s$, $s > t$, is independent of the history of M_τ^j, for $\tau \le s$, the Ito integral $\int_0^T F(s) dB_s$ is defined essentially in the same way as before (the proof of the convergence of the approximating simple integrals is the same as before). A simple example is the case where the M_s^j's are j^* independent Brownian motions, where for any given $\hat{j}$, the independence of $M_t^{\hat{j}} - M_s^{\hat{j}}$, $t > s$, on the history of the other M_τ^j's ($\tau \le s$) follows from the independence assumption. (Then B_s is assumed to be, say, M_s^1.)

In the approximation formula $\sum_i F(s^i, B_{\to s^i})(B_{s_{i+1}} - B_{s_i})$ for the Ito integal, we said that we always choose s^i equal to s_i. If instead we had chosen $s^i = (s_i + s_{i+1})/2$, we had obtained another integral, called the Stratonovich integral, a least for well-behaved integrands. There are standard formulas converting a Stratonovich integral into an Ito integral (see Karatzas and Shreve (1988)). Whether the Ito integral or the Stratonovich integral is used depends much on the interpretations these integrals have in the application at hand, but also partly on mathematical convenience. The Stratonovich integral is not a martingale, as the Ito integral is, but the Stratonovich integral has simpler calculus rules, for example, in the formula for the Stratonovich integral corresponding to Ito's formula (the latter is presented below), no second-order term appears. If the Stratonovich integral is written $\int_a^b F(t, \omega) \circ dB_t$, then for $F(t, \omega) = g(t, B_t)$, $g(t, x)$ being realvalued and C^3,

$$\int_a^b g(t, B_t) \circ dB_t = \int_a^b g(t, B_t) dB_t + \int_a^b (1/2) g_x(t, B_t) dt.$$

Note that, trivially, the Ito integral $\int_0^t dB_s$ equals B_t. We shall not give examples of how to calculate integrals until after the so-called Ito's formula has been discussed, a formula that appears in the next section.

4.3 Stochastic Differential Equations

Consider the stochastic differential equation

$$dX_t = f(t, X_t) dt + \sigma(t, X_t) dB_t, \quad X_0 = x_0 \in \mathbb{R}, \quad x_0 \text{ fixed} \tag{4.6}$$

where B_t is a given Brownian motion, f and σ are real-valued functions, piecewise and right-continuous in t, and Lipschitz continuous in x, uniformly in t (for, say, f, the Lipschitz continuity means that for some K, for all t, x, y, $|f(t,x) - f(t,y)| \leq K|x - y|$). Compared with a deterministic equation, (4.6) takes an unusual form. Actually, the equation does not have a proper meaning as written. Instead, it is really a shorthand for the corresponding integral equation:

$$X_t = x_0 + \int_0^t f(s, X_s) ds + \int_0^t \sigma(s, X_s) dB_s \tag{4.7}$$

Dividing (4.6) by dt, it seems that we get (4.4) for $W(t) = dB_t/dt$. As noted before, taken literally, this manner of stating (4.4) does not work, as (almost surely) the derivative dB_t/dt never exists (this fact is related to the fact that $E|B_{t+h} - B_t| = 2(h/2\pi)^{1/2}$, so $E|(B_{t+h} - B_t)/h| = 2/(2\pi h)^{1/2}$, which $\to \infty$ when $h \to 0$).

Let us turn instead to (4.7), which is a proper (meaningful) equation. By definition, a solution to (4.7) is a function X_s that is continuous in s, such that X_s is adapted to the given Brownian motion, and satisfies (4.7). Both integrals can be calculated, once X_s is known. The integrand in the second integral may been seen to be adapted (intuitively, it depends on past values of B_t, $(t \leq s)$, so the integral is calculable). The first integral is an ordinary one (we can calculate it for each outcome of X_s separately).

There are some general theorems securing existence of solutions to such equations. In the theorem to be presented below, allow f and X_t to be n-dimensional ($n \times 1$-matrices), $\sigma(t,x)$ to be a $n \times m$-matrix with elements $\sigma_{i,j}(x,t)$ and rows σ_j, and B_t to be an m-vector ($m \times 1$-matrix) whose components are independent Brownian motions. Then (4.6) and (4.7) become systems of n (simultaneous) equations, one for each component of X_t.

Theorem 4.1 (Existence of solutions, (Nualart (1995)). *Assume that all $f_i(t,x)$ and all $\sigma_{i,j}(t,x)$ are piecewise and right continuous in t and Lipschitz continuous in x uniformly in t, with $\sum_j |\sigma_j(t,x) - \sigma_j(t,y)| + |f(t,x) - f(t,y)| \leq K(x - y)$ for any x, y, t, and that $\sigma_j(t,x_0)$ and $f(t,x_0)$ are functions bounded by a constant K'. Then a unique solution X_t of (4.7) exists on $[0, \infty)$, adapted to (history dependent on) $B_s = (B_s^1, \dots, B_s^m)$, continuous in t and satisfying*

$$E\left[\sup_{0 \leq t \leq T} |X(t)|^2 \right] \leq \alpha K^2 T e^{\alpha K^2 T} [|X_0|^2 + ((n+1)K'T)^2],$$

for some $\alpha \geq 0$, independent of K, K', x_0 and T.   □

Some bounds on the solutions of stochastic differential equations are discussed in Exercise 4.5 below.

An approximate manner of solving such equations is "stepwise numeric integration": If X_{ih} is already calculated, then $X_{(i+1)h} = X_{ih} + f(ih, X_{ih})h + \sigma(ih, X_{ih})(B_{(i+1)h} - B_{ih})$.

Partial proof of Theorem 4.1*

Let us show, in the one-dimensional case, the existence of a solution on $[0, 1/16K^2]$, assuming that $K \geq 1$. (Iterated use of such a result yields a solution on any finite interval. When constructing the solution on $[1/16K^2, 2/16K^2]$, the initial value to be used would be $X_{1/16K^2}$, known from the previous construction. A similar remark pertains to later intervals. The below proof also works for stochastic initial values.) By the Cauchy–Schwartz inequality (see the Appendix), $\int_0^t |\phi(s)| \cdot 1 ds \leq (\int_0^t \phi(s)^2 ds)^{1/2} (\int_0^t 1 ds)^{1/2}$, or $(\int_0^t |\phi(s)| ds)^2 \leq t \int_0^t \phi(s)^2 ds$. Let $t \in [0, 1/16K^2]$, and define $Y_t^0 \equiv X_0$ and, by induction,

$$Y_t^n = X_0 + \int_0^t f(s, Y_s^{n-1}) ds + \int_0^t \sigma(s, Y_s^{n-1}) dB_s.$$

Define

$$||Y^{n-1} - Y^{n-2}||^* = \sup_{t \in [0, 1/16K^2]} E[|Y_t^{n-1} - Y_t^{n-2}|^2].$$

Using both the last inequality, $(a+b)^2 \leq 2a^2 + 2b^2$, and

$$|f(s, Y_s^{n-1}) - f(s, Y_s^{n-2})| \leq K|Y_s^{n-1}) - Y_s^{n-2}|, |\sigma(s, Y_s^{n-1}) - \sigma(s, Y_s^{n-2})|$$
$$\leq K|Y_s^{n-1}) - Y_s^{n-2}|,$$

note that

$$E|Y_t^n - Y_t^{n-1}|^2$$

$$\leq E\left[\int_0^t \{f(s, Y_s^{n-1}) - f(s, Y_s^{n-2})\} ds + \int_0^t \{\sigma(s, Y_s^{n-1}) - \sigma(s, Y_s^{n-2})\} dB_s \right]^2$$

$$\leq 2E\left[\int_0^t \{f(s, Y_s^{n-1}) - f(s, Y_s^{n-2})\} ds \right]^2 + 2E\left[\int_0^t \{\sigma(s, Y_s^{n-1}) - \sigma(s, Y_s^{n-2})\} dB_s \right]^2$$

$$\leq 2E\left[t \int_0^t \{f(s, Y_s^{n-1}) - f(s, Y_s^{n-2})\}^2 ds \right] + 2E\left[\int_0^t \{\sigma(s, Y_s^{n-1}) - \sigma(s, Y_s^{n-2})\}^2 ds \right]$$

$$\leq 2K^2 t E \int_0^t |Y_s^{n-1}) - Y_s^{n-2}|^2 ds + 2K^2 E \int_0^t |Y_s^{n-1}) - Y_s^{n-2}|^2 ds$$

$$\leq 2K^2 t^2 ||Y^{n-1} - Y^{n-2}||^* + 2K^2 t ||Y^{n-1} - Y^{n-2}||^*$$

$$\leq 4K^2 t ||Y^{n-1} - Y^{n-2}||^*,$$

(for the second inequality, the inequality $(a+b)^2 \leq 2a^2 + 2b^2$ is used, and the third one follows from the Ito isometry (4.5) and the above application of the Cauchy–Schwartz inequality).

As $4K^2 t \leq 1/4$, then $||Y^n - Y^{n-1}||^* \leq (1/4)||Y^{n-1} - Y^{n-2}||^*$. Thus the difference between Y^{n-1} and Y^{n-2}, measured by $||Y^{n-1} - Y^{n-2}|| := (||Y^{n-1} - Y^{n-2}||^*)^{1/2}$ decreases rapidly toward zero (the difference for n is $1/2$ that one for $n-1$). In fact Y_t^n is convergent in some sense (in fact, uniformly in $L^2(\Omega)$-norm), to a function Y_t, which can easily be seen to be a solution of the equation. Hence a solution exists on $[0, 1/16K^2]$. (For more details, see Exercise 4.6 below.)   □

The most important reason for including this proof is that it tells us that the above type of solutions to differential equations are history dependent on B_t: Note first that if $\alpha(s, B_{\rightarrow s})$ is history dependent, then $\int_0^t \alpha(s, B_{\rightarrow s}) ds$ and $\int_0^t \alpha(s, B_{\rightarrow s}) dB_s$ are history dependent, for given t, because the integrands $\alpha(s, B_{\rightarrow s})$, $s \leq t$, depend for each s on the values of B_τ, $\tau \leq s \leq t$. By induction, the functions Y_t^n become history dependent, and thus also, at least intuitively, the limit Y_t: None of the Y_t^n's depends on any future value of B_τ, so the limit must also be independent of such values of B_τ.

Remark 4.2. In the above theorem, f_i and σ_{ij} can be adapted functions, $f_i = f_i(t, x, \omega) = f_i(t, x, B_{\rightarrow t})$, $\sigma_{i,j} = \sigma_{i,j}(t, x, \omega) = \sigma_{i,j}(t, x, B_{\rightarrow t})$. (Then the inequality involving K must hold for all ω and K' must be a bound for all ω.) □

In particular, in (4.6), we can allow $f = f(t, x, B_{\rightarrow t})$ and $\sigma = \sigma(t, x, B_{\rightarrow t})$.

Remark 4.3. In order to define what constitutes a solution of (4.6) on $[0, T]$, the most important requirements are that, for each t, f and σ are adapted to B_t (dependent on values B_s, $s \leq t$), and piecewise and right-continuous in t, and (say) locally Lipschitz continuous in x, uniformly in t. For any function X_t, adapted to B_t, continuous in t, then $f(t, X_t, B_{\rightarrow t})$ and $\sigma(t, X_t, B_{\rightarrow t})$ become adapted to B_t and piecewise and right-continuous in t, which means that we should be able to calculate the integrals in (4.7). At least this is so if $E[\int_0^T |f(t, X_t, B_{\rightarrow t})| dt] < \infty$ and $E[\int_0^T \sigma(t, X_t, B_{\rightarrow t})^2 dt] < \infty$. Then, let us agree that X_t is a solution (sometimes called a strong solution) if it is adapted to the given Brownian motion, continuous in t, the equality in (4.7) holds for all $t \in [0, T]$, and the two just mentioned inequalities hold. □

The Ito Formula

Consider a stochastic process (sometimes called Ito process) of the form:

$$X_t = X_0 + \int_0^t u(s) ds + \int_0^t v(s) dB_s, \tag{4.8}$$

where $u(s)$ and $v(s)$ are adapted real-valued functions belonging to N^*, and $t \in J := [0, T]$. Here X_0 is a fixed number. When a process X_t is given in this manner, an informal manner of writing it is on differential form,

$$dX_t = u(t) dt + v(s) dB_s, \quad X_0 = x_0 \text{ given.} \tag{4.9}$$

Let $Y_t := g(t, X_t)$, where $g(t, x)$ has continuous second derivatives $(g : \mathbb{R}^2 \rightarrow \mathbb{R})$. Then *Ito's formula* is

$$Y_t := g(t, X_t) = g(0, X_0) + \int_0^t v(s) \left[\frac{\partial g(s, X_s)}{\partial x} \right] dB_s$$
$$+ \int_0^t \left\{ \frac{\partial g(s, X_s)}{\partial s} + u(s) \frac{\partial g(s, X_s)}{\partial x} + (1/2) v(s)^2 \left[\frac{\partial^2 g(s, X_s)}{\partial x^2} \right] \right\} ds \tag{4.10}$$

A proof of this formula is sketched below. Ito's formula is frequently informally formulated as

$$dY_t = (\partial g/\partial t)dt + (\partial g/\partial x)udt + (\partial g/\partial x)vdB_t + (1/2)v^2(\partial^2 g/\partial x^2)dt \quad (4.11)$$

or even as

$$dY_t = (\partial g/\partial t)dt + (\partial g/\partial x)dX_t + (1/2)(\partial^2 g/\partial x^2)(dX_t)^2. \quad (4.12)$$

Let us give an informal comment on (4.11) and (4.12): One often says that (4.11) is derived from (4.12), as follows: dX_t is replaced by $udt + vdB_t$, and then $(dX_t)^2 = u^2(dt)^2 + 2uvdtdB_t + v^2(dB_t)^2$ is calculated by applying the rules $dt\,dt = 0$, $dt\,dB_t = 0$, $dB_t\,dB_t = dt$, which yields $(dX_t)^2 = v^2 dt$. This calculation has no proper meaning, neither have (4.11) and (4.12).

There is a multidimensional analog of Ito's formula that will only be stated on differential form. (Again a truely meaningful formula is obtained only by first applying the calculation rules stated below and next adding integral signs.) Given the real C^2 function $g(t, x_1, \ldots, x_n)$, suppose that $dX_i(t) = u_i(t)dt + v_i(t)dB_i(t)$, where $B_i(t)$ are independent Brownian motions. Write $X_t = (X_1(t), \ldots, X_n(t))$. Then $Y_t = g(t, X_1(t), \ldots, X_n(t)) =: g(t, X_t)$ satisfies

$$dY_t = (\partial g(t, X_t)/\partial t)dt + \sum_i (\partial g(t, X_t)/\partial x_i)dX_i(t)$$
$$+ (1/2)\sum_{i,j}(\partial^2 g(t, X_t)/\partial x_i \partial x_j)dX_i(t)dX_j(t) \quad (4.13)$$

the products $dX_i(t)dX_j(t)$ being calculated by means of the rules $dt\,dt = dB_i(t)dt = 0$, $dB_i(t)dB_j(t) = 0$ if $i \neq j$, and $= dt$ if $i = j$.

To explain intuitively the content of the Ito formula, say (4.11), let us assume that dX_t, dY_t, dB_t, are not "infinitesimals," but represent changes on a small interval $[t, t+dt]$. Using that $dX_t \approx udt + vdB_t$ and Taylor's formula, including terms up to the second order, then $dY_t \approx$

$$g_t dt + g_x dX_t + (1/2)[g_{tt}dt^2 + 2g_{tx}dtdX_t + g_{xx}dX_t^2]$$
$$= g_t dt + g_x udt + g_x vdB_t + (1/2)[g_{tt}(dt)^2 + 2g_{tx}(udt + vdB_t)dt$$
$$+ g_{xx}(u^2(dt)^2 + 2uvdtdB_t + v^2(dB_t)^2)].$$

We want to drop all terms of order higher than 1. Evidently, the size of $dtdB_t$ is $(dt)^{3/2}$, $(E|dB_t|$ is proportional to $(dt)^{1/2})$, so the terms containing $(dt)^2$ and $dtdB_t$ can be dropped, compare the multiplication rules mentioned above. Then, (4.11) follows from these considerations, once we note that the term $(dB_t)^2$ cannot be dropped, but it can be replaced by dt, causing only a negligible errors. At least we have, (by the Appendix, (5.15)), that

$$E((dB_t)^2 - dt)^2 = E(dB_t)^4 - (dt)^2 = 3(dt)^2 - (dt)^2 = 2(dt)^2 \quad (4.14)$$

In fact, all (4.14) tells us is that it is slightly better to replace dB_t^2 by dt, rather than by 0, as $E((dB_t)^2 - dt)^2 = 2(dt)^2$, while $E((dB_t)^2 - 0)^2 = 3(dt)^2$. A better argument for this replacement can be found in the following proof of the integral formula (4.10).

Sketch of proof*

We use approximation by a second-order polynomial (Taylor's formula). Let $\{t_i\}, i = 1, \ldots, i^* - 1$, be a subdivision of the interval $[0, t]$ into intervals of equal lengths $h_i := t_{i+1} - t_i = h$, h small, $(t_0 = 0, t_{i^*} = t)$. Write $X_{t_{i+1}} - X_{t_i} = \Delta X_i \approx u(t_i)h_i + v(t_i)\Delta B_i$, where $\Delta B_i = B_{t_i + h_i} - B_{t_i}$. Then $g(t, X_t) - g(0, X_0) = \sum_{i=0}^{i^*-1} \{g(t_{i+h_i}, X_{t_{i+h_i}}) - g(t_i, X_t)\}$ is approximately equal to the sum

$$A := \sum_i \{g_t h_i + g_x \Delta X_i + g_{tt} h_i^2/2 + g_{tx} h_i \Delta X_i + g_{xx}(\Delta X_i)^2/2\}, \tag{4.15}$$

where the partial derivatives are evaluated at (t_i, X_{t_i}). The sum has t/h terms. Write $u_i = u(t_i)$, $v_i = v(t_i)$. We claim that this sum is approximately equal to

$$B := \sum_i \left[g_t h_i + g_x \Delta X_i + \frac{v_i^2 g_{xx} h_i}{2} \right], \tag{4.16}$$

the standard deviation of the error being of order $h^{1/2}$. Accepting this claim and letting $h \downarrow 0$ give (4.10) (the last sum is an approximation of the sum of the two integrals in (4.10)). To prove this claim, we shall show that $(E(A - B)^2)^{1/2}$ is small, where $A - B = \sum_i g_{tt} h_i^2/2 + g_{tx} h_i \Delta X_i + g_{xx}[(\Delta X_i)^2 - v_i^2 h_i])/2\}$. First, observe that $(E(A - B)^2)^{1/2} \leq$

$$\sum_i (E(g_{tt} h_i^2/2)^2)^{1/2} + \sum_i (E(g_{tx} h_i \Delta X_i)^2)^{1/2} + \Psi, \tag{4.17}$$

where $\Psi = (E(\sum_i g_{xx}[(\Delta X_i)^2 - v_i^2 h_i])/2)^2)^{1/2}$, (we have here used the general rule $(E(\sum_i a_i^2))^{1/2} \leq \sum_i (E a_i^2)^{1/2}$). Note that $(\Delta X_i)^2 = u_i^2 h_i^2 + 2h_i u_i v_i \Delta B_i + v_i^2 (\Delta B_i)^2$. Hence, $E(\Delta X_i)^2 = E u_i^2 h_i^2 + E(v_i^2)E(\Delta B_i)^2$. As $E(\Delta B_i)^2 = h_i$, this implies that $(E(\Delta X_i)^2)^{1/2}$ is of order $h^{1/2}$. So the second sum in (4.17) is of order $(t/h)h h^{1/2}$. A similar remark evidently also applies to the first sum. We claim that Ψ is of order $h^{1/2}$. Using the expression for ΔX_i above (and the above general rule),

$$\Psi \leq \sum_i (E(g_{xx} u_i^2 h_i^2/2)^2)^{1/2} + \sum_i (E(g_{xx} h_i u_i v_i \Delta B_i)^2)^{1/2}$$
$$+ \left(E\left(\sum_i v^2 g_{xx} \Delta B_i^2/2 - v_i^2 g_{xx} h_i/2 \right)^2 \right)^{1/2}.$$

The two first sums are of order (at least) $(t/h)h^{3/2}$, whereas the third term needs a little more consideration. Observe that

$$\left(\sum_i v_i^2 g_{xx} \Delta B_i^2 - v_i^2 g_{xx} h_i \right)^2$$
$$= \sum_{i,j} v(t_i)^2 g_{xx}(t_i, X_{t_i})(\Delta B_i^2 - h_i)v(t_j)^2 g_{xx}(t_j, X_{t_j})(\Delta B_j^2 - h_j).$$

Note that $v(t_i)^2 g_{xx}(t_i, X_{t_i})(\Delta B_i^2 - h_i) v(t_j)^2 g_{xx}(t_j, X_{t_j})$ and $\Delta B_j^2 - h_j$ are independent when $i < j$ (if $i > j$ interchange the roles of i and j below). As $E(\Delta B_j^2) = h_j$, taking the expectation of the last sum yields that all terms for which $i \neq j$ vanish. So by (4.14),

$$
E\left(\sum_i v_i^2 g_{xx} \Delta B_i^2 - v_i^2 g_{xx} h_i \right)^2 = E\left(\sum_i v_i^4 (g_{xx})^2 (\Delta B_i^2 - h_i)^2 \right) \sim (t/h) h^2,
$$

(here $\sim$ means proportionate to), and the square root of this entity is $\sim h^{1/2}$. Hence, the entity Ψ defined above is of order $h^{1/2}$. Thus, the whole expression in (4.17) is of order $h^{1/2}$.						□

Actually, a proof like this only works when all derivatives are bounded, when u and v are piecewise constant, and when the t_i's include the jump points of u and v (necessitating perhaps unequal h_i's) (recall that we wrote $\Delta X_i \approx u(t_i) h_i + v(t_i) \Delta B_i$, without discussing the error, which however is zero if u and v are piecewise constant!). The general case is proved by approximating general u, v-functions by piecewise constant ones, and a general g by one having bounded derivatives (perhaps even of the third order, if using the following comment). Note also that the error made by using the approximation (4.15) above has not been estimated. When g has bounded third-order derivatives, with bound K, the error term is of the form $\sum_i R_i$, where for each i, $|R_i|$ is smaller than the sum of four terms of the form $K h_i^{a_i} |\Delta B_i|^{b_i}$, $a_i + b_i = 3$, $a_i = 0, 1, 2, 3$. The square of the sum $\sum_i |R_i|$ contains $16(t/h)^2$ terms, each term being $\leq K^2 h_i^{a_i} |\Delta B_i|^{b_i} h_j^{a_j} |\Delta B_j|^{b_j}$. The "worst" of the last type of bounds are ones for which $b_i = b_j = 3$, but $E[|\Delta B_i|^3 |\Delta B_j|^3]$ is proportionate to h^3, by properties of the normal distribution. In fact all bounds are proportionate to h^k, $k \geq 3$, so $(E(\sum_i R_i)^2)^{1/2}$ is $\leq (K' 16(t/h)^2 h^3)^{1/2} = 4K'^{1/2} t h^{1/2}$, for some constant K'. Hence, the error is small when h is small.

Taking expectations in (4.10) gives the so-called *Dynkin's formula*:

$$
E[g(t, X_t)] = g(0, X_0)
$$
$$
+ E\left[\int_0^t \{ \partial g(s, X_s)/\partial s + u(s) \partial g(s, X_s)/\partial x + (1/2) v(s)^2 \partial^2 g(s, X_s)/\partial x^2 \} ds \right] \quad (4.18)
$$

The term in (4.10) containing dB_s disappears when taking expectation, for the following reason: Recalling our definition of the integral, consider it to be a sum of terms $a_i(B_{t_{i+h}} - B_{t_i})$. By our assumptions, each a_i depends only on values of B_s, for $s \leq t_i$, which are independent of $B_{t_{i+h}} - B_{t_i}$. So, therefore, is a_i. Hence $E[a_i(B_{t_{i+h}} - B_{t_i})] = E[a_i] E[B_{t_{i+h}} - B_{t_i}] = 0$. Note that $E[\int_0^b g(s) dB_s] = 0$, when $g(s)$ is any function adapted to B_s, ($g \in N^*$). In particular, this holds if $g(s)$ is replaced by $g(s) 1_{[0, \tau(\omega)]}(s)$, when both $g(s)$ and $1_{[0, \tau(\omega)]}(s)$ are adapted to B_s. This property of $\tau(\omega)$ is also expressed by saying that $\tau(\omega)$ is a *stopping time*. So Dynkin's formula also holds if t is replaced by a bounded stopping time $\tau(\omega)$.

Writing $z_t := E[g(t, X_t)]$ gives the differential form of Dynkin's formula, namely

$$
dz_t = E\{ \partial g(s, X_s)/\partial s + u(s)[\partial g(s, X_s)/\partial x] + (1/2) v(s)^2 [\partial^2 g(s, X_s)/\partial x^2] \} ds.
$$
$$
(4.19)
$$

Remark 4.4 (Additional conditions needed*). Actually, for the Ito and Dynkin formulas to hold, certain additional conditions may be needed. In the one-dimensional case, the condition $\infty > E[\int_0^T (v(t)g_x(t,X(t)))^2 dt]$ is needed, we need to "Ito-integrate" $v(t)g_x(t,X(t))$. However, this condition can be weakened (even essentially dropped), provided the Ito integral is extended to more general functions, see for example Yong and Zhou (1999), p. 36.

For the Dynkin formula, the last inequality *is* needed, together with the assumption that the expected value of the integral of the absolute value of the integrand in (4.18) is finite. $\qquad\square$

Example 4.5. Geometric Brownian motion. Let $g(t,x) = \exp\left[\left(r - \frac{\alpha^2}{2}\right)t + \alpha x\right]$, r, α given positive numbers, and let $X_t = B_t$ (i.e., $u = 0$, $v = 1$ in (4.9)). We want to find an expression for $Y_t := g(t,B_t) = \exp\left[\left(r - \frac{\alpha^2}{2}\right)t + \alpha B_t\right]$ by Ito's formula. It says that

$$
\begin{aligned}
Y_t &= 1 + \int_0^t \left(r - \frac{\alpha^2}{2}\right)\exp\left(\left(r - \frac{\alpha^2}{2}\right)s + \alpha B_s\right)ds \\
&+ \int_0^t \alpha\exp\left(\left(r - \frac{\alpha^2}{2}\right)s + \alpha B_s\right)dB_s \\
&+ (1/2)\int_0^t \alpha^2 \exp\left(\left(r - \frac{\alpha}{2}\right)s + \alpha B_s\right)ds \\
&= 1 + r\int_0^t Y_s ds + \alpha\int_0^t Y_s dB_s.
\end{aligned}
$$

Hence, Y_t is a solution of the integral equation $Y_t = 1 + r\int_0^t Y_s ds + \alpha\int_0^t Y_s dB_s$. This integral equation is also written $dY_t = rY_t + \alpha Y_t dB_t$. The solution Y_t is called geometric Brownian motion. We see that Ito's formula can sometimes be used to obtain solutions of such equations.

Can we find a formula for EY_t? This expression can be calculated by using that B_t is normally distributed. Let us wait for a moment with that, and let us instead use Dynkin's formula: Evidently, $x(t) := EY_t = 1 + Er\int_0^t Y_s ds = 1 + r\int_0^t x(s)ds$, so $dx/dt = rx, x(0) = 1$, hence $x(t) = e^{rt}$. Let us now use the first method mentioned. Now, $EY_t = \exp\left[\left(r - \frac{\alpha^2}{2}\right)t\right] E\exp(\alpha B_t)$. To calculate the integral yielding $E\exp(\alpha B_t)$, use the fact that $\alpha x - x^2/2t = -(x^2 - 2t\alpha x)/2t = -(x - t\alpha)^2/2t + (t\alpha)^2/2t$. One then gets that $E\exp(\alpha B_t)$ equals $e^{(t\alpha)^2/2t} = e^{\alpha^2 t/2}$ times the integral of the normal density function of $N(t\alpha, 2t)$, which of course is 1. Hence, again we get $EY_t = \exp(rt)$. $\qquad\square$

Example 4.6. "Integration by parts." Let us show that $\int_0^t F(s)dB_s = F(t)B_t - \int_0^t F'(s)B_s ds$, where F is deterministic and C^1. To verify this, apply Ito's formula to $g(t,x) = F(t)x$, with $X_t = B_t$. Hence, $F(t)B_t - F(0)B_0 = \int_0^t F'(s)B_s ds + \int_0^t F(s)dB_s$.

By the way, note that $\int_0^t F'(s)B_s ds$ is an "ordinary integral," for each outcome $b(s)$ of B_s, we can integrate $\int_0^t F'(s)b(s)ds$. $\qquad\square$

Example 4.7. Let us find an expression for $\int_0^t (B_t)^2 dB_t$. From ordinary integration by substitution, if h is any deterministic function, then $\int_0^t h^2 d(h(t)) = |_0^t \frac{h^3}{3}$, so we expect to obtain a term of the form $(B_t)^3$. Hence, it seems natural to try $Y_t = g(B_t)$ with $g = \frac{x^3}{3}$, to see how close we get to the original integral. Then, $Y_t = Y_0 + \int_0^t (B_s)^2 dB_s + \int_0^t B_s ds$, so $\int_0^t (B_s)^2 dB_s = Y_t - \int_0^t B_s ds = (B_t)^3/3 - \int_0^t B_s ds$. $\quad\square$

Remark 4.8 (Generalizing history dependence). Let us discuss a little more the term "adapted" used in connection with (4.8) above ($u(.)$ and $v(.)$ was said to be adapted). In the one-dimensional case, according to our conventions, this means adapted to (history dependent on) the given Brownian motion. In the multidimensional case, adapted means adapted to the vector Brownian motions $B_t = (B_1(t), \ldots, B_m(t))$, $\{B_i\}_i$ independent. More, generally, let $\{M_s^j\}_j$ be a finite collection of stochastic processes, which includes the one-dimensional B_s, and assume instead that $u(s)$ and $v(s)$ are history dependent on the M_s^j's, or adapted to ("the σ-field $\mathcal{F}_s$ generated by") the M_s^j's. Provided $B_t - B_s$, $s < t$, is independent of the history of the M_τ^j's, for $\tau \leq s$, the Ito and Dynkin formulas still hold, and the proofs are the same. $\quad\square$

The next, lengthy example is a famous result in mathematical finance that was awarded the Nobel (memorial) prize in economics.

Example 4.9 (The Black and Scholes formula for the value of a European call). A European call is a paper stating the right to buy a share of a stock at a fixed price K at a fixed date T in the future. We imagine that such papers are traded (as well as the stock). In addition, lending money at the riskless rate r is also possible. Thus a certain "completeness" of markets is assumed. At any time, the price $Y = Y_t$ of the call is a function $Y := C(t, S)$ of the price $S = S_t$ of the stock at that time. Evidently, at time T, $Y_T = S_T - K$ if $S_T \geq K, Y_T = 0$ otherwise (if $S_T < K$, nobody would exercise the right). Imagine that an investor at any given point in time operates with a portfolio of the stock and the call in relative proportions $-C_2 := -\partial C / \partial S < 0$ and 1 (a moment's reflection shows that $C_2 > 0$), such a portfolio is riskless, as we shall see. Here, we allow for the possibility of holding negative amounts of the stock or the call. (If, say, a negative amount of the stock is held, it means that the investor has borrowed the stock belonging to somebody else, and returns exactly the number of shares borrowed at the "end of time." Because he may be trading all the time in the stock, perhaps he may eventually have to buy some extra shares in the market to be able to return exactly the amount he borrowed. Each year the investor pays the owner an amount of money equal to the amount of dividends paid out according to the original number of shares he borrowed. No further compensation is paid to the owner of the stock.) The number of calls and shares of the stock the investor has are denoted $Q = Q_t$ and $P = P_t$, respectively, so all the time $P/Q = -C_2$, or $P = -C_2 Q$. The wealth is $Z = PS + YQ$. The portfolio is operated in a self-financing manner, which means that an increase in the number of calls is financed by the selling of some of the stock and vice versa. For the moment, let us operate in discrete time, with h being the unit of time, and let Δ indicate changes, e.g., $\Delta S = S_{t+h} - S_t$. If P and Q are the same at time t and $t + h$ then

$$Z_{t+h} = P_t S_{t+h} + Q_t Y_{t+h}. \tag{4.20}$$

It may well be that P and Q at time $t+h$ are changed by trading in the stock and the call, but then still $Z_{t+h} = P_{t+h}S_{t+h} + Q_{t+h}Y_{t+h} = P_t S_{t+h} + Q_t Y_{t+h}$, the last equality reflects the fact that no new money is infused or subtracted (the portfolio is self-financed). Moreover, with the proportions between P and Q being as proposed above, the portfolio is riskless: If S_{t+h} turns out to be $S_{t+h} + \delta$ rather than S_{t+h} (δ small), then, denoting the change in Z_{t+h} by δ'', we get, in the first order, that $Z_{t+h} + \delta'' = P_{t+h}(S_{t+h}+\delta) + Q_{t+h}(Y_{t+h}+\delta')$, where $\delta' := C(t+h, S_{t+h}+\delta) - C(t+h, S_{t+h}) = C_2(t+h, S_{t+h})\delta =: C_2\delta$.
As all the time $P_s = -C_2 Q_s$,

$$Z_{t+h} + \delta'' = -C_2 Q_{t+h}(S_{t+h} + \delta) + Q_{t+h}(Y_{t+h} + C_2\delta) = Z_{t+h},$$

so $\delta'' = 0$.

Subtracting $Z_t = P_t S_t + Q_t Y_t$ in (4.20), and letting h be small, we should reasonably get, using differensials, that

$$dZ_t = P_t dS_t + Q_t dY_t. \tag{4.21}$$

Using the proportions proposed above for P and Q, we get

$$dZ_t = -C_2(t, S_t)Q_t dS_t + Q_t dY_t. \tag{4.22}$$

Because the portfolio is riskless, it should earn the riskless rate of return (bank rate), r, hence, by (4.22), as $Z_t = -C_2 Q_t S_t + Y_t Q_t$, Z as a function of time satisfies

$$Q_t(-C_2(t, S_t)dS_t + dY_t) = dZ_t = rZ_t dt = rQ_t(-C_2(t, S_t)S_t + Y_t)dt. \tag{4.23}$$

It is assumed that the price of the stock follows a geometric Brownian motion, $dS_t = \alpha S_t dt + \sigma S_t dB_t$, α, β given constants. From (4.23), $-C_2 dS_t + dY_t = r(-C_2 S_t + Y_t)dt$, which yields $dY_t = C_2 dS_t - rC_2 S_t dt + rY_t dt = C_2(\alpha S_t dt + \sigma S_t dB_t) - rC_2 S_t dt + rY_t dt$. By Ito's formula,

$$dY_t = C_1 dt + \alpha S_t C_2 dt + \sigma S_t C_2 dB_t + C_{22}(\sigma^2 S_t^2/2)dt. \tag{4.24}$$

Equalizing the two different expressions for dY_t yields

$$C_1(t, S_t)dt = rC(t, S_t)dt - rS_t C_2(t, S_t)dt - (C_{22}(t, S_t)\sigma^2 S_t^2/2)dt. \tag{4.25}$$

Dividing by dt and considering S to be just an independent variable (as t), we obtain a partial differential equation

$$C_1 = rC - rSC_2 - C_{22}\sigma^2 S^2/2. \tag{4.26}$$

The boundary condition at $t = T$ is: $C = S - K$ if $S > K, C = 0$ otherwise. If we can find a solution $C(t, S)$ of this equation, the price process $Y_t = C(t, S_t)$ would satisfy

(4.25). In fact, the equation, with its boundary condition, can be solved explicitly. It has the famous solution:

$$C(t,S) = SN(x) - KN(x - \sigma(T-t)^{1/2}) \exp(-r(T-t)), \qquad (4.27)$$

where $x = \sigma^{-1}(T-t)^{-1/2}[\ln(S/K) + (r + \sigma^2/2)(T-t)]$, and $N(.)$ is the cumulative standard normal distribution. It may be tested that the partial differential equation (4.26) is satisfied by the solution in (4.27) (a lengthy calculation, so preferably see below), as well as the boundary condition (for the latter test, let $t \nearrow T$ in the solution). Thus the price Y_t equals $C(t,S_t)$, $C(t,S_t)$ as given in (4.27).

The time development of the price of the stock will be influenced by the stock's dividends; we imagine this to be taken care of by the above price equation of the stock. Now, the price of the call is as we know from the Black and Scholes formula a function of the stock price. Hence the price development of the call will also be influenced by these returns. But note that, say, Y_0 is not explicitly dependent on the drift coefficient α, only on the "volatility" coefficient σ.

Proof of the Black–Scholes formula*

That the formula (4.27) satisfies the boundary condition can be shown quite easily, as mentioned by letting $t \to T$ and using l'Hopital's rule for taking limit (the satisfaction of the boundary condition also follows from the calculations below). Also, by finding C_t, C_S, C_{SS}, one can test that the formula satisfies (4.26). However, the latter procedure is, as mentioned, quite tedious, so let us show the satisfaction by reducing stepwise the equation (4.26) to a simpler one, as done in A below. Actually we do it in the opposite order, i.e., we present the simple equation first. For the reader who wants to know what is going to happen (and perhaps does not want to go through all details), note that the first equation brings the normal distribution into the solution. The second and third equations have easily constructed modifications of the solution of the first one. Next, if $C(t,S)$ is the solution of the Black–Scholes equation, then $g(\tau,y) := C(T - \tau, e^y)$ satisfies the third equation (for suitable values of its coefficients). In the three equations, (4.28)–(4.30), a general initial condition appears, what remains is to calculate the solution for the specific initial (or here terminal) condition $C(T,S) = h(S) = \max\{0, S - K\}$. This is done in B below. Finally, C contains some additional results and comment.

A. Consider the partial differential equation (called "heat equation")

$$u_t = (\lambda/2)u_{xx}, \quad u(0,x) = f(x), \lambda > 0. \qquad (4.28)$$

where the function f in the boundary condition is a given continuous function such that, for some positive constants $A, B, \rho < 2$, $|f| \leq A\exp(B|x|^\rho)$. Write $k(\lambda,t,x) := \exp(-x^2/2\lambda t)/(2\pi\lambda t)^{1/2}$, $t > 0$, where $x \to k(\lambda,t,x)$ is actually the density of a stochastic variable Z that is normally distributed $Z \sim N(0, \lambda t)$.

Lemma. The equation (4.28) has the solution $u(t,x) = \int_{\mathbb{R}} f(\hat{z})k(\lambda,t,x - \hat{z})d\hat{z} = \int_{\mathbb{R}} f(x+z)k(\lambda,t,-z)dz = \int_{\mathbb{R}} f(x+z)k(\lambda,t,z)dz$, $t > 0$, and $\lim_{t \downarrow 0} u(t,x) = f(x)$. $\qquad \square$

Proof of the Lemma

First, note that the equation has the solution $k(\lambda,t,x)$. To see this, taking the logarithm of $k(\lambda,t,x)$ and differentiating gives $k_t/k = -1/2t + x^2/2\lambda t^2$ and $k_x/k = -x/\lambda t$, or $k_x = (-x/\lambda t)k$. Hence, $k_{xx} = \{-1/\lambda t + x^2/\lambda^2 t^2\}k$. Then, evidently, k satisfies (4.28). Next, functions $k(\lambda,t,x-y)$, y any given constant, satisfy (4.28), and sums $\sum_i \gamma_i k(\lambda,t,x-y_i)$ evidently also satisfy (4.28). Even infinite sums like $u(t,x)$ satisfy (4.5). (The growth condition on f ensures that the integral exists.) Now, $u(t,x) = Ef(x+(\lambda t)^{1/2}Z)$, $Z \sim N(0,1)$, so $\lim_{t\downarrow 0} u(t,x) = f(x)$ and the lemma is proved.

Let $u(t,x)$ now be the solution corresponding to $\lambda = 2$, and write $k(2,t,x) = k(t,x)$. Defining v by $u(t,x) = \exp(\alpha t + \beta x)v(t,x)$, it is easily calculated that v satisfies

$$v_t = v_{xx} + 2\beta v_x + (\beta^2 - \alpha)v, \qquad (4.29)$$

and $v(0,x) = u(0,x)\exp(-\beta x) = f(x)\exp(-\beta x)$, $(v(t,x) = e^{-\alpha t - \beta x}u(t,x))$.

Now, assume given a function $w(t,y)$ satisfying an equation of the form

$$w_t = aw_{yy} + bw_y + cw, \qquad (4.30)$$

with boundary condition $w(0,y) = f(y/m)\exp(-by/2m^2)$, $a > 0$, $m = a^{1/2}$. If we write $v(t,x) = w(t,mx)$, $(y = mx)$, we get that $v(0,x) = f(x)\exp(-bx/2m)$ and the equation $v_t = v_{xx} + v_x b/m + cv$.

Comparing this equation for v with the previous one, evidently these equations are equal if $b/m = 2\beta, c = \beta^2 - \alpha$, i.e., $\beta = b/2m, \alpha = b^2/4a - c$. Then, using $v = u\exp(-\alpha t - \beta x)$ gives that the solution $w(t,y)$ is

$$w(t,y) := \exp(-t(b^2/4a - c) - yb/2m^2)\int_{\mathbb{R}} k(t,y/m - \hat{z})f(\hat{z})d\hat{z}. \qquad (4.31)$$

If $C(t,S)$ satisfies the Black–Scholes equation and the boundary condition $C(T,S) = h(S)$, we get the following equation for $g(\tau,y) := C(T-\tau,e^y)$ (i.e., $g(T-t,\ln S) = C(t,S)$, where $\tau = T - t$ and $y = \ln S$): $g_\tau = (\sigma^2/2)g_{yy} + (r-\sigma^2/2)g_y - rg$. For this equation to be the same as the equation for w, of course $a = \sigma^2/2, b = r - \sigma^2/2$, $c = -r$, (so $c = -b - a = -b - m^2$ and $m = \sigma/2^{1/2}$). To have $w(\tau,y) = g(\tau,y)$, we must have that f satisfies $h(e^y) = g(0,y) = w(0,y) = f(y/m)\exp(-by/2m^2)$. Then $f(y/m) = h(e^y)\exp(by/2m^2)$, so for $\hat{z} = y/m$, we get $f(\hat{z}) = h(e^{m\hat{z}})\exp(b\hat{z}/2m)$. From this we get

$$g(\tau,y) = e^{-\tau(b^2/4a-c)-yb/2m^2}J \qquad (4.32)$$

where

$$J := \int_{-\infty}^{\infty} k(\tau,y/m - \hat{z})h(e^{m\hat{z}})e^{b\hat{z}/2m}d\hat{z} \qquad (4.33)$$

Substituting $u = y/m - \hat{z}$ in the integral gives

$$J = e^{by/2m^2}\left[-\int_{\infty}^{-\infty} k(\tau,u)h(e^{y-mu})e^{-bu/2m}du\right]$$

$$= e^{by/2m^2}\int_{-\infty}^{\infty} k(\tau,u)h(e^{y-mu})e^{-bu/2m}du. \tag{4.34}$$

B. For any γ'', note that

$$k(\tau,u)e^{-\gamma''u} = (1/2(\pi\tau)^{1/2})\exp(-u^2/4\tau)e^{-\gamma''u}$$

$$= e^{\tau\gamma''^2}(1/2(\pi\tau)^{1/2})\exp(-(u+2\tau\gamma'')^2/4\tau) = e^{\tau\gamma''^2}k(\tau,u+2\tau\gamma''). \tag{4.35}$$

The second equality was obtained by completing the square.

Consider now the case $h(S) = h^*(S) := \max\{0, S - K\}$. Then, in (4.33), we need only integrate over $\{\hat{z} : e^{m\hat{z}} - K \geq 0\}$. Equivalently, in (4.34) we need only integrate over $D = \{u : y - um \geq \ln K\} = \{u : u \leq \delta := (y - \ln K)/m\}$. From (4.34), for $h(.) = h^*(.)$, we then get

$$J = e^{by/2m^2}\left[\int_D k(\tau,u)e^{y-mu}e^{-bu/2m}du - \int_D k(\tau,u)Ke^{-bu/2m}du\right]$$

$$= e^{y+by/2m^2+\tau\gamma'^2}\int_D k(\tau,u+2\tau\gamma')mu - e^{by/2m^2+\tau\gamma^2}\int_D Kk(\tau,u+2\tau\gamma)du, \tag{4.36}$$

where $\gamma' = m + b/2m$, and $\gamma = b/2m$. Now, note that for any γ'', any δ,

$$\int_{-\infty}^{\delta} k(\tau,u+2\tau\gamma'')du = \int_{-\infty}^{\delta/(2\tau)^{1/2}+(2\tau)^{1/2}\gamma''} k(1,1,\bar{u})d\bar{u}, \tag{4.37}$$

where $\bar{u} = u/(2\tau)^{1/2} + (2\tau)^{1/2}\gamma''$, $(k(1,1,\bar{u})$ being actually the standard normal density). Using this we get

$$J = e^{y+by/2m^2+\tau\gamma'^2}N(\delta/(2\tau)^{1/2}+(2\tau)^{1/2}\gamma')$$

$$- e^{by/2m^2+\tau\gamma^2}N(\delta/(2\tau)^{1/2}+(2\tau)^{1/2}\gamma). \tag{4.38}$$

Using $\tau = T - t$, and noting that $c = -r, d = \sigma/2^{1/2}, b/2m = (r - \sigma^2/2)2^{-1/2}/\sigma$, $b/2m + m = (r + \sigma^2/2)2^{-1/2}/\sigma$ yield $g(\tau,y) =$

$$e^y N(\{y - \ln K + [r + \sigma^2/2]\tau\}/\sigma\tau^{1/2}) - Ke^{-r\tau}N(\{y - \ln K + [r - \sigma^2/2]\tau\}/\sigma\tau^{1/2}),$$

which gives the Black–Scholes formula (recall $S = e^y$). □

C. For later use, note that for a general $h(.)$, from (4.34), (4.35), for $\gamma'' = b/2m$, when $\bar{u} = u/(2\tau)^{1/2} + (2\tau)^{1/2}b/2m$, then $u = (2\tau)^{1/2}\bar{u} - \tau b/m$, and $y - mu = y - m(2\tau)^{1/2} + \tau b$, so, with $k(\tau,u+2\tau\gamma'')du = k(1,1,\bar{u})d\bar{u}$ as before,

$$J = e^{\tau(b/2m)^2+by/2m^2}\int_{-\infty}^{\infty} k(1,1,\bar{u})h(e^{y-m(2\tau)^{1/2}\bar{u}+b\tau})d\bar{u}. \tag{4.39}$$

One may prove that $C(t,S)$ is unique in the class of $C^{1,2}$-functions $\hat{C}(t,S)$ satisfying (4.26), the boundary condition and the growth condition $|\hat{C}(t,S)| \leq A\exp(B[\ln(1 + |S|)]^2)$ for all $S, t \in [0, T]$ (A and B some positive constants). The last function grows faster than any power of S but slower than any type of exponential growth. This result also follows from a corresponding uniqueness result holding for the simple heat equation (4.28), for which the growth condition is that the absolute value of the solution shall be bounded by $C\exp(Dx^2)$, C and D some constants. (See Steele (2001), from which also the above derivations are taken.)

Note that it can be shown that, for an arbitrary continuous function $h(.)$, if $C(t,S)$ satisfies (4.26) and $C(T,S) = h(S)$, then

$$C(t,S) = e^{-r(T-t)}E[h(Z_T^{T-t}S)], \tag{4.40}$$

where Z_s^t is the process starting at t, governed by $dZ_s = rZ_sds + \sigma Z_sdB_s$, $Z_t = 1$, a geometric Brownian motion with drift coefficient r (we had drift coefficient α in the process of S_t). See below.

Let an arbitrage mean a method of "speculation" that yields positive income at least in some situations but never a loss of money. Assuming that no arbitrage is possible in the market at hand and that $C(t,S)$ is the market price of the promise of getting $h(S)$ at time T, then a direct "stochastic" proof of the formula (4.40) can be given, not relying on the equation (4.26). (See Ross (1999), for an intuitive argument, at least in the case $h = \max(0, S - K)$, and the other books on finance in the References, for more formal arguments and for general $h(.)$.)

To prove the formula $C(t,S) = e^{-r\tau}E[h(Z_T^\tau S)]$, $\tau = T - t$, using the calculations above yields $e^{-r(T-t)}E[h(Z_T^{T-t}S)] = E[h(Se^{(r-\sigma^2/2)\tau+\sigma B_\tau})] =$

$$e^{-r\tau}\int_{\mathbb{R}}k(1,1,z)h(Se^{(r-\sigma^2/2)\tau+\sigma(\tau)^{1/2}z})dz$$

$$= e^{-r\tau}\int_{\mathbb{R}}k(1,1,z)h(e^{y+b\tau+(2\tau)^{1/2}mz})dz$$

$$= e^{-r\tau}\int_{\mathbb{R}}k(1,1,-\tilde{z})h(e^{y+b\tau-(2\tau)^{1/2}m\tilde{z}})d\tilde{z}$$

$$= e^{-r\tau}\int_{\mathbb{R}}k(1,1,\tilde{z})h(e^{y+b\tau-(2\tau)^{1/2}m\tilde{z}})d\tilde{z}$$

$$e^{-r\tau-\tau b^2/4a-yb/2m^2}J = g(\tau,y) = g(\tau,\ln S) = C(t,S)$$

(where (4.39) was used).

Finally, let us add the following observation. Let R_t be governed by $dR_t = rR_tdt, R_0 = 1$ and consider the portfolio $(C_2(t,S_t),(Y_t - C_2(t,S_t)S_t)/R_t)$, where $C_2(t,S_t)$ is the number of stocks held and $Y_t - C_2S_t$, $(C_2 = C_2(t,S_t))$, is the amount of money in the bank at time t, $(Y_t - C_2S_t)/R_t$ being this amount discounted back to $t = 0$. The value of this portfolio at time t is $V_t = C_2S_t + [(Y_t - C_2S_t)/R_t]R_t$. If this portfolio is operated in a self-financing manner (as we assume), then, similar to (4.21), the change in this portfolio will be $dV_t = C_2dS_t + [Y_t - C_2S_t]rdt$ (changes in the "proportions" $(C_2,(Y_t - C_2S_t)/R_t)$ may be disregarded, as we saw). Comparing this with $dY_t = C_2 dS_t - rC_2 S_t dt + rY_t dt$ gives that $dV_t = dY_t$, i.e., the value of this portfolio grows exactly as dY_t. Moreover $V_0 = Y_0$ so $V_t = Y_t$.

Hence, in the above perfect world, in a sense, the existence of the financial instrument European call option is superfluous. □

Girsanov's Theorem

Let $\psi(z)$ be a density on the real line, and let $f(x)$ be an increasing function on the real line. Let $Y = f(X)$. Can we find a density for X, such that Y gets the density $\psi(y)$? Well, using integration by substitution, for any $\bar{y}$, $\int_{-\infty}^{\bar{y}} \psi(y) dy = \int_{f^{-1}(-\infty)}^{f^{-1}(\bar{y})} \psi(f(x)) f'(x) dx$. So if X has density $\psi(f(x)) f'(x)$ on $(f^{-1}(-\infty), f^{-1}(\infty))$, then Y gets the density $\psi(y)$.

A similar property will be stated for Brownian motion.

Let P be the probability function (or "measure") on the canonical sample space $C_0([0,T])$, corresponding with a given real-valued Brownian process B_t. Hence, we have $P(B_t \in [a,b]) = N(0,t,[a,b]) = $ the probability that a $N(0,t)$ stochastic variable belongs to $[a,b]$. Let Y_t be an Ito process of the form $dY_t = a(t,\omega)dt + dB(t)$, $Y_0 = 0$, $t \in [0,T]$, where $a(t,\omega)$ is a bounded adapted function. Then there exists a probability function Q on $C_0([0,T])$ under which Y_t is a Brownian motion, in particular, $Q(Y_t \in [a,b]) = N(0,t,[a,b])$. (The explicit formula for Q is not given.)

This is one version of Girsanov's theorem. Let us apply this theorem to stochastic differential equations. Assume that X_t is a solution to the equation $dX_t = \sigma(X_t)dB_t, X_0$ given, where $\sigma(x)$ is Lipschitz continuous and $\sigma(x) > a > 0$ for all t. (By an existence theorem above, a solution exists.) Consider the equation $dY_t = c(Y_t)dt + \sigma(Y_t)dB_t, Y_0 = X_0$, where $c(.)$ is only continuous, or, say even piecewise continuous, so existence theorems do not apply (at least, not the single one we have presented). Now, from the preceding result we know that, for $u_t = -c(X_t)/\sigma(X_t)$, $\hat{B}_t = \int_0^t u_s ds + B_t$ is a Brownian motion with respect to some probability function Q, with $d\hat{B}_t = u_t dt + dB_t$. Then $dB_t = d\hat{B}_t - u_t dt$. Now, $dX_t = \sigma(X_t)dB_t = \sigma(X_t)(d\hat{B}_t - u_t dt) = c(X_t)dt + \sigma(X_t)d\hat{B}_t$. Thus the pair $(X_t, \hat{B}_t)$ satisfies the equation for Y_t, and such a pair, where the Brownian motion is not given in advance, but is "part of the construction," is called a weak solution pair. From the point of view of interpretations, a weak solution has the "weakness" that it is not necessarily history dependent (only) on the history of its "pair member" $\hat{B}_t$. Sometimes, however, we want to assume that the development of the Brownian motion is all we have to observe in order to explain – or in control problems to determine – what happens.

4.4 Stochastic Control

Assume now that the n-dimensional process $X_t \in \mathbb{R}^n$ is governed by the controlled stochastic differential equation

$$dX_t = f(t, X_t, u(t, X_t))dt + \sigma(t, X_t, u(t, X_t))dB_t, \quad X_0 = x^0. \tag{4.41}$$

The control $u(t, X_t)$ is a function taking values in a given set $U \subset \mathbb{R}^r$ and is subject to choice. In (4.41), f, σ, and x^0 are given entities, σ is an $n \times m$ matrix with
entries $\sigma_{i,j}(t, x, u)$, and B_t is a vector of m given independent Brownian motions.
We assume that the entries of f and σ are piecewise and right-continuous in t, and
(say) uniformly Lipschitz continuous in (x, u) on bounded subsets of $\mathbb{R}^{1+n+r}$. For
any given (s, x) in $[0, T] \times \mathbb{R}^n$, define

$$J(s, x, u) = E^{s,x}\left[\int_s^T f_0(t, X_t, u(t, X_t))dt + g(T, X_T)\right], \tag{4.42}$$

$u = u(.,.)$, where the symbol $E^{s,x}$ means the expected value arising from starting the
process (4.41) in state x at time s. Here $T > 0$, and f_0 and g are given continuous
functions. The optimization problem we shall consider is

$$\max_{u:=u(.,.)} J(0, x^0, u) \text{ subject to (4.41) and } u(t, x) \in U \text{ for all } (t, x). \tag{4.43}$$

In (4.43), we seek maximum among all controls of the form $u(t, x)$, so-called
Markov controls, taking values in U. We might allow more general control functions
in the problem (in (4.41)–(4.43)), namely functions $u_t := u(t, \omega)$ that are dependent
on past values of B_t (history dependent on B_t = adapted to B_t) (with $u(t, \omega) \in U$
for all (t, ω)), with $u(t, \omega)$ (say) piecewise and right-continuous in t. Then, (4.41)
changes to

$$dX_t = f(t, X_t, u_t)dt + \sigma(t, X_t, u_t)dB_t, \quad X_0 = x^0. \tag{4.44}$$

Moreover, in (4.4.2) $u(t, X_t)$ is then replaced by u_t, in which case $J(s, x, u_t)$ is obtained. Then the maximization problem consist of maximizing $J(0, x^0, u_t)$ over all
adapted controls u_t. Of course, if $u_t = u(t, X_t)$, where X_t satisfies (4.41), then (X_t, u_t)
satisfies (4.44). Most often it turns out that in order to achieve a maximum in the
set of adapted controls, it suffices to consider control functions of the form $u(t, X_t)$,
which evidently depends on past values of B_s only through X_t.

Define $J(s, x) = \sup_u J(s, x, u)$. Then (under certain conditions), in $(0, T) \times \mathbb{R}$,
J satisfies the following HJB (Hamilton–Jacobi–Bellman) equation in the one-
dimensional case ($n = m = 1$):

$$0 = J_t(t, x) + \max_{u \in U}\{f_0(t, x, u) + J_x(t, x)f(t, x, u) + (1/2)J_{xx}(t, x)[\sigma(t, x, u)]^2\} \tag{4.45}$$

In the multidimensional case ($n > 1$), for $(t, x) \in (0, T) \times \mathbb{R}^n$, the equation reads:

$$0 = J_t(t, x) + \max_{u \in U}\{f_0(t, x, u) + J_x(t, x)f(t, x, u)$$
$$+ \sum_{i,j}(1/2)J_{x_i x_j}(t, x)[\sigma(t, x, u)\sigma'(t, x, u)]_{i,j}\}. \tag{4.46}$$

Here $J_x = (J_{x_1}, \ldots, J_{x_n})$, $J_x f$ is a scalar product, σ' is the transpose of σ and $[\ ]_{i,j}$
indicate entries. The following boundary condition is evidently satisfied

$$J(T,x) = g(T,x). \tag{4.47}$$

It will be needed in order to obtain a unique solution of the HJB equation. We present more precise conditions implying (4.45), (4.46), after working out the following heuristic argument for (4.45).

Let $n = m = 1$. Consider the discrete time process X_t, $t = 0h, 1h, 2h, \ldots, i^*h = T$, h small, governed by

$$X_{t+h} = X_t + f\big(t, X_t, u(t, X_t)\big)h + \sigma(t, X_t, u(t, X_t))(B_{t+h} - B_t)$$

with criterion function

$$\tilde{J}(s, x, u) = E^{s,x}\Big[\sum_{T > t \geq s} f_0(t, X_t, u(t, X_t))h + g(T, X_T)\Big]$$

($s = i'h$ for some i', $t = ih$, i running through $i', i' + 1, \ldots, i^* - 1$), the superscript s, x indicates that the discrete process starts at (s, x). This is an approximation to the problem above. Define the optimal value function $\tilde{J}(s, x) = \sup_u \tilde{J}(s, x, u)$. Then the discrete time dynamic programming equation of Chapter 1 implies that (for any $t = ih$):

$$\tilde{J}(t,x) = \max_{u \in U}\{f_0(t, x, u)h + E^{t,x}[\tilde{J}(t + h, X^u_{t+h})]\},$$

where $X^u_{t+h} = x + f(t, x, u)h + \sigma(t, x, u)(B_{t+h} - B_t)$. Rearranging gives

$$0 = \max_u\{f_0(t, x, u)h + E^{t,x}[\tilde{J}(t + h, X^u_{t+h}) - \tilde{J}(t, x)]\}. \tag{4.48}$$

Writing J instead of $\tilde{J}$, and imagining now that X^u_s is the solution of the stochastic differential equation (4.41) on $[t, t + h]$ starting at (t, x), when using the constant control u on $[t, t + h]$, then Dynkin's formula gives, in a shorthand notation,

$$E^{t,x}[J(t + h, X^u_{t+h}) - J(t, x)]$$
$$= E^{t,x}\Big[\int_t^{t+h} J_t \, ds + \int_t^{t+h} J_x f \, ds + (1/2)\int_t^{t+h} J_{xx}\sigma^2 \, ds\Big].$$

Inserting this expression, and dividing by h, gives

$$0 = \max_u\Big\{f_0(t, x, u) + (1/h)E^{t,x}\Big[\int_t^{t+h} J_t + J_x f + (1/2)J_{xx}\sigma^2 \, ds\Big]\Big\},$$

where the entities in the integrand depend on (s, X^u_s). Denote the integrand by $\beta(s, X_s)$. When h is small, the stochastic variable $\beta(s, X^u_s)$, $s \in [t, t + h]$, is very close to the deterministic value $\beta(t, x)$, (recall that $X^u_s \to x$ as $s \to t$). So approximately, for h small,

$$(1/h)E^{t,x}\int_t^{t+h}\beta(s,X_s)ds \simeq (1/h)\int_t^{t+h}\beta(t,x)ds = \beta(t,x)$$

$$= J_t(t,x) + J_x(t,x)f(t,x,u) + (1/2)J_{xx}(t,x)\sigma(t,x,u)^2.$$

Hence, we get (4.45). Thus, a heuristic proof has been given for (a) in the theorem that follows now.

Theorem 4.10 (Necessary and sufficient conditions). *(a) (Necessary Conditions) Let J be twice continuously differentiable with respect to all variables in $(0,T) \times \mathbb{R}^n$ and be continuous on $[0,T] \times \mathbb{R}^n$. Then J satisfies the HJB equation (4.46) in $(0,T) \times \mathbb{R}^n$, together with $J(T,x) = g(T,x)$. Moreover, if $u^*(t,x)$ is an optimal control, then $u^*(t,x)$ maximizes the right-hand side of (4.46).*
(b) (Sufficient Conditions) Let $\hat{J}$ be a function that is continuous in $[0,T] \times \mathbb{R}^n$, is twice continuously differentiable in $(0,T) \times \mathbb{R}^n$, and satisfies (4.46) in $(0,T) \times \mathbb{R}^n$, together with the boundary condition $\hat{J}(T,x) = g(T,x)$ for all x. Assume that for every pair (t,x), $u^0(t,x)$ is the value of $u \in U$ that yields the maximum in the right hand side of (4.46). Then $u^0(.,.)$ is optimal. ☐

We have given a heuristic argument for (a) for $n = 1$, but no formal proof. Let us prove (b) for $n = 1$ using ideas that are more rigorous (or can easily be made so), using Dynkin's formula. (For (b) we should have added the assumption that $u^0(t,x)$ is a Markov control for which a solution of (4.41) exists for $u = u^0$.)

Proof. Let $u(t,x)$ be an arbitrary Markov control, to which there corresponds a solution X_t, and let X_t^0 correspond with $u^0(t,x)$. Let $(A^u\hat{J})(t,x)$ be the expression defined by

$$(A^u\hat{J})(t,x) = \hat{J}_t dt + \hat{J}_x f(t,x,u) + \sum_{i,j}(1/2)\hat{J}_{x_ix_j}[\sigma(t,x,u)\sigma'(t,x,u)]_{i,j}.$$

From (4.46), it follows that $(A^{u(s,X_s)}\hat{J})(s,X_s) \leq -f_0(s,X_s,u(s,X_s))$, with equality if $u = u^0$. Hence, using also the multi-dimensional version of Dynkins's formula (4.18),

$$E[\hat{J}(T,X_T)] = \hat{J}(0,X_0) + E\left[\int_0^T (A^{u(s,X_s)}\hat{J})(s,X_s)\,ds\right]$$

$$\leq \hat{J}(0,X_0) - E\left[\int_0^T f_0(s,X_s,u(s,X_s))ds\right] \qquad (4.49)$$

with equality if $u = u^0$, $X_s = X_s^0$. So, using $g(T,X_T) = \hat{J}(T,X_T)$, (the boundary condition), we get

$$\hat{J}(0,X_0) \geq E\left[g(T,X_T) + \int_0^T f_0(s,X_s,u(s,X_s))ds\right], \qquad (4.50)$$

with equality if $u = u^0$. Hence u^0 is optimal. ☐

Some four pages further on, examples showing applications of Theorem 4.10 appear.

Remark 4.11 (Solutions restricted to a subset*). Sometimes, an open set G^* in $\mathbb{R}^{n+1}$ is given, such that all solutions are required to belong to G^*, $((t,x(t,\omega)) \in G^*$, for $t \in (0,T)$, a.s.). Then if in (b) in Theorem 4.10, we only require that the HJB equation is satisfied in G^*, that $\hat{J}$ is C^2 in G^* and is continuous on $\text{cl}G^*$, that the boundary condition (4.47) holds for $(T,x) \in \text{cl}G^*$ and the solution $x^0(t,\omega)$ corresponding with $u^0(t,x)$ belongs to G^* for $t \in (0,T)$, then optimality of $u^0(.,.)$ follows again. $\square$

Remark 4.12 (Existence of a unique solutions of the HJB equation**). Let us mention that if U is compact, f,σ,f_0,g are C^3, G^* is bounded and ∂G^* is suitably smooth (defined by C^3-functions), and, in the one-dimensional case, $0 < \alpha \leq \sigma$, α independent of t,u, and x, then an existence theorem (see Fleming and Soner (1993)) ensures that a unique $C^{1,2}$-solution of the HJB equation exists. (In the multidimensional case, what is called a uniform parabolicity assumption is required, replacing the last inequality.) A Markov control $u(t,x)$ exists that yields maximum in the HJB equation, and if a corresponding solution to (4.41) exists, then $u(t,x)$ is optimal. (Unfortunately, in the general case, $u(t,x)$ can be quite "nonsmooth," measurable only.) $\square$

Remark 4.13 (T a first exit time*). Consider the case where the terminal time T is not fixed, but is a "first exit time" of a given open set G in R^{n+1} $((t,x)$-space), i.e., $T = T^{0,X_0,u}$, where, for $(s,x) \in G$, $T^{s,x,u} := \inf\{t : (t,X_t^{s,x,u}) \notin G\}$, $X_t^{s,x,u}$ being the solution starting at (s,x), corresponding with the control $u = u(.,.)$. We assume that $T^{s,x,u}$, so determined, is always $\leq T'$, for some given $T' > 0$. We assume also $(0,X_0) \in G$.

Then the following boundary condition is needed:

$$J(t,x) = g(t,x) \quad \text{for} \quad (x,t) \in \partial G, t > 0, \tag{4.51}$$

where ∂G is the boundary of G. Then, in (a) in Theorem 4.10, J satisfies the HJB equation in G, provided it is C^2 here, and the condition (4.43) should be replaced by condition (4.51) (though, strictly speaking only a slight modification of this condition is necessary, see Øksendal (2003)). In (b) replace (4.47) by $\hat{J}(t,x) = g(t,x)$ for $(t,x) \in \partial G, t > 0$ ($\hat{J}$ need only be C^2 and satisfy the HJB equation in G, but $\hat{J}$ need to be continuous on $\text{cl}G$). $\square$

Remark 4.14 (Assumptions on the controls*). In Theorem 4.10, no assumptions on the controls $u^*(.,.)$ and $u^0(.,.)$ (and on $u(.,.)$ in the proof of (b)) were specified, making the statements slightly imprecise. For (b) (the most important statement for us in this context), we can remedy this weakness somewhat by assuming of $u(t,x)$ in the proof and $u^0(t,x)$ of the theorem that they belong to the set $\mathcal{U}'$ defined as follows: The set $\mathcal{U}'$ consists of functions $u(t,x)$ for which there exists an adapted function $X_t := x(t,\omega)$, continuous in t, that satisfy the differential equation (4.44) on $[0,T]$ for $u_t := u(t,x(t,\omega))$ (as well as $E[\int_0^T |f(t,X_t,u_t)|dt] < \infty$ and $E[\int_0^T \sigma(t,X_t,u_t)^2 dt] < \infty$), $u(t,x(t,\omega))$ being assumed to be piecewise and right-continuous in t and adapted. (Usually, one also wants to say something about "well-behavior" conditions on $u(t,x)$ and $u^0(t,x)$, let us choose, say, the conditions piecewise continuity in t and in each component of x, separately. Then $u(t,x(t,\omega))$ is automatically adapted.) Moreover, for $u(.,.)$ to belong to $\mathcal{U}'$, two additional conditions are placed on $u(t,x(t,\omega))$, namely that $E \int_0^T |f_0(t,x(t,\omega),u(t,x(t,\omega)))| dt < \infty$, $E|g(T,x(t,\omega))|| < \infty$. As said above, we require $u^0(.,.) \in \mathcal{U}'$. Then Theorem 4.10 (b) essentially gives sufficient conditions for optimality in $\mathcal{U}'$. (For (a), we assume that $u^*(.,.)$ belongs to $\mathcal{U}'$.) See Remark 4.17 below for a further discussion of this point.

The roundabout method of definition of $\mathcal{U}'$ is chosen in order to allow relatively weak conditions to be placed upon $u(t,x)$. It may for example be discontinuous in x, but still, incidentally, allow a strong solution to exist. Be aware of the fact that for such controls sometimes a strong solution may fail to exists. (See Karatzas and Shreve (1988), p. 302.)

Define $U^{\hat{J}}$ to be the subset of controls in $\mathcal{U}'$ such that $E[\sup_t |\hat{J}(t,X_t^u)|] < \infty$. Let us mention that a more precise version of the sufficient conditions in (b) would contain an additional assumption, namely that $u^0(.,.) \in \mathcal{U}^{\hat{J}}$, and the statement that $u^0(.,.)$ is optimal in $\mathcal{U}^{\hat{J}}$.

A more precise proof would namely require T in (4.49) to be replaced by $T_{B(0,m)}^u = \inf\{t \leq T - 1/m : (t,X_t^u) \notin B(0,m)\}$ (if the set it empty, let $T_{B(0,m)}^u = T - 1/m$). This replacement is made to be sure that Dynkin's formula applies, which in this case yields, in analogy with (4.49),

$$\hat{J}(0,X_0) \geq E[\hat{J}(T_{B(0,m)}^u, X_{T_{B(0,m)}^u}) + \int_0^{T_{B(0,m)}^u} f_0(s,X_s,u(s,X_s))ds].$$

Then one lets $m \to \infty$ to obtain (4.50) as written, using $\hat{J}(T_{B(0,m)}^u, X_{T_{B(0,m)}^u}) \to \hat{J}(T,X_T)$ and the inequalities involved in the definitions of $\mathcal{U}'$ and $\mathcal{U}^{\hat{J}}$. For $u = u^0$, equality in (4.50) follows in the same way. $\square$

Infinite Horizon

Suppose we replace T in (4.42) by τ, with $g \equiv 0$, and let τ be an explicit argument of $J = J(\tau;s,x,u)$. Assume that we have a problem where $T = \infty$ (infinite horizon), in which case we are going to find $\max_u(.,.)J(\infty;0,x_0,u)$. This may work fine if $J(\infty;0,x_0,u)$ is finite for all u, if not it may be necessary to replace the optimality condition by another one. The control u^0 is said to be *(regularly) catching up* optimal in the infinite horizon problem if

$$\liminf_{\tau \to \infty}\{J(\tau;0,x_0,u^0) - J(\tau;0,x_0,u)\} \geq 0 \text{ for all } u(.,.). \qquad (4.52)$$

Here, only a sufficient condition for such optimality will be stated. The following inequality is needed: For any $u(.,.)$, for X_t corresponding with $u(.,.)$,

$$\liminf_{T \to \infty}\{E\hat{J}(T,X_T) - E(\hat{J}(T,X_T^0))\} \geq 0. \qquad (4.53)$$

Theorem 4.15 (Sufficient conditions, infinite horizon). *Let $\hat{J}$ be a twice continuously differentiable function in $(0, \infty) \times \mathbb{R}^n$, defined and continuous in $[0, \infty) \times \mathbb{R}^n$, satisfying (4.46) for $t > 0$. Assume that for every pair (t, x), $u^0(t, x)$ is the value of $u \in U$ that yields the maximum in the HJB equation (4.46), that a function X_t^0 corresponding with $u^0(t, x)$ exists, satisfying (4.44) for $u_t = u^0(t, X_t^0)$ (assumed to belong to $\mathcal{U}'$), and such that $E \int_0^T |f_0(s, X_s^0, u^0(s, X_s^0))| ds < \infty$ for all T. Then $u^0(.,.)$ is catching up optimal in the set $\mathcal{U}''$ of all u in $\mathcal{U}'$ for which $E \int_0^T |f_0(s, X_s^u, u(s, X_s^u))| ds < \infty$ for all T, provided $E[\sup_{t \in [0,T]} |\hat{J}(t, X_t^u)|] < \infty$ for all T, all $u \in \mathcal{U}''$.* $\qquad\square$

To prove this result, using (4.49) for arbitrary u and for $u = u^0$ (in which case (4.49) is an equality), yields

$$E \int_0^T f_0(s, X_s^0, u^0(s, X_s^0)) ds - E \int_0^T f_0(s, X_s, u(s, X_s)) ds$$
$$\geq E\hat{J}(T, X_T) - E\hat{J}(T, X_T^0). \tag{4.54}$$

It follows that if (4.53) holds, then u^0 is catching up optimal. (The more precise argument for (4.49) in Remark 4.14 can be carried over to yield a more precise proof in the present setting.)

If we use a weaker optimality criterion, obtained by replacing lim inf by lim sup in (4.52) (getting "sporadically catching up optimality"), then liminf should also be replaced by limsup in (4.53).

To obtain (4.53), one often tries to show the stronger conditions that

$$(a) \lim_{T \to \infty} E\hat{J}(T, X_T^0) = 0, \quad (b) \lim_{T \to \infty} E\hat{J}(T, X_T) \geq 0 \text{ for all admissible } X_t. \tag{4.55}$$

If admissibility imposes additional constraints on the controls (besides $u(t, x) \in U$), and assuming u^0 to be admissible, then (4.53) need only hold for such admissible u.

Autonomous Systems

Suppose now that f and σ in (4.41) are independent of t and that $f_0 = e^{-\rho t} f^0(x, u)$, $\rho \geq 0$ (formally ρ can even be negative). Such a system is called autonomous. In such an infinite horizon system, it can be expected that $J(t, x)$ is of the form $e^{-\rho t} h(x)$. The intuitive argument is the following. Let $h(x) = J(0, x)$. Then, note that the future looks the same whether we are at time $s = 0$ or $s > 0$, at least if we in the criterion in both cases discount back to the start point s (in both cases, we integrate over $[s, \infty)$), so the two optimal criterion values should then be the same. However, when $J(s, x)$ is calculated and $s > 0$, we discount back to $t = 0$, not to $t = s$, so it should then be clear that $J(t, x) = e^{-\rho t} J(0, x)$. If $e^{-\rho t} h(x)$ is inserted in the HJB equation (4.45), it becomes:

$$0 = -\rho e^{-\rho t} h + \max_u \{e^{-\rho t} f^0 + e^{-\rho t} h' f + e^{-\rho t} \sigma^2 h''/2\},$$

which gives, in this one-dimensional case, the following version of the HJB equation

$$0 = -\rho h(x) + \max_{u}\{f^0(x,u) + h'(x)f(x,u) + \sigma^2(x,u)h''(x)/2\}. \qquad (4.56)$$

This is sometimes called the *current value* form of the HJB equation (or the HJB equation with discounting). It is valid only if J has the above particular form. In the multidimensional case, the term $\sigma^2(x,u)h''(x)/2$ in (4.56) is replaced by $\sum_{i,j}(1/2)h_{x_ix_j}(x)[\sigma(x,u)\sigma'(x,u)]_{i,j}$. If $\hat{h}$ satisfies (4.56), then $\hat{J} = e^{-\rho t}\hat{h}$ satisfies (4.45).

Note that (4.56) is an ordinary differential equation.

Remark 4.16 (Existence of solutions when the horizon is finite*). Denote the set of controls $u(t,x)$ that are piecewise continuous in t and Lipschitz continuous in x with a Lipschitz constant independent of t by $\mathcal{U}'''$. If f and σ are continuous functions, with the same Lipschitz property as $u(t,x)$, then Theorem 4.1 (and the subsequent Remark) secures existence of solutions corresponding with each $u(t,x) \in \mathcal{U}'''$. Surely, if $u^0(.,.) \in \mathcal{U}'''$, then Theorems 4.10 and 4.15 yield sufficient conditions for $u^0(,.)$ to be optimal in this set. (We have $\mathcal{U}''' \subset \mathcal{U}'$, the latter set being defined in Remark 4.14.) However, stochastic optimization may require controls that are discontinuous in x. In specific examples, with specific forms of such discontinuity in any given control u, it may be possible to show that solutions to (4.41) do exist. And, as commented upon also in a previous remark, the sufficient conditions above yields optimality in $\mathcal{U}'$ (or more precisely in $\mathcal{U}^{\hat{J}}$). But general theorems on existence of (strong) solutions to (4.41) for Markov controls not in $\mathcal{U}'''$ are hard to obtain, as we cannot use Lipschitz continuity in x when $u(t,x)$ is inserted, so it is difficult to tell how much larger $\mathcal{U}'$ is compared with $\mathcal{U}'''$.

Sometimes solutions in a weaker sense can be shown to exist, as mentioned in connection with Girsanov's theorem above, but here we cannot pursue this issue any further. $\square$

Remark 4.17 (Sufficiency among adapted controls when the horizon is finite*). We can relate the above sufficient finite horizon conditions also to the following control problem. Let $\mathcal{U}^*$ be the class of controls consisting of all adapted control functions $u = u_t$ (piecewise and right continuous in t), with values in U, for which (4.44) have unique solutions X_t^u on $[0,T]$ satisfying $E[\int_0^T |f(t,X_t^u,u_t)|dt] < \infty$ and $E[\int_0^T \sigma(t,X_t^u,u_t)^2dt] < \infty$. Note that if f and σ are piecewise and right-continuous in t, continuous in u, Lipschitz continuous in x for a constant independent of t,u, and if, for any given $u_t \in \mathcal{U}^*$, $f(t,0,u_t)$ and $\sigma(t,0,u_t)$ are bounded by a deterministic constant independent of t,ω, then for any $u_t \in \mathcal{U}^*$, by Remark 4.2, (4.44) has a unique solution. If $E[\int_0^T \{|f_0(t,X_t^u,u_t) + |g(T,X_T^u)|\}] < \infty$, and $E[\sup_{t\in[0,T]} |\hat{J}(t,X_t^u)|] < \infty$ for all u_t in $\mathcal{U}^*$, let $\mathcal{U} = \mathcal{U}^*$. If the two last inequalities do not hold, let $\mathcal{U}$ be the subset of controls in $\mathcal{U}^*$, for which the first inequality is satisfied. Furthermore, let U_f to be the subset of controls in $\mathcal{U}$ such that $E[\sup_t |\hat{J}(t,X_t^u)|] < \infty$. The sufficient conditions in Theorem 4.10 ensure optimality of $u^0(t,X_t^0)$ in U_f, provided X_t^0 satisfies (4.44) for $u_t = u^0(t,X_t^0)$ and $u^0(t,X_t^0)$ belongs to U_f. Let us state this more formally: Consider the problem:

$$\max_{u_t \in \mathcal{U}} E\left[\int_0^T f_0(t, X_t, u_t)dt + g(T, X_T)\right], \text{ subject to (4.44).} \qquad (4.57)$$

Then assume that functions $u^0(t, x)$ and $x^0(t, \omega)$ exist, such that $x^0(t, \omega) = x^0(t, B_{\to t})$ is a unique solution – continuous in t – of (4.44) for $u = u^*(t, \omega) := u^0(t, x^0(t, \omega)) \in U_f$ (this function being by assumption piecewise and right continuous in t and adapted in ω), and such that $u^0(t, x)$ yields maximum in the HJB equation for $\hat{J}$ ($\hat{J}$ satisfying (4.46) and the boundary condition, C^2 in $(0, T) \times \mathbb{R}^n$, continuous in $[0, T] \times \mathbb{R}^n$). Then $u^*(t, \omega)$ is optimal in U_f. (In fact the sufficiency proof presented above for Theorem 4.10 (b) also yields this optimality.)

A similar results holds in the infinite horizon case. $\square$

Example 4.18. Consider the following problem:

$$\max_u E\left[\int_0^T -u^2 dt - x(T)^2\right], \qquad u = u(t) \in \mathbb{R}$$

subject to $dX_t = udt + \sigma dB_t$, $X_0 = x^0$, with σ a fixed number > 0, x^0 given. The HJB equation becomes

$$0 = J_t + \max_u\left\{-u^2 + J_x u + \frac{J_{xx}\sigma^2}{2}\right\}.$$

Carrying out the maximization gives $u = J_x/2$. Inserting this into the HJB equation gives

$$0 = J_t + \frac{J_x^2}{4} + \frac{J_{xx}\sigma^2}{2},$$

with $J(T, x) = -x^2$. This is a second-order nonlinear partial differential equation.

Somehow, we get the idea that perhaps a function of the form $\hat{J} = J = f(t)x^2 + g(t)$ might be a solution of the equation. Because $J(T, x) = -x^2$, then $f(T) = -1$ and $g(T) = 0$. Insert this J in the HJB equation to get $0 = (f' + f^2)x^2 + g' + f\sigma^2$, from which we try to determine f and g. Because the equation has to be satisfied for all x, including $x = 0$, $g' + f\sigma^2 = 0$, and hence also $f' + f^2 = 0$ (obtained for $x \neq 0$). Solving the last equation gives $1/f = t + C$. Because $f(T) = -1$, it follows that $f = \frac{1}{[t-T-1]}$. Integrating $g' = -f\sigma^2$ gives $g = -\sigma^2 \ln|t - T - 1| + C'$, where $C' = 0$, because $g(T) = 0$. Note that $f(t)$ is negative and decreasing and that $u = f(t)x$. (The "reason" why f is decreasing is that for small t, there is ample time left to reach a "reasonable" value of $X(T)$, whereas for t near T, a larger u is needed to reach a "reasonable" value of $X(T)$.) Note that $\sigma = 0$ gives exactly the same feedback control $u = f(t)x$ (σ does not occur in f). This is an example of the following *certainty equivalence principle* (which generally only holds in such linear-quadratic problems): "in the equations, replace each stochastic variable by its expectation and then solve the control problem." Here this means the claim that σdB_t can be replaced by its expectation ($= 0$), or equivalently, by letting $\sigma = 0$, in the equation for dX_t. Because the optimal $u = f(t)x$ does not depend on σ, $u = f(t)x$ also holds for $\sigma = 0$.

In case $\sigma = 0$, solving $dx/dt = f(t)x$ gives $x = C''(T - t + 1)$, where C'' is determined by $x_0 = C''(T + 1)$, so $C'' = x_0/(T + 1)$. So the optimal $dx/dt \ (= f(t)x)$ is actually a constant $(= -C'')$, which we also get when applying the maximum principle to the deterministic problem where $\sigma = 0$.   □

The next example stems from the theory of mathematical finance.

Example 4.19. A person's portfolio consists of one risk-free asset and one risky asset. The prices of the two assets evolve according to the equations $dp_1 = p_1 r dt$ and $dp_2 = p_2(\alpha dt + \sigma dB_t)$, respectively, $0 < r < \alpha$, r, α, σ given positive numbers. The wealth X_t equals $V_t p_1(t) + W_t p_2(t)$, where V_t is the number of risk-free assets that are held, and W_t is the number of risky assets kept. A change ΔV_t in V_t is compensated by a change ΔW_t in W_t in such a manner that $p_1 \Delta V_t = -p_2 \Delta W_t$, (so-called self-financing). We now calculate the change in X_t in a tentative manner as follows : $\Delta X_t := X_{t+\Delta t} - X_t$ equals $V_t(p_1(t + \Delta t) - p_1(t)) + [(V_{t+\Delta t} - V_t)p_1(t + \Delta t)] + W_t(p_2(t + \Delta t) - p_2(t)) + [(W_{t+\Delta t} - W_t)p_2(t + \Delta t)]$. Imagining that the changes from V_t to $V_{t+\Delta t}$ and W_t to $W_{t+\Delta t}$ take place at time $t + \Delta t$, and to prices $p_1(t + \Delta t), p_2(t + \Delta t)$, gives, by the self-financing property just discussed, that the sum of the terms in square brackets vanish. (Admittedly, this is a weak spot in the exposition!) Then, defining u_t to be the fraction of wealth held in the risky asset $(= W_t p_2(t)/X_t)$, we get $W_t p_2(t) = u_t X_t$, and $V_t p_1(t) = (1 - u_t)X_t$. Using the (reduced) expression for ΔX_t above and $p_1(t + \Delta t) - p_1(t) = p_1 r \Delta t$, $p_2(t + \Delta t) - p_2(t) = p_2(\alpha \Delta t + \sigma \Delta B_t), (\Delta B_t := B_{t+\Delta t} - B_t)$, we get $\Delta X_t = (1 - u_t) X_t r \Delta t + u_t X_t(\alpha \Delta t + \sigma \Delta B_t)$. The continuous time version of the last equation is $dX_t = (1 - u_t)X_t r dt + u_t X_t(\alpha dt + \sigma dB_t)$. If we add the fact that some of the wealth is consumed, then the wealth X_t is governed by the equation

$$dX_t = (1 - u)X_t r dt + u X_t(\alpha dt + \sigma dB_t) - C dt, \quad X_0 = x^0, \ x^0 > 0, \ x^0 \text{ given,}$$

where $C \geq 0$ is the rate of consumption per time unit.

The control problem is to maximize

$$E\left[\int_0^T C^\gamma dt + \theta(X_T)^\gamma \right]$$

with $T > 0$, $\theta > 0$, $\gamma \in (0, 1)$, T, θ, γ fixed, $(u = u(.,.)) \in [0, 1]$, $C = C(.,.) \geq 0)$. We have also the constraint $X_t > 0$ for all t. The HJB equation for the optimal value function $J(t, x)$ is

$$0 = J_t + \max_{u,C}\left[C^\gamma + J_x\{(1 - u)xr + ux\alpha - C\} + \frac{J_{xx}(ux\sigma)^2}{2} \right]$$

with $J(T, x) = \theta x^\gamma$. Ignore for the moment the constraints on u, C, and X. First-order conditions for (interior) maximum give

$$u = \frac{-(\alpha - r)J_x}{\sigma^2 x J_{xx}}, \quad C = (J_x/\gamma)^{1/(\gamma - 1)}.$$

These solutions are inserted into the HJB equation, which then becomes the nonlinear second-order partial differential equation:

$$0 = J_t + (J_x/\gamma)^{\gamma/(\gamma-1)} + J_x x r - \frac{(\alpha-r)^2 J_x^2}{2\sigma^2 J_{xx}} - (1/\gamma)^{1/(\gamma-1)} (J_x)^{\gamma/(\gamma-1)}.$$

Let us try a solution of the form $\hat{J} = \phi(t)x^\gamma$. Calculating its derivatives and inserting these into the last equation, gives

$$0 = \phi' x^\gamma + \phi^{\gamma/(\gamma-1)} x^\gamma + \phi\gamma r x^\gamma - \frac{(\alpha-r)^2 (\phi\gamma x^{\gamma-1})^2}{2\sigma^2 \gamma(\gamma-1)\phi x^{\gamma-2}}$$
$$- (1/\gamma)^{1/(\gamma-1)} \phi^{\gamma/(\gamma-1)} \gamma^{\gamma/(\gamma-1)} x^\gamma.$$

Dividing by x^γ then yields:

$$0 = \phi' + (1-\gamma)\phi^{\gamma/(\gamma-1)} + \gamma r\phi - \gamma(\alpha-r)^2 \phi/2\sigma^2(\gamma-1).$$

Put $\phi = \psi^{1-\gamma}$ (i.e., $\psi := \phi^{1/(1-\gamma)}$), and let us derive a differential equation for ψ. Inserting the expression for $\phi' = (1-\gamma)\psi^{-\gamma}\psi'$ into the above equation for ϕ, then using $\psi = \phi^{1/(1-\gamma)}$, we get $0 = \psi'(1-\gamma)\psi^{-\gamma} + (1-\gamma)\psi^{-\gamma} + \gamma r\psi^{1-\gamma} - \gamma(\alpha - r)^2 \psi^{1-\gamma}/2\sigma^2(\gamma-1)$. Multiplying by ψ^γ, and writing $\beta := r + (\alpha-r)^2/2\sigma^2(1-\gamma)$, we get $0 = (1-\gamma)\psi' + (1-\gamma) + \gamma\beta\psi$, or $\psi' = \delta\psi - 1$, $\delta := \gamma\beta/(\gamma-1)$. Because $\hat{J}(T,x) = \theta x^\gamma$, $\phi(T) = \theta$ and so $\psi(T) = \mu := \theta^{1/(1-\gamma)}$, the solution is $\psi(t) = (\mu - 1/\delta)e^{\delta(t-T)} + 1/\delta$, so $\phi(t) = ((\mu - 1/\delta)e^{\delta(t-T)} + 1/\delta)^{1-\gamma}$ and, for this ϕ, $\hat{J} = \phi(t)x^\gamma$, $C = x/\psi(t)$ and $u = -(\alpha-r)\gamma x^{\gamma-1}/\sigma^2 x\gamma(\gamma-1)x^{\gamma-2} = (\alpha-r)/\sigma^2(1-\gamma)$. Provided that this constant belongs to $(0,1)$, we have got a proper proposal for a solution. Otherwise, we have a corner solution, with either $u = 0$ or $u = 1$, C still given by $C = x/\psi(t)$.

One final comment is needed. Let $G^* = (-\infty,\infty) \times \{x \in \mathbb{R} : x > 0\}$. As we can apply Remark 4.11, this takes care of the above restriction $X_t > 0$ (x^γ is only defined for $x \geq 0$). Because C is linear in x, the solution X_t corresponding to the controls u and C stays positive all the time, and so C is positive.   □

One specific example where the sufficient conditions (and necessary conditions!) in the theorem cannot be used, is mentioned at the end of the next section, in that example the optimal value function is C^2 in $(0,T) \times R$, but not continuous on $[0,T] \times R$. In such cases, sometimes it works to replace (4.47) by (4.75) below (which is allowed).

Example 4.20. Consider the infinite horizon problem

$$\max E\left[-\int_0^\infty e^{-\rho t}(ax^2 + u^2)dt \right], \quad a > 0, \ \rho > 0, \ a, \ \rho \text{ given},$$

subject to $dX_t = u dt + \gamma X_t dB_t$, $X_0 = 1$, γ given. This is an autonomous problem, so let us write down the current value HJB equation (4.56) for this example: $0 = -\rho h + \max\{-ax^2 - u^2 + uh' + (\gamma^2 x^2/2)h''\}$. Carrying out the maximization gives $u = h'/2$, so the equation becomes $0 = -\rho h - ax^2 + h'^2/4 + (\gamma^2 x^2/2)h''$. This is a second-order differential equation in $h(x)$. In principle, we should now proceed by writing down the general solution, let us call it $h(x,C_1,C_2)$, of this equation (it would of course contain two integration constants). Next, we then should look for values of the integration constants for which the limit condition (4.53) is satisfied. Such a procedure is indicated in the solution of Exercise 4.12 below. However, in the current example, the general formula for $h(x,C_1,C_2)$ is difficult to find. Instead, let us gamble and try a solution of the form kx^2 in the HJB equation. It then becomes $0 = g(k)x^2$, where $g(k) := (\gamma^2 - \rho)k + k^2 - a$. The former equality must be satisfied for all x, so $g(k) = 0$, and this equation has two roots $k = \frac{1}{2}\{-(\gamma^2 - \rho) \pm [(\gamma^2 - \rho)^2 + 4a]^{1/2}\}$. As $g(0) = -a$, one root is positive (resulting from using the plus sign in front of the square root), and the other one is negative. The negative value of k is correct because the criterion is nonpositive. (As a hint for solutions of other problems, note that the following calculations would also reveal that we need the negative value.) Notice that $u = kx$. To check (4.55), (a), we need to calculate $z_t := E(X_t^0)^2$. This can be done, using Dynkin's formula. Using the differential form of this formula, $dz_t = 2E[X_t^0 u]dt + E(\gamma X_t^0)^2 dt = 2kE(X_t^0)^2 dt + \gamma^2 E(X_t^0)^2 dt = (2k + \gamma^2)z_t dt$. Using $z_0 = 1$, the solution is $z_t = e^{(2k+\gamma^2)t}$. Inserting the expression for k, we get $z_t = e^{(\rho-\beta)t}$, $\beta := [(\gamma^2 - \rho)^2 + 4a]^{1/2}$. Hence, $\lim_{T\to\infty} kE(X_T^0)^2 e^{-\rho t} = \lim_{T\to\infty} kz_t e^{-\rho t} = \lim_{T\to\infty} ke^{-\beta t} = 0$. Because $u = u^0 = kx$ and $e^{-\rho t}E(X_t^0)^2 = e^{-\beta t}$, it follows that $E[-\int_0^T e^{-\rho t}(a(X_t^0)^2 + u^0(t,X_t^0)^2)dt]$ has a finite limit when $T \to \infty$. It is obvious that if $\limsup_{T\to\infty} -E[e^{-\rho T}X_T^2] < 0$, i.e., if $E[-e^{-\rho t}X_t^2] \le -\alpha < 0$ for all large t (say from $t = T^*$ on), then $E\int_{T*}^\infty -e^{-\rho t}X_t^2 dt$ equals $-\infty$, so controls u giving such X_t are clearly suboptimal because we have shown that at least for one control the criterion has a finite value. We can confine the admissible controls to be of the type that the limsup above is ≥ 0. But then we have $\lim_{T\to\infty} -Eke^{-\rho T}(X_T^0)^2 = \lim_{T\to\infty} -ke^{-\beta T} = 0$, and $\limsup_{T\to\infty} Eke^{-\rho T}X_T^2 \ge 0$, for any admissible control, hence (4.53) holds for liminf replaced by limsup (we have sporadic catching up optimality). Note that here the optimal value can be calculated explicitly: Using $u^0 = kx$, it equals $E[\int_0^\infty \{-e^{-\rho t}(a+k^2)(X_t^0)^2\}dt] = -\int_0^\infty (a+k^2)e^{-\beta t}dt = -(a+k^2)/\beta$. (Note also that, above, the inequality $\rho > 0$ was not needed!) $\square$

Soft Terminal Restrictions*

In the finite horizon problem (4.42)–(4.44), let us now require that soft terminal restrictions of the form

$$Eh_i(X_T^u) = 0, i = 1,\ldots,i^*, \tag{4.58}$$

$$Eh_i(X_T^u) \ge 0, i = i^* + 1,\ldots,i^{**}, \tag{4.59}$$

have to be satisfied. (As before, $E = E^{0,X_0}$.) Assume that

$$E[|h_i(X_T^u)|] < \infty \quad \text{for all} \quad u(.). \tag{4.60}$$

Then the following procedure sometimes works. Let $q = (q_1,\ldots,q_{i^{**}})$ be an arbitrary i^{**}-vector and assume that we solve the free end problem with g replaced by

$g + qh, h = (h_1, \ldots, h_{i^{**}})$, i.e., that we solve

$$\max_u E \left[\int_0^T f_0(t, X_t^u, u_t) dt + g(T, X_T^u) + qh(X_T^u) \right], \tag{4.61}$$

subject to (4.44), u_t taking values in U.

Call this problem P_q. Denote the corresponding solution of the HJB equation by $J^q(t,x)$ and the control obtained from the maximization in this equation by $u^q(t,x)$. If we are able to find a value q^0 of q such that $X_T^{uq^0}$ satisfies (4.58), (4.59) and such that

$$q_i^0 \geq 0, q_i^0 = 0 \text{ if } Eh_i(X_T^{uq^0}) > 0, \ i > i^*, \tag{4.62}$$

then u^{q^0} is optimal in the soft end constrained problem, provided the conditions in (b) in Theorem 4.10 are satisfied for $\hat{f} = J^{q^0}$, $u^0 = u^{q^0}$, the boundary condition being $J^{q^0}(T,x) = g(T,x) + q^0 h(x)$. (To see this, simply note that for $q = q^0$, $u^{q^0}(t,x)$ is optimal in the problem P_{q^0}, and then, in particular, also in the original problem as, for any admissible u the criterion value in P_{q^0} is no less than the criterion value in the original problem, with equality if $u = u^{q^0}(t,x)$, by (4.62).)

If (4.60) is not satisfied, and in other cases where no solution to P_q can be found, one can turn instead to the stochastic maximum principle. We state a version working when σ does not depend on u, if not, a more complicated version is needed.

We need the following conditions: f, f_0, and σ are C^1 in x, and these functions, together with f_x, f_{0x}, and σ_x are piecewise and right-continuous in t and continuous in u. Moreover, f_x, f_{0x}, and σ_x are bounded functions. Finally, for some $k_1 > 0$, for all $(t,x,u), u \in U$, $|g| + |h_i| + |f_0| + |f| + |\sigma_{ij}| \leq k_1(1 + |x| + |u|)$ is satisfied. In this section, on all controls used, we will place the requirement that $E \int_0^T |u_s| ds < \infty$ (all controls u_t appearing are assumed to satisfy this requirement). So a pair (X_t, u_t) is admissible if u_t is adapted to the given Brownian motion, is piecewise and right-continuous in t and takes values in U, if the last inequality holds, and X_t is a solution corresponding to u_t of the state equation (4.44) existing on $[0,T]$, such that the terminal conditions (4.58), (4.59) are satisfied. Assume that $(X_t^*, u^*(t,x))$ is a pair such that $u^*(t,x)$ is piecewise and right-continuous in t, and in each component x_j, such that (X_t^*, u_t^{**}) satisfies the state equation (4.44) and the terminal restrictions (4.58), (4.58), for $u_t^{**} := u^*(t, X_t^*)$ (u^{**} assumed to be piecewise and right-continuous in t), and such that (X_t^*, u^{**}) is optimal in the set of all admissible pairs (X_t, u_t).

We shall make use of the following entities and equations: I is the identity matrix, $h = (h_1, \ldots, h_{i^{**}})$, $x \in \mathbb{R}^n$, $p = (p_1, \ldots, p_n) \in \mathbb{R}^n$, $\lambda_0 \geq 0$, $\lambda = (\lambda_1, \ldots, \lambda_{i^{**}})$, σ^k the k-th row of σ and

$$H(t,x,u,p) := \lambda_0 f_0 + pf. \tag{4.63}$$

In the one-dimensional case ($n = m = 1$), the function $\Phi(s,t), s \geq t$ is the solution of

$$d\Phi(s,t) = f_x(s, X_s^*, u_s^{**})\Phi(s,t)ds + \sigma_x(s, X_s^*)\Phi(s,t)dB_s, \ \Phi(t,t) = 1. \tag{4.64}$$

$(d\Phi(s,t)$ a differential with respect to s). In several dimensions, $\Phi(s,t)$ (an $n \times n$ matrix) is the solution of

$$d\Phi(s,t) = f_x(s,X_s^*,u_s^{**})\Phi(s,t)ds + \sum_k \sigma_x^k(s,X_s^*)\Phi(s,t)dB_s^k, \quad \Phi(t,t) = I. \quad (4.65)$$

$$\max_{u \in U} H(t,X_t^*,u,p^*(t)) = H(t,X_t^*,u_t^{**},p^*(t)) \quad \text{a.s.} \quad (4.66)$$

$$p^*(t) := E[\lambda h_x(X_T^*)\Phi(T,t)|X_t^*]$$

$$+E\left[\lambda_0\left\{\int_t^T f_{0x}(s,X_s^*,u_s^{**})\Phi(s,t)ds + g_x(T,X_T^*)\Phi(T,t)\right\}|X_t^*\right]. \quad (4.67)$$

$$\text{For } i > i^*, \quad \lambda_i \geq 0 \text{ and } \lambda_i = 0 \text{ if } Eh_i(X_T^*) > 0. \quad (4.68)$$

Theorem 4.21 (Maximum principle, necessary condition). *Assume that* $(X_t^*,u^*$ $(t,x))$ *is an optimal admissible pair. There exist numbers* λ_i, $i = 0,\ldots,i^{**}$, $\lambda_0 \geq 0$, *satisfying (4.68),* $(\lambda_0,\ldots,\lambda_{i^{**}}) \neq 0$, *such that (4.66) holds, for* H, $\Phi(.,)$, $p^*(.)$ *as given in (4.63), (4.64), ((4.65)), and (4.67), respectively.* $\square$

The only change needed when the optimal control is of a general adapted type (adapted to the given Brownian motion), is that in the conditional expectation in (4.67) one has to condition on the entire history of the Brownian motion. We then denote the adjoint function $p(t)$ and not $p^*(t)$, so in this case it is $p(t)$ that appears in the maximum condition (4.66). With this change, the necessary conditions can also be used to find optimal adapted controls, though we shall not give examples where optimal adapted controls come out that are not Markovian.

Variants of this result can be found for example in Kushner (1972) and Haussmann (1986). For more general conditions, see Yong and Zhou (1999). The proof is too lengthy and is omitted. When $\lambda_0 = 1$, the conditions are sufficient in the case $x \to \max_u H(t,x,u,p^*(t))$ and $\dot{x} \to g + \lambda h$ are concave and σ is linear (or affine) in x. For more general sufficient conditions (relaxing the condition on σ), see Yong and Zhou (1999) and a comment below.

For interpretations of what is going on, note that when working with adapted controls we imagine that B_t can be observed. After all, the assumption is that the person controlling the process is able to let control values be dependent on the history of B_s. It works equally well if a process uniquely related to B_t, say a corresponding geometric Brownian motion, is observable, from which the behavior of B_t can be deduced. In an economic situation, a geometric Brownian motion may be an observable price process.

(Note that even if u_t^{**} stems from a Markov control $u^*(t,x)$, to obtain the above theorem, we have assumed optimality among all (admissible) adapted controls. Recall a corresponding situation in case of the stochastic maximum principle in Chapter 1.)

Problems where only the state X_t is observable, and where we have to use only controls depending on the history of X_t, are in many ways more tricky. This case is extensively discussed in Haussmann (1986).

Sometimes, in softly end constrained problems, optimal controls in the form of functions adapted to the Brownian motion may exist that cannot be expressed as

Markov controls. When the "augmented velocity set" $\{(f_0(t,x,u)+\gamma, f(t,x,u)) : u \in U, \gamma \leq 0\}$ is convex and U is compact, in free end problems optimal *weak* solutions (X_t, u_t) do exist, for which $u_t = u(t, X_t)$ for some function $u(t,x)$ (= Markov control). The same result also holds in certain hard end constrained problems (end constraints required to be satisfied a.s.), but, as said, not necessarily in softly end constrained problems. For a precise statement, see Haussmann and Lepeltier (1990).

Let us relate the above maximum principle to the deterministic one, confining attention to the case $n = m = 1$ and $f_0 \equiv 0$: Define $E^{t,x} := E[.|X_t^* = x]$ and

$$q^{**}(t) = \lambda h_x(X_T^*)\Phi(T,t) + \lambda_0 g_x(T, X_T^*)\Phi(T,t).$$

Then $p^*(t) = E^{t, X_t^*} q^{**}(t)$, and $p(t) = E^t q^{**}(t)$, E^t indicating conditioning on the entire history of the Brownian motion up to t, and $q^{**}(t)$ satisfies

$$q^{**}(t) = \lambda h_x(X_T^*)q^*(T)^{-1}q^*(t) + \lambda_0 g_x(T, X_T^*)q^*(T)^{-1}q^*(t), \tag{4.69}$$

where $q^*(t) := \Phi(0,t)$ is an adapted solution of the following equation:

$$dq^*(t) = -q^*(t)f_x(t)dt - q^*(t)\sigma_x(t)dB_t + q^*(t)(\sigma_x(t))^2 dt, q^*(0) = 1. \tag{4.70}$$

Here $f_x(t), \sigma_x(t)$ (and later on $f_{0x}(t), f(t), \sigma(t)$) are $f_x, \sigma_x, f_{0x}, f, \sigma$ evaluated at (t, X_t^*, u_t^{**}). In the deterministic case, where $\sigma \equiv 0$, also $p(t) := (\lambda h_x(X_T^*) + \lambda_0 g_x(T, X^*(T))q^*(t)/q^*(T)$ satisfies (4.70) (we have only adjusted $q^*(t)$ by a multiplicative factor), and $p(T) = \lambda h_x(X_T^*) + \lambda_0 g_x(T, X^*(T))$ (i.e., satisfying the standard deterministic transversality condition).

Let us prove (4.70). Note that $\Phi(0,0) = I = 1$ and $\Phi(0,t)\Phi(t,0) = I = 1$, hence $\Phi(0,t) = 1/\Phi(t,0)$, so, taking differentials with respect to t, using Ito's formula and (4.64) give $d\Phi(0,t) = d(1/\Phi(t,0)) = -(1/\Phi(t,0)^2)(f_x(t)\Phi(t,0)dt + \sigma_x(t)\Phi(t,0)dB_t) + (1/\Phi(t,0)^3)\sigma_x^2(t)\Phi(t,0)^2 dt = -\Phi(0,t)f_x(t) - \Phi(0,t)\sigma_x(t)dB_t + \Phi(0,t)\sigma_x^2(t)dt$. Hence, $q^*(t) = \Phi(0,t)$ satisfies (4.70).

The maximum principle can be given another formulation, which will only be stated in the one-dimensional case. Yong and Zhou (1999) prove a maximum principle that covers the case of a u-dependent $\sigma(t,x)$-function. In that case, it has a more complicated form, but it reduces to the following result when $\sigma(t,x)$ is independent of u. (Then we may even need the further condition on the system that the second derivatives with respect to x of f_0, f, and σ exist, are piecewise and right-continuous in t, and uniformly continuous in (x,u), uniformly in t.)

The maximum condition holds for some adapted function $p(t)$ satisfying

$$dp(t) = -\lambda_0 f_{0x}(t)dt - p(t)f_x(t)dt - q(t)\sigma_x(t)dt + q(t)dB_t, \tag{4.71}$$

for some adapted function $q(t)$, with $p(T) = \lambda_0 g_x(T, x_T^*) + \lambda h_x(X_T^*)$, *the λ_i's as in* (4.68).

One might believe that by introducing the function $q(.)$ in the above equation, a plethora of solution pairs $(p(t), q(.))$ arise. (All information on $q(.)$ is what is presented above.) This is not so. Let us show that the member $p(t)$ in such a pair

$(p(t), q(t))$ is a function $p(t)$ of the type occurring in the theorem above (more precisely in the comment subsequent to the theorem). For simplicity let $f_0 \equiv 0$, and let $\psi(t) = \Phi(t, 0)$, i.e., $\psi(t)$ is the solution of $d\psi(t) = \psi(t) f_x(t) + \psi(t) \sigma_x(t) dB_t$, $\psi(0) = 1$. Let us apply Ito's formula to find $d(\psi(t)p(t))$, where $p(t)$ is as given by (4.71). We get $d(\psi(t)p(t)) =$

$$
\begin{aligned}
(d\psi(t)) & p(t) + \psi(t)(dp(t)) + (d\psi(t))(dp(t)) \\
&= \psi(t) f_x(t) p(t) dt + \psi(t) \sigma_x(t) p(t) dB_t - \psi(t) f_x(t) p(t) dt \\
&\quad - \psi(t) \sigma_x(t) q(t) dt + \psi(t) q(t) dB_t + \psi(t) \sigma_x(t) q(t) dt \\
&= \psi(t) \sigma_x(t) p(t) dB_t + \psi(t) q(t) dB_t.
\end{aligned}
$$

Letting E^t involve conditioning on the entire history of B_s up to t, taking this conditional expectation on both sides yields $d\{E^t[\psi(t)p(t)]\} = 0$, so

$$
\begin{aligned}
E^t[\psi(T)& p(T)] \\
&= E^t[\psi(t)p(t)] + E^t \left[\int_t^T d\{E^s[\psi(s)p(s)]\} ds \right] = E^t[\psi(t)p(t)] = \psi(t)p(t),
\end{aligned}
$$

hence $E^t[\psi(T)p(T)] = \psi(t)p(t)$ and

$$
p(t) = \psi(t)^{-1} E^t[\psi(T)p(T)] = \Phi(0, t) E^t[\Phi(T, 0)p(T)] = E^t \Phi(T, t)p(T).
$$

Thus, $p(t)$ is of the type occurring in (the comment subsequent to) Theorem 4.21.

Associated with the necessary conditions connected to (4.71), there is a sufficient condition that briefly stated says that, if $\lambda_0 = 1$ in the necessary conditions and $(x, u) \to H(t, x, u, p(t)) + q(t)\sigma(t, x)$ and $x \to g + \lambda h$ are concave, then the conditions are sufficient.

The discussion to follow is confined to the case where it suffices to consider Markov controls. Let us relate the equation (4.71) to an equation arising from differentiating the HJB equation (4.45) with respect to x (and using an envelope theorem): Thus, letting $w = J_x$, we then obtain the equation

$$
\begin{aligned}
0 = w_t(t, x) &+ f_{0x}(t, x, \hat{u}(t, x, w)) + w_x f(t, x, \hat{u}(t, x, w)) \\
&+ w f_x(t, x, \hat{u}(t, x, w)) + \sigma_x(t, x)\sigma(t, x)w_x(t, x) + \sigma^2(t, x)w_{xx}(t, x)/2, \quad (4.72)
\end{aligned}
$$

where $\hat{u}(t, x, w)$ gives maximum in the problem $\max_u \{f_0(t, x, u) + w f(t, x, u)\}$. Assume that $w(t, x)$ satisfies (4.72) and let $p(t) = w(t, X_t^*)$. Then $dp(t) = dw(t, X_t^*) = w_t + w_x f(t) dt + w_x \sigma(t) dB_t + \sigma^2(t)w_{xx}/2dt$. From (4.72), we get $w_t + w_x f(t) + \sigma^2(t)w_{xx}/2 = -f_{0x}(t) - w f_x(t) - \sigma_x(t)\sigma(t)w_x$, so $dp(t) = -f_{0x}(t)dt - w f_x(t)dt - \sigma_x(t)\sigma(t)w_x dt + w_x \sigma(t) dB_t$. Letting $q = w_x \sigma$, we see that $p(t)$ satisfies (4.71) for this $q(t)$ and $\lambda_0 = 1$. From a preceding argument, we know that $p(t)$ is of the form $E^t \Phi(T, t)p(T)$.

Now, if we have found a solution $\hat{J}$ of the HJB equation together with a control $\tilde{u}(t, x)$ yielding maximum in that equation, then $\hat{u}(t, x, \hat{J}_x(t, x)) = \tilde{u}(t, x)$ and $w(t, x) = \hat{J}_x(t, x)$ satisfies (4.72). This means that to find a candidate satisfying the necessary condition, one can find a solution $\hat{J}$ of the HJB equation, with $\tilde{u}(t,x)$ yielding maximum in the HJB equation, such that if X_t^* is a solution of the differential (state) equation for this control $\tilde{u}(t,x)$, which satisfies, (a), the terminal conditions, and, (b), $\hat{J}_x(T,X_T^*) = \lambda h_x(X_T^*) + g_x(T,X_T^*)$ for some λ satisfying the transversality condition (4.68), then $u^{**}(t) = \tilde{u}(t,X_t^*)$ is a candidate control: It satisfies the necessary conditions in Theorem 4.21. This holds both in one and several dimensions. Let us call this procedure P^λ. Occasionally, this closely connected procedure may work in cases where the procedure associated with the problems P_q fails (the boundary condition (b) on $\hat{J}(T,.)$ here is weaker than that on $J^{q^0}(T,.)$).

(For completeness, if one also wants to consider abnormal candidates, i.e., $\lambda_0 = 0$, then carry out the procedure in the last paragraph for f_0 and g deleted in the HJB equation and in (b).)

Note also that in cases where $w(t,x) := E^{t,x}p(t)$ is C^1 in (t,x) and the optimal control is of the form $u^*(t,X_t^*)$, the equation for $p(t)$ normally implies (4.72) for $\hat{u}(t,x,w)$ replaced by $u^*(t,X_t^*)$ and x replaced by X_t^*. A heuristic argument for this is given in B. in the Addendum to this chapter.

Example 4.22. Consider the problem

$$\max E\left[\int_0^T -u^2 dt\right] \text{ subject to } dX_t = udt + \sigma dB_t, X_0 = 0, EX_T = 1, u \in \mathbb{R}.$$

Consider first the use of the HJB equation, applied to problem P_q, i.e., to the above problem modified by replacing the end condition by the introduction of the bequest function qx (a suitable value of q is wanted). For $\hat{J} = J$, the HJB equation is $0 = J_t + \max_u\{-u^2 + J_x u + \sigma^2 J_{xx}/2\}$. Maximization yields $u = J_x/2$ and then $0 = J_t + (J_x)^2/4 + \sigma^2 J_{xx}/2$, with boundary condition $J(T,x) = qx$. Let us try a solution of the form $\phi(t) + kx$. The boundary condition gives $\phi(T) = 0, k = q$. Insertion into the HJB equation yields $0 = \phi' + q^2/4$, so $\phi(t) = q^2(T-t)/4$. The optimal control in problem P_q is $u = q/2$. When $\int_0^T udt = \int_0^T q/2dt$ equals 1, i.e., $q = 2/T$, then $X_T = 1 + \sigma B_T$ and $EX_T = 1$.

Let us solve the problem using the maximum principle (4.64), (4.66), (4.67). Maximizing the Hamiltonian gives $u = p^*(t)/2$. Now, $\Phi(s,t) \equiv 1$, in fact $p^*(t) \equiv \lambda$ for some λ. We then seek a value of λ such that $EX_T = 1$, the value is of course $\lambda = 2/T$.

Next, change the problem by replacing $X_0 = 0$ by $X_0 = x_0 > 0$ and $EX_T = 1$ by $E(X_T)^2 = k^2, k > 0$, where $k^2 - T\sigma^2 > x_0^2$. Let us use the HJB equation to solve the problem, and so first problem P_q. Let us try $J(t,x) = a(t) + b(t)x^2$, with $b(T) = q > 0, a(T) = 0$. Now, as before, $0 = J_t + J_x^2/4 + J_{xx}\sigma^2/2$, so trying the proposal for $J(t,x)$, we get $0 = a' + b'x^2 + b^2x^2 + b\sigma^2$. This gives $b' = -b^2, a' = -b\sigma^2$. Below, C and Y_0 will be integration constants. Now, $1/b = t + C, b = 1/(t+C), q = 1/(T+C)$, so $C = 1/q - T$. Moreover, $u^* = J_x/2 = b(t)X_t^*$, so $dX_t^* = (1/(t+C))X_t^* dt + \sigma dB_t$. Starting at x_0, we shall aim at a solution for which $E(X_T^*)^2 = k^2$, where k^2 is greater than $x_0^2 + T\sigma^2$, so we reasonably expect $1/(t+C)$ to be positive for all t, i.e., $C > 0$. (Whether X_t^* is positive or negative, it seems reasonable that we shall have a positive "drift" in the value of $|X_t^*|$.) Letting $Y_t = X_t^*/(t+C)$, we get $dY_t = (t+C)^{-1}\sigma dB_t$, so $Y_t = Y_0 + \int_0^t (s+C)^{-1}\sigma dB_s$. Hence, $X_t^* = x_0(t+C)/C + (t+C)\int_0^t (s+C)^{-1}\sigma dB_s$.

By Dynkin's formula,

$$dEY_t^2 = 2E[Y_t dY_t] + (t+C)^{-2}\sigma^2 dt = (t+C)^{-2}\sigma^2 dt,$$

so $EY_t^2 = Y_0^2 + \int_0^t (s+C)^{-2}\sigma^2 ds$ and

$$E(X_t^*)^2 = x_0^2(t+C)^2/C^2 + (t+C)^2 \int_0^t (s+C)^{-2}\sigma^2 ds.$$

Integrating and putting $t = T$ yield $k^2 = E(X_T^*)^2 = -(T+C)\sigma^2 + \{(x_0/C)^2 + \sigma^2/C\}(T+C)^2$. Rearranging gives $C^2\alpha - (2Tx_0^2 + T^2\sigma^2)C - Tx_0^2 = 0$ where $\alpha = k^2 - T\sigma^2 - x_0^2 > 0$, or $C = [2Tx_0^2 + T^2\sigma^2 + ((2Tx_0^2 + T^2\sigma^2)^2 + 4\alpha Tx_0^2)^{1/2}]/2\alpha$ (using the plus-sign as we need $C > 0$. The optimal control is then $u = X_t^*/(t+C)$ (optimality is provided by the comment subsequent to (4.62)). $\square$

Often the method P_q fails to work. Even for slight changes in the next to last problem, we run into some, but not insurmountable, troubles:

Example 4.23. Consider the problem:

$$\max E \int_0^T -u^2 dt \text{ subject to } dX_t = udt + t\sigma dB_t, X_0 = 0, EX_T^2 = k^2, u \in \mathbb{R},$$

where $k^2/T^2 - T\sigma^2 > 0$.

Solution. In this case, the HJB equation becomes $0 = J_t + \max_u\{-u^2 + uJ_x\} + t^2\sigma^2 J_{xx}/2$, the maximizing u equals $J_x/2$, and the HJB equation becomes $0 = J_t + J_x^2/4 + t^2\sigma^2 J_{xx}/2$. Below, C, K, and A are integration constants. A solution of the form $J = b(t)x^2 + a(t)$ now yields the equations $b' = -b^2$ and $a' = -bt^2\sigma^2$, the first one yielding $b = 1/(t+C)$. Let X_t^* be the solution corresponding to $u = b(t)x$. Now, $E(X_t^*)^2 = K^2(t+C)^2 + (t+C)^2 \int_0^t (s+C)^{-2}s^2\sigma^2 ds$ (we have written K^2 instead of x_0^2/C^2 and replaced σ^2 by $s^2\sigma^2$ in a formula for $E(X_t^*)^2$ in the previous problem). Furthermore, $0 = EX_0^2 = K^2C^2$. Here, $K = 0$ does not seem to work, at least for k large, because we are not able to lift $E(X_T^*)^2$ enough for the end condition to be satisfied if $K = 0$. So we put $C = 0$. For $Y_t := X_t^*/t$, $dY_t = d(X_t^*/t) = \sigma dB_t$, so $Y_t = K + \sigma B_t$, and $X_t^* = Kt + t\sigma B_t$ (and the K introduced here also works for the formula for $E(X_t^*)^2$ above). Now, $E(X_T^*)^2 = K^2T^2 + T^3\sigma^2$, so equating this with k^2, gives $K^2 = k^2/T^2 - T\sigma^2$. Both the positive and negative root for K work equally well, let us choose $K > 0$. Here, $b(T) = 1/T$, so $q^0 = 1/T$ in the problem P_q connected with the current problem. From now on, let $q = q^0 = 1/T$, and denote the corresponding optimal value function $J^q(t,x)$. Let us for completeness also calculate $a(t)$, which becomes $a(t) = -t^2\sigma^2/2 + T^2\sigma^2/2$. One problem that arises now is that $J^q(0,x) = x^2/0 + T^2\sigma^2/2 = \infty$, $x \neq 0$. Still however, we do retain some confidence in our candidate because we know that necessary conditions are satisfied. (Essentially, the candidate has been found using procedure P^λ, $\lambda = q$, and we know that (4.72) is satisfied, as well as (4.66).)

Let us now only consider "acceptable" admissible (adapted) controls $u(t)$ in what follows, where acceptable means $E \int_0^T u(s)^2 ds < \infty$. For $t > 0$, if $J^{q,u}(t,x)$ is the value of the free end criterion ($q = 1/T$) when the process X_s starts at (t,x) and the adapted control u is used, as always, we do have $J^q(t,x) \geq J^{q,u}(t,x)$, and so also $J^q(t,X_t^u) \geq J^{q,u}(t,X_t^u)$, X_t^u the solution starting in $(0,0)$ corresponding with u. Now, $J^{q,u}(0,0) = E[\int_0^t -u(s)^2 ds + J^{q,u}(t,X_t^u)]$, so $\lim_{t \to 0} J^{q,u}(t,X_t^u) = J^{q,u}(0,0)$.

Furthermore, by Hoelder's inequality, which implies $\int_0^t |u(s)| \cdot 1 ds \leq$

$$\left(\int_0^t u(s)^2 ds \right)^{1/2} \cdot \left(\int_0^t 1 ds \right)^{1/2} = t^{1/2} \left(\int_0^t u(s)^2 ds \right)^{1/2},$$

and Ito's isometry,

$$E(X_t^u)^2 = E \left(\int_0^t u(s) ds + \int_0^t s\sigma dB_s \right)^2$$

$$\leq E \left(\int_0^t u(s) ds \right)^2 + 2E \left[\left(\int_0^t u(s) ds \right) \left(\int_0^t s\sigma dB_s \right) \right] + E \left(\int_0^t s\sigma dB_s \right)^2$$

$$\leq E \left[t \int_0^t u(s)^2 ds \right] + 2 \left(E \left(\int_0^t u(s) ds \right)^2 \right)^{1/2} \left(E \left(\int_0^t s\sigma dB_s \right)^2 \right)^{1/2}$$

$$+ E \left(\int_0^t s\sigma dB_s \right)^2$$

$$\leq tE \int_0^t u(s)^2 ds + 2t^{1/2} \left(E \left(\int_0^t u(s)^2 ds \right) \right)^{1/2} \left(\int_0^t s^2 \sigma^2 ds \right)^{1/2} + \int_0^t s^2 \sigma^2 ds.$$

Evidently, the right-hand side divided by t goes to zero when t goes to zero, so $EJ^q(t,X_t^u) = Eb(t)(X_t^u)^2 + a(t) \to a(0)$ when $t \to 0$. Hence, $J^{q,u}(0,0) \leq a(0)$, ($J^{q,u} \leq J^q$). Now also $u^*(t) := K + \sigma B_t$ is acceptable, and if $u = u^*$, then $J^q(t,X_t^*) = J^{q,u^*}(t,X_t^*)$, ($u^* = J_x^q(t,X_t^*)/2$ where $u = J^q(t,x)/2$ yields maximum in the HJB equation for $t > 0$). Thus $J^{q,u^*}(0,0) = a(0)$, hence u^* is optimal. □

How frequently is the method associated with the problems P_q successful? We shall not be able to answer that problem in any generality. But let us consider the following problem on $J = [0,2]$:

Let $dX_t = u1_{[0,1]}(t)dt + 1_{[1,2]}(t)dB_t$, $X_0 = 0$, $E[X_2^2] = 2$, $u \geq 0$. We want to maximize $E[-X_2]$. Now $E[-X_2] = E[-Y_2] = E[-Y_1]$, where $dY_t = u1_{[0,1]}(t)dt$, $Y_0 = 0$. Any (Markov) control $u(t,x)$ leads to a deterministic control $u(t,X_t)$ on $[0,1]$ (X_t is deterministic on $[0,1]$). Then $E[X_2^2] = E[Y_1^2] + 1$, so $E[Y_1^2] = Y_1^2 = 1$, hence any deterministic control u_t on $[0,1]$, for which $\int_0^1 u_t dt = 1$ is optimal.

However, in problem P_q, where we operate with the scrap value $E(-X_2 + qX_2^2)$ ($= -Y_1 + qY_1^2 + q$), the optimal value function equals infinity if $q > 0$, which is seen by considering arbitrary large constant controls. If $q = 0$, we evidently get $u \equiv 0$, i.e., an inadmissible control (of course $E(-X_2)$ is minimized by $u \equiv 0$). So the method connected with problem P_q does not work in this case. By the way, in this problem, adapted controls are strictly better than Markov controls, to see this intuitively, we see that it pays to have $u(t, \omega)$ and (then) Y_1 large in quite improbable circumstances (i.e., $|B_t|$ large). That has only a moderate negative influence on the criterion, but lifts X_2^2 quite a lot (the cost of reaching $E[X_2^2] = 2$ should be kept down). No optimal adapted control exists.

Example 4.24. Consider the following problem:

$$\max E\left[\int_0^1 u\, dt\right], \quad \text{when} \quad dX_t = (u+Y_t)dt, \quad dY_t = \sigma dB_t, \quad EX_1^2 = 1, \quad u \in \mathbb{R}$$

$\sigma > 0, X_0 = 0, Y_0 = 0.$

Solution. Let us solve it, using necessary conditions. Let $p^x(t)$ be the multiplier corresponding with x and let $p^y(t)$ correspond with y, both governed by (4.65), (4.67), $(p^*(t) = (p^x(t), p^y(t))$. The maximization of the Hamiltonian gives no information on the maximizing u, instead it tells us that $p^x(t) \equiv -1$. The four elements of the matrix $\Phi(s,t)$ are obtained as follows: $d\Phi_{21}(s,t) = d\Phi_{22}(s,t) = 0$, so $\Phi_{21}(s,t) = 0$, and $\Phi_{22}(s,t) = 1$, hence $d\Phi_{11}(s,t) = 0$, and $d\Phi_{12}(s,t) = 1$, so $\Phi_{11}(s,t) = 1$ and $\Phi_{12}(s,t) = t - 1$. Then $p^x(t) = \lambda E^{t,X_t^*,Y_t^*}[2X_1^*]$. We already know that $p^x(t)$ is constant equal to -1, and one way to arrange this is to have X_1^* deterministic. Because the maximization of the Hamiltonian puts no restrictions on u, if we let $u^* = k - Y_t$, k a deterministic constant, then dX_t^* is deterministic, and so also X_1^*, which equals k, and if k is chosen equal to ± 1, the terminal condition is satisfied. Of the two values of k, $k = 1$ yields the highest value of the criterion. (Then $\lambda = -1/2$.)

Is the control optimal? An ad hoc argument gives that it is: Because $1 = E(X_1^u)^2 = E(X_1^u - EX_1^u)^2 + (EX_1^u)^2$, we get the largest possible value of $(EX_1^u)^2$, when there is no variance in X_1^u. Then $|EX_1^u|$ has the highest possible value. Finally, $E\int_0^1 u\, dt = EX_1^u$ and hence u^* is optimal. (Also, sufficient conditions presented earlier automatically yield optimality.)

Here it is impossible to use the P_q-procedure: The maximization in the HJB equation gives $J_x^q \equiv -1$, but at the same time $J^q(1,x,y) = qx^2$, two properties that contradict each other. (The optimal value function in the original problem actually equals $1 - x$.)

An application of the procedure P^λ would work here. Because $J_x^q \equiv -1$, we would soon come up with the simple proposal $\hat{J} = a - x$ for the solution of the HJB equation, and $-1 = \hat{J}_x(T,x^*(T)) = \lambda 2X_1^*$ would again induce us to try the possibility of a deterministic X_1^* and even dX_t^*, in fact, try $u^* = k - Y_t$, with k determined as above. (Here, we would not actually be interested in determining a, but of course $J(1,1,y) = 0$, yielding $a = 1$.)

Similarly, if in this problem we delete the soft end constraint, and instead add the scrap value $-X_1^2/2$ to the criterion, the recipe of using the HJB equation (which again yields $J_x \equiv -1$), combined with the boundary condition $J(1,x,y) = -x^2/2$ does not work. (See Exercise 4.15 below.) □

Let us finish by a comment on existence of optimal controls. Lack of existence can arise for reasons similar to those in deterministic control (unbounded control region, lack of convexity of the set of "velocity vectors" connected with the drift term). In the following problem, there is a "stochastic reason" for nonexistence: Consider the problem $\max E \int_0^1 u dt$, when $dX_t = u dt + dB_t, EX_1^2 = 1, X_0 = 0, u \in \mathbb{R}$. Using the experience from the last example, it is evident that if we were allowed to operate with an expression $k dt - dB_t$ instead of $u dt$ in the differential equation (or, informally, $u = k - dB_t/dt$), then $k = 1$ would be optimal. Such an expression would represent a too crude (volatile) control and is not allowed. Presumably, we can approximate it as closely as wanted by a smoother admissible (and "allowable") control, which then will be only approximately optimal. (Subsequent to Example 4.23, another example of nonexistence of an optimal control was indicated.)

Hard Terminal Restrictions

In the finite horizon problem (4.42)–(4.44), let us now introduce "hard" terminal restrictions of the form

$$\text{a.s.} \quad h_i(X_T^u) = 0, i = 1, \ldots, i^*, \tag{4.73}$$

$$\text{a.s.} \quad h_i(X_T^u) \geq 0, i = i^* + 1, \ldots, i^{**}. \tag{4.74}$$

We assume that h_i are C^1 and that the conditions on f, σ, f_0, and g are those stated subsequent to (4.41).

To formulate sufficient conditions, we need the following condition

$$\liminf_{t \to T} \{E \hat{J}(t, X_t^u) - E \hat{J}(t, X_t^0)\} \geq 0 \tag{4.75}$$

The following theorem holds true:

Theorem 4.25. *(Sufficient conditions, hard terminal restrictions) Let $\hat{J}(t, x)$, $t \in (0, T), x \in \mathbb{R}^n$, be a twice continuously differentiable function satisfying the HJB equation (4.45) ((4.46) in the multidimensional case), together with (4.75). Let moreover, $\hat{J}(t, x)$ be defined and be continuous even in $[0, T) \times \mathbb{R}^n$. Assume that for every pair (t, x), $u^0(t, x)$ is the value of u that yields the maximum in the HJB equation, such that $u^0 \in \mathcal{U}^{\hat{J}}$ and such that the corresponding solution $X_t^{u^0}$ of (4.41) satisfies the terminal constraints (4.73), (4.74). Then $u^0(.,.)$ is optimal in $\mathcal{U}^{\hat{J}}$.* ☐

The proof is essentially the same as that of Theorem 4.15 (just replace T by t in (4.49) and, using (4.75), let $t \to T$).

The conditions so far presented are formally sufficient, but often it is useful to formulate intuitive ideas about what sort of additional properties $\hat{J}(t, x)$ might have. Here, we shall mention only one such idea: In case $g = 0$, one should frequently expect $\lim_{t \to T} \hat{J}(t, x)$ to equal zero for all x satisfying $h_i(x) = 0, i = 1, \ldots, i^*, h_i(x) \geq 0, i = i^* + 1, \ldots, i^{**}$, and hence, for all such x, $\lim_{t \to T} \hat{J}_{x_j}(t, x)$ should be expected to equal zero in the case where the component x_j of x does not occur in the functions h_i.

Note that $\hat{J}(T, x)$ cannot reasonably be defined defined for all x (surely not for x not satisfying the h_i-constraints), which is one reason for the liminf condition in (4.75). (One might ask if not this condition could be added also in previous problems, and it could, but, for example in the free end problem, the condition $\hat{J}(T,x) = g(T,x)$ automatically entails (4.75) in case $\hat{J}$ is C^0 in $[0,T] \times \mathbb{R}^n$.)

In the above type of problems, one may need controls that are unbounded near T, but we must at least restrict the Markov controls $u(t,x)$ considered to belong to the set $\mathcal{U}^{\hat{J}}$ defined in Remark 4.14, or the adapted controls u_t considered to belong to $\mathcal{U}_f$ defined in Remark 4.17.

Example 4.26. Let us consider a very simple problem with two states. One state is free at the terminal time, and that is the only state directly influenced by B_t (i.e., having dB_t in its equation). The second state, which is fixed at the terminal time, is influenced by B_t only through the first state.

$$\max E\left[\int_0^1 -u^2 dt\right] \text{ subject to } dX_t = (u + Y_t)dt, dY_t = \sigma dB_t, X_1 = 1 \text{ a.s.,}$$

$u \in \mathbb{R}, X_0 = 0, Y_0 = 0.$

Solution. An adapted control u_s is called acceptable if $E[\int_0^1 u_s^2 ds] < \infty$. (It is admissible if also $X_1^u = 1$ a.s.) For $\hat{J} = J = J(t,x,y)$, the HJB equation is $0 = J_t + \max_u\{-u^2 + J_x(u+y)\} + J_{yy}\sigma^2/2$. The maximizing u is $J_x/2$, and the HJB equation becomes then $0 = J_t + J_x^2/4 + J_x y + J_{yy}\sigma^2/2$. Let us try the solution

$$J(t,x,y) = a(t)x + b(t)y + c(t)x^2 + d(t)xy + k(t)y^2 + h(t).$$

As $J_x = a + 2cx + dy$ and $J_{yy} = 2k$, we get

$$0 = a'x + b'y + c'x^2 + d'xy + k'y^2 + h' + (a + 2cx + dy)^2/4 + (a + 2cx + dy)y + k\sigma^2,$$

or $0 = h' + a^2/4 + k\sigma^2 + (a' + ac)x + (b' + a + ad/2)y + (c' + c^2)x^2 + (d' + cd + 2c)xy + (k' + d^2/4 + d)y^2$. This equation must hold for all x and y, so the coefficients in front of x, y, x^2, y^2, xy, y^2 and the "constant term" (i.e., purely time-dependent term) must all vanish. Hence,

(constant): $h' + a^2/4 + k\sigma^2 = 0$, (x): $a' + ac = 0$, (y): $b' + a + ad/2 = 0$
(x^2): $c' + c^2 = 0$, (xy): $d' + 2c + cd = 0$ (y^2): $k' + d + d^2/4 = 0$.

As we guess that $J(1,1,y) = 0$ and hence $J_y(1,1,y) = 0$, then, presumably, $k(1) = 0$ and $b(1) + 1 \cdot d(1) = 0$. Below, A, C, D, K, L, and M are integration constants. From (x^2), we get $c = 1/(t+C)$. This inserted in (x) and (xy) gives $a = A/(t+C)$, $d = D/(t+C) - 2$. Using $u = J_x/2 = (a + 2cx + dy)/2$, we get

$$dX_t^* = [A/2(t+C) + (1/(t+C))X_t^* + (d(t)/2 + 1)Y_t]dt.$$

Close to $t = 1$ there should be very little variability in dX_t^*, in order to obtain $X_1^* = 1$ a.s. We therefore put $d(1)/2 + 1 = 0$, which entails $D = 0$, so $dX^*(t)$ is completely deterministic. Then $X_t^* = -A/2 + K(t+C)$, or $X_t^* = Kt + M$. For arbitrary $\hat{x}$ and s, the solutions for K and M of the equations $X_s^* = \hat{x}$, and $X_1^* = 1$, gives $X_t^* = (1 - s)^{-1}(1 - \hat{x})(t - 1) + 1 = K(t - 1) + 1$, $K = (1 - s)^{-1}(1 - \hat{x}) = dX_t^*/dt$ (and then $C = -1, A = -2$, by comparing with the first formula for X_t^*). This gives

$$J(s, \hat{x}, \hat{y}) = -(1 - s)^{-1}(\hat{x} - 1)^2 - 2(\hat{x} - 1)\hat{y} - \hat{y}^2(1 - s) - \sigma^2(1 - s)^2/2,$$

as $b(t) = 2$, $k(t) = t - 1$, $h(t) = -(1 - t)^{-1} - \sigma^2(1 - t)^2/2 + L$, where L is determined by $J(1, 1, y) = 0 \Rightarrow L = 0$. Finally, for $(s, \hat{x}) = (0, 0)$, $dX_t^* = 1dt$, so the corresponding control u_t^* equals $1 - Y_t$.

To obtain (4.75), it suffices to prove that $\lim_{t \to 1}\{E[-(1 - s)^{-1}(X_t - 1)^2] - E[2(X_t - 1)Y_t]\} = 0$, for any admissible triple (u_t, X_t, Y_t) starting at $(0, 0)$ at $t = 0$. As $E|[2(X_t - 1)Y_t]| \leq (E(X_t - 1)^2)^{1/2}(E(2Y_t)^2)^{1/2}$, the second term evidently converges to zero, because below we show that $E(X_t - 1)^2 \to 0$. So let us study the first term. Note that if $v(\tau)^2$ is integrable on $[0, 1]$, then by Cauchy-Schwartz's inequality applied to $v(\tau)$ and 1 gives that $(\int_t^1 1 \cdot v(\tau)d\tau)^2 \leq (1 - t)\int_t^1 v(\tau)^2 d\tau$. Hence, as u_t is acceptable and $X_t^u := X_t$ satisfies $X_1 = 1$ a.s., then $E(X_t - 1)^2 = E(X_t - X_1)^2 =$

$$E\left[\int_t^1 (u_s + Y_s)ds\right]^2 \leq E\left[(1 - t)\int_t^1 (u_s + Y_s)^2 ds\right] = (1 - t)\int_t^1 E(u_s + Y_s)^2 ds.$$

Evidently, $E[u_s + Y_s]^2$ is integrable on $[0, 1]$, so $(1 - t)^{-1}E(X_t - 1)^2 \to 0$ when $t \to 1$. Hence (4.75) is satisfied, as also u_t^* is acceptable. $\qquad\square$

Example 4.27. Consider next the problem

$$\max E\left[\int_0^1 -u^2 dt\right] \quad \text{subject to} \quad dX_t = udt + \sigma dB_t, X_0 = 0, X_1 = 1 \text{ a.s.}, u \in \mathbb{R}.$$

In this problem, no optimal solution exists. However, let us start by trying to solve the problem, to see that trouble arises, and let us then consider a modified problem where a solution does exist.

Let us try a solution of the HJB equation of the form: $J(t, x) = a(t)x + b(t)x^2 + c(t)$. Now, as the maximizing u equals $J_x/2$, this gives $0 = J_t + J_x^2/4 + J_{xx}\sigma^2/2$, so we get $0 = a'x + b'x^2 + c' + a^2/4 + abx + b^2x^2 + b\sigma^2$. This gives $b' = -b^2$, $a' = -ab$, $c' = -a^2/4 - b\sigma^2$. Below, A, C, and D are integration constants. Evidently, we get $b = 1/(t + D)$, $a = A/(t + D)$, and $c = C + (A^2/4)(t + D)^{-1} - \sigma^2 \ln|t + D|$. Now, $dX_t^* =$

$$[A/2(t + D) + (1/(t + D))X_t^*]dt + \sigma dB_t = [(A/2 + X_t^*)/(t + D)]dt + \sigma dB_t.$$

Letting $Y_t = (A/2 + X_t^*)/(t + D)$ and using the next to last equality, we get $dY_t =$

$$[-(A/2 + X_t^*)(t + D)^{-2}]dt + (t + D)^{-1}dX_t^* = (t + D)^{-1}\sigma dB_t,$$

so $Y_t = Y_0 + \int_0^t (\tau + D)^{-1}\sigma dB_\tau$.

As $Y_0 = A/2D, X_t^* = -A/2 + (t + D)A/2D + (t + D)\int_0^t (\tau + D)^{-1}\sigma dB_\tau = tA/2D + (t + D)\int_0^t (\tau + D)^{-1}\sigma dB_\tau$. It seems that we can obtain $X_1^* = 1$ a.s. if $A/2D = 1$, i.e., $A = 2D$ and if $(1 + D) = 0$, so we put $D = -1$, and $A = -2$. Then $X_t^* = t + (t - 1) \int_0^t (\tau - 1)^{-1} \sigma dB_\tau$ and $J(t, x) =$

$$-(x - 1)^2 / (1 - t) - 1/(t - 1) + c(t) = -(x - 1)^2 / (1 - t) + C - \sigma^2 \ln |t - 1|.$$

Now, $dX_t^* =$

$$\left(1 + \int_0^t (\tau - 1)^{-1} \sigma dB_\tau\right) dt + \sigma dB_t,$$

so $u_t^* = 1 + \int_0^t (\tau - 1)^{-1} \sigma dB_\tau$. Is u_t^* "acceptable"? In fact not: Let us impose the reasonable acceptability criterion $E[\int_0^1 u_t^2 dt] < \infty$. By Ito's isometry, $E[\int_0^1 (\int_0^t (\tau - 1)^{-1} \sigma dB_\tau)^2 dt] = \int_0^1 [\int_0^t (\tau - 1)^{-2} \sigma^2 d\tau] dt = |_0^1 \{ -\sigma^2 (t + \ln |t - 1|) \}$, which is infinite. (A further discussion of this problem can be found in C in the Addendum to the chapter.)

A slight modification of the problem is to replace σ by $\sigma(1 - t)^\alpha$, $\alpha \in (0, 1]$. Now the stochastic disturbance fades away as we approach the horizon. Then an optimal u_t^* can be found using the sufficient conditions above. Let us show this only for $\alpha = 1$, in which case the differential equation is $dX_t = udt + (1 - t)\sigma dB_t$. The only change needed is then to replace σ by $\sigma(1 - t)$ everywhere above. Let us again place the following condition on u_s for u_s to be acceptable: $E[\int_0^1 u_s^2 ds] < \infty$. Then (see Example 4.26) $E[(1 - t)^{-1} (\int_t^1 |u_s| ds)^2] \leq E[\int_t^1 u_s^2 ds]$.

Now, $a(t) = -2/(t - 1)$ and $b(t) = 1/(t - 1)$ (the same as above), $c' = -a^2/4 - (1 - t)^2 b\sigma^2$, $c(t) = C + (t - 1)^{-1} - \sigma^2 (1 - t)^2 / 2$, and $J(t, x) = -(x - 1)^2 / (1 - t) - \sigma^2 (1 - t)^2 / 2 + C$, where $C = 0$, as $\lim_{t \to 1} J(t, 1) = 0$. Moreover, $Y_t = Y_0 - \int_0^t \sigma dB_\tau = Y_0 - \sigma B_t$ and $X_t^* = t + (1 - t)\sigma B_t$, which satisfies $X_1^* = 1$ a.s. For any given admissible pair X_t, u_t, using Ito's isometry, Cauchy–Schwartz's inequality, and $1 = X_1 = X_t + \int_t^1 u_s ds + \int_t^1 (1 - s)\sigma dB_s$, yield

$$E(X_t - 1)^2 = E\left(\int_t^1 u_s ds\right)^2 + 2E\left[\int_t^1 u_s ds \int_t^1 (1 - s)\sigma dB_s\right]$$

$$+ E\left(\int_t^1 (1 - s)\sigma dB_s\right)^2 \leq E\left(\int_t^1 u_s ds\right)^2$$

$$+ 2\left(E\left[\left(\int_t^1 u_s ds\right)^2\right]\right)^{1/2} \left(E\left[\left(\int_t^1 (1 - s)\sigma dB_s\right)^2\right]\right)^{1/2}$$

$$+ \int_t^1 (1 - s)^2 \sigma^2 ds$$

$$= E\left(\int_t^1 u_s ds\right)^2 + 2\left(E\left[\left(\int_t^1 u_s ds\right)^2\right]\right)^{1/2} \left(\int_t^1 (1 - s)^2 \sigma^2 ds\right)^{1/2}$$

$$+ \int_t^1 (1 - s)^2 \sigma^2 ds.$$

Then, using $\lim_{t \to 1} E[(1 - t)^{-1} (\int_t^1 u_s ds)^2] = 0$, we get that $\lim_{t \to 1} E(X_t - 1)^2 / (1 - t) = 0$. This limit implies that (4.75) is satisfied, as $u_t^* dt = dX_t^* - (1 - t)\sigma dB_t = 1 - \sigma B_t + (1 - t)\sigma dB_t - (1 - t)\sigma dB_t = 1 - \sigma B_t$ is acceptable. Thus, u_t^* is optimal. $\square$

4.5 Optimal Stopping

Consider the stochastic differential equation

$$dX_t = f(t,X_t)\,dt + \sigma(t,X_t)\,dB_t, \quad X_0 = x_0, \quad x_0 \text{ a given vector,} \qquad (4.76)$$

determining $X_t \in \mathbb{R}^n$, where f has n components, $\sigma = \{\sigma_{i,j}\}_{i=1,...,n,j=1,...,m}$ and B_t has m components, the components being independent Brownian motions. So (4.1) consists of n equations, one for each component of X_t. Let $g(t,X_t)$ be a *reward* obtained by stopping the process at (t,X_t) (or perhaps better: jumping off the process at time t), and let $f_0(t,x)$ be the reward obtained per unit of time, as long as we have not jumped off. We shall assume that f,σ,g, and f_0 are continuous and Lipschitz continuous in x, uniformly in t, with $f(t,x_0)$ and $\sigma(t,x_0)$ bounded. Furthermore, we shall assume that an open set $V \subset \mathbb{R}^{n+1}$ is given, containing $(0,x_0)$, such that we have to stop at the latest when the process X_t leaves V, and in particular we stop immediately if $(t,X_t) \in \partial V$. For simplicity, the reader may always assume that $V = (-\infty,b) \times B(0,b')$, where b and/or b' may equal ∞ (the ball $B(0,\infty)$ equals $\mathbb{R}^n$). If b is finite, it means that, at latest we have to stop at time b.

The problem is to find when it is optimal to stop, that is, we want to maximize

$$E\left[\int_0^\tau f_0(t,X_t)\,dt + g(\tau,X_\tau)\right] \qquad (4.77)$$

over the set of all "admissible" stopping rules τ. By convention, we do not get any stopping reward when we do not stop, so for $t \in [0,\infty]$, read $g(t,x)$ to mean $g(t,x)1_{[0,\infty)}(t)$. A class of stopping rules it is reasonable to consider is the one consisting of stopping rules that depend on the history of the process B_t. Such rules are called stopping times. For such a stopping rule, it may be that we stop early for some outcomes of the process, and lately for other outcomes. It may often be natural to restrict the type of history dependence we consider. In fact, a natural class of stopping times is the one consisting of Markov stopping times, which depends only on the "present state" of the process. To be more precise, Markov stopping times can be described as follows. A set $A \subset V$ is specified, such that, intuitively, if we start inside the set we stop the first time the state leaves A. The set A may be called a continuation region, and formally, we define the stopping time corresponding with A as $\tau := \tau_A = \min\{t \geq 0 : (t,X_t) \notin A\}$. If we never have $(t,X_t) \notin A$, let $\tau_A = \infty$ (and if A is empty, $\tau_A = 0$). Note that if $(0,X_0) \notin A$, then $\tau_A = 0$. In practice, we only consider open sets A, in which case the minimum always exists. Let T^M be the set of such stopping times, i.e., $T^M = \{\tau_A : A \text{ is an open subset of } V\}$. Thus, the stopping problem consists in maximizing $E[\int_0^\tau f_0(t,X_t)\,dt + g(\tau,X_\tau)]$, for $\tau \in T^M$. For any given τ, it may be that for some outcomes of history, τ equals ∞.

Define the set T^* to be the set of all history dependent, or adapted stopping times $\tau(\omega) \leq \tau_V$. (Note that $\tau(\omega)$ is adapted iff the indicator function $1_{[0,\tau(b(.))]}(t), b(.) \in C_0([0,b])$ is an adapted function in the sense we have defined earlier.) Sometimes, when maximizing the criterion, we may want to maximize the criterion over this set, which is larger than T^M. However, *optimal* stopping times most often turn out to be of the Markovian type, and if the reader wants, below he may assume that T^* equals T^M (the set of Markov stopping times).

The stopping times so far considered take values in $[0, \infty]$. Below, we also need stopping times taking values in $[t, \infty]$. The reason is that, as an aid to the solution of the optimal stopping problem, we need to consider the following situation: Given that we are in state x at time t (i.e., given that the process X_s is at the point x for $s = t$, $(t, x) \in V$), how shall we behave from then on. Then to any open set $A \subset V$, there corresponds a stopping time $\tau_{A,t,x} = \min\{s \geq t : (s, X_s) \notin A$ given that $X_t = x\}$. (Again, $\tau_{A,t,x} = \infty$ is allowed.) If $(t, x) \notin A$, then $\tau_{A,t,x} = t$. Let $T_{t,x}^M$ be the set of such Markov stopping times, and let $T_{t,x}^*$ be the set of stopping times $\leq \tau_V$ depending on the history of B_s, $s \geq t$, given that the process starts at (t, x). (If the reader wants, he may assumed that $T_{t,x}^*$ equals $T_{t,x}^M$ below.)

Again an optimal value function, denoted $J(t, x)$, is needed. It is defined as follows: Given that we start at (t, x), it is the maximal expected total reward that can be obtained when going through all possible stopping times τ. When the process starts at (t, x), the corresponding conditional expectation is denoted $E^{t,x}$ or sometimes $E[.|t, x]$. So formally, $J(t, x)$ is defined on cl$V$ by

$$J(t, x) = \sup_{\tau \in T_{t,x}^*} E^{t,x} \left[\int_t^\tau f_0(t, X_t) dt + g(\tau, X_\tau) \right].$$

Define $D := \{(t, x) \in V, t \geq 0 : g(t, x) < J(t, x)\}$. This set D is called the *optimal continuation region*, because if $(t, x) \in D$, it is optimal to continue (not stop at time t). The *optimal stopping time*, given that X_s start in (t, x), is defined by $\tau^* := \tau_{D,t,x} := \inf\{s \geq t : (s, X_s) \notin D$, given $X_t = x\}$. Evidently, if $(t, x) \in$ cl$V \setminus D$, then $\tau_{D,t,x} = t$, $J(t, x) = g(t, x)$. When $\tau_{D,0,x_0}$ is found, we have a solution to the problem of maximizing the expected total reward (4.77). Hence, when we have found D, we have solved the optimal stopping problem.

We have actually here assumed that we want to stop as rapidly as possible. Thus, by convention, in cases where we are actually indifferent, namely if the reward obtained by stopping at (t, x) equals the expected reward from some optimal stopping rule that implies that we continue at (t, x), we choose to stop immediately.

Heuristic Derivation of Necessary Conditions

We shall derive some equations that will help us to solve the stopping problem. To simplify, we put $f_0 \equiv 0$. Note that $J(t, x) = E^{t,x}[g(\tau_D, X\tau_D)]$, where we write τ_D as a shorthand for $\tau_{D,t,x}$. As before, the expectation is a conditional one: We condition on the fact that we start at (t, x). In D, the function $J(t, x)$ has a special form as will now be shown:

Assume that J is C^2. Let G be a ball around (t, x) in D. Note that if X_s crosses the boundary of G at time τ_G, then $\tau_D := \tau_{D,t,x} = \tau_{D,\tau_G,X_{\tau_G}}$. We then have

$$E[g(\tau_D, X_{\tau_D})|t, x]$$
$$= E[E[g(\tau_D, X_{\tau_D})|\tau_G, X_{\tau_G}, t, x]|t, x] = E[E[g(\tau_D, X_{\tau_D})|\tau_G, X_{\tau_G}]|t, x],$$

by the double expectation rule, and the fact that when (τ_G, X_{τ_G}) is given, the future development of X_t does not depend on (t, x), so (t, x) can be dropped in the inner expectation. (Formally, what is called the strong Markov property of X_t is used here.) Using the definition of J, this equality is the same as $J(t, x) = E[J(\tau_G, X_{\tau_G})|t, x]$. By the Dynkin formula (4.18),

$$E[J(\tau_G, X_{\tau_G})|t, x)] = J(t, x) + E^{t, x}\left[\int_t^{\tau_G} LJ(s, X_s)\, ds\right],$$

where $LJ := J_t + J_x f + (1/2)\sigma^2 J_{xx}$ (or its equivalent if the dimension $n > 1$). Because $E[J(\tau_G, X_{\tau_G})|t, x] = J(t, x)$, it follows that $E^{t, x}[\int_t^{\tau_G} LJ(s, X_s)\, ds] = 0$. When the radius of G is small, $LJ(s, X_s)$ can be replaced by $LJ(t, x)$, the error being negligible. If this is done, the integrand becomes deterministic, and the last equality simplifies to $LJ(t, x)\{E^{t, x}[\tau_G] - t\} = 0$, or $LJ(t, x) = 0$.

That the error made is negligible is shown as follows: Let $H_G := \sup\{|LJ(s, y) - LJ(t, x)| : (s, y) \in G\}$, and note that $E^{t, x}[|\int_t^{\tau_G} LJ(s, X_s) - LJ(t, x)\, ds|] \leq H_G E^{t, x}[(\tau_G - t)]$. Because $H_G \to 0$ when G decreases toward its center (t, x), the error is a small fraction of $E^{t, x}[(\tau_G - t)]$. Hence, the error is negligible.

To sum up, we have obtained the crucial equation $LJ = 0$. Hence, in D, the function $J(t, x)$ satisfies the equation

$$0 = J_t(t, x) + J_x(t, x)f(t, x) + (1/2)J_{xx}(t, x)\sigma^2(t, x), \tag{4.78}$$

for all $(t, x) \in D$. The equivalent equation in several dimensions is

$$0 = J_t + J_x f + (1/2)\sum_{i,j}(\sigma\sigma')_{i,j}J_{x_i x_j} \tag{4.79}$$

(shorthand notation). Furthermore, $J(t, x) = g(t, x)$ for $(t, x) \in \text{cl}V \setminus D$, and in particular, for $(t, x) \in \partial D$. Both properties follow from definition.

Assume that in the one-dimensional case, for some α, one has $\sigma \geq \alpha > 0$ everywhere. Then it normally follows that, for $t > 0$, $(t, x) \in \partial D \cap V$,

$$J_x(t, x) = g_x(t, x). \tag{4.80}$$

In several dimensions this equation reads

$$J_{x_i}(t, x) = g_{x_i}(t, x) \tag{4.81}$$

for all $i \notin I$, where I is the set of indices such that the corresponding equation in (4.76) does not contain any Brownian motion ($\sigma_{i,j}$ vanishes for $i \in I$, all j).

The condition (4.81) can most often only be expected to hold when further conditions are added. An example of such a condition is, if σ is simply a diagonal matrix, that for some $\delta > 0$, $\sigma_{jj}(t, x) \geq \delta > 0$ for all (t, x), $j \notin I$.

In the one-dimensional case, an informal argument for (4.80), due to A. Dixit, is as follows. We have that $J = \max\{\hat{J},g\}$, where $\hat{J}$ supposedly solves (4.78) for all (t,x) (C^2 for all (t,x)). Let $X_{\tau_D} = \hat{x}$. A kink in the graph of $x \mapsto J(t,x)$ at this boundary point $(\tau_D,\hat{x})$ of D would make the graph of $x \mapsto J(t,x)$ V-shaped near such a point. Consider a time $t = \tau_D + dt$ a little later than the stopping time τ_D. Given $(\tau_D,X_{\tau_D}) = (\tau_D,\hat{x})$, the scatter of possible points (X_t) would spread out symmetrically around $(E^{\tau_D,\hat{x}}X_t)$ with a "diameter" proportional to $(dt)^{1/2}$ (actually standard deviation $\sim (dt)^{1/2}$). For dt small (negligible as compared with $(dt)^{1/2}$), $E^{\tau_D,\hat{x}}X_t \approx \hat{x}$. Because of the V-shape, the expected value of $J(t,X_t)$, given $(\tau_D,\hat{x})$, would then be greater than $J(\tau_D,\hat{x})$: This expectation is, roughly, the average of ordinates of points in the graph of $x \mapsto J(\tau_D,x)$ spread out roughly equally on both sides of the point $(\hat{x},J(\tau_D,\hat{x}))$. (We have then used $J(t,\hat{x}) \approx J(\tau_D,\hat{x})$.) Because we *can* obtain the expected value $J(t,X_t)$ when having arrived at (t,X_t) (expectation given (t,X_t)), the expected value of $J(t,X_t)$ given $(\tau_D,\hat{x})$ (double expectation here) *can* be obtained when having arrived at $(\tau_D,\hat{x})$. As we saw, with a kink at $\hat{x}$ of $x \to J(\tau_D,x)$, this average is $> J(\tau_D,\hat{x})$, which is a contradiction. No kink means that (4.80) is satisfied.

If f_0 is nonzero, add the term f_0 on the right-hand sides of (4.78) and (4.79). When we write $\hat{J}$ instead of J, we then get

$$0 = f_0(t,x) + \hat{J}_t(t,x) + \hat{J}_x(t,x)f(t,x) + (1/2)\hat{J}_{xx}(t,x)\sigma^2(t,x). \tag{4.82}$$

The equivalent in several dimensions is (shorthand notation)

$$0 = f_0 + \hat{J}_t + \hat{J}_x f + (1/2)\sum_{i,j}(\sigma\sigma')_{i,j}\hat{J}_{x_i x_j}. \tag{4.83}$$

We can now describe a procedure for solving optimal stopping problems:

Find a function $\hat{J}$ and an open set $D^* \subset V \cap [(0,\infty) \times R^n]$ such that the following properties hold: The point $(0,x_0)$ belongs to $\mathrm{cl}D^*$,

1. $\hat{J}$ is twice continuously differentiable in $V \setminus \partial D^*$, with bounded second derivatives on bounded subsets of this set, is Lipschitz continuous on $\mathrm{cl}V$, and is C^1 in V.
2. The equality (4.82) holds in D^*, (in several dimensions (4.83)). Moreover,

$$\text{for } (t,x) \in \partial D^*, t > 0, \hat{J}(t,x) = g(t,x). \tag{4.84}$$

3.

$$\hat{J} \geq g \text{ in } V, \hat{J} > g \text{ in } D^*. \tag{4.85}$$

4.

$$\text{In } V \setminus \mathrm{cl}D^*, \ f_0(t,x) + \hat{J}_t(t,x) + \hat{J}_x(t,x)f(t,x) + (1/2)\hat{J}_{xx}(t,x)\sigma^2(t,x) \leq 0. \tag{4.86}$$

In the multidimensional case, the corresponding inequality is:

$$f_0 + \hat{J}_t + \hat{J}_x f + \sum_{i,j} (1/2) \hat{J}_{x_i x_j} [\sigma \sigma']_{i,j} \leq 0. \tag{4.87}$$

Observe that (4.84) and (4.85) actually imply (4.80) and (4.81) for all i, for $J = \hat{J}$, $D = D^*$, and (4.80) and (4.81) are useful for the construction of a solution. For $n > 1$, the C^1 condition on $\hat{J}$ in V, i.e., on $V \cap \partial D$ (it is C^2 elsewhere), may be too strong, for $i \in I$, for example $x_i \to \hat{J}(t,x)$ may have a kink, but we don't want to pursue this point.

Let $E = E^{0,X_0}$, as before.

Theorem 4.28 (Sufficient condition). *Assume that V is contained in $(-\infty, b] \times \mathbb{R}^n$, $b < \infty$, and that some function $\hat{J}$ and some open set D^* have been found, such that the properties 1. to 4. are satisfied. Moreover, assume that the process a.s. stays a negligible time on ∂D^*, formally expressed as:*

$$E \int_0^{\tau_V} 1_{\partial D^*} (s, X_s) \, ds = 0, \tag{4.88}$$

where X_s satisfies (4.76). Then $\tau_{D^*} = \inf\{t > 0 : (t, X_t) \notin D^*\}$ is an optimal stopping rule, provided $\tau_{D^*} > 0$ a.s. $\square$

Remark 4.29 (Stopping at once). Assume that $b < \infty$. Note that if some function $\hat{J}$ satisfying property 1. has been found such that $\hat{J}(0,x_0) = g(0,x_0)$, (4.87) ((4.86)) holds in $(V \backslash \mathrm{cl} D^*) \cup D^*$, and $\hat{J} \geq g$ holds in V, then we stop immediately. $\square$

Remark 4.30 (Conditions on D^*). Actually, for the theorem (and the remarks below) to hold, we also need to assume that D^* equals a set of the form $\{(s,y) \in V : \max_i \theta_i(s,y) < 0\}$, where $\{\theta_i\}_i$, is a finite collection of C^1-functions the gradients of which at each point on the boundary of the set satisfy the standard full rank condition of nonlinear programming (or the Slater condition related to these gradients). $\square$

If $\hat{J}$ is C^2 in all V, then it is not needed to test (4.88). (Often, however, $\hat{J}$ is so constructed that it equals g outside D^*. In this case, the C^2-property frequently fails on ∂D^*.) Concerning the condition $\hat{J}(t,x) = g(t,x)$ for $(t,x) \in \partial D^*$, $t > 0$, it is not necessary to require this condition on any part of ∂D^* that is never crossed. (So what we actually need is that $\hat{J}(\tau_{D^*}, X_{\tau_{D^*}}) = g(\tau_{D^*}, X_{\tau_{D^*}})$ with probability 1.)

Sketch of a proof of the theorem in the one-dimensional case

Assume that $\hat{J}$ has the properties 1. to 4. It is then easy to show that the optimal stopping time has been found: For any stopping time $\tau \in T^*$, $E[g(\tau, X_\tau)] \leq$

$$E[\hat{J}(\tau, X_\tau)] = \hat{J}(0, X_0) + E\left[\int_0^\tau L\hat{J}(s, X_s) ds\right] \leq E\left[\int_0^\tau -f_0(t, X_t) dt\right] + \hat{J}(0, X_0) \tag{4.89}$$

where we first used $\hat{J} \geq g$, then Dynkin's formula (4.18), and for the last inequality, $L\hat{J} = \hat{J}_t + \hat{J}_x f + 1/2 \hat{J}_{xx} \sigma^2 \leq -f_0$, which holds both in D^* and in $V \backslash \mathrm{cl} D^*$. Furthermore, by definition of τ_{D^*}, $\hat{J} = g$ on ∂D^* and by Dynkin's formula and (4.82), we get

$$E[g(\tau_{D^*},X_{\tau_{D^*}})] = E[\hat{J}(\tau_{D^*},X_{\tau_{D^*}})]$$

$$= \hat{J}(0,X_0) + E\left[\int_0^{\tau_{D^*}} L\hat{J}(s,X_s)ds\right] = E\left[\int_0^{\tau_{D^*}} -f_0(t,X_t)dt\right] + \hat{J}(0,X_0). \quad (4.90)$$

Hence, $E[g(\tau,X_\tau) + \int_0^\tau f_0 dt] \leq \hat{J}(0,x_0) = E[g(\tau_{D^*},X_{\tau_{D^*}}) + \int_0^{\tau_{D^*}} f_0 dt]$. So, τ_{D^*} is "best."

Here Dynkin's formula was used twice. We have sketched a proof of this formula only for C^2-functions. In the first application, the function was only C^1 on ∂D^*. When points where $\hat{J}$ is only C^1 are passed as swiftly as stated in (4.88), it may be seen that Dynkin's formula still holds. See Øksendal (2003) for the details of the arguments. (There $\hat{J}$ is approximated by functions ϕ_j being C^2 in all V, and, to be sure that Dynkins formula applies, first stopping times for which (t,X_t) stays bounded are used, which then are allowed to converge to τ and τ_{D^*}, respectively.) $\square$

Remark 4.31 (Unbounded set D^*, $b = \infty$). When D^* is not necessarily bounded and $b = \infty$, we allow for the case where we never stop ($\tau = \infty$ perhaps with positive probability). We have introduced the convention that $g(\infty,x) = 0$, more precisely we use the convention that, for any stopping time $\tau(\omega) \in T^*$, $g(\tau(\omega),X_{\tau(\omega)}) = 0$, for all ω for which $\tau(\omega) = \infty$. Assume that $E\int_0^\infty |f_0(t,X_t)|dt < \infty$. Then Theorem 4.28 holds in three cases, (a), (b), and (c):

(a) $\hat{J}$ is bounded on D^*, $\min\{0,g(t,x)\}$ is bounded on V, and $\tau_{D^*} < \infty$ a.s.

(b) Assume $n = m = 1$. Both $\lim_{t\to\infty} g(t,x_0)$ and $\lim_{t\to\infty} \hat{J}(t,x_0)$ equals zero. For some positive constants $\alpha, c_J, c_g, a_0, b_0, a, \hat{b}, a_f, b_f$, for all $(t,x) \in D^*$, $|\hat{J}_x(t,x)| \leq c_J e^{-\alpha t}$, and for all $(t,x) \in V$, $|g_x(t,x)| \leq c_g e^{-\alpha t}$, $|f_0(t,x)| \leq (a_0 + b_0|x|)e^{-\alpha t}$, $|\sigma_{ij}(t,x)| \leq a + \hat{b}|x|$, $|f(t,x)| \leq a_f + b_f|x|$, $\alpha > b_f$. (In this case, $E\int_0^\infty |f_0(t,X_t)|dt < \infty$ automatically holds.)

(c) For any given $\tau \in T^*$, there exists an increasing sequence of natural numbers n_j such that if $\tau_j := \min\{\tau, n_j, \tau_{n_j}\}$ and $\tau_j^* := \min\{\tau_{D^*}, n_j, \tau_{n_j}\}$, $\tau_{n_j} := \tau_{(-\infty,\infty)\times B(0,n_j)}$, then $\lim_{j\to\infty} E[\hat{J}(\tau_j^*(\omega),X_{\tau_j^*(\omega)})] \leq E[g(\tau_{D^*}(\omega),X_{\tau_{D^*}(\omega)})]$ and $\lim_{j\to\infty} E[g(\tau_j(\omega),X_{\tau_j(\omega)})] \geq E[g(\tau(\omega),X_{\tau(\omega)})]$ (replace limits by liminf in case the limits do not exist). If $g \geq K$ for some constant K and $\lim_{T\to\infty} g(T,X_T) = 0$ a.s., the last inequality automatically holds.

Note, finally, that if $E[\int_0^\infty |f_0(t,X_t)dt] < \infty$, and $\min\{0,g(t,x)\}$ is bounded on V, then Remark 4.29 still holds even when $b = \infty$. $\square$

We here give a proof in the case (c) of the above Remark. Given any $\tau \in T^*$. Then the proof above (cf. (4.89), (4.90)) yields $E[g(\tau_j,X_{\tau_j}) + \int_0^{\tau_j} f_0 dt] \leq \hat{J}(0,X_0) = E[\hat{J}(\tau_j^*,X_{\tau_j^*}) + \int_0^{\tau_j^*} f_0 dt]$. Taking limits on both sides as $j \to \infty$ yields $E[g(\tau,X_\tau) + \int_0^\tau f_0 dt] \leq \liminf_j E[g(\tau_j,X_{\tau_j}) + \lim_j \int_0^{\tau_j} f_0 dt] \leq E[g(\tau_{D^*},X_{\tau_{D^*}}) + \int_0^{\tau_{D^*}} f_0 dt]$. Hence, optimality of τ_{D^*} follows. A proof in case (b) is given in D in the Addendum to the chapter.

Infinite horizon, autonomous case

Assume that $V = (-\infty, \infty) \times V^*$, $f_0 = f^0(x)e^{-\alpha t}$, $\alpha > 0$, that $g(t,x) = g^0(x)e^{-\alpha t}$, $(t < \infty)$, and that f and σ are independent of t. Then we can expect that $\hat{J}$ is of the form $\hat{J} = h(x)e^{-\alpha t}$, and that D^* is of the form $(0, \infty) \times D^{**}$. In the one-dimensional case, (4.82) and (4.86) follow if $h(x)$ satisfies, respectively,

$$0 = -\alpha h + f^0 + h_x f + (1/2)h_{xx}\sigma^2 \quad \text{in } D^{**}, \tag{4.91}$$

and

$$-\alpha h + f^0 + h_x f + (1/2)h_{xx}\sigma^2 \leq 0 \quad \text{in } V^* \setminus \text{cl}D^{**}. \tag{4.92}$$

Now the solution method in the one-dimensional case consists in finding a h-function and an open set D^{**}, satisfying (4.91), $h \geq g^0$ in V^*, $h > g^0$ in D^{**}, $h = g^0$ for $x \in \partial D^{**}$, and even the equality $h_x = g_x^0$ for $x \in \partial D^{**}$ is useful for the construction of a solution. The function h must be C^2 in $V^* \setminus \partial D^{**}$, C^1 in V^* and Lipschitz continuous on $\text{cl}V^*$, and $x_0 \in \text{cl}D^{**}$.

Remark 4.32 (Negligible time on ∂D^*). Consider the case where σ is a diagonal matrix with diagonal elements σ_{ii} Then, roughly speaking, (4.88) holds if the following property is satisfied: Given the functions θ_i in Remark 4.31, for any (t,x) such that $\theta_i(t,x) = 0$, either $(\partial \theta_i(t,x)/\partial x_j)\sigma_{jj}(t,x) \neq 0$ for some j, or (if these expressions are zero), $\partial \theta_i(t,x)/\partial t + \sum_j (\partial \theta_i(t,x)/\partial x_j)f_j(t,x) + (1/2)\sum_{j \notin I}(\partial^2 \theta_i(t,x)/\partial x_j^2)\sigma_{jj}(t,x)^2 \neq 0$. This slightly heuristic result follows from considering the formula $\theta_i(t+dt) =$

$$\theta_{it}(t,x) + \theta_{ix}(t,x)(f(t,x)dt + \sigma(t,x)dB_t) + (1/2)\sum_j \theta_{ix_jx_j}(t,x)\sigma_{jj}(t,x)^2 dt$$

and noting that terms $\theta_{ix_j}(t,x)[\sigma(t,x)dB_t]_j$ such that $\theta_{ix_j}(t,x)\sigma_{jj}(t,x) \neq 0$ will dominate and the sum over j of such ones will then a.s. be nonvanishing. If on the other hand $\theta_{ix_j}(t,x_j)\sigma_{jj}(t,x) = 0$ for all j, and $\theta_{it} + \theta_{ix}f(t,x) + (1/2)\sum \theta_{ix_jx_j}\sigma_{jj}(t,x)^2 \neq 0$, then the terms corresponding with the last expression will dominate, and for dt small we will have $\theta_i(t+dt) \neq 0$. All in all, we can a.s. expect $\theta_i(t+dt) \neq 0$, i.e., we stay a negligible time on ∂D^*. $\qquad \square$

Remark 4.33 (Stochastic control with the horizon as a choice variable). Sometimes X_t, the process determined by (4.76), can be controlled, because u appears as a variable in f_0, f, and/or in σ. Assume that f, σ, g, and f_0 are continuous, and Lipschitz continuous in x, u, uniformly in t, with $|f(t,x_0,u)|, |\sigma(t,x_0,u)| \leq K$, for some K independent of (t,u). In the current case, the method of solution is analogous to the points 1. to 4. for the solution of optimal stopping problems, in fact the only changes needed in these points are that $\hat{J}$ should now be constructed as a solution in D^* of the HJB equation instead of as a solution of (4.82) and that (4.86) is modified to read

$$\hat{J}_t + \sup_u \{f_0 + \hat{J}_x f + (1/2)\hat{J}_{xx}\sigma^2\} \leq 0 \text{ in } V \setminus \text{cl}D^*, \tag{4.93}$$

when $n = 1$, i.e., in the one-dimensional case (in several dimensions use the corresponding expression). Then again D^* is the (optimal) continuation region and the optimal stopping time τ_{D^*} is again determined by D^*, and the optimal u^0 is obtained from the maximization in the HJB equation. Theorem 4.28 now holds for these modifications of conditions 1. to 4. (still condition (4.88) is needed). The smoothness conditions on $\hat{J}$ are kept.

If $b = \infty$ and D^* is unbounded, then assumptions similar to those in Remark 4.31 must also be satisfied. To present one specific such assumption, assume in (b) that the inequalities hold for all u. This modified version of (b) is sufficient for τ_{D^*} to be optimal in the current case. $\square$

In case of Remark 4.33 a sketch of the sufficiency proof is virtually the same as before (we put $f_0 \equiv 0$, and assume $b < \infty$): Now we write $L^u \hat{J}$ instead of $L\hat{J}$ because now $L\hat{J}$ depends on u. For any τ, any control function $u(.,.)$,

$$E[g(\tau, X_\tau)] \leq E[\hat{J}(\tau, X_\tau)] = \hat{J}(0, X_0) + E[\int_0^\tau L^u \hat{J}(s, X_s)\, ds] \leq \hat{J}(0, X_0),$$

where the last inequality follows from the fact that (4.93) holds for all (t, x) in $(V \setminus \mathrm{cl}D^*) \cup D^*$, $(t, x) \in V$. When $\tau = \tau_{D^*}$ and $u = u^0$ we get equalities only.

Example 4.34. Let the price of oil P_t follow a geometric (or "lognormal") Brownian motion, i.e., $dP_t = \beta P_t dt + \sigma P_t dB_t$, β and σ positive numbers. Suppose that oil is produced at a constant cost per barrel equal to C. Maximizing total expected profit from a reserve of fixed size is then equivalent to maximizing expected profit on each unit produced. In the latter case, the reward function is $(p - C)e^{-rt}$, where $r > 0$ is the discount rate. The problem is to choose the right moment in time at which to empty the field (it it is fairly obvious that it pays to empty it "in a second"). It is assumed that $\beta < r$. We want to find the time τ that gives the maximal expected reward. Let us use Theorem 4.28, i.e., let us try to construct a function $\hat{J}(t, p)$ satisfying 1. to 4. We let $V^* = (0, \infty)$, when $P_0 > 0$, P_t belongs to V^* for all t. The problem at hand is an autonomous one, so we let $\hat{J}(t, x) = h(x)e^{-rt}$ and the HJB equation to consider is (4.91), which here takes the form $0 = -rh + \beta ph' + \sigma^2 p^2 h''/2$. It is known that solutions of this equation are obtained by trying solutions of the form p^γ. Inserting this function in the differential equation, one gets $0 = \theta(\gamma)p^\gamma$, where $\theta(\gamma) = (\sigma^2/2)\gamma^2 + (\beta - \sigma^2/2)\gamma - r$. The resulting equation $\theta(\gamma) = 0$ has two solutions for γ, one defined by $\gamma = (1/\sigma^2)\{[-(\beta - \sigma^2/2) + [(\beta - \sigma^2/2)^2 + 2r\sigma^2]^{1/2}\}$ and another one, denoted γ', arising from using a minus sign in front of the square root. The general solution for h is then $h = kp^\gamma + k'p^{\gamma'}$, k, k' two arbitrary constants. We need to show that $1 < \gamma$, which follows by noting that $\theta(1) = \beta - r < 0$, $\theta(\infty) = \infty$. Moreover, $\gamma' < 0$, because $\theta(0) = -r < 0$, and $\theta(-\infty) = \infty$.

Some information about the constants k and k' should be obtained by the current value version of the boundary conditions (4.80), (4.84). But ∂D^{**}, what is that? (We need it in order to use the current value version of (4.80).) The continuation region is presumably a set of the form $\{(t, p) : p \leq p^*\}$, i.e., $D^{**} = (0, p^*)$ for some threshold or critical price p^*. (It seems reasonable to continue waiting until the price is high enough.) Here p^* is an unknown to be determined. We have $h(p^*) = p^* - C$, and from (4.80), we get one additional equation. But we have three constant to determine: k, k', and p^*. If $k' \neq 0$, $h(p)$ behaves in a suspicious way: $|J(0,0^+)| = |h(0^+)| = \infty$, which contradicts the fact that for a starting point $(t,p) = (0,0)$, the supremum of the expected reward is bounded by $-C$ and 0 (for such a starting point, $P_t = 0$ for all t). By this informal argument, we obtained a third relationship for determining the three unknowns, in fact we got $k' = 0$. Then the two former equations determine k and p^*: The equations become (by the current value versions of (4.80) and (4.84)): $p^* - C = kp^{*\gamma}$ and $1 = k\gamma p^{*\gamma-1}$. The second equation gives $k = p^{*-(\gamma-1)}/\gamma$. Inserting this in the first equation gives $p^* - C = p^*/\gamma$, or $p^* = C/(1-1/\gamma) > C > 0$. Hence $\hat{J} = (C/(1-1/\gamma))^{-(\gamma-1)}p^\gamma e^{-rt}/\gamma > 0$.

This is our proposal for $\hat{J}$. We now assume $P_0 = p_0 < p^*$. We need to make sure that the points 1. to 4. are satisfied. Now, (4.82) is satisfied in $V = (-\infty,\infty) \times (0,\infty)$. Moreover, by construction, $\hat{J}(t,p) = (p-C)e^{-rt}$ on the part of the boundary of D^* that is relevant $((-\infty,\infty) \times \{0\}$ is never crossed, see a comment subsequent to Theorem 4.28), and as a function of p, $\hat{J}$ is (strictly) convex ($\gamma > 1$) and has the line $(p-C)e^{-rt}$ as tangent at p^* by construction. Hence as a function of p, $\hat{J}$ lies above this tangent. Hence, 1. to 4. hold in the required manner.

It is needed to check that one of the conditions in Remark 4.31 applies. In fact, the conditions in (b) automatically applies. Because $\hat{J}$ in this example is C^2 in V, it is not necessary to check condition (4.88). But, actually, it is satisfied, for the following reason: For any fixed s, $\Pr[P_s = p^*] = 0$. (To reach this conclusion, we only need to know that the probability distribution of P_t is given by a density, in fact $\ln P_t$ is normally distributed.) Next, by Fubini's theorem (i.e., the fact that integrals can be interchanged) $E[\int_0^\infty 1_{(0,\infty) \times \{p^*\}}(s,P_s)ds] = \int_0^\infty E[1_{\{p^*\}}(P_s)]ds = \int_0^\infty \Pr[P_s = p^*]ds = 0$. (The fact that the expectation of the first integral is zero means that a.s. $\int_0^\infty 1_{\{p^*\}}(P_s)ds = 0$.)

Let us add the following comment. By (e.g.) looking at the graph of θ, it can be seen that γ decreases towards 1 when σ^2 increases to infinity (for fixed $\gamma > 1$, θ increases to infinity with σ). This implies that p^* increases to infinity when σ increases to infinity. $\qquad\square$

For the case $P_0 = p_0 > p^*$ see Exercise 4.19.

4.6 Controlling Diffusions with Jumps

Assume that the n-dimensional process $X_t \in \mathbb{R}^n$ is governed by the following controlled stochastic differential equation with jumps:

$$dX_t = f(t,X_t,u(t,X_t))dt + \sigma(t,X_t,u(t,X_t))dB_t, \quad X_0 = x_0, \qquad (4.94)$$

$$X_{\tau_j+} - X_{\tau_j-} = g(\tau_j, X_{\tau_j-}, V_j). \qquad (4.95)$$

Here f, g, σ, and x_0 are given entities, g is C^0, σ is an $n \times m$ matrix with entries $\sigma_{i,j}(t,x,u)$, f and σ are piecewise continuous in t, and Lipschitz continuous in x, u

on bounded subsets, uniformly in t, B_t is a vector of m independent Brownian motions, and the $V_j's$ are stochastic variables, identically and independently distributed. At certain jump time-points τ_i, $\tau_1 < \tau_2 < \dots$, the state jumps according to (4.95). The points τ_i are random variables taking values in $[0,\infty)$. By definition, the solution X_t is piecewise continuous, and continuous in (τ_j, τ_{j+1}).

The stochastic assumptions on the jump points τ_i are that they are Poisson distributed, i.e., given τ_{i-1}, τ_i, $i > 1$, is exponentially distributed in $[\tau_{i-1}, \infty)$ with parameter λ, hence the density of $\tau_i = \tau \geq \tau_{i-1}$ is $\lambda e^{-\lambda(\tau - \tau_{i-1})}$. Moreover, τ_1 is exponentially distributed in $[0,\infty)$ with parameter λ. The entities τ_i, V_i and $t \to B_t$ are independent, and the V_i's are i.i.d. Moreover $u(t,X_t)$ is a function taking values in a given set $U \subset \mathbb{R}^r$ and is subject to choice.

Define $\tau_{j,s} = s + \tau_j$ $j = 1,2,\dots$, and $J(s,x,u) =$

$$E^{s,x}\left[\int_s^T f_0(t,X_t,u(t,X_t))dt + \sum_{\tau_{j,s} \leq T} g_0(\tau_{j,s}, X_{\tau_{j,s}-}, V_j) + h_0(X_{T-})\right] \qquad (4.96)$$

where the symbol $E^{s,x}$ means the expected value arising from starting the process (4.94), (4.95) at state $x = X_{s+}$ at time s, X_t continuous and governed by (4.94) between jump points, and jumping at jump time points $\tau_{j,s}$ corresponding to (4.95), $\tau_{1,s}$ exponentially distributed in $[s,\infty)$, with intensity λ, $\tau_{2,s}$ exponentially distributed in $[\tau_{1,s}, \infty)$, given $\tau_{1,s}$, with intensity λ, and so on. Here f_0, g_0, h_0, and $T > 0$ are given entities, f_0, g_0, h_0, being continuous functions. As before, the function f_0 measures the running benefit obtained from the process, h_0 is a scrap value (or bequest function), and g_0 measures the benefit obtained at jump times. The control problem consists in maximizing $J(0,x_0,u)$, where $u = u(.,.)$ at least runs through the set of all Markov controls $u(t,x)$, or even the set of all admissible adapted $u_t(\omega)$ controls. (In the latter case, in (4.94), $u(t,X_t)$ is replaced by $u_t(\omega)$.) That $u_t(\omega)$ is adapted now means that $u_t(\omega) = u_t(B_{\rightarrow}t, \tau_1, V_1, \tau_2, V_2, \dots)$, where $u_t(\omega)$ is independent of all τ_i, V_i such that $\tau_i > t$, $(\omega = (t \mapsto B_t, \tau_1, V_1, \tau_2, V_2, \dots))$.

Define $J(s,x) = \sup_u J(s,x,u)$ (supremum over all Markov controls). Then (under certain conditions), J satisfies the following HJB equation in the one-dimensional case:

$$J_t(t,x) + \max_{u \in U}\{f_0(t,x,u) + J_x(t,x)f(t,x,u) + (1/2)J_{xx}(t,x)\sigma(t,x,u)^2\}$$
$$+ \lambda E[g_0(t,x,V) + J(t,x+g(t,x,V)) - J(t,x)] = 0, \qquad (4.97)$$

with $J(T,x) = h_0(x)$. Here, V is distributed according to the common distribution of the V_i's. In the multidimensional case, the equation reads:

$$0 = J_t(t,x) + \max_{u \in U}\{f_0(t,x,u) + J_x(t,x)f(t,x,u)$$
$$+ \sum_{i,j}(1/2)J_{x_ix_j}(t,x)[\sigma(t,x,u)\sigma'(t,x,u)]_{i,j}\}$$
$$+ \lambda E[g_0(t,x,V) + J(t,x+g(t,x,V)) - J(t,x)]. \qquad (4.98)$$

Here $J_x = (J_{x_1}, \dots, J_{x_n})$, $J_x f$ is a scalar product, and σ' is the transpose of σ.

That the HJB equation for $n = 1$ takes the form (4.97) should come as no surprise. For an equation of the form

$$X_t = x_0 + \int_0^t v(s)ds + \int_0^t \sigma(s)dB_s + \sum_{\tau_j \leq t} \hat{h}(\tau_j, V_j),$$

Dynkin's formula on differential form for $\hat{g}(t, X_t)$ in the case we condition on the start point (t, x) $(X_{t+} = x)$ takes the form $E^{t,x}\hat{g}(t + dt, X_{t+dt}) =$

$$\hat{g}(t,x) + \hat{g}_t(t,x)dt + \hat{g}_x(t,x)v(t)dt + (1/2)\hat{g}_{xx}(t,x)\sigma(t)^2 dt$$
$$+ \lambda E_V\{\hat{g}(t, x + \hat{h}(t,V)) - \hat{g}(t,x)\}dt,$$

E_V meaning expectation with respect to V. This is essentially a combination of Dynkin's formula for diffusions, and the result on the infinitesimal generator, or "Dynkin's formula," for piecewise deterministic processes (see (3.120)). A heuristic derivation of the HJB equation similar to that carried out in the case of no jumps in the state variable, see arguments subsequent to (4.47), can again be carried out in the one-dimensional case. In the case $g_0 = 0$, using the above Dynkin's formula, gives

$$E^{t,x}J(t + h, X_{t+h}) - J(t,x)$$
$$= J_t h + J_x f h + (1/2)\sigma^2 J_{xx} h + \lambda E_V\{J(t, x + \hat{h}(t,V)) - J(t,x)\}h,$$

where $\hat{h}(t,V) = g(t,x,V)$. Inserting this expression for $E^{t,x}J(t + h, X_{t+h}) - J(t,x)$ in (4.48), and dividing by h yields the HJB equation in (4.97), i.e., it brings into the HJB equation the new term $\lambda E_V\{J(t, x + g(t,x,V)) - J(t,x)\}$.

We state only a sufficient condition related to the HJB equation.

Theorem 4.35 (Sufficient conditions). *Assume that $\hat{J}$ is continuous in $[0,T] \times \mathbb{R}^n$, and is twice continuously differentiable and satisfies (4.98) in $(0,T) \times \mathbb{R}^n$, ((4.97) in the one-dimensional case), together with the boundary condition $\hat{J}(T,x) = h_0(x)$. Assume that for every pair $(t,x) \in (0,T) \times \mathbb{R}^n$, $u^0(t,x)$ is the value of u that yields the maximum on the right-hand side of (4.98), ((4.97)). Then $u^0(.,.)$ is optimal.* $\square$

Remark 4.36 (Precision added*). Also in the current situation, certain boundedness conditions are needed for the sufficient conditions to hold: The theorem yields optimality in the set S of admissible pairs (X_t^u, u_t) $(u = u_t)$, for which X_t^u is a solution of (4.94), (4.95), and for which $E[\int_0^T |f_0(t, X_t^u, u_t))|dt + |h_0(X_T^u)|] < \infty$, $E\sum|g_0(\tau_j, X_{\tau_j}^u, V_j)| < \infty$, and $E[\sup_t |\hat{J}(t, X_t^u)|] < \infty$ (u_t piecewise and right-continuous in t and adapted). It must then be assumed that a solution X_t^0 corresponding with $u^0(t,x)$ exists, such that (X_t^0, u_t^0) belongs to S, and $u_t^* = u^0(t, X^0(t))$ is piecewise and right-continuous in t and adapted (the last property follows essentially automatically).

For any $\hat{u} = u_t$, $X_t^{\hat{u}}$ is automatically known to exist, if for some numbers a_g, b_g, K and K', for all t, x, u, V, $|g(t,x,V)| \leq a_g + b_g|x|$, $|f(t,x,u) - f(t,y,u)| \leq K|x - y|$, $|\sigma_{ij}(t,x,u) - \sigma_{ij}(t,y,u)| \leq K|x - y|$, $|f(t,x_0,u)| \leq K'$, $|\sigma_{i,j}(t,x_0,u)| \leq K'$. $\square$

Example 4.37. Consider the problem:

$$\max E\left[\int_0^T -u^2/2 - 4x^2 dt\right], u \in \mathbb{R},$$

subject to

$$dX_t = udt + X_t dB_t, X_0 = x_0 \text{ given,}$$
$$X_{\tau_i+} = V_i X_{\tau_i-},$$

all τ_i, V_i and B_t independent, V_i i.i.d., with $EV_i^2 = 2$, all $|V_i| \le a$, a a given number. An unbounded number of jumps can occur, all with intensity $\lambda = 1$.

Solution. The HJB equation is

$$0 = J_t + \max[-u^2/2 - 4x^2 + J_x u] + J_{xx} x^2/2 + 1 \cdot [EJ(t, xV) - J(t, x)],$$

yielding $u = J_x$ as a maximum point, which when inserted gives the maximal value, and so

$$0 = J_t + (J_x)^2/2 - 4x^2 + J_{xx} x^2/2 + EJ(t, xV) - J(t, x).$$

Let us try a solution of the form $\hat{J} = \phi(t)x^2 + \psi(t)$. Insertion of $\hat{J}$ and its derivatives in the HJB equation gives $0 = \phi(t)'x^2 + \psi'(t) + 2\phi(t)^2 x^2 - 4x^2 + \phi(t)x^2 + E[\phi(t)(xV)^2] + \psi(t) - \phi(t)x^2 - \psi(t) = \phi'(t)x^2 + 2\phi(t)^2 x^2 - 4x^2 + \phi(t)x^2 + \phi(t)2x^2 + \psi(t) - \phi(t)x^2 - \psi(t) + \psi'(t) = x^2[\phi'(t) + 2\phi(t)^2 + 2\phi(t) - 4] + \psi'(t)$. For the right-hand expression to be identically zero, we must have $\phi'(t) = -2[\phi(t)^2 + \phi(t) - 2] = -2(\phi - 1)(\phi + 2)$, and (then) $\psi'(t) = 0$. Using $1/(\phi - 1)(\phi + 2) = (1/3)[1/(\phi - 1) - 1/(\phi + 2)]$, the first equation is a separable differential equation that gives $\ln[(\phi - 1)/(\phi + 2)] = -6t + C$, or $(\phi - 1)/(\phi + 2) = C'e^{-6t}$ (C and C' arbitrary, related, constants). Now, $\hat{J}(T, x) = 0$, which implies both $\phi(T) = 0$ and $\psi(T) = 0$, i.e., $\psi(t) \equiv 0$, and $-1/2 = C'e^{-6T}$, so $C' = -e^{6T}/2$, and (solving for ϕ), $\phi(t) = (1 - e^{6(T-t)})/(1 + e^{6(T-t)}/2) < 0$. (It can be shown that $u = u^0 = 2x\phi(t)$, with corresponding solution X_t^0, make up a pair that belongs to the set S of Remark 4.36.) □

Note that the integral version of (4.94), (4.95) (the version that has a true meaning) is as follows:

$$X_t = x_0 + \int_0^t f(s, X_s, u(s, X_s))ds$$

$$+ \int_0^t \sigma(s, X_s, u(s, X_s))dB_s + \sum_{\tau_j \le t} g(\tau_j, X_{\tau_j-}, V_j)$$

(replace $u(s, X_s)$ by u_s if general adapted controls u_s are considered).

Sometimes, more general jump processes are needed. Assume, for some constants $\alpha \ge 0, \beta \ge 0$, and $\beta' \ge 0$, that for all (t, x, V)

$$|g(t, x, V)| \le (\alpha + \beta|x|)|V|, \ |g(t, x, V) - g(t, y, V)| \le \beta'|x - y||V|. \quad (4.99)$$

Loosely speaking, we need to consider processes where we allow very frequent jumps, provided they are very small. Such processes may be needed for example to simulate better certain types of behavior of stock prices. Mathematically, such processes can be obtained by taking limits of processes of the type just studied. For simplicity, assume now that the jumps V_j are bounded by a constant K. Intuitively speaking, in the jump process now considered, small jumps as well as larger jumps can occur, small jumps can be very frequent, larger jumps are less frequent, all types of jumps occurring at time points being Poisson distributed. Let us fix a precise manner in which this happens. For some $\kappa > 1$, for each fixed $k = 1, 2, \ldots, V_1^k, V_2^k, \ldots$ are i.i.d. stochastic variables taking values in $[-K/\kappa^{k-1}, -K/\kappa^k) \cup (K/\kappa^k, -K/\kappa^{k-1}]$, and let $\tau_1^k, \tau_2^k, \ldots$ be stochastic time points at which $V_1^k, V_2^k, \ldots$ occur. The time points $\tau_1^k, \tau_2^k, \ldots$ are Poisson distributed with intensity λ_k. The $\{V_i^k\}_i$'s are distributed according to certain given densities $\pi^k(v)$. The Brownian motion and the sequences $\{\tau_i^k\}_i, \{V_i^k\}_i, k = 1, 2, \ldots$, are all independent. Now the solution X_t is given by:

$$X_t = x_0 + \int_0^t f(t, X_t, u(t, X_t)) dt +$$

$$\int_0^t \sigma(t, X_t, u(t, X_t)) dB_t + \sum_{k=1}^{\infty} \sum_{j \in \{j : \tau_j^k \le t\}} g(\tau_j^k, X_{\tau_j^k-}, V_j^k). \quad (4.100)$$

Processes of the type (4.100) are examples of what is called jump diffusions.

In the definition (4.96) of the value function $J(s, y, u)$ replace the sum by $\sum_k \sum_{\tau_{j,s}^k \le T} g_0(\tau_{j,s}^k, X_{\tau_{j,s}^k-}, V_j^k)$, where $\tau_{j,s}^k = s + \tau_j^k$ (the definition of the random variables $\tau_{j,s}^k$ corresponding with that of $\tau_{j,s}$ above). Using the name $J(s, y, u)$ on the entity obtained even after this replacement, then again, we want to maximize $J(0, x_0, u)$.

Under certain conditions on the λ_k's, if $f(t, x, u)$ and $\sigma(t, x, u)$ satisfy the same growth conditions as in Remark 4.36, then, for each admissible $u = u_t$, a unique adapted solution X_t^u of (4.100) exists a.s., being right-continuous, with left limits (these two properties for short denoted "cadlag," a French abbreviation).

The existence of X_t^u must mean, in particular, that the double sum in (4.100) exists. We do not intend to present general conditions for the double sum to exist, or general conditions for X_t^u to exist. But let us describe one case where we can have that $\lambda_k \to \infty$ and that the double sum can be expected to exist, namely the case where, for a $\mu \in (1, \kappa)$, $\lambda_k \le \mu^k$ for all k. (We now assume that the growth conditions on f and g in Remark 4.36 are satisfied.)

Assume first that $T = 1$, and that $\beta = 0$, $\alpha > 0$ in (4.99). Let m be any given natural number, let $v \in (\mu, \kappa)$, and let the number of time points $\{\tau_j^k\}_j$ in a given interval $[0, t]$ be $N^k(t)$. The double sum can be shown to exist when for all k, $N^k(t)$ belongs to the set $A_m^k := \{N^k(t) : N^k(t) \le mv^k\}$. The double sum converges because in this case each term in the outer sum is $\le \alpha(K/\kappa^{k-1}) mv^k = \alpha \kappa m K(v/\kappa)^k$, the last expression forming a convergent geometric series. Note that $E(N^k(t))^2 = E(N^k(t) - t\lambda_k)^2 + (t\lambda_k)^2 \le \lambda_k + \lambda_k^2 \le 2\mu^{2k}$. Using Chebyshev's inequality, for $\bar{A}_m^k = \{N^k(t) : N^k(t) \ge mv^k\}$, $\Pr(\bar{A}_m^k) \le E(N^k(t))^2/(mv^k)^2 \le 2m^{-2}(\mu/v)^{2k}$, so for

$\bar{A}_m = \cup_k \bar{A}_m^k$, $\Pr(\bar{A}_m) \leq 2m^{-2}/(1 - \mu/\nu)$. Now, $\Pr(\cap_m \bar{A}_m) = 0$, so the sum exists with probability 1.

Consider next the case $\beta > 0$. It can be proved, using among other things Theorem 4.1, that for some $\delta > 0$, the conditional expectation of $\sup_t |X^u(t)|$, given that the $N^k(T)$'s equal given numbers m_k, is smaller than $\delta \Pi_k (1 + (K'/\kappa^k)\beta'')^{m_k}$ where $\beta'' = \alpha + \beta$, $K' = K\kappa$, $\Pi_k = \Pi_{k=0}^\infty$ here being the product symbol. Now, $E(1 + (K'/\kappa^k)\beta'')^{m_k} \leq e^{\lambda_k T(K'/\kappa^k)\beta''}$, so $E[\delta \Pi_k(1 + (K'/\kappa^k)\beta'')^{m_k}] \leq \delta \Pi_k e^{\lambda_k T(K'/\kappa^k)\beta''} = e^{\beta'' T K' \Sigma_k \lambda_k/\kappa^k}$. As $\Sigma_k \lambda_k/\kappa^k < \infty$ (recall $\lambda_k < \mu^k$), again the double sum must exist a.s. (it exists even in "L_1-norm").

(In order to really be sure that solutions X_t^u exist, we need some convergence arguments as $m \to \infty$, for solutions $X_t^{u,m}$ to converge, where $X_t^{u,m}$ is obtained from summing to m instead of to infinity in (4.100).)

In E in the Addendum to the chapter, the double sum in (4.100) is rewritten in a form that allows one to work with weaker convergence conditions.

The HJB equation for problems of the type (4.100) reads as follows when $n = 1$:

$$J_t(t,x) + \max_{u \in U}\{f_0(t,x,u) + J_x(t,x)f(t,x,u) + (1/2)J_{x,x}(t,x)[\sigma(t,x,u)]^2\}$$
$$+ \sum_{k=1}^\infty \lambda_k E[g_0(t,x,V^k) + J(t,x+g(t,x,V^k)) - J(t,x)] = 0, \qquad (4.101)$$

again with $J(T,x) = h_0(x)$. Here, each V^k is distributed according to the common distribution of the V_j^k's, $j = 1, 2, \ldots$ A sufficient condition (verification theorem) related to this HJB equation could again be stated, it is essentially as before, but we drop stating the assumptions needed for it to hold (at least, as can be seen from what has been introduced earlier, in a shorthand notation, they would include $E[\int_0^T |f_0|dt] < \infty$, $E[h_0] < \infty$, $E[\sup_t |\hat{J}|] < \infty$, and, to operate with "sensible" solutions, $E[\int_0^T |f|dt] < \infty$ and $E[\int_0^T \sigma^2 dt] < \infty$).

Optimal stopping problems are solved by using the same tools as in the continuous diffusion case. In the pure stopping case, the HJB equation (4.82) as well as the HJB inequality (4.86) now include the term resulting from jumps, see (4.97) and (4.101). (The modification in the case where also controls are present should be obvious.) Now we must require $\hat{J} = h_0$ outside the continuation region (i.e., in $\text{cl}V \setminus D^*$), unless the process always jumps back into $\text{cl}D^*$ as long as it stays in $\text{cl}D^*$. (It is assumed that $(\tau_V, X_{\tau_V+}) \in \text{cl}V$.)

The criterion is now, (with $g_0 \equiv 0$, $h_0(\infty, .) = 0$)

$$\max_{u, \tau \leq \tau_V} E\left[\int_0^\tau f_0(t, X_t^u, u_t)dt + h_0(\tau, X_{\tau+}^u)\right]$$

(u and u_t drop out in the pure stopping case).

Further reading. For readers wanting to continue reading about the problems discussed in this chapter, the easiest book, on a more advanced level, is perhaps Øksendal (2003). Still more advanced is Yong and Zhou (1999). Books on mathematical finance include the very elementary book by Ross (1999), the (also elementary) book by Benth (2003), the book by Duffie (1996), and at a more advanced level the book by Karatzas and Shreve (1998). For the control of jump diffusions, see Øksendal and Sulem (2005).

4.7 Addendum

A. Assume that we have only risk-adverse actors in the market. Below the time is discrete, $t = kh$, $k = 0, 1, 2, \ldots$, h given. We assume a certain "constancy" through time of X_s, more precisely that $\beta_n := E[X_{t+(n+1)h} - X_{t+nh}|X_t]$ is independent of $n = 0, 1, 2, \ldots$. (This strong assumption is commented upon below.) What we shall then be able to rule out is that $\beta_n \neq 0$.

Suppose that $\beta := \beta_n \neq 0$. Imagine now that a "speculative operation" is made at each point in time $t + nh$, $n = 0, 1, 2, \ldots$: An amount $\kappa = \beta/|\beta|$ of the security is bought at time $t + nh$ for a loan of this size in the bank, and sold the next period. This operation gives, after paying back the loan at time $t + (n+1)h$, a net income $Z_{t+(n+1)h} := \kappa(X_{t+(n+1)h} - X_{t+nh})$. Define $W_{t+(k+1)h} := X_{t+(k+1)h} - X_{t+kh}$ and suppose that all $W_{t+(k+1)h}, k = 0, 1, 2, \ldots$, have the same variance σ^2. Here, and below, it may be assumed that all calculations of probabilities and (co)variances are actually conditional ones, conditional on X_t. Assume, for some given number $\mu \in (0,1)$, that $\mathrm{Covar}(W_{t+kh}, W_{t+k'h}) \leq \sigma^2 \mu^{k'-1-k}$, for all k, k', $k' > k$. Define $Q^m := \sum_{1 \leq k \leq m} Z_{t+kh}$ and let us calculate $\mathrm{Var}(Q^m)$. It equals $\sum_k \mathrm{Var}(Z_{t+kh}) + 2\sum_k \sum_{k' > k} \mathrm{Covar}(Z_{t+kh}, Z_{t+k'h}) \leq m\kappa^2\sigma^2 + 2m\sum_k \kappa^2\sigma^2/(1-\mu) \leq Km\sigma^2$, $K = 1 + 2/(1-\mu)$. Now, Chebyshev's inequality gives that, for any $\varepsilon > 0$, $\Pr[|Q^m - m|\beta|| \geq \varepsilon] = \Pr[|Q^m - EQ^m| \geq \varepsilon] \leq \mathrm{Var}(Q^m)/\varepsilon^2 \leq Km\sigma^2/\varepsilon^2$. Hence, for $\varepsilon = m|\beta|/2$, $\Pr[Q^m \leq m|\beta|/2] \leq 4K\sigma^2/m|\beta|$ and the latter probability is small when m is large.

So, for m large, the speculator is close to being sure to make a positive profit. This should not be possible in a well-functioning market, so β must be equal to zero.

The assumptions made, for example the constancy of $\beta_n := E[X_{t+(n+1)h} - X_{t+nh}|X_t]$, may be seen as unreasonable ones, but, for the last one, note that h may be, relatively speaking, very small: If β_n is (roughly) constant over weak-long periods, and trade takes place each hour (and μ is not too close to 1), our arguments may have something to say in this situation. (What is actually needed is a slow movement of β_n, if β_0 is $\geq \gamma > 0$ for some t, then "a weak after" it is still, say $\geq \gamma/2$.)

B. Proof of (4.71)$\Rightarrow$(4.72). The argument will be intuitive:

Let $s > t$, s close to t, let $\Delta t = s - t$ and $\Delta B_t = B_s - B_t$, $w(t, X_t^*) = E^{t, X_t^*} p(t)$. By Ito's formula, approximately, $w(s) - w(t) := w(s, X^*(s)) - w(t, X^*(t)) =$

$$(w_t(t) + w_x(t)f(t) + \sigma^2(t)w_{xx}(t)/2)\Delta t + w_x(t)\sigma(t)\Delta B_t,$$

and, from (4.71),

$$p(s) - p(t) = (-\lambda_0 f_{0x}(t) - p(t)f_x(t) - q(t)\sigma_x(t))\Delta t + q(t)\Delta B_t.$$

Write $E^s := E[.|s, X_s^*]$ and $E^{s,t} := E[.|t, X^*(t), s, X^*(s)]$. Note that

$$E^s[w(s) - w(t)] = E^s p(s) - E^s E^{s,t} w(t) = E^s p(s) - E^s E^t w(t)$$
$$= E^s p(s) - E^s p(t) = E^s[p(s) - p(t)].$$

Thus, the expectation E^s of the right-hand sides of the expressions for $w(s) - w(t)$ and $p(s) - p(t)$ are equal: Taking these expectations yield

$$E^s\{w_t(t) + w_x(t)f(t) + \sigma^2(t)w_{xx}(t)/2\}\Delta t + E^s\{w_x(t)\sigma(t)\}E^s[\Delta B_t]$$
$$= E^s\{-\lambda_0 f_{0x}(t)\}\Delta t - E^s\{p(t)f_x(t)\}\Delta t$$
$$-E^s\{q(t)\sigma_x(t)\}\Delta t + E^s\{q(t)\}E^s[\Delta B_t].$$

Then, presumably, the expressions in front of Δt and in front of $E^s[\Delta B_t]$ must be equal. The equality of the latter expressions gives $E^s\{w_x(t)\sigma(t)\} = E^s q(t)$, and the equality of the former expressions gives

$$E^s\{w_t(t) + w_x(t)f(t)\} + E^s\{\sigma^2(t)w_{xx}(t)/2\}$$
$$= E^s\{-\lambda_0 f_{0x}(t) - (p(t)f_x(t))\} - E^s\{q(t)\sigma_x(t)\}.$$

Letting $s \downarrow t$ gives $w_x(t)\sigma(t) = E^t q(t)$ and

$$w_t(t) + w_x(t)f(t) + \sigma^2(t)w_{xx}(t)/2$$
$$= -\lambda_0 f_{0x}(t) - (E^t p(t))f_x(t) - (E^t q(t))\sigma_x(t).$$

Inserting $w_x(t)\sigma(t) = E^t q(t)$ in the last equality gives

$$w_t(t) + w_x(t)f(t) + \sigma^2(t)w_{xx}(t)/2$$
$$= -\lambda_0 f_{0x}(t, X^*(t), u^*(t, X_t^*)) - (E^t p(t))f_x(t, X^*(t), u^*(t, X_t^*))$$
$$-w_x(t)\sigma(t)\sigma_x(t).$$

For $w(t) := w(t, X_t^*) = E^t p(t)$, $\lambda_0 = 1$, the asserted form of equation (4.72) follows.

C. It might be that u^* in an obvious sense could be catching up optimal if (4.75) holds. First note that $E(X_t^* - 1)^2 =$

$$(t-1)^2 + 2E\left[(t-1)\int_0^t (t-1)(\tau-1)^{-1}\sigma dB_\tau\right] + E\left[\int_0^t (t-1)(\tau-1)^{-1}\sigma dB_\tau\right]^2$$
$$= (t-1)^2 + E\left[\int_0^t (t-1)(\tau-1)^{-1}\sigma dB_\tau\right]^2.$$

Using the Ito isometry, $E[\int_0^t (t-1)(\tau-1)^{-1}\sigma dB_\tau]^2 =$

$$\int_0^t ((t-1)/(\tau-1))^2 \sigma^2 d\tau = (t-1)^2 \{-(t-1)^{-1}+1)\}\sigma^2.$$

Hence,

$$\lim_{t\to 1} E(X_t^* - 1)^2 = \lim_{t\to 1}\{(t-1)^2 + (t-1)(-1+(t-1))\sigma^2\} = 0,$$

so essentially $X_1^* = 1$ holds. Next, from these calculations, note that $E(X_t^* - 1)^2/(1-t) \approx \sigma^2$ when $t \approx 1$. Let $X_t = t + 2(t-1)\int_0^t (\tau-1)^{-1}\sigma dB_\tau$. Then of course, $E(X_t-1)^2/(1-t) \approx 4\sigma^2$ when $t \approx 1$. Now, $J(t,X_t) - J(t,X_t^*) =$

$$(X_t^* - 1)^2/(1-t) - (X_t - 1)^2/(1-t) \approx \sigma^2 - 4\sigma^2.$$

Evidently, $E[(X_t - 1)^2] = 0$, but condition (4.75) does not hold for this X_t. Now, X_t corresponds with an impermissibly volatile $u_t dt$, namely $(1+2\int_0^t (\tau-1)^{-1}\sigma dB_\tau)dt + \sigma dB_t$ (see the last term). But, presumably, it might be that it could be smoothened slightly to become permissible, and so that it still satisfies the terminal condition, and finally, still yields roughly $\lim_{t\to 1} E[(X_t - 1)^2]/(1-t) = 4\sigma^2$.

D. Let $n = m = 1$. Define $Y_t = (1+X_t^2)^{1/2}$ and $z_t = EY_t$, and note that $E|X_t| \le z_t \le Ke^{b_t t}$ for some K, $b_t < \alpha$, see Exercise 4.5 below. Hence, $E[e^{-\alpha t}|X_t|] \le Ke^{-(\alpha-b_t)t}$. Moreover, let $\hat\tau$ be a stopping time $\ge k$, and note that then $E|e^{-\alpha\hat\tau}X_{\hat\tau}| \le K^* e^{(b_t-\alpha)k}$, for some constant K^*. To see this, note that $e^{-\alpha\hat\tau}|X_{\hat\tau}| \le e^{-\alpha\hat\tau}Y_{\hat\tau}$. Now, $E^{k,y}[e^{-\alpha\hat\tau}Y_{\hat\tau}] = e^{-\alpha k}y + E^{k,y}[\int_k^{\hat\tau} -\alpha e^{-\alpha t}Y_t dt + \int_k^{\hat\tau} e^{-\alpha t}dY_t] \le e^{-\alpha k}y + e^{-\alpha k}E^{k,y}[\int_k^{\hat\tau} e^{-\alpha(t-k)}dY_t] =: \Gamma$. Using results from Exercise 4.5, for some constant d, for $t \ge k$, $z_t \le (z_k + d/b_t)e^{b_t(t-k)}$ and $dz_t \le [d + b_t z_k]e^{b_t(t-k)}dt$. Letting $y = Y_k$, then $E[\Gamma] = e^{-\alpha k}z_k + e^{-\alpha k}\int_k^\infty e^{-\alpha(t-k)}dz_t \le e^{-\alpha k}z_k + \int_k^\infty e^{-\alpha(t-k)}[b_t z_k + d)e^{b_t(t-k)}]dt$. Because $\int_k^\infty e^{(b_t-\alpha)(t-k)}dt = 1/(\alpha - b_t)$ and $z_k \le Ke^{b_t k}$, then $E[e^{-\alpha\hat\tau}|X_{\hat\tau}|] \le E[e^{-\alpha\hat\tau}Y_{\hat\tau}] \le E[\Gamma] \le K^* e^{(b_t-\alpha)k}$, for some K^*.

Let $\tau_k = \min\{\tau, k\}$, $k = 1, 2, \dots$. Evidently,

$$E\int_0^{\tau_k} |f_0(t,X_t)|dt \le E\int_0^\infty |f_0(t,X_t)|dt \le \int_0^\infty (a_0 + b_0 Ke^{b_t t})e^{-\alpha t}dt < \infty.$$

The bound on g_x yields

$$g(t,x) - g(t,x_0)| \le \int_0^1 |g_x(t,\theta x + (1-\theta)x_0)||(x-x_0)|d\theta \le c_g e^{-\alpha t}|x-x_0|,$$

so

$$\lim_k E[|g(\tau_k, X_{\tau_k}) - g(\tau_k, x_0) - (g(\tau, X_\tau) - g(\tau, x_0))|1_{\{\infty>\tau\}}]$$
$$= \lim_k E[|g(\tau_k, X_{\tau_k}) - g(\tau_k, x_0) - (g(\tau, X_\tau) - g(\tau, x_0))|1_{\{\infty>\tau>k\}}$$
$$\le E[c_g e^{-\alpha k}|X_k - x_0|] + E[c_g e^{-\alpha\tau}|X_\tau - x_0|1_{\{\infty>\tau>k\}}].$$

Because $g(\infty,.) = 0$, these inequalities hold even if $\infty > \tau$ is dropped, except in the very last term. The two last expectations go to zero when $k \to \infty$, see inequalities obtained above (for Γ in case of the last expectation). Because $E[|(g(\tau_k,x_0) - g(\tau,x_0))|1_{\{\tau>k\}}] \to 0$, then $E[|g(\tau_k,X_{\tau_k}) - g(\tau,X_\tau)|] \to 0$, when $k \to \infty$. Similarly, if $\tau^k = \min\{k,\tau_{D^*}\}$, then $E[|\hat{J}(\tau^k,X_{\tau^k}) - \hat{J}(\tau_{D^*},X_{\tau_{D^*}})|1_{\{\tau_{D^*}<\infty\}}] \to 0$. Now, by (4.89), $E[g(\tau_k,X_{\tau_k})] \le E[\int_0^{\tau_k} -f_0(t,X_t)dt] + \hat{J}(0,X_0)$. Letting $k \to \infty$ in the inequality, it follows that it holds for τ_k replaced by τ, by the earlier results. Next, by (4.90), $E[\hat{J}(\tau^k,X_{\tau^k})] = E[\int_0^{\tau^k} -f_0(t,X_t)dt] + \hat{J}(0,X_0)$. Note also that

$$\lim_k E[|\hat{J}(\tau^k,X_{\tau^k}) - \hat{J}(\tau^k,x_0)|1_{\{\tau_{D^*}=\infty\}}] \le \lim_k c_J e^{-\alpha\tau^k}|X_k - x_0| = 0,$$

and $E[|\hat{J}(\tau^k,x_0)|1_{\{\tau_{D^*}=\infty\}}] \to 0$, so $\lim_k E[|\hat{J}(\tau^k,X_{\tau^k})|1_{\{\tau_{D^*}=\infty\}}] = 0$. Hence, letting $k \to \infty$ in the next to last equality gives $E[g(\tau_{D^*},X_{\tau_{D^*}})] =$

$$E[g(\tau_{D^*},X_{\tau_{D^*}})1_{\{\tau_{D^*}<\infty\}}]$$
$$= E[\hat{J}(\tau_{D^*},X_{\tau_{D^*}})1_{\{\tau_{D^*}<\infty\}}]$$
$$= \lim_k E[\hat{J}(\tau^k,X_{\tau^k})1_{\{\tau_{D^*}<\infty\}}]$$
$$= \lim_k E[\hat{J}(\tau^k,X_{\tau^k})]$$
$$= E\left[\int_0^{\tau_{D^*}} -f_0(t,X_t)dt\right] + \hat{J}(0,X_0).$$

This result and the next to last inequality yield optimality of τ_{D^*}.

E. (*Rewriting the double sum in (4.100)*) Write $\mathcal{N}_+^m(t,v) = \{(\tau_j^m,V_j^m) : \tau_j \le t, 0 < V_j^m \le v\}$, and, for simplicity, assume first that all V_j^m are positive.

Then $\mathcal{N}_+^m(t,v)$ (that depends on the sequence τ_j^m,V_j^m) is a finite set a.s., let $N_+^m(t,v)$ be the number of elements in $\mathcal{N}_+^m(t,v)$. For any given $\hat{v}$ in $(K/\kappa^m, K/\kappa^{m-1}]$, the following equality holds:

$$\sum_{(\tau_j^m,V_j^m)\in\mathcal{N}_+^m(t,\hat{v})} g(\tau_j^m,X_{\tau_j^m-},V_j^m) = \int_{[0,t]}\int_{[K/\kappa^m,\hat{v}]} g(s,X_{s-},v)N_+^m(ds,dv). \quad (4.102)$$

The right-hand side is just a fancy name of the left-hand side, using essentially the notation of Stieltjes integration ($N_+^m(ds,dv)$ is the same as $dN_+^m(s,v)$): a.s., we can assume that all the points in $\mathcal{N}_+^m(t,\hat{v})$ are distinct, in which case we attach weights 1 to all points. When s and v increase, the value of $N_+^m(s,v)$ increases by 1 iff a point (τ_j^m,V_j^m) in $\mathcal{N}_+^m(t,\hat{v})$ is passed, and that change in $N_+^m(s,v)$ should be multiplied by $g(\tau_j,X_{\tau_j-},V_j^m)$. Summing over all points in $\mathcal{N}_+^m(t,\hat{v})$ yields the sum in (4.102). (See also (5.28) in the Appendix.)

Define $N_+(t,v)$ to be the function that equals $N_+^m(t,v)$ for v in the interval $(K/\kappa^m,K/\kappa^{m-1}]$. Then

$$\int_{[0,t]} \int_{(K/\kappa^k,K]} g(s,X_{s-},v)N_+(ds,dv)$$

$$= \sum_{1 \le m \le k} \int_{(K/\kappa^m,K/\kappa^{m-1}]} g(s,X_{s-},v)N_+^m(ds,dv). \quad (4.103)$$

When certain conditions on the λ_k's hold, we can let k go to infinity here and get a.s. as a limit on the left-hand side of (4.103) the expression

$$\int_{[0,t]} \int_{(0,K]} g(s,X_{s-},v)N_+(ds,dv). \quad (4.104)$$

Consider now the case where V_j^m belongs to the set $\{v : K/\kappa^m < |v| \le K/\kappa^{m-1}\}$ and can be both positive and negative. Then we define $\mathcal{N}_-^m(t,v)$ to be the set $\{(\tau_j,V_j^m) : \tau_j \le t, -K/\kappa^{m-1} \le V_j^m < -K/\kappa^m, V_j^m \le v\}$, ($\mathcal{N}_+^m(t,v)$ defined as before). Let $N_-^m(t,v)$ be the number of elements in $\mathcal{N}_-^m(t,v)$, and let $N^m(t,v) = N_-^m(t,v)$ if $v \in I_-^m := [-K/\kappa^{m-1}, -K/\kappa^m)$, $N^m(t,v) = N_+^m(t,v)$ if $v \in I_+^m := (K/\kappa^m, K/\kappa^{m-1}]$. Define also $N(v,t) = N^m(t,v)$ if $v \in I_-^m \cup I_+^m =: I^m$, $m = 1,2,\ldots$ Then for certain conditions on the λ_k's, arguments similar to the above ones would lead us to conclude that, a.s.,

$$\sum_{k=1}^{\infty} \sum_{j \in \{j:\tau_j^k \le t\}} g(\tau_j^k, X_{\tau_j^k-}, V_j^k) = \int_{[0,t]} \int_{[-K,K]} g(t,X_{t-},v)N(dt,dv). \quad (4.105)$$

The Consequences for the HJB Equation of this Reformulation

Define $v_\pm^m(t,v) = EN_\pm^m(t,v)$, and note that $v_\pm^m(t,v) = tv_\pm^m(v)$, $v_\pm^m(v) := v_\pm^m(1,v)$ (recall the expectation λt for the number of jumps in $[0,t]$ of a Poisson process with intensity λ). In fact, $v_\pm^m(v) = \lambda_\pm^m \int_{I_\pm^m \cap (-\infty,v)} \Pi^m(dv')$, where $\lambda_+^m = EN_+^m(1, K/\kappa^{m-1})$, $\lambda_-^m = EN_-^m(1, -K/\kappa^m)$, and $\Pi^m(v)$ is the common cumulative distribution of the $\{V_i^m\}_i$'s in I^m. Define also $v^m(v) = v_-^m(v)$ if $v \in I_-^m$, $v(v) = v_+^m(v)$ if $v \in I_+^m$, and let $v(v) = v^m(v)$ if $v \in I^m$, $m = 1,2,\ldots$

Note that, for any given function $\rho(t,v)$, if V^k is any of the V_j^k's, $j = 1,2,\ldots$, then $E[\rho(t,V^k)] = \int_{I^k} \rho(t,v)d\Pi^k(v)$ and hence $\lambda_k E[\rho(t,V^k)] = \int_{I^k} \rho(t,v)dv^k(v)$. So, if $\rho(t,V^k)$ is the term in square brackets in the HJB equation (4.101), then the sum in that equation equals $\sum_k \int_{I^k} \rho(t,v)dv^k(v) = \int_{[-K,K]} \rho(t,v)dv(v)$. Thus, the sum in the HJB equation (4.101) can be written as

$$\int_{[-K,K]} [g_0(t,x,V) + J(t,x+g(t,x,V)) - J(t,x)]v(dv). \quad (4.106)$$

(Here, $v(.)$ is not a cumulative distribution on $[-K,K]$, but, if such one is wanted, one can replace $v(.)$ by $v^*(.)$ defined by $v^*(v) = \int_{[-K,v]} v(dv)$.)

Certain conditions exist, not presented here, implying that such a HJB equation holds, in particular implying that the last integral is defined.

Sometimes problems appear where it is necessary to work with weaker conditions on the behavior of the jumps. Thus, for certain weaker convergence conditions,
it is possible to prove that the difference

$$\sum_{m \le k} \int_{[0,t]} \int_{[-K,K]} g(s, X_{s-}, v) N^m(ds, dv)$$

$$- \sum_{m \le k} \int_{[0,t]} \int_{[-K,K]} g(s, X_{s-}, v) v^m(dv) ds \qquad (4.107)$$

is convergent when $k \to \infty$, but that the two sums, considered separately, are not
convergent. (For example, $\lim_k \sum_{m \le k} \int_{[0,t]} \int_{[-K,K]} g(s, X_{s-}, v) v^m(dv) ds = \int_{[0,t]} \int_{[-K,K]}$
$g(s, X_{s-}, v) v(dv) ds$ may not exist.)

Let us mention that a weaker convergence condition that frequently works in
order to obtain convergence of the difference is $\int_{[-K,K]} \min\{v^2, 1\} v(dv) < \infty$.

We can always rewrite the integral equation for X_t as follows. (We shall assume
that all integrals appearing below exist.)

$$X_t = x_0 + \int_0^t f^*(s, X_s, u(s, X_s)) ds + \int_0^t \sigma(s, X_s, u(s, X_s)) dB_s$$

$$+ \int_0^t \int_{[-K,K]} g(s, X_{s-}, v) d\{N(s, v) - sv(v)\}, \qquad (4.108)$$

where the double integral is actually the limit of the expression in (4.107), and where

$$f^*(s, x, u) = f(s, x, u) + \int_{[-K,K]} g(s, x, v) v(dv). \qquad (4.109)$$

We can choose to work with equations of the form (4.108), without assuming that
f^* has the particular form (4.109), and that is sometimes done (see e.g., Øksendal
and Sulem (2005)), as weaker convergence conditions are then available.

In the case of equations of the form (4.108), then $f^*(t, x, u)$ appears in the
HJB equation instead of $f(t, x, u)$, and then the additional term $-J_x(s, x) \int_{[-K,K]}$
$g(s, x, v) v(dv)$ has to be added in the HJB equation. The reason for this may be
indicated by noting that if (4.109) is meaningful, and if f in the previous version of
the HJB equation (see (4.106)) is replaced by f^*, then that has to be compensated
for by adding the last mentioned additional term.

4.8 Exercises

Exercises for Section 4.1

4.1. Using $\text{Var}[B_t] = t$, prove that $\text{Var}[B_t - B_s] = t - s$ whenever $t > s$.

Exercises for Section 4.3

4.2. Let $g(t,x) = te^x$, and let $X_t = B_t$, i.e. $dX_t = dB_t$. Find an expression for $Y_t = g(t, B_t)$ by means of Ito's formula.

4.3. Using Ito's formula, show that $X_t = x^0/(1+t) + B_t/(1+t)$ solves the stochastic differential equation $dX_t = -1/(1+t)X_t dt + 1/(1+t)dB_t$, $X_0 = x^0$.

4.4. Let X_t be given by the stochastic differential equation $dX_t = aX_t dt + bX_t dB_t$, $b > 0$. Define $Y_t = (X_t)^\gamma$, γ positive. By means of Ito's formula, find an expression for Y_t. Define $z(t) = EY_t$, and find a differential equation for $z(t)$. What is the solution?

4.5. (A bound on the solution of stochastic differential equations.) Given the multidimensional version of the stochastic differential equation (4.6), assume that $|f| \le c + d|x|$, and $|\sigma_{i,j}| \le c' + d'|x|$. Let X_t be a solution of the equation and let $y_t = E(X_t)^2$.

(a) Use Dynkin's formula and $(a+b)^2 \le 2(a^2+b^2)$ to show that
$$dy_t \le \{E(c+d|X_t|)2|X_t| + 2(1/2)|E(c'+d'|X_t|)^2 nm\}dt \le 2E\{(c(1+|X_t|^2) + d|X_t|^2)\}dt + 2E\{(c'^2+d'^2|X_t|^2)nm\}dt \le \{2c+2nmc'^2+2(c+d+nmd'^2)y_t^2\}dt.$$
(Maybe you only want to consider the case one-dimensional case, where $n = 1$, $m = 1$.)

(b) Using (a), show that $y_t \le (x_0^2 + \alpha/\beta)e^{\beta t} - \alpha/\beta$, for some α, β. How can α and β be chosen?

Next, let $z_t = E[g(X_t)]$, where $g(x) := (1+x^2)^{1/2}$, and let $n = m = 1$.

(c) Use Dynkin's formula, $|x| \le g(x) \le 1+|x|$, and $(d^2/dx^2)(1+x^2)^{1/2} = (1+x^2)^{-3/2}$ to show that $dz_t \le E[(c+d|X_t|)|X_t|(1+|X_t|^2)^{-1/2} + (1/2)(c'+d'|X_t|)^2(1+|X_t|^2)^{-3/2}dt] \le E[c+d|X_t| + c'^2 + d'^2|X_t|^2(1+|X_t|^2)^{-3/2}dt] \le c + dz_t + c'^2 + d'^2$.

(d) Using (c), show that $E|X_t| \le z_t$, where $z_t \le -\hat{d}/d + (z_0 + \hat{d}/d)e^{dt} \le (1 + |X_0| + \hat{d}/d)e^{dt}$, for $\hat{d} = c + (c'^2 + d'^2)$.

4.6. In connection with the proof of Theorem 4.1, show that $||Y^m - Y^n|| \le ||\sum_{n \le k < m} Y^{k+1} - Y^k|| \le \sum_{n \le k} ||Y^{k+1} - Y^k|| \le \sum_{n \le k}(1/2)^{k-n}||Y^{n+1} - Y^n|| \le 2||Y^{n+1} - Y^n|| \le 2(1/2)^n||Y^1 - Y^0||$. Hence, Y_t^n is what is called a Cauchy sequence in $||\cdot||$, with a limit denoted Y_t. Now, $||Y_. - Y^n|| = ||Y_. - X_0 + \int_0^. f(s, Y_s^{n-1})ds + \int_0^. \sigma(s, Y_s^{n-1})dB_s||$. Replacing Y_t^{n-1} by Y_t in the sequence of inequalities in the proof in the text, show that $\sup_t\{E|\int_0^t(f(s, Y_s) - f(s, Y_s^{n-2}))ds + \int_0^t(\sigma(s, Y_s) - \sigma(s, Y_s^{n-2}))dB_s|^2\} \le (1/4)||Y_. - Y^{n-2}||^2$, which shows that, a.s., $Y_t = \lim_n Y_t^{n-1} = X_0 + \lim_n\{\int_0^t f(s, Y_s^{n-2}))ds + \int_0^t \sigma(s, Y_s^{n-2})dB_s\} = X_0 + \int_0^t f(s, Y_s)ds + \int_0^t \sigma(s, Y_s)dB_s$.

Exercises for Section 4.4

4.7. Solve the problem
$$\max E\left[\int_0^T -u_t^2 dt - (X_T)^2\right], \qquad u_t \in \mathbb{R}$$

subject to $dX_t = u_t dt + \sigma X_t dB_t$, $X_0 = x^0$, $\sigma > 0$, σ and x^0 fixed numbers. *Hint:* Try a solution of the HJB equation of the form $J := \phi(t)x^2$. Note that $1/(a\phi + \phi^2) = (1/a)[1/\phi - 1/(\phi + a)]$, this is useful for the solution of an integral.

4.8. Solve the problem

$$\max E\left\{ \int_0^T [-u_t^2 e^{-X_t}/2]dt + e^{X_T} \right\}, \qquad u_t \in \mathbb{R}$$

subject to $dX_t = u_t e^{-X_t} dt + \sigma dB_t$, $X_0 = x^0$, $\sigma > 0$, σ and x^0 fixed numbers. *Hint:* Try a solution of the HJB equation of the form $J := \phi(t)e^x + \psi(t)$. See also the hint in the preceding exercise.

4.9. Solve the problem:

$$\max E\left\{ \int_0^\infty [\beta(X_t)^\gamma + (c_t)^\gamma]e^{-t}dt \right\},$$

subject to $dX_t = \alpha X_t dt + \sigma X_t dB_t - c_t dt$, $X_0 = x^0 > 0$, c_t the control, $c_t > 0, \alpha, \sigma$, positive, $\gamma \in (0,1)$, $\alpha\gamma < 1$, $(X_t > 0)$.

(a) Try an optimal value function of the form: $h(x)e^{-t}$, where $h(x) = ax^\gamma + b$ then is going to be the solution of the current value HJB equation. Equations obtained for a and b can be solved for $\gamma = 1/2$. What are the solutions?

(b) For $\gamma = 1/2$, is the condition (4.53) satisfied?

4.10. Solve the problem:

$$\max E\left\{ \int_0^\infty [-\beta(X_t)^2 - (u_t)^2]e^{-\rho t}dt \right\},$$

subject to $dX_t = \alpha X_t dt + \gamma X_t dB_t + u_t dt$, $X_0 = x^0$, u_t the control, $u_t \in \mathbb{R}$, β, γ, ρ positive. Try a solution of the current value HJB of the form $ax^2 + bx + c$.

4.11. Solve the problem

$$\max E\left\{ \int_0^\infty -e^{-\rho t}[X_t^2 + u_t^2]dt \right\}, \qquad u_t \in \mathbb{R}$$

subject to $dX_t = u_t dt + \sigma dB_t$, $X_0 = x^0$, with $\sigma > 0$, $\rho > 0$. Try a solution of the current value HJB of the form $ax^2 + bx + c$.

4.12. The following problem has the solution $u \equiv 1$, a glance at the problem will give the reader a strong suspicion that this is so:

$$\max E\left\{ \int_0^\infty e^{-t}X_t dt \right\}, \qquad u_t \in [0,1]$$

subject to $dX_t = (1/2)u_t X_t dt + \sqrt{2X_t}dB_t$, $X_0 = x^0 = 1$. Below, we only consider $x > 0$. Write down the HJB equation, show that it has the general solution $h =$

$2x + C_1 x^{r_+} + C_2 x^{r_-}$, where $r_+ = (1/4)[1 + (17)^{1/2}]$ and $r_- = (1/4)[1 - (17)^{1/2}]$. Find values of the integration constants such that (4.55) holds, in fact (4.55) (a) is sufficient for determining constants. *Hint*: Use Dynkin's formula to find $z_t^{(r)} := E(X_t^0)^r$.

4.13. Solve the problem

$$\max E\left\{ \int_0^1 -u_t^2 dt \right\}, \text{ subject to } dX_t = u_t dt + \sigma x dB_t, X_0 = 1, E(X_1)^2 = k,$$

where $u_t \in \mathbb{R}, k > e^{\sigma^2}$. *Hint*: Try $J = z(t)x^2, z(1) = q$ in (4.71).

4.14. Solve the problem

$$\max E\left[\int_0^1 -(u_t X_t)^2 dt \right] \text{ subject to } dX_t = u_t dt + \sigma X_t dB_t, EX(1)^4 = k,$$

$X_0 = x^0 > 0, u_t \in \mathbb{R}$, where σ and k are constants, k suitable large. *Hint*: Try $J = b(t)x^4$.

4.15. Using the maximum principle, solve the problem:

$$\max E\left[\int_0^1 u_t dt - X_1^2/2 \right], \text{ when } dX_t = (u + Y_t)dt, dY_t = \sigma dB_t,$$

$u_t \in \mathbb{R}, \sigma > 0, X_0 = 1, Y(0) = 0$.

Hint: The maximum condition gives $p^x \equiv -1$, (p^x the costate corresponding with x), so $-1 = p^x(1) = -X_1^*$, i.e., X_1^* is deterministic. One way to obtain that is to let $u_t^* = k - Y_t$, k a deterministic constant. (To be sure that an optimal control has been found, prove that the criterion equals $1/2 - E[(X_1 - 1)^2/2]$.)

Exercises for Section 4.5

4.16. Solve the stopping problem

$$\max_\tau E[(X_\tau)e^{-\rho\tau}], \tau \text{ subject to choice in } [0,\infty), dX_t = dB_t, X_0 = 0.$$

Hint: $\hat{J}(t,x) = h(x)e^{-\rho t}$, and the continuation region is presumably of the form $(-\infty, a)$. To determine constants, also use that $J(0,x)$ is non-negative, but presumably is not large, when x is near $-\infty$ (non-negativity because we know that sometime X_t becomes zero, with prob. 1).

4.17. (Øksendal). Solve the stopping problem

$$\max_\tau E[(B_\tau)^2 e^{-\rho t}], \tau \text{ subject to choice in } [0,\infty).$$

Hint: $\hat{J}(t,x) = h(x)e^{-\rho t}$, $dX_t = dB_t$, $(X_t = B_t)$, and the continuation region is presumably of the form $(-a,a)$. Find an (unsolvable) equation for a.

4.18. Solve the problem:

$$\max_{\tau} E[(e^{X_\tau/4} - 1)e^{-\tau}], \tau \text{ subject to choice in } [0,\infty),$$

where $dX_t = 1dt + 2dB_t$, $X_0 = 0$. *Hint*: The continuation region is of the form $(-\infty, a)$. $J(t,x) = h(x)e^{-t}$.

4.19. Consider Example 4.34, still with $r > \beta$. (a) Show that when $\beta \geq \sigma^2/2$, then we stop a.s. (b) Assume that $p_0 > p^*$. Using $V = (p^*, \infty)$ (if $p_t = p^*$, we surely stop) and $\hat{J} := p - C$, show by means of Remark 4.29 that it is optimal to stop immediately.

Exercises for Section 4.6

4.20. Solve the problem:

$$\max E\left[\int_0^T (-u_t^2/2 - X_t^2)dt\right], dX_t = u_t dt + dB_t, u_t \in \mathbb{R}, X_0 = x^0 \text{ given,}$$

where X_t can have an unbounded number of jumps all of intensity 1, given by $X_{\tau_j+} = X_{\tau_j-} + V_j$. Here, all V_j are i.i.d., $E[V_j] = 0$, $E[V_j^2] = 1$.

Hint: Try a solution $\hat{J} = \phi(t)x^2 + \psi(t)$.

Chapter 5
Appendix: Probability, Concepts, and Results

Certain concepts and results from probability theory are presented. The reader is advised to look up in the Appendix definitions, results, or explanations related to themes appearing in the chapter he is currently reading and not read the Appendix in its entirety.

5.1 Elementary Probability

A discrete sample space is a "countable" set Ω with typical element ω. The set Ω is countable if it is possible to label the various elements in Ω such that $\Omega = \{\omega_1, \omega_2, \ldots, \omega_i, \ldots\}$; it may be finite or infinite. The set Ω includes all possible outcomes that can occur, and on Ω we imagine that there is given a function $P(\omega)$ that tells for each ω in Ω how probable it is that ω will occur. Think of Ω as a sort of box from which draws are made and that $P(\omega)$ is the probability that ω is drawn. The probabilities $P(\omega)$, $\omega \in \Omega$, are non-negative numbers, and these probabilities must satisfy

$$\sum_{\omega \in \Omega} P(\omega) := \sum_i P(\omega_i) = 1. \tag{5.1}$$

The function $P(\omega)$ is often called a *(probability) mass function* or a *probability distribution*.

In some cases, we can think of $P(\omega)$ as indicating how frequently ω occurs. That is, if we repeatedly are in the same situation (we repeatedly make draws out of Ω, which are put back again), then $P(\omega)$ indicates the relative frequency of occurrences of ω. When applying the theory of probability in concrete situations, then somehow, by (long) experience, or by careful design of the situation, we "know" the probabilities. For example, in situations where 10 balls numbered 1 to 10 are thrown into an urn, and the urn is thoroughly shaken, then if a ball is drawn from the urn, there is a probability of 1/10 of drawing a ball with number (say) 2. Another situation occurs if I have carefully constructed a biased die that turns up 6 with a probability (or frequency) 1/3, shown in lengthy trials. Then I believe that the probability of getting

A. Seierstad, *Stochastic Control in Discrete and Continuous Time*,
DOI: 10.1007/978-0-387-76617-1_5,
© Springer Science+Business Media, LLC 2009

6 is 1/3, which is twice what an uninformed opponent would believe, namely 1/6, as is the case for an unbiased die.

Frequently, a subset A of Ω is called an *event*. (We may think of it as the event that the outcome ω happens to belong to the set A.) We shall sometimes use the word (sub)set and sometimes the word event. For any ω, the set $A = \{\omega\}$ is sometimes called an elementary event. For any event A in Ω, by definition, the probability of A equals $\Pr(A) := P(A) := \sum_{\omega_i \in A} P(\omega_i)$. We often write $\Pr(A)$ and not $P(A)$, similarly, $\Pr(\omega_i) := P(\omega_i)$. From the definition of $P(A)$, it follows that if the sets A_j, $j = 1, 2, \ldots$ are (pairwise) disjoint, then

$$\Pr(\cup_j A_j) = \sum_j \Pr(A_j) \qquad (A_j \text{ disjoint}) \qquad (5.2)$$

(a property called *countable additivity* of $P(.)$). If the A_j's are disjoint and have a union equal to Ω, this family of sets (events) is called a *partition* of Ω. Evidently then $\sum_j \Pr(A_j) = 1$. Note, moreover that

(α) $\qquad\qquad\qquad A \subset B \Rightarrow \Pr(A) \le \Pr(B)$

(β) $\qquad\qquad\qquad\Pr(A) + \Pr(\complement A) = 1$

(γ) $\quad \Pr(A \cup B) = \Pr(A) + \Pr(B) - \Pr(A \cap B)$ for arbitrary events $A, B \subset \Omega$

(δ) $\qquad\qquad\qquad \Pr(\Omega) = 1, \Pr(\emptyset) = 0. \qquad\qquad\qquad (5.3)$

All these properties follow easily from the definitions.

An example of a probability mass function is the *Poisson mass function*

$$p(k; \gamma) = e^{-\gamma} \frac{\gamma^k}{k!} \qquad (\text{expectation } \gamma, \text{ variance } \gamma) \qquad (5.4)$$

k running in $\{0, 1, 2, \ldots\}$, $(k! = 1 \cdot 2 \cdots \cdot k, 0! = 1)$. (For the definition of expectation and variance, see below.) Then $\Pr(k) = p(k; \gamma)$. The number k of people calling an office in a given time interval is sometimes assumed to be Poisson distributed, for some given parameter (or "intensity") γ.

Another probability distribution is the *binomial distribution*, which has the mass function

$$b(k; n, p) = \binom{n}{k} p^k (1-p)^{n-k}, \qquad (\text{expectation } np, \text{ variance } np(1-p)) \qquad (5.5)$$

where $\binom{n}{k} = n!/r!(n-r)!$, k running in the set $\{0, 1, 2, \ldots n\}$, $(\binom{n}{0} = 1)$. This mass function gives the probability of k "successes" in the following situation: There are n trials, each one can turn out to be a success or not, the probability of success in any given trial being p.

The following rule is frequently useful.

Theorem 5.1 (Simple law of total probability). *Let A_1, A_2, ... be disjoint events such that $\Omega = \cup_i A_i$ (i.e., $\{A_i\}$ is a partition of Ω). Then for any event B in Ω,*

$$\Pr[B] = \sum_i \Pr[B \cap A_i]. \qquad\qquad (5.6)$$
$\square$

Cumulative Distribution

If Ω is a countable set of real numbers, and $f(\omega)$ is a given mass function on Ω, then

$$F(\omega) := \Pr[\omega' \le \omega] = \sum_{\omega' \le \omega} f(\omega')$$

is the *cumulative distribution* for f.

We apply the term *random variable* to any function defined on Ω. In particular, the function $Y = \omega$ (or more formally $Y := \omega \to \omega$), is a random variable. The image set of a random variable X can be any set Ω''. We can always take it to be $\Omega' := X(\Omega)$, which is a countable subset of Ω''. Most often one encounters real-valued random variables ($\Omega'' = \mathbb{R}$), or random variables taking values in $\mathbb{R}^n$.

Function of Random Variables

Let f be a given mass function on Ω, and let Y be a random variable on Ω with values in Ω''. We assume that $\Omega'' = \Omega_Y := \{Y(\omega) : \omega \in \Omega\}$ and furnish this countable set with the mass function $f_Y(\omega') := \sum_{\omega \in \{\omega : Y(\omega) = \omega'\}} f(\omega)$. This is the mass function of the random variable Y. For any subset A in Ω'' we have that $\Pr(A) = \Pr\{Y \in A\}$, where $\Pr(A) := \sum_{\omega' \in A} f_Y(\omega')$ and $\Pr\{Y \in A\}$ means $\sum_{\omega \in \{\omega : Y(\omega) \in A\}} f(\omega)$.) Because $\Pr(A)$ is calculated by means of f_Y, we often write $\Pr(A) = \Pr_{f_Y}(A)$. When working with the random variable Y, we frequently forget about the original space Ω and its mass function f, and work only with the mass function $f_Y(\omega')$ in Ω'', once $f_Y(\omega')$ has been found. Note that $\{Y \in B\} = \{\omega \in \Omega : Y(\omega) \in B\}$, B any subset of Ω''.

Joint Mass Functions

Let $\Omega = \Omega_1 \times \Omega_2$, where Ω_1 and Ω_2 are countable. Then Ω is also countable. On Ω we assume that a mass function $f(\omega, \omega')$ is defined, $(\omega, \omega') \in \Omega$. If $\Omega \subset \mathbb{R}^2$, the cumulative distribution corresponding to $f(\omega, \omega')$ is

$$F(\omega, \omega') := \Pr[\omega_1 \le \omega, \omega_2 \le \omega'] = \sum_{\omega_1 \le \omega, \omega_2 \le \omega'} f(\omega_1, \omega_2)$$

The so-called *marginal mass functions* are defined by $\Pr(\omega) = \sum_{\omega' \in \Omega_2} f(\omega, \omega') =: f_1(\omega)$ and $\Pr(\omega') = \sum_{\omega \in \Omega_1} f(\omega, \omega') =: f_2(\omega')$, respectively.

These definitions are extended to random variables in general. Let X and Y be two random variables defined on Ω, with values in Ω' and Ω'', respectively, where we assume $\Omega' = X(\Omega)$ and $\Omega'' = Y(\Omega)$. Let $f_{X,Y}(\omega', \omega'')$ be the mass function of the random variable (X, Y), i.e.,

$$f_{X,Y}(\omega', \omega'') = \Pr[(X, Y) = (\omega', \omega'')] = \sum_{\omega \in \{\omega : (X(\omega), Y(\omega)) = (\omega', \omega'')\}} f(\omega)$$

(f being the mass function on Ω). Then the marginal mass function of X, $f_X(\omega')$ is defined by $f_X(\omega') := \Pr(X = \omega') := \sum_{\omega \in \{\omega:X(\omega)=\omega'\}} f(\omega) = \sum_{\omega'' \in \Omega''} f_{X,Y}(\omega', \omega'')$ and the marginal distribution of Y, $f_Y(\omega'')$ is defined by $f_Y(\omega'') := \Pr(Y = \omega'') := \sum_{\omega \in \{\omega:Y(\omega)=\omega''\}} f(\omega) = \sum_{\omega' \in \Omega'} f_{X,Y}(\omega', \omega'')$. We repeat:

$$f_Y(\omega'') = \sum_{\omega' \in \Omega'} f_{X,Y}(\omega', \omega''), \quad f_X(\omega') = \sum_{\omega'' \in \Omega''} f_{X,Y}(\omega', \omega''). \tag{5.7}$$

Combinatorics

Only a few formulas are presented here. Out of a list of n different symbols, n^r different ordered r-tuples of r symbols can be constructed, when "replacement" is allowed (when each symbol can be repeated). A different formula holds when replacement is not allowed, i.e., in the case where a symbol that has been used once in the ordered r-tuple can never be used again. In this case, the number of different ordered r-tuples is $(n)_r := n(n-1) \cdot \ldots \cdot (n-r+1)$. In the last case, assume that order is of no significance, i.e., that two r-tuples are counted as one and the same if they contain the same symbols but in different order. Then we speak of r-subsets instead of r-tuples. Without replacement, $\binom{n}{r} := n!/r!(n-r)!$ different r-subsets can be constructed, where $n! = (n)_n$.

5.2 Conditional Probability

If A and B are given subsets of Ω, then, by definition, the conditional probability of A given B equals

$$P(A|B) := \Pr[B|A] := P(B \cap A)/P(A) \text{ provided } P(A) > 0, \tag{5.8}$$

An example of the use of conditional probability is the following. In an urn there are two white and two red balls. Two draws are made. If the first ball drawn is red, what is the probability that the second one is also red? Well, in the urn three balls remain, one red, so because a red ball was drawn at the first draw, the probability is 1/3 for drawing a red ball at the second draw. This is the conditional probability of drawing a red ball at the second draw, given that we drew a red ball at the first draw. Here, the conditional probability was not determined by using (5.8), but by considering instead the information available when the event "Drawing a red ball at the first draw" has already happened. In such cases, and they frequently occur, we have an intuitive feeling for what the conditional probability is. Using the formal definition, we can argue as follows. Number the balls from 1 to 4, 1 and 2 being red. Generally, the number of the first ball drawn is denoted i_1, the second one drawn has the number i_2, with $i_1, i_2 \in \{1, \ldots 4\}$. All together, there are 12 ordered pairs that can be drawn. (For example, if $i_1 = 1$, there are three remaining possible values of i_2, i.e., pairs start with 1. Similarly, three pairs start with 2, 3, and 4. Compare formulas with the Combinatorics section above.) All pairs are equally possible, and only two consist of red ones, namely $(1,2)$ and $(2,1)$. So the probability of drawing a red pair is $2/12 = 1/6 = P(A \cap B)$, where A denotes the event of drawing "red" at the first draw, and B the event of drawing red at the second draw. Evidently, $P(A) = 1/2$, so $\Pr(B|A) = (1/6)/(1/2) = 1/3$.

In this case, we were able to find the conditional probability much more quickly, not using the formal definition. In concrete situations, the fact that an event has occurred, changes the "stochastic situation" in ways that we can intuitively exploit. We can also often deduce probabilities of "composite" events, using conditional probabilities that we somehow feel are correct ones. In fact, we often use the definition of conditional probability rather as a theorem (axiom!). Consider, e.g., the example above: To derive the probability of a red pair, we could use the relation $P(A \cap B) = P(B|A)P(A)$ and argue as follows: It is obvious that $P(B|A) = 1/3$ and that $P(A)$ is 1/2, so $P(A \cap B) = 1/3 \cdot 1/2 = 1/6$.

Two formulas involving conditional probabilities are useful: From the simple law of total probability, we get the following *law of total probability*:

Theorem 5.2. *Let* $A_1, A_2, \ldots$ *be a partition of* Ω, *and let B be any event in* Ω. *Then*

$$\Pr(B) = \sum_i \Pr[B|A_i] \cdot \Pr(A_i). \tag{5.9}$$

$\square$

We also have the following result (a consequence of (5.9)):

Theorem 5.3. *(Bayes' formula) Let* $A_1, A_2, \ldots$ *be a partition of* Ω *and let B be any event in* Ω. *Then*

$$\Pr(A_j|B) = \Pr(B|A_j)\Pr(A_j) / \left\{ \sum_i \Pr(B|A_i)\Pr(A_i) \right\} \tag{5.10}$$

$$(= \Pr(A_j \cap B)/\Pr(B)).$$

$\square$

Conditional Mass Functions

Given two random variables $X : \Omega \to \Omega'$ and $Y : \Omega \to \Omega''$, with joint mass function $f_{X,Y}(\omega, \omega')$. Then the conditional mass function $f_{X|Y}(\omega'; \omega'')$ equals $f_{X,Y}(\omega', \omega'')/f_Y(\omega'')$. It is the same as $\Pr[X = \omega'|Y = \omega'']$. It is only defined for ω''-s for which $f_Y(\omega'') > 0$.

Independence

The events A and B are said to be *independent* if

$$P(A \cap B) = P(A)P(B). \tag{5.11}$$

Another manner of expressing this is evidently $P(A|B) = P(A)$. This equality says that it does not change our belief in how likely the event A is when we hear that B has happened. Also concerning independence, we often have an intuition about it, without apparently needing to test it against the definition. However, as with conditional probabilities, if the situation is slightly complex, a test using the definition may be needed. Sometimes it seems trivial. If balls are replaced in the urn considered above (again the urn is thoroughly shaken), then drawing a red ball at the second draw intuitively is independent of drawing a red ball at the first draw. We may test this using the definition. Because of replacement, the number of possible pairs are now 16. (If ball no 1 is drawn, four different balls can be drawn at the second draw, and thus we get 4 pairs. That we also get if the first ball is number 2,3, and 4.) So the probability $P(A \cap B) = 4/16 = 1/4$ (pairs (1,1),(1,2),(2,1) and (2,2) are all red), so the equality $1/4 = P(A \cap B) = P(A)P(B) = 1/2 \cdot 1/2$ holds, showing independence.

We say that two random variables X and Y defined on Ω, taking values in Ω' and Ω'', respectively, are *independent* if the events $\{X = \omega'\}$ and $\{Y = \omega''\}$ are independent for all $(\omega', \omega'') \in \Omega' \times \Omega''$. For two random variables X and Y, we have the following result:

Theorem 5.4. *X and Y are independent iff we have the factorization $f_{X,Y}(\omega', \omega'') = f_X(\omega') \cdot f_Y(\omega'')$ of the mass function $f_{X,Y}$ of (X,Y).* □

Another property equivalent to independence, is: $f_X(\omega') = f_{X|Y}(\omega'; \omega'')$.

A family F of events is called *independent* if

$$\Pr(A_1 \cap \cdots \cap A_n) = \Pr(A_1) \cdot \cdots \cdot \Pr(A_n)$$

for all finite collections of events A_i, $i = 1, \ldots, n$, from F.

Pairwise independence does not imply independence: A family F consisting of three events A_1, A_2, A_3 may well exhibit pairwise independence, (A_i, A_j independent for all pairs (i, j)), without being independent.

A family of random variables $X_1, X_2, \ldots$ from Ω into Ω' is *independent* if the family of sets (events) $A_{i,\omega'} := \{X_i = \omega'\}$, $\omega' \in \Omega$, is independent.

If $\Omega' = \Omega'' = \{\text{red, white}\}$, Ω is the urn, and X is the outcome in the first draw, Y in the second draw in the example above, then X and Y are independent variables in the replacement case (not in the nonreplacement case).

Product Probability

If Ω_1 and Ω_2 are two sample spaces, with mass functions $f_1(\omega_1)$ and $f_2(\omega_2)$, then on $\Omega = \Omega_1 \times \Omega_2$ we can introduce the mass function $f(\omega_1, \omega_2) := f_1(\omega_1) \cdot f_2(\omega_2)$. For such a mass function f, the outcomes ω_1 and ω_2 are independent. Similarly, if X is a random variable on Ω_1 with values in Ω^1, and Y is a random variable on Ω_2 with values in Ω^2, then the mass function of (X,Y) on $\Omega^1 \times \Omega^2$ is $f_{X,Y}(\omega^1, \omega^2) := f^X(\omega^1) \cdot f^Y(\omega^2)$, where $f^X(\omega^1)$ (respectively, $f^Y(\omega^2)$) is the mass function of X (resp. Y). For the moment they are written with superscripts instead of subscripts in order to express the following fact: We evidently have $f^X(\omega^1) = f_X(\omega^1)$, $f^Y(\omega^2) = f_Y(\omega^2)$, where subscript indicate marginal distributions (relative to $f_{X,Y}$ just defined).

5.3 Expectation

Let Y be a random variable on Ω into Ω', with mass function $f_Y(\omega')$. Then the expectation of Y, denoted $E(Y)$, is defined by

$$E(Y) := \sum_{\omega \in \Omega} Y(\omega)f(\omega) = \sum_{\omega' \in Y(\Omega)} \omega' f_Y(\omega') \qquad (5.12)$$

where f, as usual, is the mass function on Ω.

The expectation of Y is the weighted average of the values of Y, the weights being $f_Y(\omega')$. Or, it is the maximum amount a gambler should be willing to pay for getting access to a lottery with stochastic reward Y.

(The last idea is sometimes made the basis for the introduction of subjective probability. Imagine that $Y = 1_A$, the indicator function of the set $A \subset \Omega$. Then $P(A) = E(1_A)$. Suppose that the gambler is able to connect "subjective probabilities" $P(A)$ to all events A, the number $P(A)$ having the property that he is indifferent between getting $P(A)$ surely or getting 1 dollar iff A occurs. It can be shown that if the gambler fixes the numbers $P(A)$ in a (so-called coherent) manner in which it is impossible for an opponent to earn money by exploiting the above-mentioned indifference, then it can be shown that the numbers $P(A)$ satisfy the axioms of a probability distribution.)

The expectation $E(Y)$ satisfied a number of relations (below a.s. =: "almost surely" means the same as "with probability 1").

$(\alpha) \qquad\qquad E(X) = 0$ if $X = 0$ a.s.,

$(\beta) \qquad\qquad E(a) = a$, (expectation of a constant),

$(\gamma) \quad E(aX + bY) = aE(X) + bE(Y)$ (a and b are constants),

$(\delta) \quad E(XY) = (E(X)) \cdot (E(Y))$ when X and Y are independent.

$$(5.13)$$

All properties are easy consequences of the definition.

The variance of Y is defined by

$$\mathrm{Var}(Y) := E[(Y - E(Y))^2] \qquad (5.14)$$

We have

$$\mathrm{Var}(Y) = E(Y^2) - (E(Y))^2 \qquad (5.15)$$

and

$$\mathrm{Var}(aY + b) = a^2 \mathrm{Var}(Y) \quad (a \text{ and } b \text{ are constants}) \qquad (5.16)$$

and

$$\mathrm{Var}\left(\sum_{1 \le i \le n} X_i\right) = \sum_i \mathrm{Var}(X_i) \text{ when } X_1, \ldots, X_n \text{ are independent.} \qquad (5.17)$$

Without independence, $\mathrm{Var}(\sum_{1 \le i \le n} X_i) = \sum_{1 \le i \le n} \sum_{1 \le j \le n} \mathrm{Cov}(X_i, X_j)$, where

$$\text{Cov}(X,Y) := E[(X - EX)(Y - EY)] = E[XY] - E(X)E(Y).$$

If $\psi(\omega')$ is any function on Ω', Y is a random variable on Ω, taking values in Ω' with mass function $f_Y(\omega)$, then for $Z = \psi(Y)$, we have

$$E(Z) = \sum_{\omega \in \Omega} \psi(Y(\omega))f(\omega) = \sum_{\omega' \in Y(\Omega)} \psi(\omega')f_Y(\omega') = \sum_{\omega'' \in \psi(Y(\Omega))} \omega'' f_{\psi(Y)}(\omega'') \tag{5.18}$$

where $f_{\psi(Y)}(\omega'')$ is the mass function of the random variable $\omega'' = \psi(Y(\omega))$.

Conditional Expectation

The conditional expectation of X given Y is defined by

$$E[X|Y] := \sum_{\omega' \in X(\Omega)} \omega' f_{X|Y}(\omega'; \omega''). \tag{5.19}$$

Similarly,

$$E[\phi(X)|Y] = \sum_{\omega' \in X(\Omega)} \phi(\omega')f_{X|Y}(\omega'; \omega''). \tag{5.20}$$

Note that $E[\phi(X)|Y]$ is a random variable on the image set Ω'' of Y. The following double expectation rule is of great use:

$$E[E[\phi(X)|Y]] = E[\phi(X)]. \tag{5.21}$$

More generally we have

$$E[E[\phi(X)|Y,Z]|Z] = E[\phi(X)|Z]. \tag{5.22}$$

We also have $E[\psi(Y)\phi(X)] = E[\psi(Y)E[\phi(X)|Y]]$. Furthermore,

$$E[\phi(X)|Y] = E[\phi(X)] \text{ if } X \text{ and } Y \text{ are independent.} \tag{5.23}$$

Finally the so-called Chebyshev's inequality holds: For $a > 0$,

$$\Pr[|X| \geq a] \leq E(X^2)/a^2. \tag{5.24}$$

5.4 Uncountable Sample Space

Frequently, the sample space is an uncountable set, say a subset of $\mathbb{R}^n$. Formally, probability theory can be developed using an "abstract," perhaps uncountable, sample space Ω. That approach is sketched in the next section. Here, we shall assume that $\Omega \subset \mathbb{R}^n$, for the current purpose we can actually assume $\Omega = \mathbb{R}^n$ by simply setting $\Pr(\mathbb{R}^n \backslash \Omega) = 0$.

Consider first the case $n = 1$. For $\Omega = \mathbb{R}$, for any x, a probability $F(x)$ is assigned to the event $\{\omega : \omega \leq x\}$. $F(x)$ is called a *(cumulative) distribution function*. By definition, a cumulative distribution function is nondecreasing, $F(x) \leq 1$, $F(\infty) := \lim_{x \to \infty} F(x) = 1$, and $F(-\infty) = 0$. The probability of $\{\omega : \omega > x\}$ is $1 - F(x)$ and the probability of $\{\omega : x < \omega \leq y\}$ is $F(y) - F(x)$. The probability of $\{\omega : \omega < x\}$ is $F(x-)$, the left limit of F at x. F is always taken to be right-continuous, so $F(x) = F(x+)$. (In fact, F has to be so, due to property (6) in Section 5.6 below.) Finally, note that $\Pr\{x \leq \omega < y\} = F(y-) - F(x-)$.

Frequently, $F(x)$ has further smoothness properties: $F(x)$ may be a definite integral, $F(x) = \int_{-\infty}^{x} f(\omega)d\omega$, for all x, where $f(x)$ piecewise continuous. Then $f(x)$ is called the *density* of $F(x)$. $F(x)$, as a function of the upper limit of such a definite integral, gets a property named absolute continuity, which we do not define here. (It is equivalent to being such a definite integral for a $f(x)$ perhaps only integrable, see below.) When $F(x)$ is such an integral, $F(x)$ is definitely continuous in the ordinary sense and $\Pr\{\omega : x < \omega < y\} = \Pr\{\omega : x \leq \omega \leq y\} = \int_{x}^{y} f(\omega)d\omega$. Moreover, $E[\phi(\omega)] := \int_{-\infty}^{\infty} \phi(\omega)f(\omega)d\omega$, so in particular, $E(\omega) := \int_{-\infty}^{\infty} \omega f(\omega)d\omega$, and $\mathrm{Var}(\omega) := \int_{-\infty}^{\infty} (\omega - E\omega)^2 f(\omega)d\omega$.

Two frequently occurring densities are the *normal* or *Gaussian* density

$$f(x) = 1/(2\pi\sigma^2)^{1/2} \exp(-(x-\mu)^2/2\sigma^2) \quad \text{(expectation } \mu, \text{ variance } \sigma^2) \quad (5.25)$$

and the *(negative) exponential density*, for $\Omega = [0, \infty)$,

$$f(x) = \lambda e^{-\lambda x}, \ x \geq 0 \quad \text{(expectation } 1/\lambda, \text{ variance } 1/\lambda^2). \quad (5.26)$$

If $\Omega = \mathbb{R}^n$ and $\mathbf{x} \leq \mathbf{y}$ is the standard vector ordering ($x_i \leq y_i$ for all i), probabilities for events in Ω are obtained by specifying a cumulative distribution function $F(\mathbf{x})$, $\mathbf{x} \in \mathbb{R}^n$ (such that $\Pr\{\omega \leq \mathbf{x}\} = F(\mathbf{x})$). Again $F(\mathbf{x})$ is nondecreasing in $\mathbf{x}$ (nondecreasing in each component of $\mathbf{x}$) and $F(\mathbf{x})$ is right-continuous in the sense that $F(\mathbf{x}) = \lim_n F(\mathbf{y}_n)$ if $\mathbf{y}_n \downarrow \mathbf{x}$, and $F(\mathbf{x}) \leq 1$. Furthermore, $F(n1') \to 1$ when $n \to \infty$, where $1'$ is the vector whose components all equal 1, and $F(-n1') \to 0$ as $n \to \infty$. Sometimes F is generated by a bounded density $f(\mathbf{x})$, which is, separately, piecewise continuous in each variable x_i (we then call it separately piecewise continuous). Then, by definition, $F(\mathbf{x})$ equals the n-multiple integral over $\{\mathbf{z} \leq \mathbf{x}\}$ of $f(\mathbf{z})$ and $F(\mathbf{x})$ becomes continuous.

Marginal Densities, Conditional Densities

If $f(x, y)$, defined on $\Omega = \mathbb{R}^2$ is a density, then $f_X(x) = \int_{-\infty}^{\infty} f(x, y)dy$ is the *marginal density* of $X : x \to x : \mathbb{R} \to \mathbb{R}$, with $f_Y(y)$ having a similar definition, $Y : y \to y : \mathbb{R} \to \mathbb{R}$. Of course, on the basis of $f_X(x)$, we can calculate probabilities like $\Pr\{a \leq X \leq b\}$ ($= \int_a^b f_X(x)dx$.) The *conditional density* of X given Y is $f_{X|Y}(x; y) = f(x, y)/\int_{-\infty}^{\infty} f(x, y)dx$. (If the integral equals zero for some value of y, for this value of y, any value, for example 0, can be assigned to $f_{X|Y}(x; y)$. For the standard applications of such conditional densities, the values for such y's do not matter. A similar remark could have been made also in the discrete case, we did not.) The conditional density of X given Y is used to calculate conditional probabilities like $\Pr[a \leq X \leq b|Y = y]$ $(= \int_a^b f_{X|Y}(x;y)dx)$. Now, the conditional expectation $E(\phi(X)|Y)$ is given by $\int_{-\infty}^{\infty} \phi(x)f_{X|Y}(x;y)dx$ and the property (5.21) holds also in this case.

Here, in the definition of marginal and conditional densities, x and y could even be n'-vectors and n''-vectors, respectively, with $n' + n'' = n$, $(\Omega = R^n)$, the only change needed is that the integrals above are multiple integrals.

5.5 General Probability Distributions

So far, probability distributions have either been discrete or continuous. Often a common notation for the calculation of expectations and probabilities is useful. Given any cumulative distribution $F(x)$ on $\mathbb{R}$, we shall write $E(\phi(x)) = \int_{-\infty}^{\infty} \phi(x)\,dF(x)$, and similarly, $\Pr\{a \leq X \leq b\} = \int_{[a,b]} dF(x)$. So let us give a meaning to an integral of the form $\int_{[a,b]} \phi(x)\,dF(x)$. If $F(x)$ stems from a piecewise continuous density f, we evidently let $dF(x)$ have the meaning $dF(x) = f(x)\,dx$, so inserting this for $dF(x)$ in $\int_{[a,b]} \phi(x)\,dF(x)$ yields an ordinary integral. If $F(x)$ is discrete with mass function $f(x)$, by definition

$$\int_{[a,b]} \phi(x)\,dF(x) := \sum_{a \leq x \leq b} \phi(x)f(x) = \sum_{a \leq x \leq b} \phi(x)(F(x) - F(x^-)).$$

The last equality stems from the fact that $f(x) = F(x) - F(x^-)$ in this case. In fact, this formula holds for all x, with $F(x) - F(x^-)$ nonzero only for a discrete set of x's at which we have nonzero mass. (The sum needs only extend over this set. We allow this set to be countable.)

Suppose we have a probability distribution that is a combination of a distribution given by a piecewise continuous density and a discrete distribution. More precisely, suppose the cumulative distribution $F(x)$ is a sum $F(x) = F_1(x) + F_2(x)$, such that F_1 is discrete and F_2 has a piecewise continuous density, $F(\infty) = 1$, F_1, F_2 nondecreasing, and $F_i(-\infty) = 0$, $i = 1, 2$, Then, for any piecewise continuous function $\phi(x)$, by definition, we let

$$\int_{[a,b]} \phi(x)\,dF(x) := \int_{[a,b]} \phi(x)\,dF_1(x) + \int_{(a,b)} \phi(x)\,dF_2(x).$$

A more general definition of an integral of the form $\int_I \phi(x)\,dF(x)$ where I is an interval, can be given. It is useful also outside probability theory. We take $F(x)$ to be any real-valued nondecreasing and right-continuous function on $\mathbb{R}$. Given any interval $[a,b]$, let $F(b) - F(a-) = M$. Now, $F(x) - F(x-) \geq 1/n$ for only a finite number of points in $[a,b]$, the number in fact is $\leq nM$. (Evidently, the sum of these "jump increases" of F cannot be larger than the total increase $F(b) - F(a-)$.) So F has at most a countable number of jump points in any interval $[a,b]$. In fact, F has a countable number of jump points in all $(-\infty, \infty)$. Let $F_1 = \sum_{y \leq x} F(y) - F(y-)$ (the sum in effect is over all jump points $\leq x$). Then F_1 has the same jumps as F. Hence, $F_2 = F - F_1$ becomes continuous. Moreover, F_2 is nondecreasing, because F_1 is constant between jumps. Unfortunately, F_2 cannot always be expressed as the integral of a density. So let us explain what we mean by an integral over $[a,b]$ of a piecewise continuous function ϕ with respect to the integrator F_2, $\int_{[a,b]} \phi(x) dF_2(x)$. As usual for integrals, we first state the definition for piecewise constant functions: Let $\phi(x) = \sum_i \alpha_i 1_{I_i}(x)$, where I_i is a partition of $[a,b]$ into intervals I_i with endpoints $b_{i-1} < b_i$, and where $1_{I_i}(x)$ is the indicator function of I_i, (equal to 1 if $x \in I_i$, equal to 0 otherwise). Then

$$\int_{[a,b]} \phi(x) dF_2(x) := \sum_i \alpha_i (F_2(b_i) - F_2(b_{i-1})).$$

The integral with respect to F_2 of a piecewise continuous function ϕ is defined as a limit of integrals of piecewise constant functions. More precisely, it is done as follows. Define, for each n, $\phi_n(x) = \sum_i \phi((b_{i-1} + b_i)/2) 1_{(b_{i-1}, b_i]}(x)$, where $0 < b_i - b_{i-1} \leq 1/n$ (we let b_{i-1}, b_i depend on n), moreover $\{b_i\}$ includes all discontinuity points of ϕ. Then we define $\int_{[a,b]} \phi(x) dF_2(x) = \lim_{n \to \infty} \int_{[a,b]} \phi_n(x) dF_2(x)$. It can be proved that the limit exists. Note that for any sequence $\phi_n(x)$ of the above type, $\phi_n(x) \to \phi(x-)$ for all $x > a$. In fact, we get the same limit for the integrals of all bounded sequences of piecewise constant functions converging to $\phi(x)$ for all $x > a$, x a continuity point of $\phi(x)$.)

Again, by definition, for any interval I,

$$\int_I \phi(x) dF(x) = \sum_{x \in I} \phi(x)(F_1(x) - F_1(x^-)) + \int_I \phi(x) dF_2(x).$$

The integral with respect to F over a one point set can be nonzero. In fact, $\int_{[a,a]} \phi(x) ds = \phi(a)(F(a) - F(a-))$. Moreover,

$$\int_{[a,b]} \phi(x) dF(x) = \phi(a)(F(a) - F(a-)) + \phi(b)(F(b) - F(b-)) + \int_{(a,b)} \phi(x) dF(x).$$

In this connection, note that an integral with respect to F_2 over an interval does not change when we omit one or the other (or both) endpoints from the interval. However, the entity F_1 may cause changes. Note, finally that if $\hat{F}$ is a nondecreasing function, perhaps only defined on an interval $[a,b]$, not on all $\mathbb{R}$, then by definition,

$$\int_{[a,b]} \phi(x) d\hat{F}(x) := \int_{[a,b]} \phi(x) dF(x),$$

where $F(x) = \hat{F}(x+)$, for $x \in [a,b)$, $F(x) = \hat{F}(b)$, $x \geq b$, and $F(x) = \hat{F}(a)$, $x < a$. The integral presented here is called the *Stieltjes integral*.

So far we have defined the Stieltjes integral for nondecreasing "integrator" functions F. We can also define it for functions F^* that are differences of two nondecreasing functions F^1 and F^2, $F^* = F^1 - F^2$, by defining

$$\int_{[a,b]} \phi(x) \, dF^*(x) = \int_{[a,b]} \phi(x) \, dF^1(x) - \int_{[a,b]} \phi(x) \, dF^2(x).$$

Functions F^* being such differences have "bounded variation" (a property equivalent to being such a difference), a term we do not define.

It is easily checked that $\int_{[c,d]} dF(x) = F(d) - F(c-)$ and that $\int_{(a,d]} dF(x) = F(d) - F(a)$, from which it follows that $\lim_{a \uparrow c} \int_{(a,b]} dF(x) = \int_{[c,b]} dF(x)$. Even $\lim_{a \uparrow c} \int_{(a,b]} \phi(x) dF(x) = \int_{[c,b]} \phi(x) dF(x)$, ($\phi(x)$ piecewise continuous), and similarly, $\lim_{d \downarrow b} \int_{[c,d)} \phi(x) dF(x) = \int_{[c,b]} \phi(x) dF(x)$. (Here $(a,b]$ and $[c,d)$ can be replaced by $[a,b]$ and $[c,d]$, respectively.)

The definition of the integral $\int_{(a,b]} \phi(x) \, dF(x)$ could also have been carried out in the following manner. Define $\int_I dF(x) = F(b) - F(a)$ if $I = (a,b]$. Let $\phi(x) = \sum_i \alpha_i 1_{(b_{i-1}, b_i]}$, with $b_{i-1} < b_i$, $i = 1, 2, \ldots, m$, $b_0 = a, b_m = b$. Then

$$\int_{(a,b]} \phi(x) \, dF(x) := \sum_i \alpha_i \int_{I_i} dF(x) = \sum_i \alpha_i (F(b_i) - F(b_{i-1})).$$

Next, for any bounded function ϕ that is a pointwise limit of such piecewise constant functions ϕ_n, define $\int_{(a,b]} \phi(x) dF(x) = \lim_n \int_{(a,b]} \phi_n(x) dF(x)$. In particular, piecewise and right-continuous functions are such pointwise limits.

The point of mentioning this alternative approach is that it generalizes more easily to the case where $\mathbf{x} \in \mathbb{R}^n$, for $n = 2, 3, \ldots$.

So let us sketch such a generalization, confining the discussion mainly to $\mathbb{R}^2$. Then $(\mathbf{a}, \mathbf{b}]$ has to be understood as a generalized rectangle, denoting the set $\{\mathbf{a} \ll \mathbf{x} \leq \mathbf{b}\}$ ($\mathbf{a} \ll \mathbf{x}$ means $a_i < x_i$ for all i). Let $F(\mathbf{x}) := F(x_1, x_2)$ be a function of $\mathbf{x} = (x_1, x_2) \in \mathbb{R}^2$, nondecreasing and continuous from above ($f(\mathbf{y}) \downarrow f(\mathbf{x})$ when $\mathbf{y} \downarrow \mathbf{x}$). Now, let $(\infty, \mathbf{b}] = \{\mathbf{x} : \mathbf{x} \leq \mathbf{b}\}$. Write $\mathbf{a} = (a_1, a_2)$, $\mathbf{b} = (b_1, b_2)$. As $(\mathbf{a}, \mathbf{b}] = (\infty, \mathbf{b}] - (\infty, (a_1, b_2)] - (\infty, (a_2, b_1)] + (\infty, \mathbf{a}]$ (draw a figure !), we define the integral $\int_{(\mathbf{a}, \mathbf{b}]} dF(\mathbf{x})$ over $(\mathbf{a}, \mathbf{b}]$ to be equal to

$$F(b_1, b_2) - F(a_1, b_2) - F(a_2, b_1) + F(a_1, a_2). \tag{5.27}$$

We make the assumption that for any $\mathbf{a}, \mathbf{b}, \mathbf{a} \ll \mathbf{b}$, the expression in (5.27) is non-negative. We can again define integrals of simple functions, namely functions of the form $\phi = \sum_i \alpha_i 1_{I_i}$, $I_i = (\mathbf{a}_i, \mathbf{b}_i] \subset (\mathbf{a}, \mathbf{b}]$, the I_i's disjoint, by the formula $\int_{(\mathbf{a}, \mathbf{b}]} \phi(\mathbf{x}) dF(\mathbf{x}) = \sum_i \alpha_i \int_{I_i} dF(\mathbf{x})$. Next, any bounded function ψ that is a limit of such functions $\phi = \phi_n$ (for each $\mathbf{x}$) can be "integrated" in the sense that we define $\int_{(\mathbf{a}, \mathbf{b}]} \psi(\mathbf{x}) dF(\mathbf{x}) = \lim_{n \to \infty} \int_{(\mathbf{a}, \mathbf{b}]} \phi_n(\mathbf{x}) dF(\mathbf{x})$. (The integral will not depend on which sequence ϕ_n we choose.) In particular, continuous functions are pointwise limits of such sequences, so we can integrate continuous functions. Again, we have (for example) that $\int_{(\mathbf{a}, \mathbf{b}]} \psi(\mathbf{x}) dF(\mathbf{x})$ often differs from $\int_{(\mathbf{a}, \mathbf{b}]} \psi(\mathbf{x}) dF(\mathbf{x})$, the former integral being defined by $\lim_{n\to\infty} \int_{(\mathbf{a}-1'/n,\mathbf{b}]} \psi(\mathbf{x}) dF(\mathbf{x})$, $1' = (1,1)$. We can also define an integral over $[\mathbf{a},\mathbf{b})$ by a suitable limit. Then we can integrate simple functions defined by such intervals, and then also pointwise limits of simple functions of this type. These extension makes it possible to integrate various types of discontinuous functions.

If $F(\mathbf{x}) = \sum_{\mathbf{x}_j \leq \mathbf{x}} a_{\mathbf{x}_j}$, where $a_{\mathbf{x}_j}$, $j = 1, \ldots, j'$ are given positive entities, then for any continuous $\psi(\mathbf{x})$,

$$\int_{(\mathbf{a},\mathbf{b}]} \psi(\mathbf{x}) dF(\mathbf{x}) = \sum_{\mathbf{x}_j \in (\mathbf{a},\mathbf{b}]} \psi(\mathbf{x}_j) a_{\mathbf{x}_j}. \tag{5.28}$$

To see this, just let $\phi_n(\mathbf{x}) = \psi(\mathbf{b}_i) 1_{I_i^n}$, where I_i^n is a partition of $(\mathbf{a},\mathbf{b}]$, so fine that each $I_i^n := (\mathbf{a}_i^n, \mathbf{b}_i^n]$ contains at most one point $\mathbf{x}_j$, and if it does, $\mathbf{b}_i^n = \mathbf{x}_j$. Note that in this case, $\int_{I_i^n} dF(x)$ equals $a_{\mathbf{x}_j}$ if $\mathbf{x}_j \in I_i^n$, and equals zero otherwise, (this actually follows from (5.27)). So, for fine partitions, $\int_{(\mathbf{a},\mathbf{b}]} \phi_n(\mathbf{x}) dF(\mathbf{x})$ is independent of n, and equals the right hand side of (5.28). Finally, $\phi_n(\mathbf{x}) \to \psi(\mathbf{x})$ when $\max_i(b_i^n - a_i^n) \to 0$ as $n \to \infty$.

The above discussion can be generalized to functions $F(\mathbf{x})$ defined on $\mathbb{R}^n$, being separately nondecreasing in each component of $\mathbf{x}$ and continuous from above (same definition as above). In fact the above discussion can be carried over word by word, except for the formula (5.27) for the integral $\int_{(\mathbf{a},\mathbf{b}]} dF(\mathbf{x})$, (now $\mathbf{a},\mathbf{b} \in \mathbb{R}^n$), which is now given by a more complicated formula involving the values of F at all the various "corners" of $(\mathbf{a},\mathbf{b}]$. (In this connection again a crucial non-negativity assumption, a generalization of the assumption on (5.27), is needed.)

5.6 Abstract Measures

Measures are non-negative functions $\mu(A)$ defined on subsets A of a given "universal set" Ω. (Now, Ω may be any type of set, no "structure" of Ω is assumed beforehand. In abstract measure theory, the structure described below is simply postulated.) In probability theory, a given probability measure μ gives probabilities attached to subsets of Ω, and $\mu(\Omega) = 1$. (Given $A \subset \Omega$, the probability that an outcome belongs to A is $\mu(A)$.) In measure theory generally, we do not necessarily have that $\mu(\Omega)$ is finite, we allow that $\mu(A)$ takes value in $[0,\infty]$.

The domain of definition of a measure is always a specified family $\mathcal{F}$ of subsets of Ω. $\mathcal{F}$ is required to be a σ-algebra (or σ-field), which, by definition, has the following properties:

(a) The empty set and Ω belongs to $\mathcal{F}$.
(b) $A \in \mathcal{F} \implies \complement A \in \mathcal{F}$.
(c) $A_1, A_2, \ldots$ all belong to $\mathcal{F} \implies \cup_n A_n \in \mathcal{F}$.

As a consequence of these properties, note that a finite or countable intersections $\cap_n A_n$ also belongs to $\mathcal{F}$ if all A_n belong to $\mathcal{F}$.

Thus, for μ to be a *measure*, by definition, it is defined on a given σ–field $\mathcal{F}$, and, furthermore, satisfies the following three properties:

1. $\mu(\emptyset) = 0$.
2. $\mu(A) \geq 0$ for all $A \in \mathcal{F}$.
3. If $A_1, A_2, \cdots \in \mathcal{F}$ are disjoint, then $\mu(\cup_n A_n) = \sum_n \mu(A_n)$ (a property called *countable additivity*).

Some further properties easily follow from (1)–(3):

4. $A \subset B$ with $A, B \in \mathcal{F}$ imply $\mu(A) \leq \mu(B)$.
5. If $A_1, A_2, \cdots \in \mathcal{F}$ is an increasing sequence of sets (i.e., $A_i \subset A_{i+1}$), then $\mu(\cup_i A_i) = \sup_i \mu(A_i) = \lim_i \mu(A_i)$.
6. If $\mu(\Omega) < \infty$, and $A_1, A_2, \cdots \in \mathcal{F}$ is a decreasing sequence of sets (i.e., $A_i \supset A_{i+1}$), then $\mu(\cap_i A_i) = \inf_i \mu(A_i) = \lim_i \mu(A_i)$.

We allow ∞ as values of sums like $\sum_n \mu(A_n)$ in (3). The properties (1)–(3) should be compared with the ones listed for probabilities above.

A null set (or μ-null set) A is any set $A \in \mathcal{F}$ such that $\mu(A) = 0$. (In probability theory, the event A has then zero probability of occurring.)

A real-valued function $\phi(\omega)$ on Ω is *measurable* (or $\mathcal{F}$-measurable) if $\{\omega : \phi(\omega) \leq r\} \in \mathcal{F}$ for all $r \in \mathbb{R}$. A vector valued function is measurable if all its components are measurable. A *simple measurable* real-valued function is of the form $\psi = \sum_i a_i 1_{A_i}$, $A_i \in \mathcal{F}$, $a_i \in \mathbb{R}$, $i = 1, \ldots, m$, A_i disjoint. Note that a bounded measurable real-valued function ϕ can always be approximated by simple measurable functions ψ. Indeed, simply put $\psi(\omega) = \psi_n(\omega) = \sum_i (iM/2^n) 1_{B_i}(\omega)$, where

$$B_i = \{\omega : iM/2^n \leq \phi(\omega) < (i+1)M/2^n\}, \; i = -2^n, -2^n + 1, \ldots, 0, \ldots, 2^n$$

where $M = \sup_\omega |\phi(\omega)|$. We then have $|\phi(\omega) - \psi_n(\omega)| \leq M/2^n$ for all ω, and $\psi_n(\omega) \leq \psi_{n+1}(\omega) \leq \psi(\omega)$ for all ω. (We do have $B_i \in \mathcal{F}$, as a consequence of the above properties of a σ–field.)

The *integral* (or μ-integral) of a simple non-negative function $\sum_i a_i 1_{A_i}(\omega)$ on Ω (a_i given positive numbers, A_i given disjoint sets from $\mathcal{F}$), is defined by $\int_\Omega \psi(\omega) d\mu(\omega) := \sum_i a_i \mu(A_i)$ (perhaps the value of the sum is ∞). The integral of a non-negative measurable function ϕ is defined as the limit $\lim_{n \to \infty} \int_\Omega \psi_n(\omega) d\mu(\omega) = \sup_n \int_\Omega \psi_n(\omega) d\mu(\omega)$, perhaps the limit is ∞ (ψ_n defined above, here becoming non-negative). The limit is unique, i.e., the limit also equals the limit $\lim_n \int_\Omega \psi^n(\omega) d\mu(\omega)$, where ψ^n is any other sequence of non-negative simple functions converging to $\phi(\omega)$ pointwise (i.e., for each ω), with $\psi^n(\omega) \leq \psi^{n+1} \leq \phi(\omega)$ for all n, ω.

The integral is next extended to measurable real-valued function ϕ that are not non-negative: $\int_\Omega \phi(\omega) d\mu(\omega) := \int_\Omega \phi_+(\omega) d\mu(\omega) - \int_\Omega \phi_-(\omega) d\mu(\omega)$, where $\phi_+ := \max(0, \phi) \geq 0$ and $\phi_- := -\min(0, \phi) \geq 0$. Here we agree that the integral is only defined if both the integrals on the right-hand side in the definition are finite. In this case ϕ is called *integrable* (or μ-integrable). (When the term integrable is used, it includes being measurable.) Non-negative function are only called integrable if their integrals are finite. A measurable real-valued function $\phi(\omega)$ is *integrable* iff $\int_\Omega |\phi(\omega)| d\mu(\omega) < \infty$. Note that, if $A \in \mathcal{F}$, then $\int_A \phi(\omega) d\mu(\omega) := \int_\Omega 1_A(\omega) \phi(\omega) d\mu(\omega)$.

If $f \geq 0$, and $\int_\Omega f d\mu(\omega) = 0$, then $f = 0$ a.e. (or μ-a.e.), a.e. here meaning for all ω, except possible in a μ-null set.

The reason why we operate with the general concept of measurable functions, is partly because of the next theorem. (In the results below, let μ be any given measure defined on a σ-field $\mathcal{F}$ of subsets of a given set Ω.)

Theorem 5.5. *If $\phi_n(\omega)$ is any sequence of measurable functions converging to $\phi(\omega)$ for all $\omega \in \Omega$, except perhaps for ω in a null set (so-called convergence a.e.), then ϕ is also measurable.* □

Piecewise continuous functions on $\Omega = \mathbb{R}$ do not satisfy such a property: A pointwise limit of such functions need not be piecewise continuous.

Moreover, the following theorem holds:

Theorem 5.6 (Dominated convergence). *In case of the previous theorem, if there exists a $\psi(\omega)$ such that $|\phi_n(\omega)| \leq \psi(\omega)$ for all ω, except for ω in a null set perhaps dependent on n, and $\int_\Omega \psi(\omega) d\mu(\omega) < \infty$, ($\psi$ is integrable), then $\lim_n \int_\Omega \phi_n(\omega) d\mu(\omega) = \int_\Omega \phi(\omega) d\mu(\omega)$.* □

Two more results can be mentioned. The *monotone convergence theorem* says that if $0 \leq f_n \uparrow f$ a.e., then $\int_\Omega f_n \uparrow \int_\Omega f$ (shorthand notation). The *Fatou's lemma* says that for $f_n \geq 0$, we have $\int_\Omega \liminf_{n \to \infty} f_n \leq \liminf_{n \to \infty} \int_\Omega f_n$.

So far, we have only discussed real-valued functions. If $\phi(\omega)$ takes values in $\mathbb{R}^n$, then $\int_\Omega \phi(\omega) d\omega := (\int_\Omega \phi_1(\omega) d\omega, \ldots, \int_\Omega \phi_n(\omega) d\omega)$. The two theorems above also hold for $\mathbb{R}^n$-valued functions. (Then $|a|$ means norm of $a \in \mathbb{R}^n$.)

Some inequalities are of frequent use: If f and g are measurable, and $|f|^p$ and $|g|^q$ are integrable, where $p > 1$, and $1/p + 1/q = 1$, then $|fg|$ is integrable and (shorthand notation)

$$\int_\Omega |fg d\mu| \leq \left(\int_\Omega |f|^p d\mu\right)^{1/p} \left(\int_\Omega |g|^q d\mu\right)^{1/q} \quad \text{(Hoelders inequality)} \quad (5.29)$$

(The expression $(\int_\Omega |f|^p d\mu)^{1/p}$ is called the L_p-norm of f (L_p-μ-norm.) For $p = q = 2$, the inequality furnishes (a particular case of) the Cauchy–Schwartz inequality. If f and g are measurable, and $|f|^p$ and $|g|^p$ are integrable, where $p \geq 1$, then

$$\left(\int_\Omega |f+g|^p d\mu\right)^{1/p} \leq \left(\int_\Omega |f|^p d\mu\right)^{1/p} + \left(\int_\Omega |g|^p d\mu\right)^{1/p} \quad \text{(Minkowski's inequality)}$$

$$(5.30)$$

Theorem 5.7 (Interchange of integral and differentiation). *Assume that $f(x, \omega): \mathbb{R}^n \times \Omega \to \mathbb{R}^m$ is measurable in ω and C^1 in x, that $f(x_0, \omega)$ is integrable with respect to ω, and that $|f_x(x, \omega)| \leq \phi(\omega)$ for all x in some ball $B(x_0, \delta)$, where $\phi(\omega)$ is some integrable function. Then the derivative $[(d/dx) \int_\Omega f(x, \omega) d\mu(\omega)]_{x=x_0}$ exists and equals $\int_\Omega [(d/dx) f(x, \omega)]_{x=x_0} d\mu(\omega)$.* □

Proof. Let $x_j = x_0 + a/j$, a any unit vector in $\mathbb{R}^n$. Note that, for each component i, for each ω, for $\psi_j(\omega) := (f_i(x_j, \omega) - f_i(x_0, \omega))$, $|j\psi_j(\omega) - [(d/dx)f_i(x, \omega)]_{x=x_0} a| \to 0$, when $j \to \infty$, so by Theorem 5.2, and the bound ϕ, we have that $(1/j^{-1}) \int_\Omega (f_i(x_j, \omega) - f_i(x_0, \omega))d\mu(\omega) = j \int_\Omega \psi_j(\omega)d\mu(\omega) \to \int_\Omega [(d/dx)f_i(x, \omega)]_{x=x_0} a \, d\mu(\omega)$, when $j \to \infty$, (note that by Theorem 5.5, $[(d/dx)f_i(x, \omega)]_{x=x_0} a$ is measurable.) $\qquad\square$

Particular Examples of Measures

Consider the family $\mathcal{F}_0$ of intervals in $\mathbb{R}$ of any given type (open, half-open, closed), including the empty set and $(-\infty, \infty)$. We can enlarge this family by including all countable unions of sets from $\mathcal{F}_0$, obtaining a family $\mathcal{F}_1$. An even bigger family, $\mathcal{F}_2$, is obtained by including in the family all complements of sets in $\mathcal{F}_1$. An enlargement, $\mathcal{F}_3$ of $\mathcal{F}_2$ is obtained by including all countable unions of sets in $\mathcal{F}_2$. And so we can continue the enlargements by taking complements and countable unions, constructing $\mathcal{F}_4, \mathcal{F}_5, \ldots$. The *Borel-field* $\mathcal{F}^*$ on $\mathbb{R}$ is, by definition, the smallest σ-field containing $\mathcal{F}_0$ ($\mathcal{F}^*$ is "generated by" $\mathcal{F}_0$). Evidently, it contains all the $\mathcal{F}_i$'s described above. We can think of it as the union of all the $\mathcal{F}_i$'s (but actually it is strictly larger). Now, in particular, $\mathcal{F}^*$ is generated by the family of all intervals of the type $(a, b]$. Similarly, the Borel-field on $\mathbb{R}^n$ is the smallest σ-field containing all sets of the form $\{x \in \mathbb{R}^n : a \ll x \le b\}$, a, b arbitrary vectors. A *Borel measurable* function ϕ (or Borel function for short) is a real-valued function with the property that $\{x : \phi(x) \le r\}$ is a Borel set (i.e., a member of $\mathcal{F}^*$), for all $r \in \mathbb{R}$. Piecewise continuous functions on $\mathbb{R}$ are Borel functions. All continuous functions in $\mathbb{R}^n$ are Borel functions, even all functions being, separately, piecewise continuous in each of the n variables are Borel functions.

It can easily be shown that, for a given σ-field $\mathcal{F}$ in Ω, if $f(\omega) : \Omega \to \mathbb{R}^n$ is $\mathcal{F}$-measurable, and g is a Borel function from $\mathbb{R}^n$ into $\mathbb{R}$, then $g(f(\omega))$ is $\mathcal{F}$-measurable.

Let us next define the term Lebesgue null set. A set $N \subset \mathbb{R}$ is called a *Lebesgue null set* if for each $\varepsilon > 0$, a countable collection of disjoint intervals $\{I_i\}$ can be found, for which $N \subset \cup_i I_i$, and for which $\sum_i \text{length}(I_i) \le \varepsilon$, ($\text{length}(I_i) = $ length of I_i).

Define the *Lebesgue measure* $\gamma(I)$ of an interval I to be length(I). The (Lebesgue) measure of a finite or countable union of disjoint intervals I_i is defined as $\sum_i \gamma(I_i)$. For real-valued piecewise constant functions $\psi(x) = \sum_i a_i 1_{Ii}$, where the I_i's are bounded disjoint intervals, we define the Lebesgue integral over $\mathbb{R}$ by $\int_\mathbb{R} \psi \, dx = \sum_i a_i \cdot \text{length}(I_i)$. It turns out that for any bounded Borel measurable function ψ, there exists a sequence of piecewise constant functions ϕ_n such that $\phi_n \to \psi$ for all ω, except perhaps for ω in a Lebesgue null set N, with $|\phi_n| \le |\psi|$. For a bounded Borel function ψ, we then define the *Lebesgue integral* $\int_{[a,b]} \psi(x) \, dx$ by $\lim_{n\to\infty} \int_{[a,b]} \phi_n dx$. It turns out that the limit exists and is independent of which sequence ϕ_n we choose, as long as all sequences we use are bounded, i.e., $\sup_{n,x} |\phi_n(x)| < \infty$. If $\phi \ge 0$ is an unbounded Borel function, we define

$$\int_{[a,b]} \phi \, dx = \lim_{n\to\infty} \int_{[a,b]} \min\{n, \phi(x)\} \, dx.$$

For general Borel functions ϕ, the integral is again defined by $\int_{[a,b]} \phi(x)\,dx = \int_{[a,b]} \phi_+(x)\,dx - \int_{[a,b]} \phi_-(x)\,dx$, the integral existing when both terms on the right-hand side are finite, and then ϕ is called *(Lebesgue) integrable*.

For piecewise continuous functions the integral so defined is the same as the elementary integral in calculus. By the method above, the elementary integral is extended to a larger class of functions. The *Lebesgue measure* of a Borel set B, is defined as $\lambda(B) := \int_{\mathbb{R}} 1_B(x)\,dx := \lim_{n\to\infty} \int_{[-n,n]} 1_B(x)\,dx$.

Above we defined a Stieltjes integral, for a given (nondecreasing) distribution function F. (We shall here assume F to be any nondecreasing right continuous real-valued function on $\mathbb{R}$.) Recall that we defined the Stieltjes integral for piecewise constant functions, and then for piecewise continuous functions. A μ_F-null set N is any set, for which, for any $\varepsilon > 0$, there exist a countable union $J = \cup_j J_j$ of disjoint intervals $J_j = (a_j, b_j]$, for which $\sum_j F(b_j) - F(a_j) \leq \varepsilon$, and for which $N \subset J$. Now, it may be proved that for any bounded Borel function ψ there exists a sequence of piecewise constant functions ϕ_n converging to ψ for all x, except perhaps for x in a μ_F-null Borel set N. For any bounded Borel function ψ, a Lebegue-Stieltjes integral can be defined by letting $\int_{[a,b]} \psi\,dF(x) = \lim_{n\to\infty} \int_{[a,b]} \phi_n\,dF(x)$, where $\phi_n(x)$ is any sequence of piecewise constant functions converging to ψ for all x, except perhaps for x in a μ_F-null Borel set B, $\sup_{x,n} |\phi_n(x)| < \infty$. (The limit can be seen to be the same for all choices of such sequences for which $\sup_{x,n} |\phi_n(x)| < \infty$.) The extension of the integral to non-negative unbounded functions ϕ, and then to unbounded Borel function ϕ not necessarily non-negative, is done exactly as for the Lebesgue integral. For any Borel set B, define its *Lebesgue-Stieltjes measure* by $\mu_F(B) = \lim_{n\to\infty} \int_{[-n,n]} 1_B\,dF(x)$.

Often the Borel-field is enlarged to a larger σ-field, namely to the smallest σ-field which also includes all null sets, both in the case the measure is λ (Lebesgue measure) or μ_F. (This larger σ-field simply equals $\{B \cup N : B$ a Borel set, N a null set$\}$, N any Lebesgue – or μ_F – null set, as the case may be.) In the former case we then get the σ-field of Lebesgue measurable sets, in the latter case the σ-field of "Lebesgue-Stieltjes measurable set," in the latter case the σ-field will depend on the function $F(x)$.

5.7 Intuition

What do mathematical artifacts as the σ-algebra $\mathcal{F}$ and measurability with respect to $\mathcal{F}$, mean, intuitively speaking? Very often, the sets in $\mathcal{F}$ represent the information it is possible to have: We know that an outcome ω has occurred at the time considered, but often it is the case that what we can know about ω is whether ω belongs to A for each set A in $\mathcal{F}$. This information will not always be enough for knowing the outcome ω exactly. Frequently, for example in stochastic control, the rules specifying how to act, is, perhaps by necessity, based on such "limited" knowledge. To be specific, consider, e.g., the simple $\mathcal{F}$-measurable function $\phi := a1_A + b1_B, A, B \in \mathcal{F}$. If this happens to be our action rule, we "do" $a + b$ if $\omega \in A$ and $\omega \in B$, if $\omega \in A$

and $\omega \notin B$, we do a, etc. We see that to act according to this rule, the knowledge needed is not exactly what ω is, but only whether it belongs to A and B. A similar remark pertains to general simple functions $\sum_i a_i 1_{A_i}, A_i \in \mathcal{F}$, all we need to calculate its value is to know if $\omega \in A_i$ or not, for all i. If a general $\mathcal{F}$-measurable function ϕ specifies our action rule, then, as it can be approximated arbitrarily closely by simple functions, intuitively we can say that to act, we only need to know, for any $A \in \mathcal{F}$, if $\omega \in A$. We see this clearly in the following manner: For any $x \in \mathbb{R}$, define $A_x := \{\omega : \phi(\omega) = x\}$, ($A_x$ does belong to $\mathcal{F}$). We "go with speed x" (or do x) iff ω belongs to A_x.

A slightly more sophisticated setting is the following: Think of the function ϕ instead as a measuring instrument, the readings of which depend on ω, in such a way that the function ϕ becomes $\mathcal{F}$-measurable. Our rule of action is based on these readings. The readings are outcomes of the stochastic variable $X = \phi(\omega)$. Let us agree that we can always know if X satisfies $a < X \le b$, for any a, b. Let G be the σ-field generated by the sets $\{\omega : a < X(\omega) \le b\}$. This is a sub ($\sigma$-)field of $\mathcal{F}$, and, at most, it is only for members A of this subfield G that we can know if any given ω belongs to A or not. Now, by necessity, we must require that our action rule is measurable with respect to G (more demanding than $\mathcal{F}$-measurability). Then it is nice to have the following (easily proved) theorem, in which we have n measuring instruments X_i :

Theorem 5.8. *Let $\phi_1, \ldots, \phi_n$ be real-valued measurable functions with respect to a given σ-field $\mathcal{F}$ of sets in Ω. Let $\mathcal{P}$ be the smallest σ-field generated by all the sets $\{x : \phi_i(x) \le a_i\}$, a_i arbitrary real numbers. Assume that ψ is a real-valued function that is measurable with respect to $\mathcal{P} \subset \mathcal{F}$. Then $\psi = f(\phi_1, \ldots, \phi_n)$, for some Borel function f.* $\qquad\square$

Thus, that our given action rule ψ is measurable with respect to G (quite an abstract or perhaps incomprehensible property) means that our action rule can be written $\psi = f(\phi)$, where f is a Borel function, it means nothing else than that our actions are a function of the readings of the instrument.

5.8 General Conditional Expectation and Probability

A probability measure is a measure μ such that $\mu(\Omega) = 1$. It is fully specified when the triple $(\Omega, \mathcal{F}, \mu)$ is specified.

A *random variable* $\Omega \to \mathbb{R}^n$ is defined to be the same as a $(\mathcal{F}-)$ measurable function from Ω into $\mathbb{R}^n$. The (cumulative) distribution function of a given random variable $\mathbf{X}(\omega)$ is defined by $F_X(\mathbf{x}) = \mu\{\omega : \mathbf{X}(\omega) \le \mathbf{x}\}$.

Note that for any real-valued continuous function (or even Borel) function ψ on $\mathbb{R}^n$, $\int_\Omega \psi(X(\omega)) d\mu(\omega) = \int_{\mathbb{R}^n} \psi(x) dF_X(x)$.

Let P be any probability measure defined on the σ-field $\mathcal{F}$ of sets in Ω, and let Y be any simple random variable, i.e., $Y = \sum_i a_i 1_{A_i}$, $i = 1, \ldots, m$, $A_i \in \mathcal{F}$ and disjoint. Then, note that if H is any set in $\mathcal{F}$, by definition $E[Y|H] = \sum_i a_i P[A_i|H]$, where, as before, the conditional probability $P(A|H)$ is defined as $P(A \cap H)/P(H)$, ($P(A|H)$ having any value, e.g., 0, if $P(H) = 0$). Evidently, then $\int_H E[Y|H]dP(\omega) = E[Y|H]P(H) = \{\sum_i a_i P[A_i|H]\}P(H) = \sum_i a_i P(A_i \cap H) = \int_H Y dP(\omega)$. Hence, we have $E[Y|H]P(H) =$

$$\int_H E[Y|H]dP(\omega) = \int_H Y dP(\omega). \tag{5.31}$$

For an arbitrary bounded measurable function Y, we can use $E[Y|H]P(H) = \int_H Y dP(\omega)$ to define $E[Y|H]$ (again only when $P(H) > 0$). Next, let $\mathcal{H}^* := \{H_j\}$ be any finite partition of Ω, $H_j \in \mathcal{F}$, and let $E[Y|\mathcal{H}^*](\omega)$ be the simple function defined by $E[Y|\mathcal{H}^*](\omega) = E[Y|H_j]$, for $\omega \in H_j$. Then,

$$\int_H E[Y|\mathcal{H}^*](\omega)dP(\omega) = \int_H Y dP(\omega) \tag{5.32}$$

evidently holds for each $H \in \mathcal{H}^*$. More generally, if $\mathcal{H}^*$ is any given σ-field contained in $\mathcal{F}$, it is possible to define a function $E[Y|\mathcal{H}^*]$ being measurable relative to $\mathcal{H}^*$ such that (5.32) holds for all $H \in \mathcal{H}^*$. In other words, the conditional expectation of a bounded random variable Y with respect to a sub-σ-field $\mathcal{H}^*$ of $\mathcal{F}$ is definable. We cannot give the precise construction, but one can think of it in the manner that we can find say an increasing family of finite partitions $\mathcal{H}^i = \{H_{i,j} : j = 1, \ldots, k_j\}$ of Ω, all $\mathcal{H}^i \subset \mathcal{H}^*$, for which the functions $E[Y|\mathcal{H}^i]$ are convergent to some $\mathcal{H}^*$-measurable function, which we denote $E[Y|\mathcal{H}^*]$, and for which we obtain "in the limit" (5.32) for all $H \in \mathcal{H}^*$, using that $\int_H E[Y|\mathcal{H}^i]dP(\omega) = \int_H Y dP(\omega)$ holds for all $H \in \mathcal{H}^i$, for each i.

Conditional probability with respect to any sub-σ-field $\mathcal{H}^*$ can be introduced in a similar manner, or more simply, as follows: $P(A|\mathcal{H}^*) := E[1_A|\mathcal{H}^*]$. Hence, as any conditional expectation, this is a function of $\omega \in \Omega$, measurable with respect to $\mathcal{H}^*$.

If $\mathcal{H}^*$ is generated by a real random variable X (i.e., $\mathcal{H}^*$ is generated by the sets $\{\omega : X(\omega) \leq r\}$, we write $E[Y|X]$ and $P(A|X)$ instead of $E[Y|\mathcal{H}^*]$ and $P(A|\mathcal{H}^*)$, respectively.

The above abstract definitions can be compared (or applied) to the definition of conditional densities. Let $f(x,y)$ be a given joint bounded density on $\mathbb{R}^2$, for simplicity defined on $[a,b] \times [c,d]$. The family $\mathcal{F}$ is now the family of Borels set in $[a,b] \times [c,d]$. Let now $\mathcal{H}^*$ be the family of sets $[a,b] \times B$, B a Borel set in $[c,d]$. Let us copy the arguments above for calculating conditional probabilities in this case (conditioning on y). Above we first considered finite subfamilies $\mathcal{H}^*$ and then indicated a limit argument. Let us do the same here: Thus we now let the finite family $\mathcal{H}^*$ be a fine subdivision of $[c,d]$ into intervals of the form $[y_i, y_i + \Delta y]$. Let A be any interval (or Borel set) in $[a,b]$. Then, for any y_i, $\Pr[X \in A | Y \in [y_i, y_i + \Delta y]]$

$$= \left\{ \int_{A \times [y_i, y_i + \Delta y]} f(x,y) \, dy \, dx \right\} / \iint_{[a,b] \times [y_i, y_i + \Delta y]} f(x,y) \, dy \, dx$$

$$\approx \Delta y \int_A f(x, y_i) dx / \Delta y \int_{[a,b]} f(x, y_i) dx = \int_A f_{X|Y}(x; y_i) \, dx.$$

The next to last equality is an approximative one, the smaller Δy is the better is the approximation. Approximately, $\int_A f_{X|Y}(x, y_i)dx$ is the conditional probability of A, given $y \in [y_i, y_i + \Delta y_i]$, which depends on y_i. Next, if $\mathcal{H}^*$ is generated by all sets of the form $\{y : y \leq r\}$, as the subdivision gets finer and finer, we get $P(A|H^*) = P(A|Y) = \int_A f_{X|Y}(x, y)dx$, a function of $y \in \mathbb{R}$.

Going back to the general conditional probability $P(A|\mathcal{H}^*)$, note that we "almost" have that for each given ω, $A \rightarrow P(A|\mathcal{H}^*)$ is a probability measure on Ω. We do have $P(\emptyset|\mathcal{H}^*) = 0$, and $P(\Omega|\mathcal{H}^*) = 1$. Some problems remain: First, conditional expectation and conditional probability, both functions of ω, are not uniquely determined, but only up to a null set in $\mathcal{H}^*$. We hence speak of versions of, say conditional probability, that can differ in this sense. However, for most purposes the versions are equivalent (it doesn't matter which version is used). Most often, it is the property (5.32) (or properties directly following from it) that is used, and then the nonuniqueness does not cause problems. Related to this nonuniqueness is the fact that if A_i, $i = 1, 2, \ldots$ are disjoint, then the equality $\Sigma_i P(A_i|\mathcal{H}^*) = P(\cup_i A|\mathcal{H}^*)$ holds unfortunately not for each and every ω, but only for all ω, except a null set in $\mathcal{H}^*$. A version of conditional probability, for which for *each* ω, $A \rightarrow P(A|\mathcal{H}^*)$ is a probability measure, is called a *regular conditional probability*. In the general case, such ones does not always exist. For probability measures defined on the Borel-field in $\mathbb{R}^n$, regular versions of conditional probabilities exist for any given sub-σ-field $\mathcal{H}^*$ of the Borel field.

Note that if $P(A|\mathcal{H}^*)$ is a regular probability measure, we have, for any bounded random variable Y that

$$E[Y|\mathcal{H}^*] = \int_{\Omega} Y(\omega)dP(\omega|\mathcal{H}^*).$$

A conditional distribution function of a real-valued random variable X with respect to $\mathcal{H}^*$ is defined by $F(\omega; x) := P(X \leq x|\mathcal{H}^*)$. Though perhaps no regular version of conditional probability exists, at least there exists a version of conditional probability such that, if this version is used in the definition, $F(\omega; x)$ becomes regular in the following sense: $F(\omega; x)$ is, for each ω, a distribution function on $\mathbb{R}$. Corresponding to F there is, of course, a conditional probability measure $Q(\omega; B)$ on the Borel field $\mathcal{B}$ of $\mathbb{R}$, such that $Q(\omega; B) = \int_B dF(\omega; x)$, and $Q(\omega; B)$ becomes regular.

5.9 Stochastic Processes

A *stochastic process* X_t, $t \in [0, \infty)$, $X_t \in \mathbb{R}$, is simply a collection of random variables, one for each t, defined on a sample space Ω with σ-field $\mathcal{F}$, X_t $\mathcal{F}$-measurable. Sometimes an increasing family (a so-called filtration) of sub-σ-fields $\mathcal{F}_t$ of $\mathcal{F}$ appear together with a given stochastic process X_t. If then X_t is $\mathcal{F}_t$-measurable for each $t \in [0, \infty)$, X_t is called $\mathcal{F}_t$-*adapted*. If, for any s, $(t, \omega) \rightarrow X_t$ is simultaneously Borel$\times \mathcal{F}_s$-measurable on $[0, s] \times \Omega$, then the process is called $\mathcal{F}_t$-*progressively measurable* (sometimes the word $\mathcal{F}_t$-adapted is used even for this property). A stochastic process X_t is a *Markov process* with respect to an increasing family of σ-algebras $\mathcal{F}_t \subset \mathcal{F}$ if X_t is adapted to $\mathcal{F}_t$ and if $E[f(X_{t+h})|\mathcal{F}_t] = E[f(X_{t+h})|X_t]$ for any piece-wise continuous (or even bounded Borel) function f, for any t and any $h > 0$. In particular, for $f = 1_A$, this equality says that the conditional probability of finding X_{t+h} in A, given the history up to t, is the same as the conditional probability of finding X_{t+h} in A, given X_t. In other words, we can make the same predictions for X_{t+h} whether we know the entire prehistory up to t (know whether ω belongs to B or not, for each $B \in \mathcal{F}_t$), or whether we only know $X_t(\omega)$ (know whether ω belongs to B or not, for each B in the σ-field generated by the sets $\{\omega : X_t(\omega) \leq a\}$, a any real number). Solutions of stochastic differential equations are Markov processes.

A *martingale* is a stochastic process adapted to $\mathcal{F}_t$ such that $E[X_t|\mathcal{F}_s] = X_s$ for all s, t, $s < t$. (It is also assumed that $E|X_t| < \infty$ for all t.) For martingales for which $\sup_t E|1_B X_t|$ is small when meas(B) is small (this is called uniform integrability), there exists a random variable X with $E|X| < \infty$, such that $X_t = E[X|\mathcal{F}_t]$.

Given an increasing sequence of σ-fields $\mathcal{F}_t$ in $\mathcal{F}$, a *stopping time* $\tau(\omega)$ relative to $\mathcal{F}_t$ is a random variable with the additional property that $\{\omega : \tau(\omega) \leq t\}$ for any t, not only belongs to $\mathcal{F}$ but to $\mathcal{F}_t$, for each t.

A *Brownian motion* B_t is both a martingale and a Markov process. It has even the following property: A Markov process X_t (adapted to some $\mathcal{F}_t$) has the "strong Markov property" if $E[f(X_{\tau+h})|\mathcal{F}_\tau] = E[f(X_{\tau+h})|X_\tau]$ for any piecewise continuous (or bounded Borel) function f where τ is now no longer any fixed time point, but is a (stochastic) stopping time a.s. $< \infty$, and $\mathcal{F}_\tau := \{M \in \mathcal{F} : M \cap [\omega : \tau(\omega) \leq t] \in \mathcal{F}_t$ for all $t\}$.

Note that the solution of a stochastic differential equation (more precisely, the solution appearing in Theorem 4.1) has the strong Markov property with respect to the family $\mathcal{F}_t$ of σ–algebras generated by the Brownian motions, i.e., in the one-dimensional case generated by the sets $H_{s,r} = \{b(.) : b(s) \leq r\}$, $s \leq t$, $r \in \mathbb{R}$.

Further reading. From a vast number of books, let us mention the classic book by Feller (1957, sec. ed.), the relatively elementary book by Grimmett and Stirzaker (1982, first ed.). More advanced books are Billingsley (1979, first ed.), Fristedt and Grey (1997), and Kallenberg (1997, first ed.), all making full use of measure theory.

5.10 Exercises

5.1. Let $F(x)$ be defined by $F(x) = 0$, $x < 1$, $F(x) = 1 + x$, $x \in [1,2)$, $F(x) = 5 + x^2$ for $x \in [2, \infty)$. Let $\phi = x^3$ and calculate: $\int_{(0,3)} \phi(x)dF(x)$, $\int_{[0,3]} \phi(x)dF(x)$, $\int_{[0,2)} \phi(x)dF(x)$, $\int_{[0,2]} \phi(x)dF(x)$.

5.2. A coin is tossed repeatedly, and we want to find the probability that the first head appears after an even number of tosses. Show that a nonfinite sample space is needed, and that it is useful to have (5.9) holding for a countable set of A_i's to calculate the probabilities.

5.3. Show that if F is a cumulative distribution with a countable number of jumps, then

$$\int_{(c,d]} dF(x) \to \int_{[a,d]} dF(x) \quad \text{when } c \uparrow a.$$

5.4. Let μ be a finite measure on Ω. Assume that A_n, $n = 1, 2 \ldots$ is a decreasing family of measurable subsets of Ω, with $\cap_n A_n = \emptyset$. Show that $\mu(A_n) \downarrow 0$. (*Hint:* Use countable additivity and (with $A_0 = \Omega$), $\Omega = \cup_{n \geq 0}(A_n \setminus A_{n+1})$, and $\cup_{n \geq m}(A_n \setminus A_{n+1}) = A_m$.)

5.5. Prove Theorem 5.6. (*Hint:* Let N be the μ-null set, and let $\Omega' = \Omega \setminus N$. Then the set $\{\omega \in \Omega' : \phi(\omega) > r\} = \cup_k \cup_m \cap_{n \geq m} \{\omega \in \Omega' : \phi_n(\omega) > r + 1/k\}$, k, n, m natural numbers.)

Solutions

Exercises of Chapter 1

1.2 $u_{T-1} = 2/3x, u_{T-2} = 1$.

1.3 $\alpha_T = \delta, \alpha_t = 2(\alpha_{t+1}K)^{1/2}, u_t = x - (1/2\gamma)\ln(\alpha_{t+1}K)$.

1.4 The optimal controls are $C_t = A_t/k_t$ and w_t, where w_t is the solution of $E[(r_t - V_t)(1 + V_t + \{r_t - V_t\}w_t)^{-1}] = 0$, no explicit formula is available. Here $k_t = 1 + \frac{k_{t+1}}{1+\theta}$, $k_T = k$. The value function is $J(t, A_t) = (1+\theta)^{-1}k_t \ln A_t + b_t$, where for $t < T$, $b_t = -\ln k_t + (1+\theta)^{-1}k_{t+1}[\ln\{k_{t+1}(1+\theta)^{-1}\} - \ln\{1 + k_{t+1}(1+\theta)^{-1}\} + d_t] + (1+\theta)^{-1}b_{t+1}, b_T = 0, d_t = E\ln[(1+r_t)w_t + (1+V_t)(1-w_t)]$.

1.5 $J(t, x_t) = k_t + a_t x_t, a_t = a2^{t-T}, k_{t-1} = k_t + 2/a_t, k_T = 0, u_{t-1} = 4/a_t^2$.

1.6 $J(T, x) = x, J(t, x) = 2^{T-t}x + b_t, b_{t-1} = 2^{T-t}/\lambda + b_t, u_T = 0, b_T = 0, u_{T-1}$ arbitrary, $u_t = 1, t < T - 1$.

1.7 $J(t, x) = \gamma^{-1}\alpha^t H_t(S_t + x)^\gamma, H_{t-1}^{1/(1-\gamma)} = 1 + (\alpha a H_t)^{1/(1-\gamma)}, H_T = K, S_T = 0, S_{t-1} = (S_t + y_{t-1})/r, c_t = (S_t + x_t)H_t^{1/(\gamma-1)}$. Guess that $J(t, x) = I(t, x + \sum_{s=t}^{T-1} r^{t-s}y_s)$, where $I(t, w) = \alpha^t H_t w^\gamma/\gamma$. Using (*), the optimality equation becomes $\alpha^{t-1}H_{t-1}(x + \sum_{s=t-1}^{T-1} r^{t-1-s}y_s)^\gamma/\gamma = \max_c\{\alpha^{t-1}c^\gamma/\gamma + \alpha^t H_t a(x - c + \sum_{s=t}^{T-1} r^{t-s}y_s)^\gamma/\gamma\}$.

1.8 $J(T, x) = 2x^2$, for $t < T : J(t, x) = \max\{x^2, a_t + 2x^2)\}, a_{t-1} = -1 + a_t/4, a_{T-1} = -1, u_t = 0$ if $a_t + x_t^2 \leq 0, u_t = 1$ if $a_t + x_t^2 > 0$.

1.9 $u_{t-1} = x/(1 + a_t^2), a_{t-1} = (1/2)(1 + a_t)^{1/2}, a_T = a/2$.

1.10 $u_t = \bar{u} = (\hat{q} - \hat{p})/(\hat{q} + \hat{p}), \hat{q} = q^{-1/\alpha}, \hat{p} = p^{-1/\alpha}, J(t, x) = A_t x^{1-\alpha}, A_{t-1} = A_t[p(1 + \bar{u})^{1-\alpha} + q(1 - \bar{u})^{1-\alpha}]$.

1.11 $a_T = b_T = 1, a_{t-1} = 9a_t^2 c_t^2 + (3a_t/4)(4 + b_t)^2 c_t^2 + (9b_t/4)a_t^2 c_t^2, b_{t-1} = a_t^2 d_t^2 + (a_t/4)(4 + 3b_t)^2 d_t^2 + (3b_t/4)a_t^2 d_t^2, c_t = 1/(4 + 3a_t + b_t), d_t = 1/(4 + a_t + 3b_t), u_{t-1}(x, 1) = -3a_t c_t x, u_{t-1}(x, 0) = -a_t d_t x$.

277

1.12 (a) $J(x) = -\hat{a}x^2 - \hat{c}, \hat{a} = -(2\alpha)^{-1}(1-2\alpha-(1+4\alpha^2)^{1/2}), \hat{c} = \alpha\hat{a}d/(1-\alpha)$. Solution of (b): $J(t,x) = -\alpha^t(a_t x_t^2 + b_t), b_{t-1} = \alpha a_t d + \alpha b_t, a_{t-1} = 1 + \alpha a_t/(1 + \alpha a_t), b_T = 0, a_T = 1$.

1.13 $J(x) = a\ln x + b$, where $a = (2-\beta)^{-1}, b = [\beta a d + \beta a\ln(\beta a) - (1+\beta a)\ln(1+\beta a)](1-\beta)^{-1}$, and $d = E\ln V, u = (1-\alpha)(1+\alpha)^{-1}x$.

1.14 By the summation formula for geometric series, $\sum_{t=0}^{\infty} l^t = 1/(1-l)$ provided that $|l| < 1$. Now, for $x = x_0, X_t = V_t \rho V_{t-1} \cdot \ldots \cdot V_2 \rho V_1 \rho x$, so $EX_t^{1-\gamma} = D^t(\rho^{1-\gamma})^t x^{1-\gamma}$. Hence, $E\sum_{t=0}^{\infty} \beta^t X_t^{1-\gamma} u_*^{1-\gamma} = x^{1-\gamma}(1-\rho)^{1-\gamma}\sum_{t=0}^{\infty} l^t$ for $l = \beta D\rho^{1-\gamma} = \rho < 1$, so

$$J^{u_*} = (1-\rho)^{1-\gamma}x^{1-\gamma}\frac{1}{1-\rho} = (1-\rho)^{-\gamma}x^{1-\gamma}.$$

This is precisely $J(x)$ in the text.

1.15 $J(T,x) = x, J(T-1,x) = 3x, J(T-2,x) = 3^2 x$ and so on, $u_T = 0, u_{T-1}$ arbitrary, $u_t = 1, t < T-1$.

1.16 $x_1 = x_0 + 1/2 - v_1, x_2 = x_1 + 1/2 - v_2, x_3 = x_2 + 1/2 - v_3$.

1.18 See Example 1.18.

1.19 Solution by means of Remark 1.30: $J(t,x_t) = k_t + a_t x_t, a_t = a2^{t-T}, k_{t-1} = k_t + 2/a_t, k_T = 0, u_{t-1} = 4/a_t^2$, where now $a > 0$ is determined by $Ex_T = 0$. Now, $0 = EX_T = x_0/2^T - \sum_{t=0}^{T-1} u_t/2^{T-t} = x_0/2^T + 4(1-2^T)/a^2$, which determines a.

1.20 $u_t^* = p - q$ for all t

1.21 Solution: See solution to Exercise 1.3.

1.24 (a) Evidently, $X_2 \in \{0,2\}, \Pr[X_2 = 2] = 1/4$. If $u_2 = u_2(x_2)$, then $EX_3 = E[X_2 u_2(X_2)] + Eu_2(X_2) = [2u_2(2) + u_2(2)](1/4) + u_2(0)(3/4) \neq 1$ for any combination of values $u_2(2) = \pm 1, u_2(0) = \pm 1$. If $u_2 = u_2(x_1)$, then $EX_3 = E[X_2 u_2(X_1) + u_2(X_1)] = E[u_2(X_1)E[X_2 + 1|X_1]] = E[u_2(X_1)(X_1 + 1)] = 2u_2(1)(1/2) + u_2(0)(1/2) = 1$ if $u_2(1) = 1, u_2(0) = 0$.

(b) Essentially repeating the above calculations, for a $u_2(X_2, V_2) = u_2(V_2), EX_3 = E[V_2 u_2(V_2) + u_2(V_2)] = E[E[V_2 u_2(V_2) + u_2(V_2)|V_1]] = \{(2u_2(2) + u_2(2))(1/2) + u_2(0)(1/2)\}\Pr[V_1 = 1] + u_2(0)\Pr[V_1 = 0] = [2u_2(2) + u_2(2)](1/4) + u_2(0)(3/4) \neq 1$ for all choices of $u_2(v_2)$. For $u_2 = u_2(V_1), EX_3 = E[V_2 u_2(V_1) + u_2(V_1)] = E[u_2(V_1)E[V_2 + 1|V_1]] = E[u_2(V_1)(V_1 + 1)] = 2u_2(1)(1/2) + u_2(0)(1/2) = 1$, if $u_2(1) = 1, u_2(0) = 0$.

1.25 Absorbing problem: Stop iff $y_t \leq 2t - 1/2, t < T$.

1.26 For $x \geq 0$ trivially stop at once. For $x < 0, J(T-k,x) = \max\{x, 3^{-k}(x-k/2)\}$. Stop iff $x_{T-k} \geq (-k/2)/(1-3^{-k})$.

1.27 At the second day, Jane stops iff $s_2 \geq 2$. At the first day, Jane stops iff $s_1 = 2$. Note that if $s_1 = 1$ it is strictly better to continue than to stop, the criterion values are, respectively, 25 and 24.

1.28 Iff $n > n^*$, John continues even if place n is vacant.

1.29 Time 2: stop iff $x \geq 1$, time 1: stop iff $x \geq 3/2$, time 0: stop iff $x \geq 7/4$.

1.30 Note that $x_t \geq 0, y_t \geq 0$ for all t. $B_t := \{(x,y) : x \geq E[X_{t+1}|(x,y)], y \geq 0\} = \{(x,y) : x \geq (1/2)[x/2+y] + (1/2)x/2, y \geq 0\} = \{(x,y) : x \geq y, y \geq 0\}$, $B_T = R^2$. Evidently B_t is absorbing.

1.31 For any $k \in [0,1], E[\max\{k,X^2\}] = \int_0^{k^{1/2}} k dx + \int_{k^{1/2}}^1 x^2 dx = k^{3/2} + (1/3)(1 - k^{3/2}) = 1/3 + 2k^{3/2}/3 = \phi(k)$. Let $a_{T-1} = Ex^2 = 1/3$, and, generally, define $a_t = \phi(a_{t+1})$. The optimality equation yields at time t: stop iff $x_t^2 \geq a_t$.

1.32 Stop as soon as $x \geq \mu a e^{-b}/(1 - e^{-b})$. Solution by one stage look ahead gives stop when x is in $B = \{x : x e^{-bt} \geq (1-\mu)xe^{-b(t+1)} + \mu(x+a)e^{-b(t+1)} = xe^{-b(t+1)} + \mu a e^{-b(t+1)}\}$. Evidently, B is absorbing: If $x_t \in B$, then $x_{t+1} \geq x_t$ is in B.

1.33 Stop when the maximum of job offers so far received, $\tilde{v}_t$ satisfies $c \geq \int_{\tilde{v}_t}^R (v - \tilde{v}_t)\varphi(v)dv$.

1.35 The equation for k is $k = k(\alpha/2)^k/(1 - \alpha/2) + (\alpha/2)(k+1)$.

1.36 When starting in $k_0 = 2$, the probability of stopping is the sum of probabilities of stopping when reaching 1 after one step, after two steps, after three steps, after four steps, etc., which equals $(1/2)[1 + 1/4 + (3/4)(1/16) + (3/4)(15/16)(1/64) + \ldots] < (1/2)[1 + 1/4 + (1/16) + (1/64) + \ldots] = 2/3$. When starting in $k_0 = 3, 4, \ldots$, a similar calculation yields that (at least) the same bound holds.

1.39 We have $g_t(\theta) = g_t(v_{\to t-1}, \theta)$. By means of the relations $\psi_t(v_t|v_{\to t-1}) \sim \psi_{t-1}(v_{t-1})|v_{\to t-2})g_t(v_{\to t-1}, \theta_1) + \psi_{t-1}(v_{t-1})|v_{\to t-2})g_t(v_{\to t-1}, \theta_2)$, $\psi_0(v_0) \sim \phi(v_0, \theta_1)g_0(\theta_1) + \phi(v_0, \theta_1)g_0(\theta_2)$, find the densities $\psi_t(v_t|v_{\to t-1})$ recursively forwards. Next, define $v^T(v_{\to T-1}) = -c + \int_0^R v\psi_T(v|v_{\to T-1})$, and $v^t(v_{\to t-1}) := -c + E[J(t, V_{\to t})|v_{\to t-1}]$, where $J(t, v_{\to t}) = \max\{v_t, v^{t+1}(v_{\to t})\}$. Calculate recursively backwards the functions $v^t(v_{\to t-1})$ by $v^{t-1}(v_{\to t-2}) =$

$$-c + E[\max\{V_{t-1}, v^t(v_{\to t-1})\}|v_{\to t-2}]$$

$$= -c + \int_0^R \max\{v_{t-1}, v^t(v_{\to t-1})\}\psi_{t-1}(v_{t-1}|v_{\to t-2})dv_{t-1}.$$

Stop iff $v_t \geq v^{t+1}(v_{\to t})$, where $v_{\to t}$ is the actual history up to time t.

1.40 At time t stop iff $0 \geq -1 + 10(1 - \gamma_t)$, $\gamma_t = \int_0^1 F(q)\psi_t(q)dq$, ψ_t the update of $\psi_0(q)$, according to $\psi_{t+1}(q) = F(q)\psi_t(q)/\gamma_t$. To show absorption, one needs $\beta(s) := \int_0^s [\psi_t(q) - F(q)\psi_t(q)/\gamma_t]dq \geq 0$, which follows from $\beta(0) = \beta(1) = 0$, and $\beta'(q) = \psi_t(q)(1 - F(q)/\gamma_t)$ being first ≥ 0 and next (for greater q), ≤ 0.

1.41 $x_0 = 1/2$, the updating: $x_{t+1} = 0$ if a black ball is drawn at time t, otherwise $x_{t+1} = 2x_t/(x_t + 1)$. A one-stage look-ahead policy is: Stop iff $x_t = 0$ or $x_t - 0.5 \geq -1/32 + (2x_t/(x_t + 1) - 0.5)(1/2 + x_t/2)$, i.e., $(1/4)x_t \geq 1/2 - 1/4 - 1/32 = 7/32$, or $x_t \geq 7/8$.

1.50 $B = \{(x, \tau^0) : xe^{-b\tau^0} \geq E[(x+V_1)e^{-b(\tau+\tau^0)}]\} = \{(x, \tau^0) : xe^{-b\tau^0} \geq e^{-b\tau^0}(x + a)\lambda/(b+\lambda)\}$. Stop at the first jump at which x has become $x \geq a\lambda/b$. Evidently, B is absorbing.

1.51 For some K', the criterion is $\leq EK'[\sum_{j=0}^{\infty} e^{-\alpha\tau_j}] \leq \sum_{i=0}^{\infty} KK'e^{-\delta i}$.

Exercises of Chapter 2

2.1 $\phi = (T+1-t)^{1/2}$

2.2 $u^* \equiv 1$, see *Hint*.

2.3 $V(0,0,x^0) = \max\{0, x^0\}$.

2.4 (a): $x(t;s,y) = 5/4 - C/2 + Ct/2 - t^2/4$, $C = \alpha(s,y) := (2y + s^2/2 - 5/2)/(s-1)$, $p(t;s,y) = C - t$. (b) C instead given by $\max\{1, \alpha(s,y)\}$, not differentiable with respect to (s,y) at (s,y) such that $p^{(a)}(1;s,y) = 0$.

2.5 (a): $p(t;s,y) = p(s,y) = 4y/(2s+1)$, $x(t;s,y) = p(s,y)t/2 + p(s,y)/4$. (b) $p(t;s,y) = p(s,y) = y/(s-3/2)$, $x(t;s,y) = p(s,y)(t-3/2)$, they fail to exist for $s = 3/2$.

2.6 (a) We use characteristic solutions to obtain a proposal for W: Whenever a switch point $\sigma \in (0,1)$ occurs, where we switch from using $u = 1$ to $u = 0$, the solutions are $x(t;s,y) = y + t - s$, for $t < \sigma$, $x(t;s,y) = y + \sigma - s$ for $t > \sigma$, $p(t;s,y) = 1 - t, t > \sigma, p(t;s,y) = 1 - \sigma$ for $t < \sigma, \sigma = (1 - y + s)/2$. Then $W(s,y) = (y + \sigma - s)(1 - \sigma) = (y/2 - s/2 + 1/2)(1 + y - s)/2$, if $\sigma \in (s,1)$, $W(s,y) = y(1-s)$ (and $p(t;s,y) = 1 - t$) if $\sigma \leq s$ (we need only consider $y > 0$, in which case $\sigma < 1$). Then $Z^* := \{(s,y) : \sigma = s\} = \{(s,y) : s + y = 1\}$ is slim.
(b) Put $\phi = x - p$. Then $\phi_x \dot{x} + \phi_p \dot{p} = 1 > 0$, so (2.38) holds. Now, $y = p(s;s,y)$ implies $s + y = 1$, test both for $p(s;s,y) = 1 - s$ ($s \geq \sigma$) and for $p(s;s,y) = 1 - \sigma$ ($s < \sigma$). For (s,y) such that $s + y \neq 1$, we have continuous differentiability.

2.7 Insertion of the proposal in the current value HJB equation yields the equation $-\alpha a + 2a^2 + 1 = 0$ for a, i.e., $a = \alpha/4 \pm (\alpha^2 - 8)^{1/2}/4$, where the following argument only works when using the minus sign. For this proposal of $h(x)$, the optimal $x^*(t) = x^0 e^{2at}$, and then $\lim_{t\to\infty} W(t, x^*(t)) = 0$, because $W(t, x^*(t)) = a(x^0)^2 e^{4at} e^{-\alpha t} = a(x^0)^2 e^{-(\alpha^2-8)^{1/2}t}$. So (2.47) is satisfied ($S \equiv 0$). Finally (2.46) is trivially satisfied as $a > 0$.

2.8 $T = 1, u \equiv 1$.

Exercises of Chapter 3

3.1 (a) No change. (b) For scrap value $ax(T)$, $u(t;s,y,0) = a$. For scrap value $ax(T)^2/2$, $p(t;s,y,0)$ and $x(t;s,y,0)$ are determined by $\dot{p} = -\lambda[\alpha(t) + a(x+2)/(1+$

$a(t-T))]+\lambda p$ and $\dot{x}=p(t)$, with $x(s;s,y)=y$ and $p(T)=ax(T)$, where $\alpha(t)=$ $-ae^{-\lambda T}e^{\lambda t}/(1+a(t-T))$. With $w(1+a(t-T))=x+2$, $\ddot{w}=-2a\dot{w}/(1+a(t-$ $T))+\lambda\dot{w}-\lambda\alpha(t)/(1+a(t-T))$, which has the solution $\dot{w}(s)=\psi(s)+K\gamma(s)^{-1}$, where $\gamma(t)=(1+a(t-T))^2e^{-\lambda t}$ and $\psi(s)=-\gamma(s)^{-1}\int_T^s\lambda\gamma(r)\alpha(r)/(1+a(r-T))dr$. So $x(t;0,0,0)=(\int_T^t\psi(s)+K\gamma(s)^{-1}ds+H)(1+a(t-T))-2$, K and H two integration constants determined by $x(t)=0$ and $\dot{x}(T)=ax(T)$.

3.2 $\hat{u}=1/4p^2$, all $p(t,s,y,j)$ independent of (s,y), $p(t;2)=e^{T-t}$, $p(t;1)=2e^{T-t}-e^{(1-\lambda)(T-t)}$, $p(t;0)=4e^{T-t}-3e^{(1-\lambda)(T-t)}+2\lambda e^{(1-\lambda)(T-t)}(t-T)$.

3.3 $x(t;0,0,0)=1/2+Ae^{r_+t}+Be^{r_-t}$, $r_\pm=(1/2)[\lambda\pm(\lambda^2+8\lambda)^{1/2}]$, A, B determined by $A+B+1/2=0$, $r_+Ae^{r_+T}+r_-Be^{r_-T}=1$.

3.4 Generally, $u(t,j)=1/p(t,j)$. (Neither p nor u depend on the starting point (s,y)). Now, $p(t,N)=1$. Consider $\dot{p}(t,N-1):=\lambda p-\lambda 2$. Evidently, $p(t,N-1)=2-e^{\lambda(t-1)}$ (recall $p(1)=1$). Next, $\dot{p}(t,N-2):=\lambda p-\lambda 2(-e^{\lambda(t-1)}+2)$ so, $p(t,N-2)=4-3e^{\lambda(t-1)}+2\lambda e^{\lambda(t-1)}(t-1)$. Writing $p(t,N-k)=a_k+a_{k,0}e^{\lambda(t-1)}+e^{\lambda(t-1)}a_{k,1}(t-1)+\ldots+e^{\lambda(t-1)}a_{k,k-1}(t-1)^{k-1}$, one easily calculates $p(t,N-(k+1))$, and hence the difference equations for the coefficients. In fact, $a_{k+1}=2a_k$, $a_{k,0}=1-a_k$, and for $i>0$, $a_{k+1,i}=-2\lambda a_{k,i-1}/i$, $a_0=1$.

3.5 $u(t;s,y,1)=a=p(t;s,y,1)$, $u(t;s,y,0)=[a-\lambda a(\lambda-1)^{-1}]e^{(\lambda-1)(t-T)}+\lambda a(\lambda-1)^{-1}=p(t;s,y,0)$.

3.6 After a jump: $y>1\Rightarrow u(s,y,1)=1$, $y<1\Rightarrow u(s,y,1)=0$. Before a jump, for $(s,y)=(0,1/2)$: $u=0$ after t^*, $u=1$ before t^*, $p(t)=Ke^{\lambda t}+1/2-(t-t^*)+2-1/\lambda$, $p(1)=x(1)=1/2-(t-t^*)$ giving $K=(1/\lambda-2)e^{-\lambda}$ for $t>t^*$, $p(t)=1/2+2+He^{\lambda t}$ for $t<t^*$, $p(t^*)=1$ giving $H=-3e^{-\lambda t^*}/2$. t^* is the solution of $1=(1/\lambda-2)e^{\lambda(t^*-1)}+1/2+2-1/\lambda$, i.e. $t^*=\lambda^{-1}\ln[(3/2-1/\lambda)/(2-1/\lambda)]+1\in(0,1)$. (Existence theory says that an optimal solution exists, we have found one candidate satisfying the necessary conditions. For sufficient conditions, one might turn to Remark 3.45.)

3.7 If we start in (s,y), the after-a-jump-adjoint function $p(t;s,y,1)=(3+s-y)/2-t$, $u=-1$ in $(\sigma',2)$, where $\sigma'=(3+s-y)/2)$, $u=1$ in (s,σ'). Before a jump, we only seek the solution $x(t)=x(t;0,0,0)$. Now, $p(t)=p(t;0,0,0)$ satisfies $\dot{p}(t)=-1-\lambda p(t,t,x(t)+1,1)+\lambda p=-1-\lambda(3-x(t)-1-t)/2+\lambda p$, so $p(t)=he^{\lambda t}+1/\lambda+e^{\lambda t}\int_t^2\lambda e^{-\lambda z}[(3-x(z)-1-z)/2]dz$, where h is a constant such that $he^{2\lambda}+1/\lambda=p(2)\leq 0$, (so $h<0$). Now, $x(z)\leq 2-z$, ($x(z)$ cannot come above a line through $(2,0)$ with slope -1, given by $2-z$). Hence, $(3-x(z)-1-z)/2\geq(3-(2-z)-1-z)/2=0$. Now, $\dot{p}\leq-1$ when $p\leq 0$, when going backwards $p(t)$ is first negative and then positive, with $p(\rho)=0$ for some ρ. Now, $u=1$ for $t<\rho$, $u=-1$ for $t>\rho$. Using $x(0)=0$, $x(2)=0$ we must have $\rho=1$. The solution is reasonably enough independent of λ.

3.8 (i) $\hat{u}=(p/\gamma)^{1/(\gamma-1)}$, $p(t;1)=ae^{k(T-t)}$, $p(t;0)=ae^{k(T-t)}$.
(ii) Let $\mu=1/(\gamma-1)$. $x(t;s,y,1)=D(y,s)e^{kt}+B(y,s)e^{-\mu kt}$, $B(y,s)=y/(e^{-\mu ks}-e^{-\mu kT-kT+ks})$, $D(y,s)=-B(y,s)e^{-kT-\mu kT}$. Moreover, $p(t;s,y,1)=C(y,s)e^{-kt}$,

where $C(y,s) = \gamma B(y,s)^{1/\mu} (\mu k + k)^{1/\mu} e^{-kT}$. For $j = 0$, $\dot{p}(t;0,c,0) = -pk + \lambda p - \lambda C(x+b,t)e^{-kt}$, so $\ddot{x} = k\dot{x} - \gamma^{-\mu}\mu p^{\mu-1}\dot{p} = k\dot{x} - \gamma^{-\mu}\mu p^{\mu-1}[-pk + \lambda p - \lambda \gamma(B(x+b,t)(\mu k + k))^{1/\mu}e^{k(T-t)}] = k\dot{x} + \gamma^{-\mu}\mu(k - \lambda)p^{\mu} - \lambda \gamma B(x + b,t)(\mu k + k)e^{k(T-t)}$, where $p = \gamma(kx - \dot{x})^{\gamma-1}$, so $\ddot{x} = k\dot{x} + \mu(k - \lambda)(kx - \dot{x}) - \lambda \gamma(B(x+b,t)(\mu k + k))^{1/\mu}e^{k(T-t)}$.

3.9 $\ddot{w} = 2\dot{w}/(1-t) + \lambda \dot{w} - \lambda/(1-t)^2 + \lambda/(1-t)$, $\dot{w} = (t-1)^{-2}\{Ce^{\lambda t} - (1 - 1/\lambda)/(t-1)^2 + 1/(t-1) =: \phi(C,t)$, for $y_2 = 0$, $x_1(t;0,0,0) = (1-t)\{C' + \int_0^t \phi(C,r)dr\}$, where C is determined by $x_1(0;0,0,0,0) = 0$, $x_1(1;0,0,0,0) = 1$, so $C' = 0$. Correspondingly, for arbitrary (s,y_1,y_2), $C' = C(s,y_1,y_2)$ and $C = C(s,y_1,y_2)$ are determined by $x_1(s;s,y_1,y_2,0) = y_1, x_1(1;s,y_1,y_2,0) = 1$. The discussion of strong semiadmissibility is as in Example 3.29, if one want to use Theorem 3.27. Alternatively, "concave sufficient" conditions can be used to show optimality, in which case admissibility of the nonanticipating candidate $x^*(t,\omega)$ corresponding to the characteristic solutions obtained can be shown as in Example 3.29.

3.10 $u(t;s,y,1) = y/(s-T)$, $p(t;s,y,1) = 1$, $x(s;t,s,y,0) = D + t + Ce^{\lambda(t-s)}$, $C = C(s,y) = (T+y-s)/(1-e^{\lambda(T-s)})$, $D = y - C - s$. Note, for the question of strong semiadmissibility, that $C(s,\hat{x}(s))$ and so $|u(t;s,\hat{x}(s),0)|$ is bounded (before a jump) for any admissible pair $x(t,\omega), u(t,\omega)$.

3.11 $T(s,y,1) = \max\{s,y\}$, $T(0,0,0) = \lambda^{-1}[(1+\lambda) - ((1+\lambda)^2 - \lambda^2)^{1/2}] \in (0,1)$.

3.12 Always, $u \equiv 1$. $\eta(T,s,y,1) = -T/2 + 1/2$, $T(s,y,1) = \max\{1,s\}$, $T(0,0,0) = (2/\lambda)(1/2 - [1/4 - \lambda(1/2 - 5\lambda/4)]^{1/2}) \in (0,1)$, as $\eta(T,0,0,0) = 1/2 - T/2 + \lambda T^2/4 - 5\lambda/4$, $T \leq 1$, $\eta(T,0,0,0) = -T/2 + 1/2 - \lambda$ for $T > 1$ (note that $\eta(1,0,0,0) = -\lambda$, $\eta(0,0,0,0) = 1/2 - 5\lambda/4 > 0$).

Exercises of Chapter 4

4.1 $E(B_t - B_s)^2 = E(B_t^2 - 2B_t B_s + B_s^2) = E(B_t^2 - 2(B_t - B_s)B_s - 2B_s^2 + B_s^2) = EB_t^2 - EB_s^2$.

4.2 $dY_t = e^{B_t}dt + te^{B_t}dB_t + (te^{B_t}/2)dt$.

4.4 $dY_t = \gamma(X_t)^{\gamma-1}(aX_t dt + bX_t dB_t) + (1/2)\gamma(\gamma-1)(X_t)^{\gamma-2}b^2(X_t)^2 dt = \gamma Y_t a dt + \gamma b Y_t dB_t + 2^{-1}\gamma(\gamma-1)b^2 Y_t dt$, so $dEY_t = a\gamma EX_t^{\gamma} + \gamma(\gamma-1)b^2(1/2)E(X_t)^{\gamma} = kEY_t$, $k = a\gamma + \gamma(\gamma-1)b^2/2$. Hence $EY_t = X_0^{\gamma}e^{kt}$.

4.5 (b) $\alpha = 2c + 2n^2m^2c'^2$, $\beta = 2(c + d + n^2m^2d'^2)$.

4.7 $\phi(t) = \sigma^2/(De^{\sigma^2 t} - 1)$, $D = (1 - \sigma^2)e^{-\sigma^2 T}$.

4.8 $\psi(t) = 0$, $\phi(t) = \sigma^2/(Ce^{t\sigma^2/2} - 1)$, $C = (1+\sigma^2)e^{-T\sigma^2/2}$.

4.9 Let $d = a^{\gamma/(\gamma-1)}$. The optimization yields $c = a^{1/(\gamma-1)}x$. a and b must satisfy the equations: $-b = 0$, i.e., $b = 0$, and $0 = -a + \beta + d(1 - \gamma) + \gamma\alpha a + a\gamma(\gamma-1)\sigma^2/2$. For $\gamma = 1/2$, $0 = -a + \beta + 1/2a + \alpha a/2 - a\sigma^2/8$, or (as $a = 0$ is not possible), after multiplying by a and solving, we get $a = [-\beta - (\beta^2 - 2\delta)^{1/2}]/2\delta$, $\delta = (\alpha/2 - \sigma^2/8 - 1) < 0$. As the criterion is positive, - must be used. Now, the optimal c equals $a^{-2}x$, so the optimal solution X_t^0 satisfies $dX_t^0 = \alpha X_t^0 dt - a^{-2}X_t^0 dt + \sigma X_t^0 dB_t$. Moreover, $E(X_T^0)^{1/2} = (X_0)^{1/2}e^{kt}$, where $k = \alpha/2 - a^{-2}/2 - \sigma^2/8$. As $\alpha\gamma = \alpha/2 < 1$, surely $E(X_T^0)^{1/2}e^{-T} \to 0$ when $T \to \infty$. Hence (4.55), (a) holds. Because $X_t > 0$ for all admissible X_t, (4.55), (b) automatically holds, (or at least $\liminf_T E\hat{J}(T, X_T) \ge 0$).

4.10 For $k = -\rho + 2\alpha + \gamma^2$, $a = 2^{-1}(-k - (k^2 + 4\beta)^{1/2})$, $b = 0$, $c = 0$. Moreover, $u^0 = u = ax$, and $\delta(t) := E(X_t^0)^2$ satisfies $\dot\delta = \mu\delta$, where $\mu = 2\alpha + 2a + \gamma^2 = \rho - (k^2 + 4\beta)^{1/2}$, so $Ee^{-\rho t}E(X_t^0)^2 \to 0$ when $t \to \infty$ from which (4.55) (a) follows. (4.55) (b) needs only be tested for solutions X_t for which $\limsup_{T\to\infty} Ee^{-\rho T}X_T^2 = 0$. At least sporadic catching up optimality follows.

4.11 $a = 2^{-1}[\rho - (\rho^2 + 4)^{1/2}]$, $b = 0$, $c = a\sigma^2/\rho$. (4.53) needs only be tested for solutions X_t for which $\limsup_{T\to\infty} Ee^{-\rho T}X_T^2 = 0$, compare Example 4.20. Use Dynkin's formula to show that $\eta(t) := E(X_t^0)^2$ satisfies $\eta' = 2a\eta + \sigma^2$, hence $\lim_{T\to\infty} Ee^{-\rho T}(X_T^0)^2 = 0$. The last calculation also yields finiteness of the criterion for the proposed candidate for the optimal control. At least sporadic catching up optimality holds, see previous exercise.

4.12 HJB: $0 = -h + x + (1/2)xh' + x^2h''$, $z_t^{(r_\pm)} = e^t$, and $r_\pm = (1/2)[1/2 \pm (1/4 + 4)^{1/2}]$, $r_+ > 1$, $r_- < 0$. The criterion is positive (X_t never gets negative), so both $C_1 \ge 0$ and $C_2 \ge 0$ (look at very large values of x and values close to zero). Finally, $E[e^{-t}h(X_t^0)] = E[e^{-t}2X_t^0] + C_1 + C_2 \to C_1 + C_2$ when $t \to \infty$ ($(X_t^0)^{r_\pm} = e^t$, $EX_t^0 = e^{t/2}$). From (4.55) $C_1 = C_2 = 0$. (Actually, using sufficient conditions, once we got the "proposal" $h = 2x + C_1 x^{r_+} + C_2 x^{r_-}$, we could immediately put $C_1 = C_2 = 0$, because what we need is just to find one proposal for h that works.)

4.13 $z(t) = \sigma^2/(De^{\sigma^2 t} - 1)$, $u = xz(t)$, defining $y(t) = E(X_t^0)^2$, we get $y(t) = K[(De^{\sigma^2 t} - 1)^2 e^{-\sigma^2 t}]$, where $D > 1$, D and K are so chosen that $y(1) = k$, $y(0) = x_0^2$, i.e., $D = (k^{1/2} - e^{-\sigma^2/2})/(k^{1/2} - e^{\sigma^2/2})$, $K = 1/(D-1)^2$.

4.14 $b(t) := 6\sigma^2(Ce^{6\sigma^2 t} - 4)^{-1}$, $y(t) := E(X_t^0)^4 = Ke^{\int_0^t (8b(s) + 6\sigma^2)ds} = K[1 - e^{-6\sigma^2 t}/C]^8 e^{6\sigma^2 t}$, K and C determined by $y(0) = x_0^4$, $y(1) = k$. Now, $J^q(0, x_0) = \infty$, at least for $q > 1/3$ for small σ (check for $\sigma = 0$), hopefully the value $q^0 = b(1)$ is so small that J^{q^0} is finite, hence sufficient conditions applies.

4.15 $k = 1$.

4.16 $a = (2\rho)^{-1/2}$.

4.17 a is given by the equation $a = (2/\rho)^{1/2}(e^{2(2\rho)^{1/2}a} + 1)(e^{2(2\rho)^{1/2}a} - 1)^{-1}$ and $h(x) = a^2(e^{(2\rho)^{1/2}x} + e^{-(2\rho)^{1/2}x})/(e^{(2\rho)^{1/2}a} + e^{-(2\rho)^{1/2}a})$.

4.18 The second-order equation for h becomes $-h + h' + 2h'' = 0$. The general solution is $be^{r_+x} + ce^{r_-x}$, $r_\pm = (-1\pm 3)/4$. Because we can expect $J(t,x) \ge 0$ for all x, both $b \ge 0$, $c \ge 0$. When x is large negative, the solution will be decreasing in x if $c \neq 0$, so we must have $c = 0$. Furthermore, a and b are determined by smooth pasting, i.e., the use of (4.80) (with $r_+ = 1/2$): $be^{r+a} = be^{a/2} = e^{a/4} - 1$ and $be^{a/2}/2 = e^{a/4}/4$, which yields $a = 4\ln 2$, and $b = 1/4$. Evidently, (4.82) is satisfied in $(-\infty, 4\ln 2)$ and by the fact that $be^{x/2} - e^{x/4}$ is convex in $(0, \infty)$, we get $be^{x/2} \geq e^{x/4} - 1$ for $x \geq 0$. For $x < 0$, $e^{x/4} - 1 < 0 < (1/4)e^{x/2}$.

4.19 (a) Recall that $M_t = \sup_{0 \leq s \leq t} B_s$ has a density two times the density of a $N(0,t)$–distributed variable. Note first that for any positive constants c and d, a number s exists such that $\Pr[M_s < c] < d$, (when $t \to \infty$, the graph of the distribution of the $N(0,t)$– variable is uniformly pushed down towards zero). Next, observe that $P_t = p_0 \exp[(\beta - \sigma^2/2)t + \sigma B_t]$, so $P_t < p^* \Leftrightarrow \sigma B_t < \ln p^* - \ln p_0 - (\beta - \sigma^2/2)t$, the last number $\leq \ln p^* - \ln p_0 =: a$. Now, B_t reaches sometimes even $c := a/\sigma$, with probability 1: By contradiction, if not, there is a positive probability ε that $B_t < c$ for all t, and then even more for $t \leq s$, s the number above corresponding to c and $d = \varepsilon/2$. A contradiction has arisen.
(b) In V, $\hat{f}$ satisfies (4.86).

4.20 The HJB equation becomes $0 = \psi' + \phi'x^2 + 2\phi^2x^2 - x^2 + \phi + E[\phi(x+V)^2 + \psi] - \phi x^2 - \psi = \psi' + \phi'x^2 + 2\phi^2x^2 - x^2 + 2\phi$, which yields $\psi' = -2\phi$, $\phi' = 1 - 2\phi^2$ with solution $\phi = (Ce^{2^{3/2}t} - 1)/(Ce^{2^{3/2}t} + 1)$, where $\phi(T) = 0$, $\psi(T) = 0$, so $C = e^{-2^{3/2}T}$, $(u = 2\phi x)$.

References

Arkin, V.I., Evstigneev, L.V. (1987) Stochastic Models of Control and Economic Dynamics, Academic Press, London.

Benth, F.E. (2003) Option Theory and Stochastic Analysis - An Introduction to Mathematical Finance, Springer-Verlag, New York.

Bertzekas, D.P. (1976) Dynamic Programming and Stochastic Control, Academic Press, New York.

Bertsekas, D.P., Shreve, S.E. (1978) Stochastic Optimal Control: The Discrete Time Case. Academic Press, New York.

Billingsley, P. (1979) Probability and Measure (First ed.), J. Wiley, New York.

Bliss, G.A. (1946) Lectures on the Calculus of Variations, University of Chicago Press, Chicago.

Boltyanskii, V.G. (1971) Mathematical Methods of Optimal Control, Holt, Rinehart, Winston, New York.

Costa, O.L.V, Raymundo, C.A.D., Dufor, F. (2000) Optimal stopping with continuous control of piecewise deterministic Markov processes, *Stochastic Reports* 70, 41–73.

Costa, O.L.V, Raymundo, C.A.D. (2000) Impulse and continuous control of piecewise deterministic Markov processes, *Stochastic Reports* 70, 75–107.

Dempster, M.A.H., Ye, J.J. (1995) Impulse control of piecewise deterministic Markov processes, *Ann. Appl. Probab.* 5, 399–423.

Davis, M.H.A (1993) Markov Models and Optimization, Chapman and Hall, London.

Davis, M.H.A., Farid, M. (1999) Piecewise deterministic processes and viscosity solutions, McEneaney, W.M. et al. (ed) Stochastic Analysis, Control, Optimization and Applications, A Volume in Honour of W.H.Fleming, on Occasion of His 70th Birthday. Birkäuser, 249–268.

Duffie, D. (1996) Dynamic Asset Pricing Theory (Second ed.), Princeton Univ. Press, Princeton, N.J.

Feller, W. (1957) An introduction to Probability Theory and Its Applications, (vol I), (Second ed.) J. Wiley, New York.

Fleming, W.H., Rishel, R.W. (1975) Deterministic and Stochastic Control, Springer-Verlag, New York.

Fristedt, B., Gray, L. (1997) A Modern Approach to Probability Theory, Birkhäuser, Boston.

Hernandez-Lerma, O., Lasserre, J.B. (1996) Discrete - Time Markov Control Processes, Springer-Verlag, New York.

Grimmett, G.R., Stirzaker, D.R. (1982) Probability and Random Processes, (First ed.) Oxford University Press, Oxford, UK.

Haussmann, U.G. (1986) A Stochastic Maximum Principle for Optimal Control of Diffusions, Pitman Research Notes in Mathematics Series 151, Longman Scientific and Technical, Harlow, Essex, UK.

Haussmann, U.G., Lepeltier, J.P. (1990) On the existence of optimal controls, *SIAM J. Control Optimization* 28: 851–902.

Kallenberg, O. (1997) Foundations of Modern Probability (First ed.) Springer-Verlag, New York.

Karatzas, I., Shreve, S.E. (1988) Brownian Motion and Stochastic Calculus, Springer-Verlag, New York.

Karatzas, I., Shreve, S.E. (1998) Methods of Mathematical Finance, Springer-Verlag, New York.

Kushner, H.J. (1972) Necessary conditions for continuous parameter stochastic optimization problems, *SIAM J. Control Optimization* 10: 550–565.

Nualart, D. (1995) The Malliavin Calculus and Related Topic, Springer-Verlag, New York.

Puterman, M.L. (1994) Markov Decision Processes: Discrete Stochastic Programming, J. Wiley, New York.

Ross, S.M. (1983) Introduction to Stochastic Dynamic Programming, Academic Press, New York.

Ross, S.M. (1992) Applied probability and Optimization Applications, Dover Publications, Mineola, New York.

Ross, S.M. (1999) An Introduction to Mathematical Finance, Cambridge Press, Cambridge, UK.

Seierstad, A. (2001) Necessary conditions and sufficient conditions for optimal control of piecewise deterministic control systems, *Memorandum No 5/2001, Dept. of Economics*, University of Oslo, Norway.

Seierstad, A. (2002) Maximum principle for stochastic control in continuous time with hard end constraints. *Memorandum No 24/2002, Dept. of Economics*, University of Oslo, Norway.

Seierstad, A. (2008) Existence of optimal controls in piecewise deterministic systems, *Memoradum from Dept. of Economics*, University of Oslo (forthcoming).

Seierstad, A., Sydsaeter, K. (1987) Optimal Control Theory with Economic Applications, North-Holland, Amsterdam.

Steele, J.M. (2001) Stochastic Calculus and Financial Applications, Applications of Mathematics 45, Springer-Verlag, New York.

Stokey, N.L., Lucas, R.E., with Prescott, E.C. (1989) Recursive Methods in Economic Dynamics, Harvard University Press, Cambridge, Massachusetts.

Sydsaeter, K. et al. (2005) Further Mathematic for Economic Analysis, Prentice Hall, Harlow, Essex, UK.

Ye, J.J. (1996) Dynamic programming and the maximum principle of piecewise deterministic Markov processes, AMS Lectures in Applied Mathematics: Yin, G.G, Zhang, G. *The Mathematics of stochastic manufacturing systems*, vol.33, 365–383.

Yong, J., Zhou, X.Y. (1999) Stochastic Controls, Springer-Verlag, New York.

Øksendal, B. (1998) Stochastic Differential Equations (Fifth ed.) Springer-Verlag, New York.

Øksendal, B., Sulem, A. (2006) Applied Stochastic Control of Jump Diffusions, Springer-Verlag, New York.

Seierstad's memoranda can be retrieved from http://www.oekonomi.uio.no/memo/

Index

A

absorbing set, 41
adapted function, 189
adapted process, 274
admissible, 87
admissible set, 87
autonomous case for diffusions, 237
autonomous system, 213

B

Bellman equation, 10
binomial distribution, 256
Black and Scholes formula, 203
Borel measurable, 270
Borel-field, 270
Brownian motion, 186

C

candidate control, 118
catching up optimal, 212
Cauchy–Schwartz inequality, 269
characteristic pairs, 118
characteristic solutions, 96, 118
conditional density, 263
conditional expectation, 262, 273
conditional mass functions, 259
conditional probability, 258, 273
control of diffusions with jumps, 239
controllable set, 85
countable additivity, 256, 268
criterion containing an integral, 92
cumulative distribution, 257, 263
cumulative distribution function, 263

D

discrete sample space, 255
dominated convergence, 269
double expectation rule, 262
dynamic programming equation, 4
Dynkin's formula, 199

E

equilibrium optimality equation, 10
Euler equation, 24
event, 256
expectation, 261
exponential density, 263
extremal method, 117, 134, 138
extremal method for free terminal time, 157
extremals, 96

F

Fatou's lemma, 269
feedback control, 88
free terminal time problems, 108

G

Gaussian distribution, 263
general distribution of jump times, 132
geometric Brownian motion, 200
Girsanov's theorem, 207
growth conditions, 126

H

Hamilton–Jacobi–Bellman equation, 86, 208
Hamiltonian, 117
hard end constraints, 36, 137
hard terminal constraints, 137
hard terminal restrictions for diffusions, 227